# Microchip
# Fabrication

# Microchip Fabrication

## A Practical Guide to Semiconductor Processing

## Peter Van Zant

**Fifth Edition**

**McGraw-Hill**

New York   Chicago   San Francisco   Lisbon   London   Madrid
Mexico City   Milan   New Delhi   San Juan   Seoul
Singapore   Sydney   Toronto

**Library of Congress Cataloging-in-Publication Data**

Van Zant, Peter.
    Microchip fabrication / Peter Van Zant.—5th ed.
      p. cm.
    Includes index.
    ISBN 0-07-143241-8
      1. Semiconductors—Design and construction. 2. Integrated circuits—Design
    and construction. I. Title.

    TK7871.85.V36    2004
    621.3815'2—dc22                    2004040287

1  2  3  4  5  6  7  8  9  0  DOC/DOC  0  1  0  9  8  7  6  5  4

ISBN 0-07-143241-8

*The sponsoring editor for this book was Stephen S. Chapman and the production supervisor was Pamela Pelton. It was set in Century Schoolbook by J. K. Eckert & Company, Inc.*

Printed and bound by R. R. Donnelley.

McGraw-Hill books are available at special quantity discounts to use as premiums and sales promotions, or for use in corporate training programs. For more information, please write to the Director of Special Sales, McGraw-Hill Professional, Two Penn Plaza, New York, NY 10121-2298. Or contact your local bookstore.

This book is printed on recycles, acid-free paper containing a minimum of 50% recycled, de-inked fiber.

*This edition is dedicated to two exceptional women, Marilyn (Van Zant) O'Connor and Anne Miller. Marilyn is my lovely and loving sister. She is also my good friend, enthusiastic supporter, and a wise confidant. Thanks, sis.*

*For over twenty years Anne has been a collaborator, business partner, and friend. Her wise business counsel and contributions to this text are greatly appreciated.*

# Contents

Preface    xiii

## Chapter 1.  The Semiconductor Industry                                        1

Overview                                                                         1
Objectives                                                                       1
Birth of an Industry                                                             1
The Solid-State Era                                                              3
Integrated Circuits (ICs)                                                        4
Process and Product Trends                                                       5
Industry Organization                                                           13
Stages of Manufacturing                                                         14
The Junction Transistor                                                         17
Five Decades of Industry Development                                            19
The Nano Era                                                                     21
Review Questions                                                                23
References                                                                      24

## Chapter 2.  Properties of Semiconductor Materials and Chemicals              25

Overview                                                                        25
Objectives                                                                      25
Atomic Structure                                                                26
The Periodic Table of the Elements                                              27
Electrical Conduction                                                           29
Dielectrics and Capacitors                                                      30
Intrinsic Semiconductors                                                        31
Doped Semiconductors                                                            32
Electron and Hole Conduction                                                    33
Semiconductor Production Materials                                              37
Semiconducting Compounds                                                        37
Silicon Germanium                                                               39
Engineered Substrates                                                           39
Ferroelectric Materials                                                         40
Diamond Semiconductors                                                          40
Process Chemicals                                                               40
States of Matter                                                                42

Properties of Matter                                    43
Pressure and Vacuum                                     45
Acids, Alkalis, and Solvents                            46
Chemical Purity and Cleanliness                         48
Review Questions                                        48
References                                              49

## Chapter 3. Crystal Growth and Silicon Wafer Preparation    51

Overview                                                51
Objectives                                              51
Introduction                                            52
Semiconductor Silicon Preparation                       52
Crystalline Materials                                   53
Crystal Orientation                                     55
Crystal Growth                                          56
Crystal and Wafer Quality                               59
Wafer Preparation                                       62
Wafer Slicing                                           64
Wafer Marking                                           65
Rough Polish                                            65
Chemical Mechanical Polishing (CMP)                     67
Backside Processing                                     67
Double-Sided Polishing                                  68
Edge Grinding and Polishing                             68
Wafer Evaluation                                        68
Oxidation                                               69
Packaging                                               69
Engineered Wafers (Substrates)                          69
Review Questions                                        69
References                                              70

## Chapter 4. Overview of Wafer Fabrication                   71

Overview                                                71
Objectives                                              71
Goal of Wafer Fabrication                               72
Wafer Terminology                                       72
Basic Wafer-Fabrication Operations                      73
Example Fabrication Process                             83
Chip Terminology                                        87
Wafer Sort                                              87
Packaging                                               89
Summary                                                 89
Review Questions                                        90
References                                              90

## Chapter 5. Contamination Control                           91

Overview                                                91
Objectives                                              91
Introduction                                            92
Contamination Sources                                   96
Cleanroom Construction                                  106
Cleanroom Materials and Supplies                        119

Cleanroom Maintenance                                      119
Wafer Surface Cleaning                                     120
Review Questions                                           135
References                                                 136

## Chapter 6.  Productivity and Process Yields                139

Overview                                                   139
Objectives                                                 139
Yield Measurement Points                                   140
Accumulative Wafer-Fabrication Yield                       140
Wafer-Fabrication Yield Limiters                           142
Assembly and Final Test Yields                             155
Overall Process Yields                                     155
Review Questions                                           157
References                                                 157

## Chapter 7.  Oxidation                                       159

Overview                                                   159
Objectives                                                 159
Silicon Dioxide Layer Uses                                 160
Thermal Oxidation Mechanisms                               163
Oxidation Processes                                        191
Postoxidation Evaluation                                   193
Review Questions                                           195
References                                                 195

## Chapter 8.  The Ten-Step Patterning Process—Surface Preparation
to Exposure                                                197

Overview                                                   197
Objectives                                                 197
Introduction                                               198
Overview of the Photomasking Process                       199
Ten-Step Process                                           201
Basic Photoresist Chemistry                                202
Photoresist Performance Factors                            207
Physical Properties of Photoresists                        214
Photomasking Processes                                     217
Surface Preparation                                        218
Photoresist Application (Spinning)                         221
Soft Bake                                                  227
Alignment and Exposure                                     231
Advanced Lithography                                       243
Review Questions                                           243
References                                                 244

## Chapter 9.  The Ten-Step Patterning Process—Developing
to Final Inspection                                        245

Overview                                                   245
Objectives                                                 245
Hard Bake                                                  252
Integrated Image Processing                                254

Etch                                                                      258
Wet Etching                                                               259
Dry Etch                                                                  264
Resist Stripping                                                          273
Final Inspection                                                          277
Mask Making                                                               278
Summary                                                                   280
Review Questions                                                          280
References                                                                281

### Chapter 10.  Advanced Photolithography Processes                     283

Overview                                                                  283
Objectives                                                                283
Issues of VLSI/ULSI Patterning                                            284
Other Exposure Issues                                                     292
Surface Problems                                                          300
Antireflective Coatings                                                   301
Photoresist Process Advances                                              304
Improving Etch Definition                                                 319
Self-Aligned Structures                                                   320
Etch Profile Control                                                      321
Review Questions                                                          321
References                                                                322

### Chapter 11.  Doping                                                   325

Overview                                                                  325
Objectives                                                                325
Introduction                                                              325
Formation of a Doped Region by Diffusion                                  326
Formation of a Doped Region and Junction by Diffusion                     328
Diffusion Process Steps                                                   333
Deposition                                                                333
Drive-in Oxidation                                                        341
Introduction to Ion Implantation                                          345
Concept of Ion Implantation                                               346
Ion Implantation System                                                   346
Dopant Concentration in Implanted Regions                                 355
Evaluation of Implanted Layers                                            359
Uses of Ion Implantation                                                  361
The Future of Doping                                                      362
Review Questions                                                          363
References                                                                363

### Chapter 12.  Layer Deposition                                        365

Overview                                                                  365
Objectives                                                                365
Introduction                                                              365
Chemical Vapor Deposition Basics                                          369
CVD Process Steps                                                         372
CVD System Types                                                          372
Atmospheric-Pressure CVD Systems                                          373
Low-Pressure Chemical Vapor Deposition (LPCVD)                            377

Atomic Layer Deposition (ALD)                                   381
Vapor Phase Epitaxy (VPE)                                       382
Molecular Beam Epitaxy (MBE)                                    383
Metalorganic CVD (MOCVD)                                        384
Deposited Films                                                385
Deposited Semiconductors                                       385
Epitaxial Silicon                                              386
Polysilicon and Amorphous Silicon Deposition                  392
SOS and SOI                                                   394
Gallium Arsenide on Silicon                                    395
Insulators and Dielectrics                                     395
Conductors                                                    399
Review Questions                                              399
References                                                    400

Chapter 13.  Metallization                                    401

Overview                                                      401
Objectives                                                    401
Introduction                                                  402
Conductors—Single-Level Metal                                 402
Conductors—Multilevel Metal Schemes                           403
Conductors                                                    405
Electrochemical Plating (ECP)                                 413
Chemical Mechanical Processing                                414
Metal Film Uses                                               415
Deposition Methods                                            416
Vacuum Pumps                                                  426
Summary                                                       431
Review Questions                                              433
References                                                    433

Chapter 14.  Process and Device Evaluation                    435

Overview                                                      435
Objectives                                                    435
Introduction                                                  436
Wafer Electrical Measurements                                 437
Process and Device Evaluation                                 439
Physical Measurement Methods                                  442
Layer Thickness Measurements                                  443
Junction Depth                                                448
Contamination and Defect Detection                            454
General Surface Characterization                              462
Contamination Identification                                  465
Device Electrical Measurements                                467
Review Questions                                              477
References                                                    478

Chapter 15.  The Business of Wafer Fabrication                481

Overview                                                      481
Objectives                                                    481
Wafer Fabrication Costs                                       484
Equipment                                                     491

Automation                                                              494
Factory-Level Automation                                                497
Equipment Standards                                                     499
Statistical Process Control (SPC)                                       501
Inventory Control                                                       506
Quality Control and Certification—ISO 9000                              508
Line Organization                                                       508
Review Questions                                                        510
References                                                              510

## Chapter 16.  Introduction to Devices and Integrated Circuit Formation    513

Overview                                                                513
Objectives                                                              513
Semiconductor-Device Formation                                          513
Alternative (Scaled) Transistor Designs                                 531
Integrated-Circuit Formation                                            533
Bi-MOS                                                                  543
Superconductors                                                         544
Review Questions                                                        550
References                                                              552

## Chapter 17.  Introduction to Integrated Circuits                         553

Overview                                                                553
Objectives                                                              553
Introduction                                                            553
Circuit Basics                                                          554
Integrated Circuit Types                                                556
The Next Generation                                                     566
Review Questions                                                        567
References                                                              567

## Chapter 18.  Packaging                                                    569

Overview                                                                569
Objectives                                                              569
Introduction                                                            569
Chip Characteristics                                                    571
Package Functions and Design                                            572
Overview of Packaging Operations                                        574
Packaging Processes                                                     580
Alternative Process                                                     589
Transfer to Packaging Area                                              590
Package Process Flows                                                    600
Package/Bare Die Strategies                                             600
Package Design                                                          602
Package Type/Technology Summary                                         608
Review Questions                                                        608
References                                                              609

Glossary    611
Index    631

# Preface
# to the Fifth Edition

Despite recessions, the microchip industry continues its evolutionary march to the physical limits of silicon-based ICs. Fortunately, the end seems always just over the hill, and the industry keeps chugging along. Unfortunately, keeping a textbook current with the advances in microchip fabrication means frequent updates. Hence this fifth edition.

This edition follows the same chapter sequence as the previous editions. Hopefully, this will assist instructors in upgrading their course curriculums. Fortunately, the basics of semiconductor device operation and wafer processing remain the same and will be found in this edition.

My thanks go to Steve Chapman, my editor at McGraw-Hill. His guidance and patience with my writing schedule are appreciated.

Many thanks to Anne Miller and Michael Heynes of Semiconductor Services for their consultation and input. Alex Braun, of Semiconductor International, and Nikki Wood, of Future Fab International, were most helpful with securing permission to reproduce material from their fine publications. Jeff Eckert, of J. K. Eckert & Co., did a fine job on organizing the over 600 figures and editing the manuscript. Mark Hall, Mark Hall Design, and David Wellner did yeoman's work transforming my hand drawings into understandable illustrations.

Last, but not least, thanks to my wife Mary Dewitt for enduring my 5:30 A.M. writing sessions and her unending support.

# About the Author

Peter Van Zant is an internationally known semiconductor professional with an extensive background in process engineering, training, consulting, and writing. Principal of Peter Van Zant Associates, a firm that supplies writing, training, and consulting services to business and industry, he is the author of *Semiconductor Technology Glossary, Third Edition; Integrated Circuits Text; Safety First Manual;* and *Chip Packaging Manual.* His books and training materials are used by chip manufacturers, industry suppliers, colleges, and universities. Peter Van Zant Associates' customers include Intel, National Semiconductor, Applied Materials, Air Products and Chemicals, SCP Global Inc., and a number of educational institutions. Mr. Van Zant is also the elected District 1 Supervisor in his home county of Nevada in California.

# The Semiconductor Industry

## Overview

In this chapter, you will be introduced to the semiconductor industry with a description of the historic product and process developments and the rise of semiconductors into a major world industry. The major manufacturing stages, from material preparation to packaged product, are introduced along with the mainstream product types, transistor building structures, and the different integration levels. Industry product and processing trends are identified.

## Objectives

Upon completion of this chapter, you should be able to:

1. Describe the difference between discrete devices and integrated circuits.
2. Define the terms "solid-state, "planar processing" and "N-type" and "P-type" semiconducting materials.
3. List the four major stages of semiconductor processing.
4. Explain the integration scale and at least three of the implications of processing circuits of different levels of integration.
5. List the major process and device trends in semiconductor processing.

## Birth of an Industry

The electronics industry got its jump start with the discovery of the audion vacuum tube in the 1906 by Lee Deforest.[1] It was made possible the radio, television, and other consumer electronics. It also was the brains of the world's first electronic computer, named the Electronic Numeric Integrator and Calculator (ENIAC), first demonstrated at the Moore School of Engineering in Pennsylvania in 1947.

This ENIAC hardly fits the modern picture of a computer. It occupied some 1500 square feet, weighed 30 tons, generated large quantities of heat, required the services of a small power station, and cost $400,000 in 1940 dollars. The ENIAC was based on 19,000 vacuum tubes along with thousands of resistors and capacitors (Fig. 1.1).

A vacuum tube consists of three elements: two electrodes separated by a grid in a glass enclosure (Fig. 1.2). Inside the enclosure is a vacuum, required to prevent the elements from burning up and to allow the easy transfer of electrons.

Tubes perform two important electrical functions: *switching* and *amplification*. Switching refers to the ability of an electrical device to turn a current on or off. Amplification is a little more complicated. It is the ability of a device to receive a small signal (or current) and amplify it while retaining its electrical characteristics.

Vacuum tubes suffer from a number of drawbacks. They are bulky and prone to loose connections and vacuum leaks, they are fragile, they

| Size, ft | 30 × 50 |
|---|---|
| Weight, tons | 30 |
| Vacuum Tubes | 18,000 |
| Resistors | 70,000 |
| Capacitors | 10,000 |
| Switches | 6000 |
| Power Requirements, W | 150,000 |
| Cost (in 1940) | $400,000 |

**Figure 1.1** ENIAC statistics. *(Source:* Foundations of Computector Technology, *J. G. Giarratano, Howard W. Sams & Co., Indianapolis, IN, 1983.)*

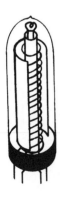

**Figure 1.2** Vacuum tube.

require relatively large amounts of power to operate, and their elements deteriorate rather rapidly. One of the major drawbacks to the ENIAC and other tube-based computers was a limited operating time due to tube burn-out. However, the world did not recognize the potential of computers early on. IBM Chairman, Thomas Watson, in 1943, ventured that, "I think there is a worldwide market for maybe five computers."

These problems were the impetus leading many laboratories around the country to seek a replacement for the vacuum tube. That effort came to fruition on December 23, 1947, when three Bell Lab scientists demonstrated an electrical amplifier formed from the semiconducting material germanium (Fig. 1.3).

This device offered the electrical functioning of a vacuum tube but added the advantages of being solid state (no vacuum), being small and lightweight, and having low power requirements and long life-time. First named a *transfer resistor*, the new device soon became known as the *transistor*.

The three scientists, John Bardeen, Walter Brattin, and William Shockley were awarded the 1956 Nobel Prize in physics for their invention.

## The Solid-State Era

That first transistor was a far distance from the high-density integrated circuits of today. But it was the component that gave birth to the solid-state electronics era with all its famous progeny. Besides transistors, solid-state technology is also used to create diodes, resistors, and capacitors. Diodes are two-element devices that function in a circuit as an on/off switch. Resistors are monoelements devices that serve to limit current flow. Capacitors are two-element devices that store charge in a circuit. In some integrated circuits, the technology is used to create fuses. Refer to Chapter 14 for an explanation of these concepts and an explanation of how these devices work.

**Figure 1.3**  The first transistor.

These devices, containing only one device per chip, are called *discrete* devices (Fig. 1.4). Most discrete devices have less-demanding operational and fabrication requirements than integrated circuits. In general, discrete devices are not considered leading-edge products. Yet, they are required in most sophisticated electronic systems. In 1998, they accounted for 12 percent of the dollar volume of all semiconductor devices sold.[2] The semiconductor industry was in full swing by the early 1950s, supplying devices for transistor radios and transistor based computers.

## Integrated Circuits (ICs)

The dominance of discrete devices in solid-state circuits came to an end in 1959. In that year, Jack Kilby, a new engineer at Texas Instruments in Dallas, Texas, formed a complete circuit on a single piece of the semiconducting material germanium. His invention combined several transistors, diodes, and capacitors (five components total) and used the natural resistance of the germanium chip (called a *bar* by Texas Instruments) as a circuit resistor. This invention was the *integrated circuit*, the first successful integration of a complete circuit in and on the same piece of a semiconducting substrate.

The Kilby circuit did not have the form that is prevalent today. It took Robert Noyce, then at Fairchild Camera, to furnish the final piece of the puzzle. In Fig. 1.5 is a drawing of the Kilby circuit. Note that the devices are connected with individual wires.

Earlier, Jean Horni, also at Fairchild Camera, had developed a process of forming electrical junctions in the surface of a chip to create a solid-state transistor with a flat profile (Fig. 1.6). The flattened profile was the outcome of taking advantage of the easily formed natural oxide of silicon, which also happened to be a dielectric (electrical insulator). Horni's transistor used a layer of evaporated aluminum, that was patterned into the proper shape, to serve as wiring for the device. This technique is called *planar technology*. Noyce applied this technique to

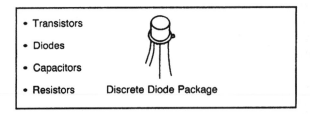

* Transistors

* Diodes

* Capacitors

* Resistors        Discrete Diode Package

**Figure 1.4**  Solid-state discrete devices.

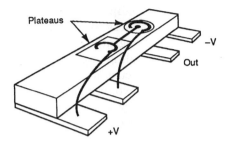

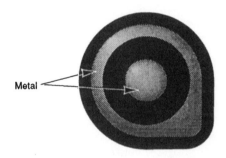

**Figure 1.5** Kilby integrated circuit from his notebook.

**Figure 1.6** Horni "teardrop" transistor.

"wire" together the individual devices previously formed in the silicon wafer surface (Fig. 1.7).

The Noyce integrated circuit became *the* model for all integrated circuits. The techniques used not only met the needs of that era, but contained the seeds for all the miniaturization and cost-effective manufacturing that still drives the industry. Kilby and Noyce shared the patent for the integrated circuit.

## Process and Product Trends

Since 1947, the semiconductor industry has seen the continuous development of new and improved processes. These process improvements have in turn led to the more highly integrated and reliable circuits that have, in their turn, fueled the continuing electronics revolution. These process improvements fall into two broad categories: process and structure. Process improvements are those that allow the fabrication of the devices and circuits in smaller dimensions, in ever higher density, quantity, and reliability. The structure improvements are the invention of new device designs allowing greater circuit performance, power control, and reliability.

Device component size and the number of components in an IC are the two common trackers of IC development. Component dimensions are characterized by the smallest dimension in the design. This is

# NOYCE PATENT U.S. PATENT No. 2,981,877

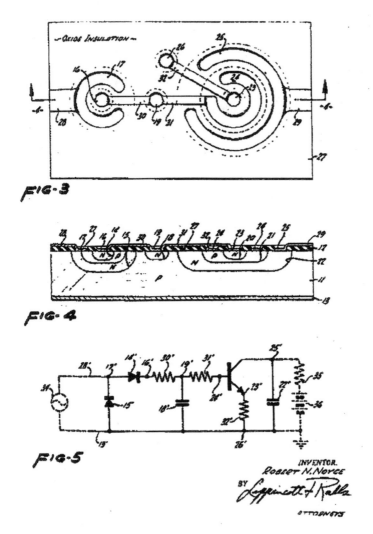

**Figure 1.7** Noyce IC patent. *(Courtesy of* Semiconductor Reliability News, *June 2003.)*

called the *feature size* and is usually expressed in microns or nanometers. A micron is 1/1,000,000 of a meter or about 1/100 the diameter of a human hair. A nanometer is 1/1,000,000,000 of a meter. A more specific tracker of semiconductor devices is *gate width*. Transistors are composed of three parts, one of which acts to allow the passage of current. In today's technology, the most popular transistor is the metal-oxide-semiconductor (MOS) structure. The controlling part is called the *gate*. Smaller gate widths drive the industry by producing smaller and faster transistors and more dense circuits. Currently, the industry is driving to the 90-nm gate width, with projections in the *International Technology Roadmap* for semiconductors projecting 22 nm size in 2016.[3]

In 1965, Gordon Moore, a founder of Intel, noted that the number of transistors on a chip were doubling every 18 months. He published the observation, which was immediately dubbed *Moore's law*. Industry observers have used this law to predict the future density of chips. Over the years, it has proven very accurate and now drives technical advances (Fig. 1.8). It is the basis of the *International Technology Roadmap for Semiconductors*, developed by the Semiconductor Industry Association.

Circuit density is tracked by the *integration level*, which is the number of components in a circuit. Integration levels (Fig. 1.9) range from small scale integration (SSI) to ultra large scale integration (ULSI). ULSI chips are sometimes referred to as very very large scale integration (VVLSI). The popular press calls these newest products *megachips*.

| Year | Transistor count |
|------|------------------|
| 1978 | 29,000 |
| 1982 | 275,000 |
| 1985 | 1,200,000 |
| 1991 | 3,100,000 |
| 1993 | 7,500,000 |
| 1997 | 9,500,000 |
| 2001 | 55,000,000 |

Figure 1.8  *Source: Intel Corporation.*

| Level | Abbreviation | # Components per Chip |
|-------|--------------|----------------------|
| Small Scale Integration | SSI | 2 - 50 |
| Medium Scale Integration | MSI | 50 - 5000 |
| Large Scale Integration | LSI | 5000 - 100,000 |
| Very Large Scale Integration | VLSI | Over 100,000 - 1,000,000 |
| Ultra Large Scale Integration | ULSI | > 1,000,000 |

Figure 1.9  IC integration table.

In addition to the integration scale, memory circuits are identified by the number of memory bits contained in the circuit (a four-meg memory chip can store four million bits of memory). Logic circuits are often rated by their number of "gates." A gate is the basic operational component of a logic circuit.

**Decreasing feature size**

The journey from small scale integration to today's megachips has been driven primarily by reductions in the feature size of the individual components. This decrease has been brought about by dramatic increases in the imaging process, known as lithography, and the trend to multiple layers of conductors. Actual and projected feature sizes are shown in Fig. 1.10. The Semiconductor Industry Association (SIA) has projected feature sizes decreasing to 22 nm (0.0022 µm) by the year 2016.[3] Along with the ability to make components on the chip smaller comes the benefit of crowding them closer together, further increasing density.

An analogy used to explain these trends is the layout of a neighborhood of single-family homes. The density of the neighborhood is a function of the house size, lot size, and the width of the streets. Accommodating a higher population could come by increasing size of the neighborhood (increasing the chip area). Another possibility is to reduce the size of the individual houses and place them on smaller lots. We can also reduce the street size to increase density. However, at some point, the streets cannot be reduced anymore in size or they won't be wide enough for autos. Furthermore, at some point, the houses cannot be further reduced in size and still function as dwelling units. At this point, an option is to build up by building multidecked freeways and/or replacing individual homes with apartment buildings. All of these concepts are used in semiconductor technology.

| Year | DRAM pitch (nm) |
|------|-----------------|
| 2001 | 130 |
| 2004 | 90 |
| 2007 | 65 |
| 2010 | 45 |
| 2013 | 32 |
| 2016 | 22 |

**Figure 1.10** Lithography DRAM pitch size. *(Source: SIA,* International Technology Roadmap for Semiconductors.)

There are several benefits to the reduction of the feature size and its attendant increase in circuit density. At the circuit performance level, there is an increase in circuit speed. With less distances to travel and with the individual devices occupying less space, information can be put into and gotten out of the chip in less time. Anyone who has waited for their personal computer to perform a simple operation can appreciate the effect of faster performance. These same density improvements result in a chip or circuit that requires less power to operate. The small power station required to run the ENIAC has given way to powerful lap top computers that run on a set of batteries.

### Increasing chip and wafer size

The advancement of chip density from the SSI level to ULSI chips has driven larger chip sizes. Discrete and SSI chips average about 100 mils (0.1 in) on a side. ULSI chips are in the 500 to 1000 mil (0.5 to 1.0 in) per side, or larger, range. ICs are manufactured on thin disks of silicon (or other semiconductor material, see Chapter 2) called *wafers*. Placing square or rectangular chips on a round wafer leaves unavailable areas around the edge (see Fig. 6.6). These unavailable areas can become large as the chip size increases (Fig. 1.11). The desire to offset the loss of usable silicon has driven the industry to larger wafers. As the chip size increases, the 1-in diameter wafers of the 1960s have given way to 200- and 300-mm (8-in and 12-in) sized wafers. Production efficiency increases, because the area of a circle increases as the mathematical square of the radius. Thus, doubling the wafer diameter from 6 to 12 in increases the area available for chip fabrication by four times.

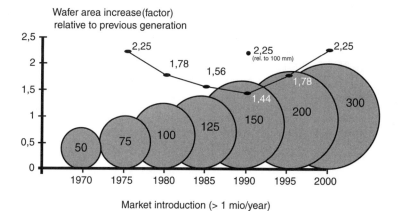

**Figure 1.11**  Wafer size history. *(Courtesy of Future Fab International.)*

### Reduction in defect density

As feature sizes have decreased, the need for reduced defect density and defect size on the chips (and in the manufacturing process) has become critical. A 1-micron piece of dirt on a 100-micron sized transistor may not be a problem. On a one-micron sized transistor, it becomes a killer defect that can render the component inoperable (Fig. 1.12). Contamination control needs has driven the cost of building an IC manufacturing facility into the multibillion dollar range.

### Increase in interconnection levels

The component density increase has led to a "wiring" problem. In the neighborhood analogy, reducing street widths was one strategy to increase density. But, at some point, the streets become too narrow to allow cars to travel. The same thing happens in IC design. The increased component density and close packing rob the surface space needed on the surface to connect the components. The solution is multiple levels of "wiring" stacked (Fig. 1.13) above the surface components in layers of insulators and conducting layers (Chapter 13).

### The SIA roadmap

These major IC parameters are interrelated. Moore's law predicts the future of component density, which triggers the calculation of the integration level (component density), chip size, defect density (and size), and the number of interconnection levels required. The Semiconductor Industry Association has made these projections into the future in a series of "roadmaps" covering these and other critical device and production parameters (Fig. 1.14).

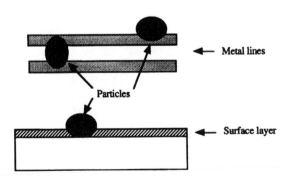

**Figure 1.12**  Relative size of airborne particles and wafer dimensions.

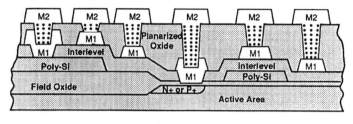

:: Via Plugs          M1 = First Metal          M2 = Second Metal

**Figure 1.13** Cross section of typical planarized two-level metal VLI structure showing range of via depths after planarization. (*Courtesy of Solid State Technology.*)

| Year of Production | 2001 | 2006 | 2012 |
|---|---|---|---|
| Line width (nm) | 150 | 100 | 50 |
| Memory size | 1 Gb | 16Gb | 64 Gb |
| Logic Bits/cm$^2$ | 380M | 2.2B | 17B |
| Chip Size-DRAM (mm$^2$) | 445 | 790 | 1580 |
| Max wiring levels | 7 | 7-8 | 9 |
| Mask layers | 23 | 24/26 | 28 |
| Defect density-DRAM (D/m) | 875 | 490 | 250 |
| Chip conections-I/Os | 1195 | 1970 | 3585 |
| Wafer diameter (mm) | 300 | 300 | 450 |

**Figure 1.14** Projection of wafer and chip parameters.

## Chip cost

Perhaps the most significant effect of these process and product improvements is the cost of the chips. Figure 1.15 shows the year-by-

| Year | Transistor cost ($) |
|---|---|
| 1986 | 1.00 |
| 1972 | 0.01 |
| 1976 | 0.001 |
| 1985 | 0.0001 |
| 1992 | 0.00001 |
| 2000 | 0.000001 |

**Figure 1.15** Declining transistor cost. (*Source: Intel Corporation.*)

year drop in transistors through the 1990s. The reductions are typical for any maturing product. Prices start high and, as the technology is mastered and manufacturing efficiencies increase, the prices drop and eventually become stable. These chip prices have constantly declined even as the performance of the chips has increased. In its first 30 years, the semiconductor industry had 2 to 5 times the economic impact in the U.S. that the railroads had in a similar period.[4] The factors affecting chip cost are discussed in Chapter 15.

The two factors, increased performance and less cost, have driven the explosion of products using solid-state electronics. By the 1990s, an auto had more computing power on-board than the first lunar space shots. Even more impressive is the personal computer. Today, for a moderate price, a desktop computer can deliver more power than a IBM mainframe manufactured in 1970. Major industry use of chips is shown in Fig. 1.16. By 2008, the global chip industry will be producing a billion transistors per person worldwide.[5]

### Semiconductor industry growth

Overall, the semiconductor industry has experienced worldwide continuous growth. From its birth in the 1950s, it has grown to worldwide sales of over $200 billion dollars a year, supported by a supplier industry of over $30 billion.[6] The millions of chips are supplied by factories located throughout the world. Interestingly, even as the industry shows signs of maturing, it is still growing faster than other "mature" industries, indicating that microchips still have a lot of growth potential (Fig. 1.17).

An example of increasing chip power is shown in Fig. 1.18, which indicates the number of volumes of the *Encyclopaedia Britannica* that can be stored on larger capacity DRAM memory chips.

The history of the semiconductor industry is one of continual developments and advances emerging to world dominance in the mid-

| CHIP USES | 1996 |
|-----------|------|
| Computer | 48.0% |
| Consumer | 21.2% |
| Telecom | 14.7% |
| Industrial | 9.8% |
| Automotive | 4.4% |
| Military | 1.9% |
| Total | 100.0% |

Figure 1.16 Semiconductor chip uses. *(Courtesy of In-Stat-1995 SEMI ISS seminary.))*

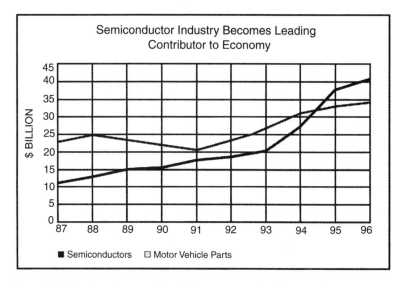

Figure 1.17   Semiconductor and vehicle parts growth. *(Courtesy Semiconductor Industry Association.)*

| Year | 1993 | 1996 | 1999 | 2002 | 2005 |
|---|---|---|---|---|---|
| DRAM Capacity | 16Mb | 64Mb | 256Mb | 1Gb | 4Gb |
| Feature Size (micron) | 0.5 | 0.35 | 0.25 | 0.18 | 0.08 |
| Volumes of Enc. Britannica | 1/4 Vol | 1 Vol. | 4 Vol. | 16 Vol. | 2 Sets |

Figure 1.18   Future DRAM capacity. *(Source:* Business Week, *July 1994.)*

1990s. In that year, the semiconductor industry became the nation's leading value added industry, outperforming the auto industry (Fig. 1.19).

## Industry Organization

The electronics industry is divided into two major segments: semiconductors and systems (or products). "Semiconductors" encompasses the material suppliers, circuit design, chip manufacturers, and all of the equipment and chemical suppliers to the industry. The systems segment encompasses the industry that designs and produces the vast number of semiconductor device based products, from consumer electronics to space shuttles. The electronics industry includes the manufacturers of printed circuit boards.

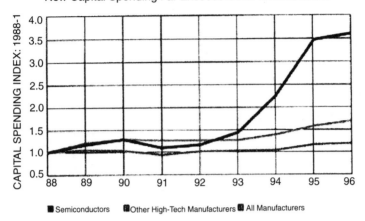

**Figure 1.19**  Growth of semiconductor industry capital spending. *(Courtesy Semiconductor Industry Association.)*

The semiconductor segment is composed of two major subsegments. One is the firms that actually make the semiconductor solid-state devices and circuits. The manufacturing process is named *wafer fabrication*. Within this segment there are three types of chip suppliers. Integrated device manufacturers (IDMs) design, manufacture, package, and market chips. Foundry companies build circuit chips for other chip suppliers. Waferless (or fabless) companies design and market chips, buying finished chips from chip foundries. Chips are fabricated by both merchant and captive producers. Merchant suppliers manufacture just chips and sell them on the open market. Captive suppliers are firms whose final product is a computer, communications system, or other product, and they produce chips in house for their own products. Some firms produce chips for in-house use and also sell on the open market, and others produce specialty chips in house and buy others on the open market. Since the 1980s, the trend has been to a greater percentage of chips being fabricated in captive fab areas.

**Stages of Manufacturing**

Solid-state devices are manufactured in the following five distinct stages (Fig. 1.20):

1. Material preparation
2. Crystal growth and wafer preparation
3. Wafer fabrication and sort

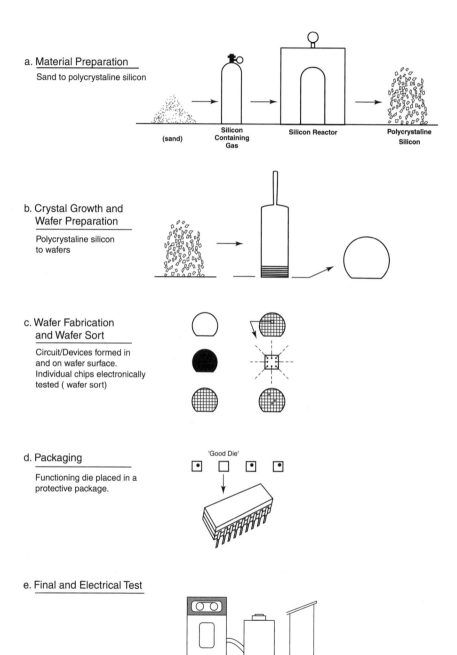

a. Material Preparation

Sand to polycrystaline silicon

(sand)    Silicon Containing Gas    Silicon Reactor    Polycrystaline Silicon

b. Crystal Growth and Wafer Preparation

Polycrystaline silicon to wafers

c. Wafer Fabrication and Wafer Sort

Circuit/Devices formed in and on wafer surface. Individual chips electronically tested ( wafer sort)

d. Packaging

Functioning die placed in a protective package.

'Good Die'

e. Final and Electrical Test

Tester    Test Head    Handler

**Figure 1.20**   Stages of semiconductor production.

4. Packaging

5. Final and electrical test

In the first stage, material preparation (see Chapter 2), the raw semiconducting materials are mined and purified to meet semiconductor standards. For silicon, the starting material is sand, which is converted to pure silicon with a polysilicon structure (Fig. 1.20a).

In stage two, the material is formed into a crystal with specific electrical and structural parameters. Next, thin disks called *wafers* are cut from the crystal and surface treated (Fig. 1.20b) in a process called crystal growth and wafer preparation (see Chapter 3). The industry also makes devices and circuits from germanium and compounds of different semiconductor materials.

In stage three (Fig. 1.20c), wafer fabrication, the devices or integrated circuits are actually formed in and on the wafer surface. Up to several thousand identical devices can be formed on each wafer, although two to three hundred is a more common number. The area on the wafer occupied by each discrete device or integrated circuit is called a *chip* or *die*. The wafer fabrication process is also called fabrication, fab, chip fabrication, or microchip fabrication. While a wafer fabrication operation may take several thousand individual steps, there are two major activities. In the front end of the line (FEOL), the transistors and other devices are formed in the wafer surface. In the back end of the line (BEOL), the devices are wired together with metallization processes, and the circuit is protected with a final sealing layer.

Following wafer fabrication, the devices or circuits on the wafer are complete, but untested and still in wafer form. Next comes an electrical test (called *wafer sort*) of every chip to identify those that meet customer specifications. Wafer sort may be the last step in the wafer fabrication or the first step in the *packaging* process.

Packaging (Fig. 1.20d) is the series of processes that separate the wafer into individual die and place them into protective packages. This stage also includes final testing of the chip for conformance to customer specifications. The industry also refers to this stage as *assembly and test* (A/T). A protective chip package is necessary to protect the chip from contamination and abuse, and to provide a durable and substantial electrical lead system to allow connection of the chip onto a printed circuit board or directly into an electronic product. Packaging takes place in a different department of the semiconductor producer and quite often in a foreign plant.

The vast majority of chips are packaged in individual packages. But a growing percentage are being incorporated into hybrid circuits, in multichip modules (MCMs), or mounted directly on printed circuit boards (chip-on-board, COB). An integrated circuit is an electrical circuit formed entirely by semiconductor technology on a single chip. A hybrid circuit combines semiconductor devices (discretes and ICs) with thick or thin film resistors and conductors and other electrical components on a ceramic substrate. These techniques are explained in Chapter 18.

## The Junction Transistor

While the tremendous advantages of solid-state electronics was recognized early on, the advancements possible from miniaturization were not realized until two decades later. During the 1950s, engineers set to work and defined many of the basic processes and materials still used today.

The structure that makes semiconductor devices function is the *junction* (Fig. 1.21). It is formed by creating a structure that is rich in electrons (negative polarity or *N-type*) next to a region rich in *holes* (locations with missing electrons that act electrically positive or *P-type*) (see Chapter 11).

A transistor requires two junctions to work (see Chapter 16). Early commercial transistors were of the bipolar type (see Chapter 14), which dominated production well into the 1970s. The term *bipolar* refers to a transistor structure that operates on both negative and positive currents. The other major method of building a solid-state transistor is the *field effect transistor* (FET). William Shockley published the operational basics of a FET in 1951. These transistors operate with only one type of current and are also called *unipolar devices*. The FET came to the marketplace in volume in a structure known as the *metal oxide semiconductor* (MOS) transistor.

William Shockley and Bell Labs get much of the credit for the spread of semiconductor technology. Shockley left Bell Labs in 1955 and formed Shockley Laboratories in Palo Alto, California. While his company did not survive, it established semiconductor manufacturing on the West Coast and provided the beginning of what eventually became known as Silicon Valley. Bell Labs helped the fledgling industry with the decision to license its semiconductor discoveries to a host of companies.

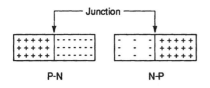

**Figure 1.21**  P-N and N-P junctions.

The early semiconductor devices were made in the material germanium. Texas instruments changed that trend with the introduction of the first silicon transistor in 1954. The issue over which material would dominate was settled in 1956 and 1957 by two more developments from Bell Labs: diffused junctions and oxide masking.

It was the development of oxide masking that ushered in the "silicon age." Silicon dioxide ($SiO_2$) grows uniformly on silicon and has a similar index of expansion, which allows high-temperature processing without warping. Silicon dioxide is a dielectric material, which allows it to function on the silicon surface as an insulator. Additionally, $SiO_2$ is an effective block to the dopants that form the N and P regions in silicon.

The net effect of these advances was planar technology (Fig. 1.22), introduced by Fairchild Camera in 1960. With the above-named techniques, it was possible to form (diffusion) and protect (silicon dioxide) junctions during and after the wafer fabrication process. Also, the development of oxide masking allowed two junctions to be formed through the top surface of the wafer (Fig. 1.22); that is, in one *plane*. It was this process that set the stage for the development of thin film wiring.

Bell Labs conceived of forming transistors in a high purity layer of semiconducting material deposited on top of the wafer (Fig. 1.23). Called an *epitaxial layer*, this discovery allowed higher speed devices and provided a scheme for the closer packing of components in a bipolar circuit.

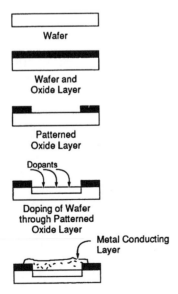

**Figure 1.22** Basics of silicon planar processing.

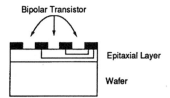

Bipolar Transistor

Epitaxial Layer

Wafer

**Figure 1.23** Double diffused bipolar transistor formed in epitaxial layer.

The 1950s was indeed the golden age of semiconductor development. During this incredibly short time, most of the basic processes and materials were discovered. The decade opened with the knowledge of how to manufacture small volumes of crude devices in germanium and ended with the first integrated circuit and silicon firmly established as the semiconductor of the future.

### Five Decades of Industry Development

In the 1950s, basic and crude products and processes launched an industry that has grown into a major world manufacturing sector. The 1960s was the decade the industry started growing into a sophisticated industry, driven by new products that demanded new fabrication processes, which demanded new materials and new production equipment. The chip price erosion trend of the industry, well established in the 1950s, also was an industry driver.

Technology spread as engineers changed companies in the industry clusters in Silicon Valley, Route 128 around Boston, and in Texas. By the 1960s, the number of fab areas had grown sufficiently, and processes were approaching a level of commonalty that attracted semiconductor specialty suppliers.

On the company front, many of the key players of the 1950s formed new companies. Robert Noyce left Fairchild to found Intel (with Andrew Grove and Gordon Moore), and Charles Sporck also left Fairchild to grow National Semiconductor into a major player. Signetics became the first company dedicated exclusively to the fabrication of ICs. New device designs were the usual driver of start-up companies. However, the ever present price erosion was a cruel trend that drove both established and new companies out of business.

Price dropping was accelerated by the development of a plastic package for silicon devices in 1963. Also in that year, RCA announced the development of the *insulated field effect transistor* (IFET), which paved the way for the MOS industry. RCA also pioneered the first complementary MOS (CMOS) circuits.

At the start of the 1970s, the industry was manufacturing ICs primarily at the MSI level. The move to profitable, high-yield LSI devices was being somewhat hampered by mask-caused defects and the damage inflicted on the wafers by the contact aligners. The mask and aligner defect problem was solved with the development of the first practical projection aligner by Perkin and Elmer Company.

The decade also saw the improvement of cleanroom construction and operation, the introduction of ion implantation machines, and the use of e-beam machines for high-quality mask generation, and mask steppers began to show in fab areas for wafer imaging.

Automation of processes started with spin/bake and develop/bake systems. The move from operator control to automatic control of the processes increased both wafer throughput and uniformity.[20]

Once the processes were integrated into the equipment, the stage was set for dissemination throughout the world. Along with the process improvements came a more detailed understanding of the physics of solid-state devices, which allowed the mastering of the technology by student engineers worldwide.

The focus in the 1980s was automation of all phases of wafer fabrication and packaging and elimination of operators from the fab areas. Automation increases manufacturing efficiency, minimizes processing errors, and keeps the wafer fabrication areas cleaner by limiting the number of operators, who are one of the major sources of contamination in the process. These issues are examined in more detail in Chapter 4.

One feature of automation is flexibility. As in automobile industry automation, especially in the area of design, manufacturers began to design more complicated chips. The new designs, in turn, presented new manufacturing challenges that led to the development of new processes. At these sophisticated levels, machine automation is required to achieve the process control and repeatability.

The 1980s started with American and European dominance and ended as a worldwide industry. Through the 1970s and 1980s, the one-micron feature size barrier loomed as both opportunity and challenge. The opportunity was a new era of megachips with vastly increased speeds and memory. The challenge was the limitations of conventional lithography, additional layers, more step height variation on the wafer surface, and increasing wafer diameters, to mention a few. The one-micron barrier was crossed in the early 1990s when 50 percent[7] of microchip fabrication lines were working at the micron or submicron level.

The industry matured into more traditional focuses on manufacturing and marketing issues. Early on, the profit strategy was to ride the innovation curve. That meant always being first (or close to first) with the latest and greatest chip that could be sold with enough profit to pay for the R&D and finance new designs. The profit potential of this strategy overcame manufacturing yield problems and lower efficiencies. The spread of the technology (competition) and improvements in process control, however, moved the industry to greater emphasis on the production issues. The primarily productivity factors are automation, cost control, process characterization and control, and worker efficiency.

Strategies to control the cost have included detailed analysis of equipment cost of ownership, new fab layouts (such as cluster tools), robotic automation, wafer isolation technology (WIT), computer integrated manufacturing (CIM), sophisticated statistical process control, advanced metrology instruments, just-in-time inventory schemes, and others (see Chapter 15).

Technical driving factors (feature size reduction, wafer diameter increases, and yield improvement) all have physical or statistical limits. But productivity improvement, which incorporates many factors (see Chapter 15), is the source of continuing profit. The pressures are enormous.

Wafer fab facilities are in the gigabuck ($2 to 3 billion and increasing) level, and equipment and process development are equally expensive. Manufacturing chips with features sizes below 0.35 microns will require extensive and expensive development of conventional lithography or X-ray and deep UV (DUV) lithography.

The challenge of the *SIA Roadmap* (IRTS) is that many of the processes required to produce the next generations of chips are unknown or in very primitive states of development. However, the good news is that the industry is moving forward along an evolutionary curve rather than relying on revolutionary breakthroughs. Engineers are wringing every bit of productivity out of the processes before looking for a big technology jump to solve problems. This is another sign of a maturing industry.

Perhaps the major technological change of the decade was copper wiring. Aluminum wiring ran into limitations in several areas, notably in contact resistance with silicon. Copper has always been a better conductor but was difficult to deposit and pattern. It was also a killer of circuit operation if it got into the silicon. IBM[8] developed usable copper processes (Chapters 10 and 13), which gained almost instant acceptance for wiring together advanced chips.

### The Nano Era

Microtechnology in the popular sense means "small." In the science world, it refers to one-billionth. Thus, feature sizes and gate widths are expressed in microns (micrometers), as in 0.018 μm. It is becoming more common to use nanometers ($1 \times 10^{-9}$ meters), thus making the above gate width 180 nm (Fig. 1.24).[9]

The way to the nano future is sketched out in the Semiconductor Industry Association's *International Technology Roadmap for Semiconductors (ITRS)*. Gate widths of 22 nm or less are predicted by 2016. At these levels, the operational parts of devices consist of only a few atoms or molecules.

Getting there will not be easy. There is a predictable train of events that happen as devices are scaled to smaller dimensions. Advantages are faster operating transistors and higher-density chips. However, smaller dimensions require more sophisticated processes and equipment.

A gate area is the critical working part of an MOS transistor. Smaller gates are more vulnerable to contamination, which drives the development of cleaner chemicals and processes. Detecting lower levels of contamination requires more sensitive measurement techniques.

Surface roughness becomes a parameter requiring control. As the devices get closer together, they drive the need for a superstructure of metallization layers stacked on top of the surface. These put pressure on planarization techniques to keep the surfaces flat enough to allow patterning. More metallization layers bring with them higher electrical resistances, which drive the need for new metallization materials, such as copper. As the number of electrical functions on a chip increases, so does the internal temperature, driving the need for heat dissipation techniques.

All of these advances will have to take place in ever cleaner wafer fabrication facilities with super-clean materials and chemicals and with process tools clustered to minimize exposure to contamination and to increase process efficiency.

Wafer diameters will move into the 450+ mm range, and factory automation will be at the tool-to-tool level with on-board process moni-

| | Meters (m) |
|---|---|
| Meter (m) | 1 |
| Centimeter (cm) | 1/100 |
| Millimeter (mm) | 1/1,000 |
| Micrometer (micron) (μm) | 1/1,000,000 |
| Nanometer (nm) | 1/1,000,000,000 |
| Angstrom (Å) | 1/10,000,000,000 |

**Figure 1.24** Comparative length units.

toring. More processes at higher levels of detail will require higher-volume wafer fabrication plants with more sophisticated process automation and factory management. Price tags for these mega-plants are headed to the $10 billion level.[10] This level of investment will pressure faster R&D activities and quick factory startups.

By 2016, the industry and circuits will be far different from what they are now, and the industry will be near the end of the basic physics of silicon transistors. Post-silicon production materials have yet to be identified, but the industry will grow. Not all IC uses have to be state of the art. It is unlikely that toasters, refrigerators, and automobiles will require cutting-edge devices. New base materials are in R&D labs. Compound semiconductors, such as gallium arsenide (GaAs) are candidates. Technologies such as molecular beam epitaxy (MBE) (Chapter 12) may be employed to build entirely new materials one atom at a time.

Another use of the term "nano" is a new way to build very small structures, called *nanotechnology*. It is based on the discovery of a structure of carbon flat crystals shaped like a hollow tube (*nanotube*). These structures have promise for a number of uses. In semiconductor technology, it appears that these nets of carbon atoms can be doped to act as electronic devices and, eventually, electronic circuits.

It is safe to say that the semiconductor industry will continue to be the dominant industry as it continues to push the limits of material and manufacturing technology. It is also safe to predict that the use of ICs will continue to shape our world in ways yet unknown.

### Review Questions

1. List the four types of discrete devices.

2. Describe the advantages of solid-state devices over vacuum tubes.

3. A VLSI circuit has more components than a ULSI circuit (true or false).

4. Describe the difference between a hybrid and integrated circuit.

5. State the stage of processing in which wafers are produced.

6. State the stage of processing that processes "chips."

7. Describe an N-P junction.

8. Describe what is meant by the term "feature size."

9. List three trends that have driven the semiconductor industry.

10. Describe the functions of a semiconductor package.

# References

1. E. Antebi, *The Electronic Epoch* (New York: Van Nostrand Reinhold), p. 126.
2. Economic Indicator, *Semiconductor International,* January 1998, p. 176.
3. Semiconductor Industry Association, *International Technology Roadmap for Semiconductors,* 2001/2003 update, www.semichips.org.
4. K. Flamm, "More for Less: The Economic Impact of Semiconductors," Dec. 1997.
5. D. Hatano, "Making a Difference: Careers in Semiconductors," Semiconductor Industry Association, Matec Conference, August 1998.
6. Economic Indicator, *Semiconductor International,* January 1998, pp. 176–177.
7. Rose Associates, 1994 Semiconductor Equipment and Materials International (SEMI) Information Seminar.
8. P. Singer, "Copper Goes Mainstream: Low k to Follow," *Semiconductor International,* November 1997, p. 67.
9. J. Baliga, Ed., *Semiconductor International,* January 1998, p. 15.
10. C. Skinner and G. Gettel, *Solid State Technology,* February 1998, p. 48.

Chapter

# 2

# Properties of Semiconductor Materials and Chemicals

## Overview

Semiconductor materials possess electrical and physical properties that allow the unique functions of semiconductor devices and circuits. These properties are examined along with the basics of atoms, electrical classification of solids, and intrinsic and doped semiconductors.

Wafer fabrication is a long series of steps that include many cleaning operations using ordinary and specialty chemicals. The basic properties of gases, acids, bases, and solvents are discussed.

## Objectives

Upon completion of this chapter, you should be able to:

1. Identify the parts of an atom.

2. Name the two unique properties of a doped semiconductor.

3. List at least three semiconducting materials.

4. Explain the advantages and disadvantages of gallium arsenide compared with silicon.

5. Explain the difference in composition and electrical functioning of N- and P-type semiconducting materials.

6. Describe the properties of resistivity and resistance.

7. Identify the differences between acids, alkalis, and solvents.

8. List the four states of nature.

9. Give the definition of an atom, a molecule, and an ion.

10. Explain four or more basic chemical handling safety rules.

## Atomic Structure

### The Bohr atom

The understanding of semiconductor materials requires a basic knowledge of atomic structure.

Atoms are the building blocks of the physical universe. Everything in the universe (as far as we know) is made from the 96 stable materials and 12 unstable ones known as *elements*. Each element has a different atomic structure. The different structures give rise to the different properties of the elements.

The unique properties of gold are due to its atomic structure. If a piece of gold is divided into smaller and smaller pieces, one eventually arrives at the last piece that exhibits the properties of gold. That last piece is the atom.

Dividing that last piece further will yield the three parts that compose individual atoms. They are called the *subatomic particles*. These are *protons, neutrons,* and *electrons*. Each of these subatomic particles has its own properties. A particular combination and structure of the subatomic particles are required to form the gold atom. The basic structure of the atom most used to understand physical, chemical, and electrical differences between different elements was first proposed by the famous physicist Niels Bohr (Fig. 2.1).

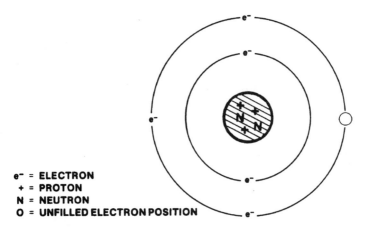

e⁻ = ELECTRON
+ = PROTON
N = NEUTRON
O = UNFILLED ELECTRON POSITION

**Figure 2.1**  Bohr atom model.

The Bohr atom model has the positively charged protons and neutral neutrons located together in the nucleus of the atom. The negatively charged electrons move in defined orbits about the nucleus, similar to the movement of the planets about the sun. There is an attractive force between the positively charged protons and the negatively charged electrons. However, this force is balanced by the outward centrifugal force of the electrons moving in their orbits. The net result is a structurally stable atomic structure.

Each orbit has a maximum number of positions available for electrons. In some atoms, not all of the positions are filled, leaving a "hole" in the structure. When a particular electron orbit is filled to the maximum, additional electrons must go into the next outer orbit.

## The Periodic Table of the Elements

The elements differ from each other in the number of electrons, protons, and neutrons in their atoms. Fortunately, nature combines the subatomic particles in an orderly fashion. An examination of some of the rules governing atomic structure is helpful in understanding the properties of semiconducting materials and process chemicals. Atoms (and therefore the elements) range from the simplest, hydrogen (with one electron) to the most complicated one, lawrencium (with 103 electrons).

Hydrogen consists of only one proton in the nucleus and only one electron. This arrangement illustrates the first of the following rules of atomic structure.

1. In each atom, there is an equal number of protons and electrons.

2. Each element contains a specific number of protons, and no two elements have the same number of protons. Hydrogen has one proton in its nucleus, while the oxygen atom has eight.

   This fact leads to the assignment of numbers to each of the elements. Known as the *atomic number,* it is equal to the number of protons (and therefore electrons) in the atom. The basic reference of the elements is the periodic table (Fig. 2.2). The periodic table has a box for each of the elements, which is identified by two letters. The atomic number is in the upper left hand corner of the box. Thus, calcium (Ca) has the atomic number 20, so we know immediately that calcium has 20 protons in its nucleus and 20 electrons in its orbital system.

   Neutrons are electrically neutral particles that, along with the protons, make up the mass of the nucleus.

   Figure 2.3 shows the atomic structure of elements no. 1, hydrogen; no. 3, lithium; and no. 11, sodium. When constructing the dia-

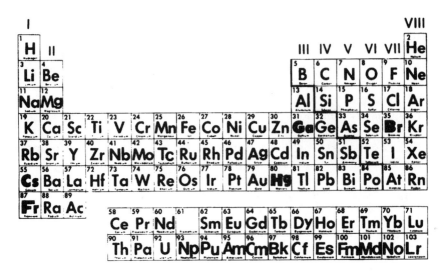

**Figure 2.2**  Periodic table of the elements.

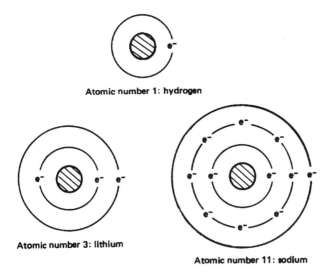

Atomic number 1: hydrogen

Atomic number 3: lithium

Atomic number 11: sodium

**Figure 2.3**  Atomic structures of hydrogen, lithium, and sodium.

grams, several rules were observed in the placement of the electrons in their proper orbits. The rule is that each orbit (n) can hold $2n^2$ electrons. Solution of the math for orbit no. 1 dictates that the first electron orbit can hold only two electrons. This rule forces the third electron of lithium into the second ring. The rule limits the number of electrons in the second ring to 8 and that of the third ring to 18. So, when constructing the diagram of the sodium atom, with 11 protons and electrons, the first two orbits take up 10 electrons, leaving the 11th in the third ring.

These three atoms have a commonalty. Each has an outer ring with only one electron in it. This illustrates another observable fact of elements.

3. Elements with the same number of outer-orbit electrons have similar properties. This rule is reflected in the periodic table. Note that hydrogen, lithium, and sodium appear on the table in a vertical column labeled with the Roman numeral one (I). The column number represents the number of electrons in the outer ring and all of the elements in each column share similar properties.

It is no accident that the three of the best electrical conductors (copper, silver, and gold) all appear in the same column (Ib) (Fig. 2.4) of the periodic table.

There are two more rules of atomic structure relevant to the understanding of semiconductors.

4. Elements are stable with a filled outer ring or with eight electrons in the outer ring. These atoms tend to be more chemically stable than atoms with partially filled rings.

5. Atoms seek to combine with other atoms to create the stable condition of full orbits or eight electrons in their outer ring.

Rules 4 and 5 influence the creation of N- and P-type semiconductor materials, as explained in the section on doped semiconductors.

## Electrical Conduction

### Conductors

An important property of many materials is the ability to conduct electricity or support an electrical current flow. An electrical current is

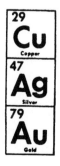

**Figure 2.4**  The three best electrical conductors.

simply a flow of electrons. Electrical conduction takes place in elements and materials where the attractive hold of the protons on the outer ring electrons is relatively weak. In such a material, these electrons can be easily moved, which sets up an electrical current. This condition exists in most metals.

The property of materials to conduct electricity is measured by a factor known as *conductivity*. The higher the conductivity, the better the conductor. Conducting ability is also measured by the reciprocal of the conductivity, which is *resistivity*. The lower the resistivity of a material, the better the conducting ability.

$$C = 1/\rho$$

where  $C$ = conductivity

$\rho$ = resistivity in ohm-centimeters ($\Omega$-cm)

## Dielectrics and Capacitors

At the opposite end of the conductivity scale are materials that exhibit a large attractive force between the nucleus and the orbiting electrons. The net effect is a great deal of resistance to the movement of electrons. These materials are known as *dielectrics*. They have low conductivity and high resistivity. In electrical circuits and products, dielectric materials such as silicon dioxide (glass) are used as insulators.

An electrical device known as a *capacitor* is formed whenever a dielectric layer is sandwiched between two conductors. In semiconductor structures, capacitors are formed in MOS gate structures, between metal layers and silicon substrates separated by dielectric layers, and other structures (see Chapter 16). The practical effect of a capacitor is that it stores electrical charges. Capacitors are used for information storage in memory devices to prevent unwanted charges to build up in conductors and silicon surfaces, and to form the working parts of field effect (MOS) transistors. The capacitance ability of a film is relative to the area and thickness and a property parameter known as the *dielectric constant*. Semiconductor metal conduction systems need high conductivity and, therefore, low-resistance and low-capacitance materials. These are referred to as *low-k dielectrics*. Dielectric layers used as insulators between conducting layers need high capacitances or *high-k dielectrics*.

$$C = \frac{kE_0 A}{t}$$

where  $C$ = capacitance
$k$ = dielectric constance of material
$E_0$ = permittivity of free space (free space has the highest "capacitance")
$A$ = area of capacitor
$t$ = thickness of dielectric material

### Resistors

An electrical factor related to the degree of conductivity (and resistivity) of a material is the electrical resistance of a specific volume of the material. The resistance is a factor of the resistivity and dimensions of the material. Resistance to electrical flow is measured in ohms as illustrated in Fig. 2.5.

The formula defines the electrical resistance of a specific volume of a specific material (in this illustration, the volume is a rectangular bar with dimensions X, Y, and Z). The relationship is analogous to density and weight, *density* being a material property and *weight* being the force exerted by a specific volume of the material.

Electric current flow is analogous to water flowing in a hose. For a given hose diameter and water pressure, only a given amount of water will flow out of the hose. The resistance to flow can be reduced by increasing the hose diameter, shortening the hose, and/or increasing the pressure. In an electrical system, the electron flow can be increased by increasing the cross section of the material, shortening the length of the piece, increasing the voltage (analogous to pressure), and/or decreasing the resistivity of the material.

### Intrinsic Semiconductors

Semiconducting materials, as the name implies, are materials that have some natural electrical conducting ability. There are two elemental semiconductors (silicon and germanium), and both are found in col-

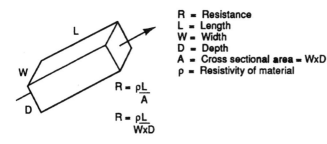

**Figure 2.5**  Resistance of rectangular bar.

umn IV (Fig. 2.6) of the periodic table. In addition, there are some tens of material compounds (a compound is a material containing two or more chemically bound elements) that also exhibit semiconducting properties. These compounds come from elements found in columns III and V, such as gallium arsenide, indium gallium phosphide (InGaP), and gallium phosphide (GaP). Others are compounds from elements from columns II and VI of the periodic table.

The term *intrinsic* refers to these materials in their purified state and not contaminated with impurities or dopants purposely added to change properties.

**Doped Semiconductors**

Semiconducting materials, in their intrinsic state, are not useful in solid-state devices. However, through a process called *doping,* specific elements can be introduced into intrinsic semiconductor materials. These elements increase the conductivity of the intrinsic semiconductor material. The doped material displays two unique properties that *are* the basis of solid-state electronics. The two properties are

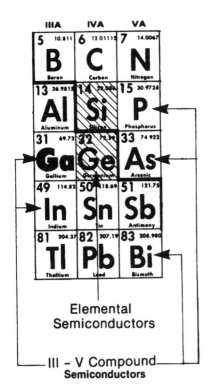

**Figure 2.6**  Semiconductor materials.

1. Precise resistivity control through doping

2. Electron and hole conduction

**Resistivity of doped semiconductors.**  Metals have a conductivity range limited to $10^4$ to $10^6$/ohm-cm. The implications of this limit are illustrated by an examination of the resistor represented in Fig. 2.5. Given a specific metal with a specific resistivity, the only way to change the resistance of a given volume is to change the dimensions. In a semiconducting material, the resistivity can be changed, giving another degree of freedom in the design of the resistor. Semiconductors are such a material. Their resistivity can be extended over the range of $10^{-3}$ to $10^3$ by the addition of dopant atoms.

Semiconducting materials can be doped into a useful resistivity range by elements that make the material either electron rich (N-type) or hole rich (P-type).

Figure 2.7 shows the relationship of the doping level to the resistivity of silicon. The x-axis is labeled the carrier concentration because the electrons or holes in the material are called *carriers*. Note that there are two curves: N-type and P-type. That is due to the different amount of energies required to move an electron or a hole through the material. As the curves indicate, it takes less of a concentration of N-type dopants than P-type dopants to create a given resistivity in silicon. Another way to express this phenomenon is that it takes less energy to move an electron than to move a hole.

It takes only 0.000001 to 0.1 percent of a dopant to bring a semiconductor material into a useful resistivity range. This property of semiconductors allows the creation of regions of very precise resistivity values in the material.

### Electron and Hole Conduction

Another limit of a metal conductor is that it conducts electricity only through the movement of electrons. Metals are permanently N-type. Semiconductors can be made either N- and P-type by doping with specific dopant elements. N- and P-type semiconductors can conduct electricity by either electrons or holes. Before examining the conduction mechanism, it is instructive to examine the creation of free (or extra) electrons or holes in a semiconductor structure.

To understand the situation of N-type semiconductors, consider a piece of silicon (Si) doped with a very small amount of arsenic (As) as shown in Fig. 2.8. Assuming even mixing, each of the arsenic atoms would be surrounded by silicon atoms. Applying the rule from the "Periodic Table of the Elements" section that atoms attempt to stabilize by having eight electrons in their outer ring, the atom is shown shar-

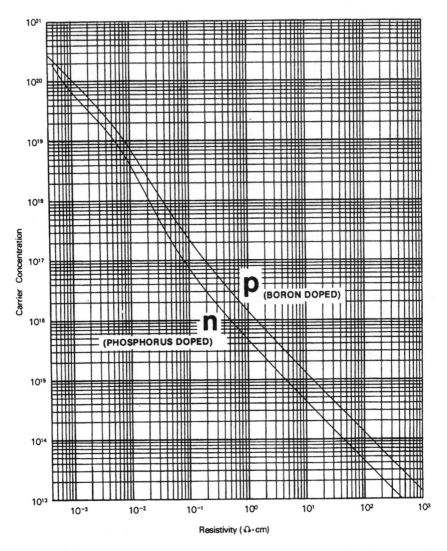

**Figure 2.7**  Silicon resistivity versus doping (carrier) concentration. *(After Thurber et al., Natl. Bur. Standards Spec. Publ. 400-64, May 1981, Tables 10 and 14.)*

ing four electrons from its neighboring silicon atoms. However, arsenic is from column V, which means it has five electrons in its outer ring. The net result is that four of them pair up with electrons from the silicon atoms, leaving one left over. This one electron is available for electrical conduction.

Considering that a crystal of silicon has millions of atoms per $cm^3$, there are lots of electrons available to conduct an electrical current. In

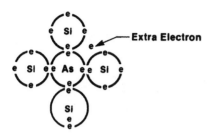

Extra Electron

**Figure 2.8** N-type doping of silicon with arsenic.

silicon, the elements arsenic, phosphorus and antimony create N-type conditions.

An understanding of P-type material is approached in the same manner (Fig. 2.9). The difference is that only boron, from column III of the periodic table, is used to make silicon P-type. When mixed into the silicon, it too borrows electrons from silicon atoms. However, having only three outer electrons, there is a place in the outer ring that is not filled by an electron. This unfilled position is defined as a *hole*.

Within a doped semiconductor material, there is a great deal of activity: holes and electrons are constantly being created. The electrons are attracted to the unfilled holes, in turn leaving an unfilled position, which creates another hole.

How the electrons contribute to electrical conduction is illustrated in Fig. 2.10. When a voltage is applied across a piece of conducting or semiconducting material, the negative electrons move toward the positive pole of the voltage source, such as a battery.

In P-type material (Fig. 2.11), an electron will move toward the positive pole by jumping into a hole along the direction of route ($t_1$).

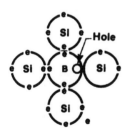

**Figure 2.9** P-type doping of silicon with boron.

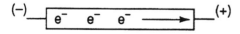

**Figure 2.10** Electron conduction in N-type semiconductor material.

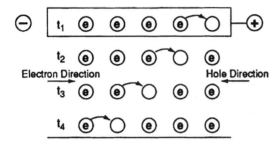

**Figure 2.11**  Hole conduction in P-type semiconductor material.

Of course, when it leaves its position, it leaves a new hole. As it continues toward the positive pole, it creates a succession of holes. The effect to someone measuring this process with a current meter is that the material is supporting a positive current, when actually it is a negative current moving in the opposite direction. This phenomenon is called *hole flow* and is unique to semiconducting materials.

The dopants that create a P-type conductivity in a semiconductor material are called *acceptors*. Dopants that create N-type conditions are called *donors*. An easy way to keep these terms straight is that acce*p*tor has a p and do*n*or is spelled with an n.

The electrical characteristics of conductors, insulators, and semiconductors are summarized in Fig. 2.12. The particular characteristics of doped semiconductors are summarized in Fig. 2.13.

N- and P-type conditions are also created in germanium and compound semiconductors with specific dopant elements.

| Classification | Electrons | Examples | Conductivity |
|---|---|---|---|
| 1. Conductor | Free to Move | Gold<br>Copper<br>Silver | $10^4$–$10^6$<br>/ohm-cm |
| 2. Insulator<br>(Dielectric) | Bound | Glass<br>Plastic | $10^{-22}$–$10^{-10}$<br>/ohm-cm |
| 3. Semiconductor<br>a. Intrinsic | Some Available | Germanium<br>Silicon<br>III-IV | $10^{-9}$–$10^3$<br>/ohm-cm |
| b. Doped | Controlled<br>Amount Available | N-type Semiconductor<br>P-type Semiconductor | |

**Figure 2.12**  Electrical classification of materials.

| | N-TYPE | P-TYPE |
|---|---|---|
| 1. Conduction | Electrons | Holes |
| 2. Polarity | Negative | Positive |
| 3. Dopant Term | Donor | Acceptor |
| 4. Doping Elements in Silicon | Arsenic Phosphorus Antimony | Boron |

**Figure 2.13**   Characteristics of doped semiconductors.

### Carrier mobility

It was mentioned previously that it takes "less energy" to move an electron than a hole through a piece of semiconducting material. In a circuit we are interested in both the energy required to move these carriers (holes and electrons) and the speed at which they move. The speed of movement is called the *carrier mobility,* with holes having a lower mobility than electrons. This factor is an important consideration in selecting a particular semiconducting material for a circuit.

## Semiconductor Production Materials

### Germanium and silicon

Germanium and silicon are the two elemental semiconductors. The first transistor was made with germanium, as were the initial devices of the solid-state era. However, germanium presents problems in processing and in device performance. Its 937°C melting point limits high-temperature processing. More importantly, its lack of a natural occurring oxide leaves the surface prone to electrical leakage.

The development of silicon/silicon dioxide planar processing solved the leakage problem of integrated circuits, flattened the surface profile of the circuits and allowed higher temperature processing due to its 1415°C melting point. Consequently, silicon represents over 90 percent of the wafers processed worldwide.

### Semiconducting Compounds

There are many semiconducting compounds formed from elements listed in columns III and IV and II to VI of the periodic table. Of these compounds, the ones most used in commercial semiconductor devices are gallium arsenide (GaAs) and gallium arsenide-phosphide (GaAsP), indium phosphide (InP), gallium aluminum arsenic (GaAlAs), and indium gallium phosphide (InGaP).[1] These compounds have special properties.[2] Diodes made from GaAs and GaAsP give off visible and laser light when activated with an electrical current. They are the materials used to make the light-emitting diodes (LEDs) used in electronic panel displays.

An important property of gallium arsenide is its high (electrical) carrier mobility. This property allows a gallium arsenide device to react to high-frequency microwaves and effectively switch them into electrical currents in communications systems faster than silicon devices.

This same property, carrier mobility, is the basis for the excitement over gallium arsenide transistors and ICs. Devices of GaAs operate two to three times faster than comparable silicon devices and find applications in super-fast computers and real-time control circuits such as airplane controls.

GaAs has a natural resistance to radiation-caused leakage. Radiation, such as that found in space, causes holes and electrons to form in semiconductor materials. It gives rise to unwanted currents that can cause the device or circuit to malfunction or cease functioning. Devices that can perform in a radiation environment are known as *radiation hardened*. GaAs is naturally radiation hardened.

GaAs is also semi-insulating. In an integrated circuit, this property minimizes leakage between adjacent devices, allowing a higher packing density, which in turn results in a faster circuit because the holes and electrons travel shorter distances. In silicon circuits, special isolating structures must be built into the surface to control surface leakage. These structures take up valuable space and reduce the density of the circuit.

Despite all of the advantages, GaAs is not expected to replace silicon as the mainstream semiconducting material. The reasons reside in the trade-offs between performance and processing difficulty. While GaAs circuits are very fast, the majority of electronic products do not require their level of speed. On the performance side, GaAs, like germanium, does not possess a natural oxide. To compensate, layers of dielectrics must be deposited on the GaAs, which leads to longer processing and lower yields. Also, half of the atoms in GaAs are arsenic, an element that is very dangerous to human beings. Unfortunately, the arsenic evaporates from the compound at normal process temperatures, requiring the addition of suppression layers (caps) or pressurized process chambers. These steps lengthen the processing and add to its cost.

Evaporation also occurs during the crystal growing stage, resulting in nonuniform crystals and wafers. The nonuniformity produces wafers that are very prone to breakage during fab processing. Also, the production of large-diameter GaAs wafers has lagged behind that of silicon (see Chapter 3).

Despite the problems, gallium arsenide is an important semiconducting material that will continue to increase in use and will probably have a major influence on computer performance of the future.

### Silicon Germanium

Competitors to GaAs are silicon/germanium (SiGe) structures. The combination increases transistor speeds to levels that allow ultra-fast radios and personal communication devices.[3] Device/IC structures feature a layer of germanium deposited by ultra-high vacuum/chemical vapor deposition (UHV/CVD).[4] Bipolar transistors are formed in the Ge layer. Unlike the simpler transistors formed in silicon technology, SiGe required transistors with *hetrostructures* or *heterojunctions*. These are structures with several layers and specific dopant levels to allow high-frequency operations (see Chapter 16).

A comparison of the major semiconducting production materials and silicon dioxide is presented in Fig. 2.14.

### Engineered Substrates

A bulk wafer was the traditional substrate for fabricating microchips. Electrical performance demands new substrates, such as silicon on an

|  | Ge | Si | GaAs | SiO$_2$ |
|---|---|---|---|---|
| Atomic Weight | 72.6 | 28.09 | 144.63 | 60.08 |
| Atoms/cm$^3$ or Molecules | $4.42 \times 10^{22}$ | $5.00 \times 10^{22}$ | $2.21 \times 10^{22}$ | $2.3 \times 10^{22}$ |
| Crystal Structure | Diamond | Diamond | Zinc–Blends | Amorphous |
| Atoms/Unit Cell | 8 | 8 | 8 | — |
| Density | 5.32 | 2.33 | 5.65 | 2.27 |
| Energy Gap | 0.67 | 1.11 | 1.40 | 8 (approx.) |
| Dielectric Constant | 16.3 | 11.7 | 12.0 | 3.9 |
| Melting Point (°C) | 937° | 1415° | 1238° | 1700° (approx.) |
| Breakdown Field (V/ ) | 8 (approx.) | 30 (approx.) | 35 (approx.) | 600 (approx.) |
| Linear Coefficient of Thermal Expansion $\frac{\Delta L}{LT} \quad \frac{1}{C}$ | $5.8 \times 10^{-6}$ | $2.5 \times 10^{-6}$ | $5.9 \times 10^{-6}$ | $0.5 \times 10^{-6}$ |

**Figure 2.14**  Physical properties of semiconductor materials.

insulator (SOI) such as sapphire, and silicon on diamond (SOD). Diamond dissipates heat better than silicon. Another structure is a layer of "strained" silicon deposited on a wafer of silicon-germanium. Strained silicon occurs when silicon atoms are deposited on a Si/Ge (sSOI) layer previously deposited on an insulator. Si/Ge atoms are more widely spaced than normal silicon. During the deposition, the silicon atoms "stretch" to align to the SI/Ge atoms, staining the silicon layer. The electrical effect is to lower the silicon resistance, allowing electrons to move up to 70 percent faster. This structure brings performance benefits to MOS transistors (see Chapter 16).

### Ferroelectric Materials

In the ongoing search for faster and more reliable memory structures, ferroelectrics have emerged as a viable option. A memory cell must store information in one of two states (on/off, high/low, 0/1), be able to respond quickly (read and write), and be capable of changing states reliably. Ferroelectric material capacitors such as $PbZr_{1-x}T_xO_3$ (PZT) and $SrBi_2Ta_2O_9$ (SBT) exhibit these desirable characteristics. They are incorporated into SiCMOS (see Chapter 16) memory circuits known as ferroelectric random access memories (FeRAMs).[5]

### Diamond Semiconductors

Moore's law cannot go indefinitely into the future. One end point is when the transistor parts become so tiny that the physics governing transistor action no longer work. Another limit is heat dissipation. Bigger and denser chips run very hot. Unfortunately, high heat also degrades the electrical operations and can render the chip useless. Diamond is a crystal material that dissipates heat much faster than silicon. Despite this positive aspect, diamond as a semiconductor wafer has faced barriers of cost, uniformity, and finding a supply of large diamonds. However, there is new research into making synthetic diamonds using vapor deposition techniques. Doping diamond is the next barrier. This material is being explored and may find its way into fabrication areas of the future.[6]

### Process Chemicals

It should be fairly obvious that extensive processing is required to change the raw semiconducting materials into useful devices. The majority of these processes use chemicals. In fact, microchip fabrication is primarily a chemical process or, more correctly, a series of chemical processes. Up to 20 percent of all process steps are cleaning or wafer surface preparation.[6]

Great quantities of acids, bases, solvents, and water are consumed by a semiconductor plant. Part of this cost is due to the extremely high purities and special formulations required of the chemicals to allow precise and clean processing. Larger wafers and higher cleanliness requirements need more automated cleaning stations and the cost of removal of spent chemicals is rising. When the costs of producing a chip are added up, process chemicals can be up to 40 percent of all manufacturing costs.

The cleanliness requirements for semiconductor process chemicals are explored in Chapter 4. Specific chemicals and their properties are detailed in the process chapters.

### Molecules, compounds, and mixtures

At the beginning of this chapter, the basic structure of matter was explained by the use of the Bohr atomic model. This model was used to explain the structural differences of the elements that make up all the materials in the physical universe. But it is obvious that the universe contains more than 103 (the number of elements) types of matter.

The basic unit of a nonelemental material is the molecule. The basic unit of water is a molecule composed of two hydrogen atoms and one oxygen atom. The multiplicity of materials comes about from the ability of atoms to bond together to form molecules.

It is inconvenient to draw diagrams such as in Fig. 2.15 every time we want to designate a molecule. The more common practice is to write the molecular formula. For water, it is the familiar $H_2O$. This formula tells us exactly the elements and their number in the material. Chemists use the more precise term *compound* in describing different combinations of elements. Thus, $H_2O$ (water), NaCl (sodium chloride or salt), $H_2O_2$ (hydrogen peroxide), and $As_2O_3$ (arsine) are all different compounds composed of aggregates of individual molecules.

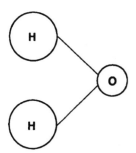

**Figure 2.15**  Diagram of water molecule.

Slurries are used in polishing operations such as *chemical mechanical polishing* (CMP). Typical slurries have fine pieces of silica (glass) suspended in a mild base solution such as ammonium hydroxide.

Some elements combine into *diatomic molecules*. A diatomic molecule is one composed of two atoms of the same element. The familiar process gases (oxygen, nitrogen, and hydrogen), in their natural state, are all composed of diatomic molecules. Thus, their formula's are $O_2$, $N_2$, and $H_2$.

Materials also come in two other forms: mixtures and solutions. Mixtures are composed of two or more substances, but the substances retain their individual properties. A mixture of salt and pepper is the classic example.

Solutions are mixtures of a solid dissolved in a liquid. In the liquid, the solids are interspersed, with the solution taking on unique properties. However, the substances in a solution do not form into a new molecule. Saltwater is an example of a solution. It can be separated back into its starting parts: salt and water.

### Ions

The term *ion* or *ionic* is used often in connection with semiconductor processing. This term refers to any atom or molecule that exists in a material with an unbalanced charge. An ion is designated by the chemical symbol of the element or molecule followed by a superscripted positive or negative sign ($Na^+$, $Cl^-$). For example, one of the serious contamination problems is mobile ionic contamination such as sodium ($Na^+$). The problem comes from the positive charge carried by the sodium when it gets into the semiconductor material or device. In some processes, such as the ion implantation process, it is necessary to create an ion, such as boron ($B^+$), to accomplish the process.

### States of Matter

#### Solids, liquids, and gases

Matter is found in four different states. They are solids, liquids, gases, and plasma (Fig. 2.16).

- Solid
- Liquid
- Gas
- Plasma

**Figure 2.16**  Four states of nature.

- Solids are defined as having a definite shape and volume, which is retained under normal conditions of temperature and pressure.

- Liquids have definite volume but a variable shape. A liter of water will take the shape of any container in which it is stored.

- Gases have neither definite shape nor volume. They too will take the shape of any container but, unlike liquids, they will expand or can be compressed to entirely fill the container.

The state of a particular material has a lot to do with its pressure and temperature. Temperature is a measure of the total energy incorporated in the material. We know that water can exist in three states (ice, liquid water, and steam or water vapor) simply by changing the temperature and/or pressure. The influence of pressure is more complicated and beyond the scope of this text.

### Plasma state

The fourth state of nature is plasma. A star is an example of a plasma state. It certainly does not meet the definitions of a solid, liquid, or gas. A plasma state is defined as a high-energy collection of ionized atoms or molecules. Plasma states can be induced in process gases by the application of high-energy radio-frequency (RF) fields. They are used in semiconductor technology to cause chemical reactions in gas mixtures. One of their advantages is that energy can be delivered at a lower temperature than in convention systems, such as convection heating in ovens.

### Properties of Matter

All materials can be differentiated by their chemical compositions and the properties that arise from those compositions. In this section, several key properties required to understand and work with properties of semiconductor materials and chemicals are defined.

### Temperature

The temperature of a chemical exerts great influence on its reactions with other chemicals, whether in an oxidation tube or in a plasma etcher. Additionally, safe use of some chemicals requires knowledge and control of their temperatures. Three temperature scales are used to express the temperature of a material. They are the Fahrenheit, Centigrade (or Celsius) and the Kelvin scale (Fig. 2.17).

The Fahrenheit scale was developed by Gabriel Fahrenheit, a German physicist, using a water and salt solution. He assigned to the so-

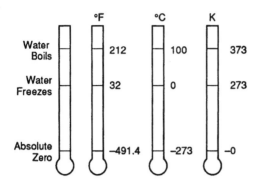

Figure 2.17  Temperature scales.

lution's freezing temperature the value of zero degrees Fahrenheit (0°F). Unfortunately, the freezing temperature of pure water is more useful, and we have ended up with the Fahrenheit scale having a water freezing point at 32°F and a boiling point of 212°F, with 180° between the two points.

The Celsius or centigrade scale is more popular in scientific endeavors. It more sensibly sets the freezing point of water at 0°C and boiling at 100°C. Note that there are exactly 100 degrees Celsius between the two points. This means that a one-degree change in temperature as measured on the centigrade scale requires more energy than a one-degree change on the Fahrenheit scale.

The third temperature scale is the Kelvin scale. It uses the same scale factor as the centigrade scale but is based on absolute zero. Absolute zero is the theoretical temperature at which all atomic motion would cease. This value corresponds to –273°C. On the Kelvin scale, water freezes at 273 K and boils at 373 K.[*]

### Density, specific gravity, and vapor density

An important property of matter is density. When we say that something is *dense*, we refer to its mass or weight per unit volume. A cork has a lower density than an equal volume of iron. Density is expressed as the weight, in grams, per cubic centimeter of the material. Water is the standard, with one cubic centimeter (at 4°C) weighing 1 g. The densities of other substances are expressed as a ratio of their density to that of a comparable volume of water. Silicon has a density of 2.3. Therefore, a piece of silicon one cubic centimeter (1 cm$^3$) in volume will weigh 2.3 g.

*Specific gravity* is a term used to reference the density of liquids and gases at 4°C. It is the ratio of the density of the substance to that of

---

[*]The degree symbol is omitted when using the Kelvin scale.

water. Gasoline has a specific gravity of 0.75, which means it is 75 percent as dense as water.

*Vapor density* is a density measurement of gases under certain conditions of temperature and pressure. The reference is air, with one cubic centimeter having an assigned density of one (1). Hydrogen has a vapor density of 0.60 which makes it 60 percent the density of a similar volume of air. The contents of a container of hydrogen will weigh 60 percent less than a similar container of air.

## Pressure and Vacuum

Another important aspect of matter is pressure. Pressure, as a property, is usually used in connection with liquids and gases. It is defined as the force per unit area exerted against the surface of the container. It is the gas pressure in a cylinder that forces the gas out into a process chamber. All processes machines using gases must have gauges to measure and control the pressure.

Pressures are expressed in pounds per square inch of area (psia), in atmospheres or in torrs. An atmosphere (A) is the pressure exerted by the atmosphere surrounding the Earth at a specific temperature. Thus, a high-pressure oxidation system operated at 5 atmospheres contains a pressure 5 times that of the atmosphere.

One atmosphere of air has a pressure of 14.7 psia. Pressures inside gas tanks are measured in psig units or pounds per square inch gauge. This means that the gauge reading is absolute; it does not include the pressure of the outside atmosphere.

*Vacuum* is also a term and condition encountered in semiconductor processing. It is actually a condition of low pressure. Generally, pressures below standard atmospheric pressures are referred to as vacuums. But a vacuum condition is measured in units of pressure.

Low pressures tend to be expressed in torrs. The unit is named after the Italian scientist, Torricelli, who made many of the early discoveries in the field of gases and their properties. A torr is defined as the equivalent of one millimeter of mercury in a pressure measuring device known as a *manometer.*

Imagine the effect on the column of mercury in the manometer in Fig. 2.18*a* of increasing the pressure above atmospheric pressure. As the pressure goes up, it pushes down the mercury in the dish and raises the mercury in the column. Now imagine what happens as air is extracted from the system (Fig 2.18*b*) below atmospheric pressure creating a vacuum. As long as there are any gas molecules or atoms in the manometer, some small pressure will be exerted on the mercury in the dish, and the mercury in the column will rise some small but finite

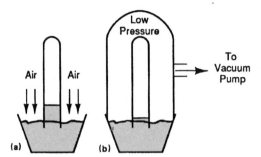

**Figure 2.18** Pressure vacuum measurement.

amount. The amount of the rise as measured in millimeters (mm) is relative to the pressure or, in this case, the vacuum.

Vacuum systems for evaporation, sputtering, and ion implantation are operated at vacuums (pressures) of $10^{-6}$ to $10^{-9}$ torrs. Translated into a vacuum system containing a simple manometer, this means that the column of mercury would rise only 0.000000001 ($1 \times 10^{-9}$) to 0.000001 ($1 \times 10^{-6}$) millimeters—a *very* short length! In actual practice, a mercury manometer cannot measure these extremely low pressures. Other, more sensitive gauges are used.

## Acids, Alkalis, and Solvents

### Acids and alkalis

Semiconductor processing requires large amounts of liquid chemicals to etch, clean, and rinse the wafers and packages. These chemicals are identified by chemists into three major classifications.[7]

- Acids
- Alkalis
- Solvents

Acids and alkalis differ from each other due to the presence of specific ions in the liquid. Acids contain *hydrogen ions,* while alkalis (also called *bases*) contain *hydroxide ions.* An examination of the water molecule explains the differences.

The chemical formula for water normally is written as $H_2O$. It can also be written in the form HOH. When separated into its parts, we find that water is made up of a positively charged hydrogen ion ($H^+$) and a negatively charged hydroxide ion ($OH^-$).

When water is mixed with other elements, either the hydrogen or hydroxyl ion combines with other substances (Fig. 2.19). Liquids that contain the hydrogen ion are called *acids*. Liquids that contain the hydroxyl ion are called *alkalis* or *bases*. Acids and bases are commonly

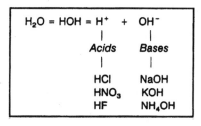

$H_2O = HOH = H^+ + OH^-$

| Acids | Bases |
|-------|-------|
| HCl | NaOH |
| HNO$_3$ | KOH |
| HF | NH$_4$OH |

**Figure 2.19**  Acid and base solutions.

found in the home: lemon juice and vinegar are acids, and ammonia and baking soda in a solution of water are bases.

Acids are further divided into two categories: organic and inorganic. Organic acids are those that contain hydrocarbons, whereas inorganic acids do not. Sulfonic acid is an organic acid, and hydrogen fluoride (HF) is an inorganic acid.

The strength and reactivity of acids and bases are measured by the pH scale (Fig. 2.20). This scale ranges from 0 to 14, with 7 being a neutral point. Water is neutral, neither an acid or a base; therefore, it has a pH of 7. Strong acids, such as sulfuric acid ($H_2SO_4$), will have low pH values of 0 to 3. Strong bases, such as sodium hydroxide (NaOH), have pH values greater than 7.

Both acids and bases are reactive with skin and other chemicals and should be stored and handled with all of the prescribed safety precautions.

### Solvents

Solvents are liquids that do not ionize; they are neutral on the pH scale. Water is a solvent; in fact, it is the solvent with the greatest

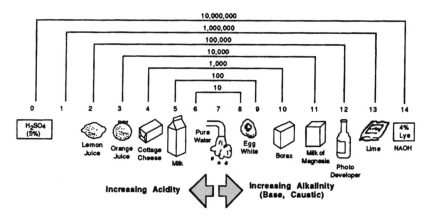

**Figure 2.20**  pH scale.

ability to dissolve other substances. It is also the most commonly used solvent in semiconductor processing. Alcohol and acetone are other common solvents in the wafer fabrication process.

Most of the solvents in fab processing are volatile, flammable, or combustible. It is important to use them in properly exhausted stations and observe prescribed precautions in their storage and use.

## Chemical Purity and Cleanliness

While the names of chemicals used in fabrication areas sound familiar, there is an entire supply industry dedicated to producing the highest-quality chemicals to meet semiconductor processing demands. Chemicals must meet very high purity requirements. In general the target is "six nines" purity. This translates to 99.9999 percent pure. Physical contamination such as particles are also controlled. Typical chemical specifications limit particles to parts per billion (PPB)/liter and microns. These and other specifications are established in the 2001 *International Technology Roadmap for Semiconductors* (ITRS). Specific chemicals used are identified in the chapter on contamination control and the individual process chapters.

### Safety issues

The storage, use, and disposal of chemicals and electrical and other risks are present in semiconductor process areas. Companies address these risks by developing employee knowledge, skill, and awareness through training programs and safety inspections.

### The Material Safety Data Sheet

For every chemical brought into a manufacturing site, the supplier must provide a form called the Material Safety Data Sheet (MSDS). It is required by the federal Occupational, Safety, and Health Administration (OSHA). The form is also called OSHA Form 20. This form contains chemical, storage, health, first aid, and usage information about the chemical. Under current regulations, the MSDSs must be filed on the site and available to employees.

## Review Questions

1. Describe the electrical difference between a conductor, a dielectric and a semiconductor.

2. Why are doped semiconductors required for solid-state devices?

3. Which has a higher resistivity, a metal or an intrinsic semiconductor?

4. Give two reasons silicon is the most common semiconducting material.

5. Name one element that will make silicon "N" type.

6. A "P"-type semiconductor exhibits a (negative or positive) current.

7. Indicate which classification (acid, base, solvent) is described below.

   a. Contains $OH^-$ ions
   b. Contains $H^-$ ions
   c. Is neutral on the pH scale

8. Acids have a pH between _____ and _____.

9. Bases have a pH between _____ and _____.

10. What is the pH of water?

## References

1. Fujitsu Quantum Devices Limited, website.
2. R. E. Williams, *Gallium Arsenide Processing Techniques,* Artech House, Inc., Dedham, MA, 1984.
3. "Industry News," *Semiconductor International,* March 1994, p. 22.
4. W. Conrad Holton, *Silicon Germanium: Finally for Real, Solid State Electronics,* November 1997, p. 119.
5. Robert E. Jones, "Integration of Ferroelectrics into Nonvolatile Memories," *Solid State Technology,* October 1997, p. 201.
6. Joshua Davis, The Diamond Wars Have Begun, Wired Magazine, Sept. 2003, p. 96.
7. R. Allen et al., "MNST Wafer Cleaning," *Solid State Technology,* January 1994, p. 61.

# 3

# Crystal Growth and Silicon Wafer Preparation

## Overview

In this chapter, the preparation of semiconductor-grade silicon from sand, its conversion into crystals and wafers (material preparation stage), and the processes required to produce polished wafers (crystal growth and wafer preparation) are explained.

## Objectives

Upon completion of this chapter, you should be able to:

1. Explain the difference between crystalline and noncrystalline materials.
2. Explain the differences between a polycrystalline and a single crystalline material.
3. Draw a diagram of the two major wafer crystal orientations used in semiconductor processing.
4. Explain the Czochralski, float zone and liquid crystal encapsulated Czochralski methods of crystal growing.
5. Draw a flow diagram of the wafer preparation process.
6. Explain the use and meaning of the flats or notches ground on wafers.
7. Describe the benefits in the wafer fab process that come from edge-rounded wafers.

8. Describe the benefits in the wafer fab process that come from flat and damage-free wafer surfaces.

## Introduction

The evolution of higher-density and larger-size chips has required the delivery of larger diameter wafers. Starting with 1-in diameter wafers in the 1960s, the industry is now introducing 300-mm (12-in) diameter wafers into production lines. According the International Technology Roadmap for Semiconductors, 300-mm wafers will be the standard diameter until about 2007, and 400- or 450-mm diameter wafers are predicted for the far future (Figure 3.1). Larger-diameter wafers are necessary to accommodate increasing chip sizes with cost effective wafer fabrication processes (see Chaps. 6 and 15). The challenges in wafer preparation are formidable. In crystal growth, the issues of structural and electrical uniformity and contamination become challenges. In wafer preparation, flatness, diameter control, and crystal integrity are issues. Larger diameters are heavier, which requires more substantial process tools and, ultimately, full automation. A production lot of 300-mm diameter wafers weighs about 20 lb (7.5 kg) and can have a production value of half a million dollars or more.[1] These challenges coexist with ever tightening specifications for almost every parameter. Keeping abreast of these challenges and providing ever larger diameter wafers is a key to continued microchip evolution.

## Semiconductor Silicon Preparation

Semiconductor devices and circuits are formed in and on the surface of wafers of a semiconductor material, usually silicon. Those wafers must have very low levels of contaminants, be doped to a specified resistivity level, have a specific crystal structure, be optically flat, and meet a host of other mechanical and cleanliness specifications. Manufacture of IC grade silicon wafers proceeds in four stages.

| Year of production | 2001 | 2006 | 2012 |
|---|---|---|---|
| Wafer diameter (mm) | 300 | 300 | 450 |

**Figure 3.1**  Wafer diameters. *(Courtesy of SIA.)*

**Silicon wafer preparation stages**

- Conversion of ore to a high-purity gas
- Conversion of gas to polysilicon silicon
- Conversion of polysilicon silicon to a single crystalline, doped crystal ingot
- Preparation of wafers from the crystal ingot

The first stage of semiconductor manufacturing is the extraction and purification of the raw semiconductor material(s) from the earth. Purification starts with a chemical reaction. For silicon, it is the conversion of the ore to a silicon-bearing gas such as silicon tetrachloride or trichlorosilane. Contaminants, such as other metals, are left behind in the ore remains. The silicon bearing gas is then reacted with hydrogen (Fig. 3.2) to produce semiconductor grade silicon. The silicon produced is 99.9999999 percent pure, one of the purist materials on Earth.[2] It has a crystal structure known as polycrystalline or *polysilicon.*

## Crystalline Materials

One way that materials differ is in the organization of their atoms. In some materials, such as silicon and germanium, the atoms are arranged into a very definite structure that repeats throughout the material. These materials are called *crystals.*

Materials without a definite periodic arrangement of their atoms are called noncrystalline or *amorphous.* Plastics are examples of amorphous materials.

## Unit cells

There are actually two levels of atomic organization possible for crystalline materials. First is the organization of the individual atoms. The atoms in a crystal arrange themselves at specific points in a structure known as a *unit cell.* The unit cell is the first level of organization in a crystal. The unit cell structure is repeated everywhere in the crystal.

$$2SiHCl_3 \text{ (gas)} + 3H_2 \text{ (gas)} \rightarrow 2Si \text{ (solid)} + 6HCl \text{ (gas)}$$

**Figure 3.2**  Hydrogen reduction of trichlorosilane.

   Another term used to reference crystal structures is *lattice*. A crystalline material is said to have a specific lattice structure and that the atoms are located at specific points in the lattice structure.

   The number of atoms, relative positions, and binding energies between the atoms in the unit cell gives rise to many of the characteristics of the material. Each crystalline material has a unique unit cell. Silicon atoms have 16 atoms arranged into a diamond structure (Fig. 3.3). GaAs crystals have 18 atoms in a unit cell configuration called a zincblend (Fig. 3.4).

**Poly and single crystals**

   The second level of organization within a crystal is related to the organization of the unit cells. In intrinsic semiconductors, the unit cells are *not* in a regular arrangement to each other. The situation is similar to a disorderly pile of sugar cubes, with each cube representing a unit cell. A material with such an arrangement has a polycrystalline structure.

   The second level of organization occurs when the unit cells (sugar cubes) are all neatly and regularly arranged relative to each of the others (Fig. 3.5). Materials thus arranged have a single (or mono-) crystalline structure.

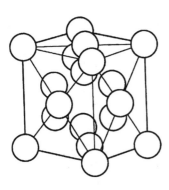

**Figure 3.3**  Unit cell of silicon.

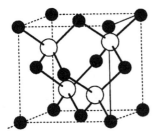

**Figure 3.4**  GaAs crystal structure.

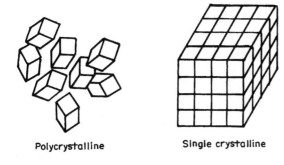

**Polycrystalline**          **Single crystalline**

**Figure 3.5**   Poly- and single-crystal structures.

Single-crystal materials have more uniform and predictable proper-ties than polycrystalline materials. The structure allows a uniform and predictable electron flow in semiconductors. At the end of the fab process, crystal uniformity is essential for separating the wafer into die with non-ragged edges (see Chapter 18).

## Crystal Orientation

In addition to the requirement of a single-crystal structure for a wafer, there is the requirement of a specific *crystal orientation*. This concept can be visualized by considering slicing the single crystalline block in shown in Fig. 3.5. Slicing it in the vertical plane would expose one set of *planes,* while slicing it corner-to-corner would expose a different plane. Each plane is unique, differing in atom count and binding ener-gies between the atoms. Each has different chemical, electrical, and physical properties that are imparted to the wafers. Specific crystal orientations are required for the wafers.

Crystal planes are identified by a series of three numbers known as Miller Indices. Two simple cubic unit cells nestled into the origin of a XYZ coordinate system are shown in Fig. 3.6. The two most common orientations used for silicon wafers are the ⟨100⟩ and the ⟨111⟩ planes. The plane designations are verbalized as the one-oh-oh plane and the one-one-one plane. The brackets indicate that the three numbers are Miller indices.

⟨100⟩ oriented wafers are used for fabricating MOS devices and cir-cuits, while the ⟨111⟩ oriented wafers are used for bipolar devices and circuits. GaAs wafers are also cut along the ⟨100⟩ planes of the crystal.

Note that the ⟨100⟩ plane in Fig. 3.6 has a square shape, while the ⟨111⟩ plane is triangular in shape. These orientations are revealed when wafers are broken as shown in Fig. 3.7. The ⟨100⟩ wafers break

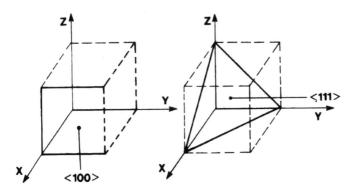

**Figure 3.6**   Crystal planes.

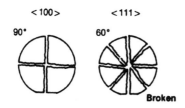

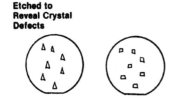

**Figure 3.7**   Wafer orientation indicators.

into quarters or with right angle (90°) breaks. The ⟨111⟩ wafers break into triangular pieces.

## Crystal Growth

Semiconductor wafers are cut from large crystals of the semiconducting material. These crystals, also called *ingots*, are grown from chunks of the intrinsic material, which have a polycrystalline structure and are undoped. The process of converting the polycrystalline chunks to a large crystal of single-crystal structure, with the correct orientation and the proper amount of N- or P-type, is called *crystal growing.*

Three different methods are used to grow crystals: the Czochralski, liquid encapsulated Czochralski, and float zone techniques.

## Czochralski (CZ) method

The majority of silicon crystals are grown by the CZ method (Fig. 3.8). The equipment consists of a quartz (silica) crucible that is heated by surrounding coils that carry radio frequency (RF) waves or by electric heaters. The crucible is loaded with chunks of polycrystalline of the semiconductor material and small amounts of dopant. The dopant material is selected to create either an N-type or P-type crystal. First, the poly and dopants are heated to the liquid state at 1415°C (Fig. 3.9). Next, a seed crystal is positioned to just touch the surface of the liquid material (called the *melt*). The seed is a small crystal that has the same crystal orientation required in the finished crystal. Seeds can be

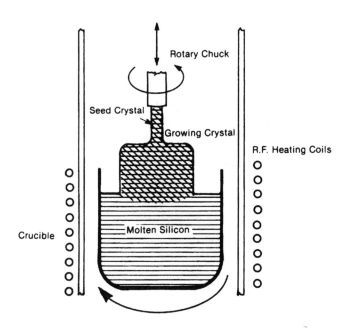

**Figure 3.8**   Czochralski crystal-growing system.

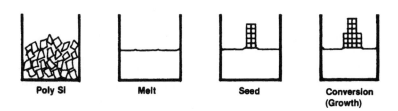

**Figure 3.9**   Crystal growth from a seed.

produced by chemical vapor techniques. In practice, they are pieces of previously grown crystals and are reused.

Crystal growth starts as the seed is slowly raised above the melt. The surface tension between the seed and the melt causes a thin film of the melt to adhere to the seed and then to cool. During the cooling, the atoms in the melted semiconductor material orient themselves to the crystal structure of the seed. The net effect is that the crystal orientation of the seed is propagated in the growing crystal. The dopant atoms in the melt become incorporated into the growing crystal, creating an N- or P-type crystal.

To achieve doping uniformity, crystal perfection, and diameter control, the seed and crucible (along with the pull rate) are rotated in opposite directions during the entire crystal-growing process. Process control requires a complicated feedback system integrating the parameters of rotational speed, pull speeds, and melt temperature.

The crystal is pulled in three sections. First a thin neck is formed, followed by the body of the crystal ending with a blunt tail. The CZ method is capable of producing crystals several feet in length and with diameters up to 12 or more inches. A crystal for 200-mm wafers will weigh some 450 lb (168 kg) and take three days to grow.

### Liquid encapsulated Czochralski (LEC)

LEC crystal growing[3] is used for the growing of gallium arsenide crystals. It is essentially the same as the standard CZ process but with a major modification for gallium arsenide. The modification is required because of the evaporative property of the arsenic in the melt. At the crystal growing temperature, the gallium and arsenic react, and the arsenic can evaporate, resulting in a nonuniform crystal.

Two solutions to the problem are available. One is to pressurize the crystal growing chamber to suppress the evaporation of the arsenic. The other is the LEC process (Fig. 3.10). LEC uses a layer of boron trioxide ($B_2O_3$) floating on top of the melt to suppress the arsenic evaporation. In this method, a pressure of about one atmosphere is required in the chamber.

### Float zone

Float zone crystal growth is one of several processes explained in this text that were developed early in the history of the technology and are still used for special needs. A drawback to the CZ method is the inclusion of oxygen from the crucible into the crystal. For some devices, higher levels of oxygen are intolerable. For these special cases, the

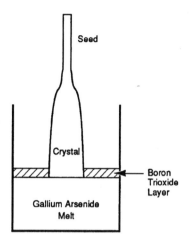

**Figure 3.10**  LEC system of crystal growth.

crystal might be grown by the float zone technique, which produces a lower oxygen content crystal.

Float zone crystal growth (Fig. 3.11) requires a bar of the polysilicon and dopants that have been cast in a mold. The seed is fused to one end of the bar and the assemblage placed in the crystal grower. Conversion of the bar to a single-crystal orientation starts when an RF coil heats the interface region of the bar and seed. The coil is then moved along the axis of the bar, heating it to the liquid point a small section at a time. Within each molten region, the atoms align to the orientation started at the seed end. Thus, the entire bar is converted to a single crystal with the orientation of the starting seed.

Float zone crystal growing cannot produce the large diameters that are obtainable with the CZ process, and the crystals have a higher dislocation density. But the absence of a silica (silicon) crucible yields higher-purity crystals with lower oxygen content. Lower oxygen allows crystals with higher that find use in semiconductor devices such as power thyristors and rectifiers. The two methods are compared in Fig. 3.12.

## Crystal and Wafer Quality

Semiconductor devices require a high degree of crystal perfection. But even with the most sophisticated techniques, a perfect crystal is unobtainable. The imperfections, called *crystal defects*, result in process problems by causing uneven silicon dioxide film growth, poor epitaxial film deposition, uneven doping layers in the wafer, and other problems. In finished devices, the crystal defects cause unwanted current

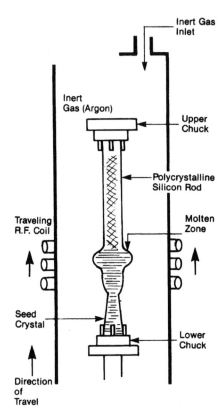

**Figure 3.11** Float zone crystal-growing system.

| PARAMETER | CZ | FLOAT ZONE |
|---|---|---|
| Large Crystal | Yes | Difficult |
| Cost | Lower | |
| Dislocations | $0 - 10^4/cm^2$ | $10^3 - 10^5/cm^2$ |
| Resistivity | Up to 100 ohm-cm | 2000 ohm-cm Max. |
| Radial | | |
| Resistivity | 5 – 10% | 5 – 10% |
| Oxygen Content | $10^{16} - 10^{18}$ atoms/cm³ | 0 – Very Low |

**Figure 3.12** Comparison of CZ and float crystal-growing methods.

leakage and may prevent the devices from operating at required voltages. There are three major categories of crystal defects:

1. Point defects
2. Dislocations
3. Growth defects

## Point defects

Point defects come in two varieties. One comes when contaminants in the crystal become jammed in the crystal structure, causing strain. The second is known as a vacancy. In this situation, there is an atom missing from a location in the structure (Fig. 3.13).

Vacancies are natural phenomena that occur in every crystal. Unfortunately, vacancies occur whenever a crystal or wafer is heated and cooled, such as in the fabrication process. The minimization of vacancies is one of the driving forces behind the desire for low-temperature processing.

## Dislocations

Dislocations are misplacements of the unit cells in a single crystal. They can be imagined as a orderly pile of sugar cubes with one of the cubes slightly out of alignment with the others.

Dislocations occur from growth conditions and lattice strain in the crystal. They also occur in wafers from physical abuse during the fab process. A chip or abrasion of the wafer edge serves as a lattice strain site that can generate a line of dislocations that progresses into the wafer interior with each subsequent high-temperature processing of the wafer. Wafer dislocations are revealed by a special etch of the surface. A typical wafer has a density of 200 to 1000 dislocations per square centimeter.

Etched dislocations appear on the surface of the wafer in shapes indicative of their crystal orientation. ⟨111⟩ wafers etch into triangular dislocations, and ⟨100⟩ wafers show "squarish" etch pits (Fig. 3.7).

## Growth defects

During crystal growth, certain conditions can result in structural defects. One is *slip,* which refers to the slippage of the crystal along crystal planes (Fig. 3.14). Another problem is *twinning.* This is a situation in which the crystal grows in two different directions from the same interface. Both of these defects are cause for rejection of the crystal.

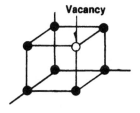

**Vacancy**

**Figure 3.13** Vacancy crystal defect.

Figure 3.14   Crystal slip.

## Wafer Preparation

### End cropping

After removal from the crystal grower, the crystal goes through a series of steps that result in the finished wafer. First is the cropping off of the crystal ends with a saw.

### Diameter grinding

During crystal growth, there is a diameter variation over the length of the crystal (Fig. 3.15). Wafer fabrication processing, with its variety of wafer holders and automatic equipment, requires tight diameter control to minimize warped and broken wafers.

Diameter grinding is a mechanical operation performed in a centerless grinder. This machine grinds the crystal to the correct diameter without the necessity of clamping it into a lathe-type grinder with a fixed center point—although lathe-type grinders are used.

### Crystal orientation, conductivity, and resistivity check

Before the crystal is submitted to the wafer preparation steps, it is necessary to determine whether it meets orientation and resistivity specifications.

The crystal orientation (Fig. 3.16) is determined by either X-ray diffraction or collimated light refraction. In both methods, an end of the crystal is etched or polished to remove saw damage. Next, the crystal is mounted in the refraction apparatus and the X-rays or collimated light reflected off the crystal surface onto a photographic plate (X-rays) or screen (collimated light). The pattern formed on the plate or

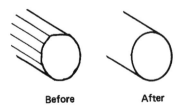

**Before**          **After**

Figure   3.15   Crystal   diameter grinding.

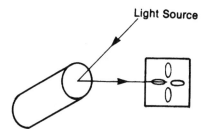

**Figure 3.16** Crystal orientation determination.

screen is indicative of the crystal plane (orientation) of the grown crystal. The pattern shown in Fig. 3.16 is representative of a ⟨100⟩ orientation.

Most crystals are purposely grown several degrees off the major ⟨111⟩ or ⟨100⟩ plane. This off-orientation provides several benefits in wafer fabrication processing, particularly ion implantation. The reasons are covered in the applicable process chapters.

The crystal is positioned on a slicing block to ensure that the wafers will be cut from the crystal in the correct orientation.

Because each crystal is doped, an important electrical check is the conductivity type (N or P) to ensure that the right dopant type was used. A hot-point probe connected to a polarity meter is used to generate holes or electrons (depending on the type) in the crystal. The conductivity type is displayed on the meter.

The amount of dopant put into the crystal is determined by a resistivity measurement using a four-point probe. See Chapter 13 for a description of this measurement technique. The curves presented in Chapter 2 (Fig. 2.7) show the relationship between resistivity and doping concentration for N- and P-type silicon.

The resistivity is checked along the axis of the crystal due to dopant variation during the growing process. This variation results in wafers that fall into several resistivity specification ranges. Later in the process, the wafers will be grouped by resistivity range to meet customer specifications.

### Grinding orientation indicators

Once the crystal is oriented on the cutting block, a flat is ground along the axis (Fig. 3.17). This flat will show up on each of the wafers and is called the *major flat*. The position of the flat is along one of the major crystal planes, as determined by the orientation check.

In the fabrication process, the flat functions as a visual reference to the orientation of the wafer. It is used to place the first pattern mask on the wafer so that the orientation of the chips is always to a major crystal plane.

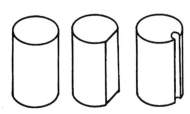

**Figure 3.17** Crystal flat grinding.

On most crystals, there is a second, smaller, secondary flat ground on the edge. The position of the secondary flat to the major flat is a code that tells the wafer fabrication department both the orientation and conductivity type of the wafer. The code is shown in Fig. 3.18.

For larger wafer diameters, a notch is ground on the crystal to indicate the wafer crystal orientation (Fig. 3.17). In some cases, a simple notch is ground into the crystal to act as a production orientation locator.

## Wafer Slicing

The wafers are sliced from the crystal with the use of diamond-coated inside diameter saws (Fig. 3.19). These saws are thin circular sheets of steel with a hole cut out of the center. The inside of the hole is the cutting edge and is coated with diamonds. An inside diameter saw has rigidity, but without being very thick. These factors reduce the kerf (cutting width) size which in turn prevents sizable amounts of the crystal being wasted by the slicing process.

For 300-mm diameter wafers, wire saws are used to ensure flat surfaces with little tapering and with a minimal amount of "kerf" loss.

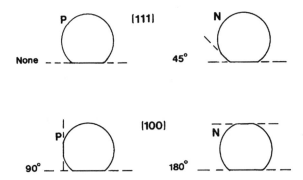

**Figure 3.18** Wafer flat locations.

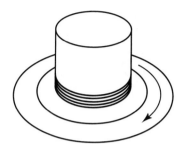

a. Inside Diameter Diamond Saw

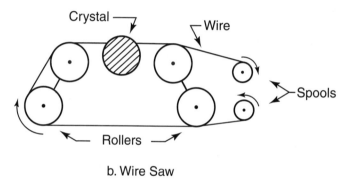

b. Wire Saw

**Figure 3.19**   Inside-diameter saw wafer slicing.

## Wafer Marking

Large-area wafers represent a high value in the wafer fabrication process. Identifying them is necessary to prevent misprocessing and to maintain accurate tracking. Laser marking, using bar codes and a data matrix code[4] (Fig. 3.20), is used. Laser dots are the agreed upon method for 300-mm wafers.

## Rough Polish

The surface of a semiconductor wafer has to be free of irregularities and saw damage and must be absolutely flat. The first requirement comes from the very small dimensions of the surface and subsurface layers making up the device. They have dimensions in the 0.5 to 2 micron range. To get an idea of the relative dimensions of a semiconductor device, imagine the cross section in Fig. 3.21 as tall as a house

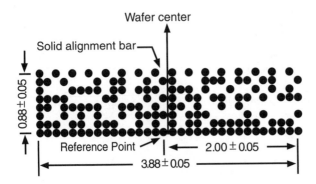

**Figure 3.20** Laser dot coding. *(Reprinted from the Jan. 1998 edition of* Solid State Technology, *copyright 1998 by Penn-Well Publishing Company.)*

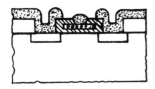

**Figure 3.21** Cross section of a MOS transistor.

wall, about 8 ft (2.4 m). On that scale, the working layers on the top of the wafer would all exist within the top 1 or 2 in (25 or 50 mm) or less.

Flatness is an absolute requirement for small dimension patterning (see Chapter 11). The advanced patterning processes project the required pattern image onto the wafer surface. If the surface is not flat, the projected image will be distorted just as a film image will be out of focus on a non-flat screen.

The flatting and polishing process proceeds in two steps: rough polish and chemical/mechanical polishing (Fig. 3.22). Rough polishing is a conventional abrasive slurry lapping process, but it is fine tuned to semiconductor requirements. A primary purpose of the rough polish is to remove the surface damage left over from the wafer slicing process.

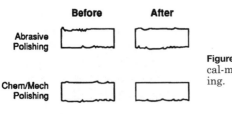

**Figure 3.22** Abrasive and chemical-mechanical surface polishing.

## Chemical Mechanical Polishing (CMP)

The final polishing step is a combination of chemical etching and mechanical buffing. The wafers are mounted on rotating holders and lowered onto a pad surface rotating in the opposite direction. The pad material is generally a cast and sliced polyurethane with a filler or a urethane coated felt. A slurry of silica (glass) suspended in a mild etchant, such as potassium or ammonium hydroxide, is dropped onto the polishing pad.

The alkaline slurry chemically grows a thin layer of silicon dioxide on the wafer surface. The buffing action of the pad mechanically removes the oxide in a continuous action. High points on the wafer surface are removed until an extremely flat surface is achieved. If a semiconductor wafer surface was extended to 10,000 ft (3,048 m—the length of a typical airport runway), it would vary about plus or minus 2 in (50 mm) over its entire length.

Achieving the extreme flatness requires the specification and control of the polishing time, the pressure on the wafer and pad, the speed of rotation, the slurry particle size, the slurry feed rate, the chemistry (pH) of the slurry, and the pad material and conditions.

Chemical/mechanical polishing is one of the techniques developed by the industry that has allowed production of larger wafers. CMP is used in the wafer fabrication process to flatten wafer surfaces after the buildup of new layers creates uneven surfaces. In this application, CMP is the abbreviation for chemical mechanical *planarization.* A detailed explanation of this use of CMP appears in Chapter 10.

## Backside Processing

In most cases, only the front side of the wafer goes through the extensive chem/mech polishing. The backs may be left rough or etched to a bright appearance. For some device use, the backs may receive a special process to induce crystal damage, called *backside damage.* Backside damage causes the growth of dislocations that radiate up into the wafer. These dislocations can act as a trap of mobile ionic contamination introduced into the wafer during the fab process. The trapping phenomenon is called *gettering* (Fig. 3.23). Sandblasting of the back-

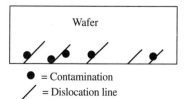

Figure 3.23   Trapping.

● = Contamination
/ = Dislocation line

side is another standard technique. Other methods include the deposition of a polysilicon layer or silicon nitride layer on the back.

## Double-Sided Polishing

One of the many demands on larger diameter wafers is flat and parallel surfaces. Most manufacturers of 300-mm diameter wafers employ double-sided polishing to achieve flatness specifications of 0.25 to 0.18 μm over 25 × 25 mm sites.[5] A downside is that all further processing must employ handling techniques that do not scratch or contaminate the backside.

## Edge Grinding and Polishing

Edge grinding is a mechanical process that leaves the wafer with a rounded edge (Fig. 3.24). Chemical polishing is employed to further create an edge that minimizes edge chipping and damage during fabrication that can result in wafer breakage or serve as the nucleus for dislocation lines.

## Wafer Evaluation

Before packing, the wafers (or samples) are checked for a number of parameters as specified by the customer. Figure 3.25 illustrates a typical wafer specification.

Primary concerns are surface problems such as particulates, stain, and haze. These problems are detected with the use of high-intensity lights or automated inspection machines.

**Before**        **After**

Figure 3.24   Wafer edge grinding.

| Typical 200 mm Wafer Specification | |
|---|---|
| Diameter | 200mm |
| Diam. Tolerance | 0.25mm |
| Thickness | 725 +/- μ |
| Crystal Orientation | <100> +/- 1 degree |
| Resistivity | 2.7-4 Ω-cm |
| Res. Gradient | 10% |
| Oxygen | 25-29 ppma |
| Oxygen Gradient | 5% |
| Carbon | 0.3 ppma |

Figure 3.25   Typical 200-mm wafer specification.

## Oxidation

Silicon wafers may be oxidized before shipment to the customer. The silicon dioxide layer serves to protect the wafer surface from scratches and contamination during shipping. Most companies start the wafer fabrication process with an oxidation step, and buying the wafers with an oxide layer saves a manufacturing step. Oxidation processes are explained in Chapter 7.

## Packaging

While much effort goes into producing a high-quality and clean wafer, the quality can be lost during shipment to the customer or, worse, from the packaging method itself. Therefore, there is a very stringent requirement for clean and protective packaging. The packaging materials are of nonstatic, nonparticle-generating materials, and the equipment and operators are grounded to drain off static charges that attract small particles to the wafers. Wafer packaging takes place in cleanrooms.

## Engineered Wafers (Substrates)

Increasingly, wafer fabrication companies are asking for wafer manufacturers to supply wafers with deposited top-side layers, such as epitaxial silicon. Other wafer products include silicon deposited on insulators (SOI and SOS) such as sapphire or diamond (Chapter 12).

## Review Questions

1. In a polycrystalline structure, the atoms are not arranged (true or false).

2. In a single-crystal structure, the unit cells are not arranged (true or false).

3. Draw a cubic unit cell and identify the $\langle 100 \rangle$ plane.

4. $\langle 111 \rangle$ oriented wafers are used for (bipolar, MOS) devices.

5. What is the orientation of a semiconductor crystal if the seed has a $\langle 100 \rangle$ orientation.

6. Draw a diagram of a CZ crystal grower and identify all the major parts.

7. During crystal growth, the molten material is changed from single crystal structure to a polycrystal structure (true or false).

8. Why are wafers edge rounded?

9. Draw a flow diagram of the wafer preparation process.

10. Give two reasons why semiconductor wafers require a flat surface.

## References

1. Russ Arensman, "One-Stop Automation," *Electronic Business*, July 2002, p. 54.
2. Sumitomo Sitix product brochure.
3. R. E. Williams, *Gallium Arsenide Processing Techniques*, Artech House Inc., Dedham MA, 1984, p. 37.
4. S. J. Brunkhorst and D. W. Sloat, "The impact of the 300-mm transition on silicon wafer suppliers," *Solid State Technology*, January 1998, p. 87.
5. *Ibid.*

# 4

# Overview of Wafer Fabrication

## Overview

This chapter will introduce the four basic processes used in the wafer fabrication to form the electrical elements of an integrated circuit in and on the wafer surface. Circuit design is traced from the functional diagram to the production of photomasks and reticles. Wafer and chip features and terminology are detailed. Finally, a flow diagram for the fabrication of a simple semiconductor device is presented.

## Objectives

Upon completion of this chapter, you should be able to:

1. Identify and explain the four basic wafer operations.
2. Identify the parts of a wafer.
3. Draw a flow diagram of the circuit-design process.
4. Explain the definition and use of a composite drawing and mask set.
5. Draw cross sections showing the doping sequence of basic operations.
6. Draw cross sections showing the metallization sequence of basic operations.
7. Draw cross sections showing the passivation sequence of basic operations.
8. Identify the "parts" of an integrated circuit chip.

## Goal of Wafer Fabrication

The four stages of microchip manufacturing are materials preparation, crystal growth and wafer preparation, wafer fabrication, and
packaging. The first two stages have been explored in Chapter 3. In
this chapter, the fundamentals of stage 3, wafer fabrication, are explained.

*Wafer fabrication* is the manufacturing processes used to create the
semiconductor devices in and on the wafer surface. The polished starting wafers come into fabrication with blank surfaces and exit with the
surface covered with hundreds of completed chips (Fig. 4.1).

## Wafer Terminology

A completed wafer is illustrated in Fig. 4.2. The regions of a wafer surface are:

1. Chip, die, device, circuit, microchip, or bar. All of these terms are
   used to identify the microchip patterns covering the majority of the
   wafer surface.

2. Scribe lines, saw lines, streets, and avenues. These areas are
   spaces between the chips that allow separation of the chip from the
   wafer. Generally, the scribe lines are blank, but some companies
   place alignment targets, or electrical test structures (see photomasking) in them.

3. Engineering die, test die. These chips are different from the regular device or circuit die. They contain special devices and circuit elements that allow electrical testing during the fabrication
   processing for process and quality control.

4. Edge chips. The edges of the wafer contain partial chip patterns
   that are wasted space. Larger chips on the same diameter wafer

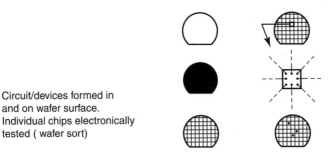

Circuit/devices formed in
and on wafer surface.
Individual chips electronically
tested ( wafer sort)

**Figure 4.1**   Wafer fabrication and wafer sort.

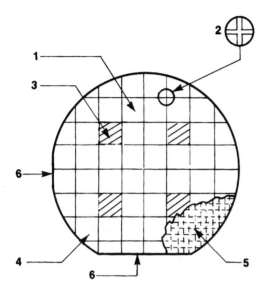

Figure 4.2   Wafer terminology.

result in large numbers of partial chips. Larger-diameter wafers minimize the amount of wasted space from larger chips.

5. Wafer crystal planes. The cutaway section illustrates the crystal structure of the wafer under the circuit layers. The diagram shows that the chip edges are oriented to the wafer crystal structure.

6. Wafer flats/notches. Wafers come from the wafer preparation stage with flats that indicate information about the crystal orientation and doping polarity of the wafer. Longer flats are called *major flats,* and shorter flats are called *minor flats.* The depicted wafer has a major and minor flat, indicating that it is a P-type ⟨100⟩ oriented wafer (see Chapter 3 for the flat code). Three-hundred-millimeter diameter wafers use notches as crystal orientation indicators. The flats and notches also assist alignment of the wafer in a number of the wafer fabrication processes.

## Basic Wafer-Fabrication Operations

There are hundreds of thousands of different microchip types and functions. However, they are made with a small number of basic structures (primarily bipolar or MOS structures, see Chapter 16) and manufacturing processes. An analogy is the auto industry. This industry produces a wide variety of products, from sedans to bulldozers. Yet the processes of metal forming, welding, painting, and so on are common to all plants. Within the plant, these basic processes are applied in different ways to produce the desired product.

The same is true in microchip fabrication. Companies use four basic operations in an infinite number of sequences and variations to produce specific microchips. They are layering, patterning, doping, and heat treatments (Fig. 4.3). Figure 4.4 illustrates how the four operations could be used to form a doped region in the wafer surface and contact it with a conductive layer of metal. A cross section of a simple metal oxide substrate (MOS) silicon gate transistor illustrates how these basic operations are used to create a real-life semiconductor device.

### Layering

Layering is the operation used to add thin layers to the wafer surface. An examination of the simple MOS transistor structure in Fig. 4.4 shows a number of layers that have been added to the wafer surface. These layers could be insulators, semiconductors, or conductors. They are of different materials and are grown or deposited by a variety of processes.

The major techniques are grown silicon dioxide layers and deposition (Fig. 4.5) of a variety of materials. Common deposition techniques are physical vapor deposition (PVD), chemical vapor deposition (CVD), evaporation, and sputtering. Electroplating is used to deposit gold metallization on high density integrated circuits. Figure 4.6 lists common layer materials and layering processes. The details of each are explained in the process chapters. The role of the different layers in the structures is explained in Chapter 16.

### Patterning

Patterning is the series of steps that results in the removal of selected portions of the added surface layers (Fig. 4.7). After removal, a *pattern* of the layer is left on the wafer surface. The material removed may be in the form of a hole in the layer or just a remaining island of the material.

The patterning process is known by the names photomasking, masking, photolithography, and microlithography. During the wafer fabrication process, the various physical *parts* of the transistors, diodes, capacitors, resistors, and metal conduction system are formed in and on the wafer surface. These parts are created one layer at a time by the combination of putting a layer on the surface and removing portions, with a patterning process, to leave a specific shape. The goal of the patterning operation is to create the desired shapes in the exact dimensions (feature size) required by the circuit design, and to locate them in their proper location on the wafer surface and in relation to the other *parts*.

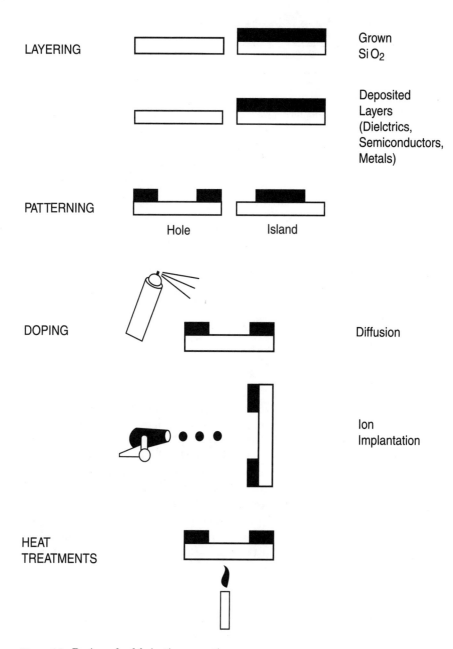

**Figure 4.3**  Basic wafer-fabrication operations.

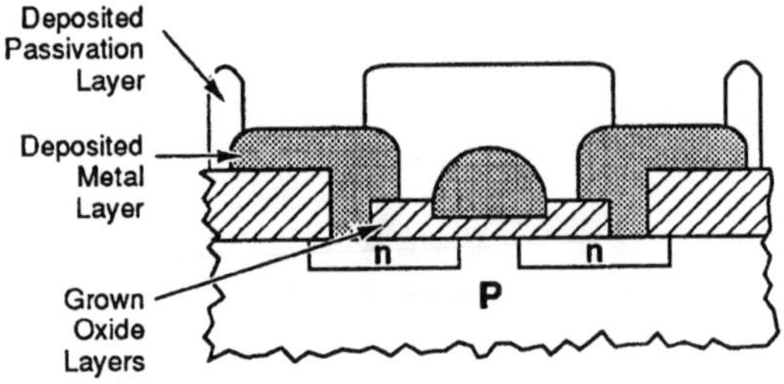

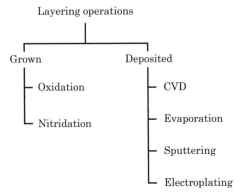

**Deposited Passivation Layer**

**Deposited Metal Layer**

**Grown Oxide Layers**

n    n

P

**Figure 4.4** Cross section of completed metal gate MOS transistor with grown and deposited layers.

Layering operations

Grown

- Oxidation

- Nitridation

Deposited

- CVD

- Evaporation

- Sputtering

- Electroplating

**Figure 4.5**  Layering operations.

Patterning is the most critical of the four basic operations. This operation sets the critical dimensions of the devices. Errors in the patterning process can cause distorted or misplaced patterns that result in changes in the electrical functioning of the device/circuit. Misplacement of the pattern can have the same bad results. Another problem is defects. Patterning is a high-tech version of photography but performed at incredibly small dimensions. Contamination in the process steps can introduce defects. This contamination problem is magnified by the fact that patterning operations are performed on the wafer from 5 to 20 or more times in the course of the wafer-fabrication process.

## Doping

Doping is the process that puts specific amounts of electrically active dopants in the wafer surface through openings in the surface layers

| Layers | Thermal oxidation | Chemical vapor deposition | Evaporation | Sputtering | Electroplating |
|---|---|---|---|---|---|
| Insulators | silicon dioxide | silicon dioxide<br>silicon nitride | | silicon dioxide<br>silicon monoxide | |
| Semiconductors | | epitaxial silicon<br>polycrystalline silicon | | | |
| Conductors | | | aluminum<br>aluminum alloys<br>nichrome<br>gold | aluminum<br>aluminum alloys<br>tungsten<br>titanium<br>molybdenum | gold<br>copper |

**Figure 4.6**  Table of layers, processes, and materials.

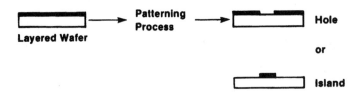

**Figure 4.7**  Patterning.

(Fig. 4.8). The two techniques are *thermal diffusion* and *ion implanta-tion*, which are detailed in Chapter 11.

Thermal diffusion is a chemical process that takes place when the wafer is heated to the vicinity of 1000°C and is exposed to vapors of the proper dopant. A common example of diffusion is the spreading de-odorant vapors into a room after being released from a pressurized can. Dopant atoms in the vapor move into the exposed wafer surface through the chemical process of diffusion to form a thin layer in the wafer surface. In microchip applications, diffusion is also called solid-state diffusion, since the wafer material is a solid. Diffusion is a chem-ical process.

Ion implantation is a physical process. Wafers are loaded in one end of an implanter and dopant sources (usually in gas form) in the other end. At the source end, the dopant atoms are ionized (given an electri-cal charge), accelerated to a high speed, and swept across the wafer surface. The momentum of the atoms carries them into the wafer sur-face, much like a ball shot from a cannon lodges in a wall.

The purpose of the doping operation is to create pockets in the wafer surface (Fig. 4.9) that are either rich in electrons (N-type) or rich in electrical holes (P-type). These pockets form the electrically active re-gions and N-P junctions required for operation of the transistors, di-odes, capacitors, and resistors in the circuit.

**Heat treatments**

Heat treatments are the operations in which the wafer is simply heated and cooled to achieve specific results (Fig. 4.10). In the heat

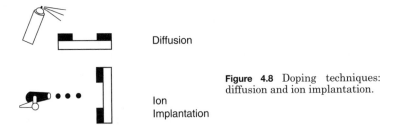

Diffusion

Ion
Implantation

**Figure 4.8** Doping techniques: diffusion and ion implantation.

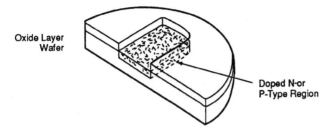

**Figure 4.9**   Formation of N- or P-type region in wafer surface.

| Operation | Heat treatment |
|---|---|
| Patterning | Soft bake |
| | Hard bake |
| | Post exposure bake (develop) |
| Doping | Post ion implant anneal |
| Layering | Post metal deposition and patterning anneal |

**Figure 4.10**   Table of major heat treatments.

treatment operations, no additional material is added or removed from the wafer, although contaminates and vapors may evaporate from the wafer surface.

An important heat treatment takes place after ion implantation. The implantation of the dopant atoms causes a disruption of the wafer crystal structure that is repaired by a heat treatment, called *anneal*, at about 1000°C. Another takes place after the conducting stripes of metal are formed on the wafer. These stripes carry the electrical current between the devices in the circuit. To ensure good electrical conduction, the metal is "alloyed" to the wafer surface by a heat treatment, which takes place at 450°C. A third important heat treatment is the heating of wafers with photoresist layers to drive off solvents that interfere with accurate patterning.

### Circuit design

Circuit design is the first step in creation of a microchip. A circuit designer starts with a block functional diagram of the circuit such as the logic diagram in Fig. 4.11. This diagram lays out the primary functions and operation required of the circuit. Next, the designer translates the functional diagram to a schematic diagram (Fig. 4.12). This diagram identifies the number and connection of the various circuit

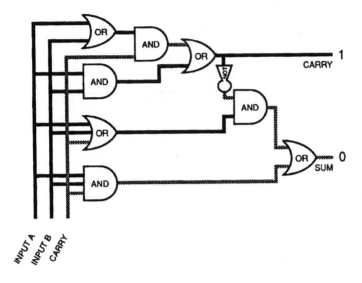

Figure 4.11  Example functional logic design of a simple circuit.

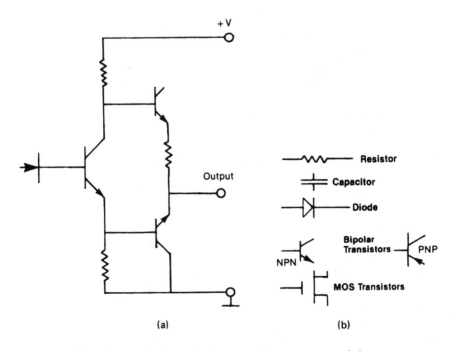

Figure 4.12  Example circuit schematic diagram with component symbols.

components. Each component is represented by a symbol. Accompanying the schematic diagram are the electrical parameters (voltage, current, resistance, and so forth) required to make the circuit work.

The third step, circuit layout, is unique to semiconductor circuits. Circuit operation is dependent on a number of factors, including material resistivity, material physics, and the physical dimensions of the individual component "parts." The placement of the parts relative to each other is another factor. All these considerations dictate the physical layout and dimensions of the part/device/circuit. Layout starts with using sophisticated computer-aided design (CAD) systems to translate each circuit component into its physical shape and size. Through the CAD system, the circuit is built, exactly duplicating the final design. The result is a composite picture of the circuit surface showing all of the sublayer patterns. This drawing is called a *composite* (Fig. 4.13). The composite drawing is analogous to the blueprint of a multistory office building as viewed from the top and showing all of

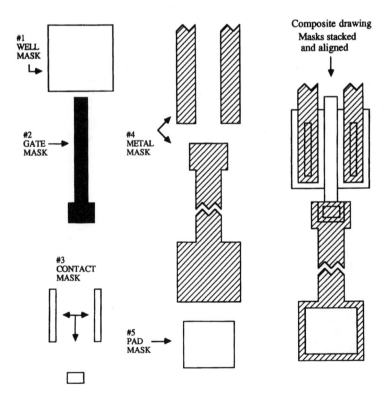

**Figure 4.13** Composite and layer drawings for five-mask silicon gate transistor.

the floors. However, the composite is many times the dimensions of the final circuit.

Both buildings and semiconductor circuits are built one layer at a time. Therefore, it is necessary to separate the composite drawing into the layout for each individual layer in the circuit. Figure 4.13 illustrates the composite and individual layer patterns for a simple silicon gate MOS transistor.

Each layer drawing is digitized (digitizing is the translation of the layer drawings to a digital data base) and plotted on a computerized $X–Y$ plotting table.

### Reticle and masks

The patterning process is used to create the required layer pattern and dimensions in and on the wafer surface. Getting the pattern from the digitized pattern to the wafer surface requires several steps. For the photo processes, there is an intermediate step called a reticle. A reticle is a "hard copy" of the individual drawing recreated in a thin layer of chrome deposited on a glass or quartz plate (Fig. 4.14a). The

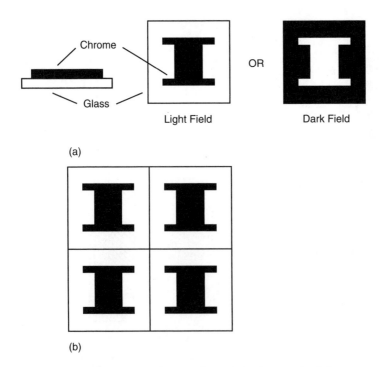

**Figure 4.14**  (a) Chrome on glass reticle and (b) photomask of the same pattern.

reticle may be used directly in the patterning process or may be used to produce a photomask. A photomask is also a glass plate with a thin chrome layer on the surface. After production, it is covered with many copies of the circuit pattern (Fig. 4.14b). It is used to pattern a whole wafer surface in one pattern transfer. (Reticle and mask-making processes are detailed in Chapter 11.)

Reticles and masks are produced in a separate department or are purchased from outside vendors. They supply the fabrication area with a separate set of reticles or set of masks (*mask set*) for each circuit type.

## Example Fabrication Process

The manufacture of a circuit starts with a polished wafer. The cross section sequence in Fig. 4.15 shows the basic operations required to form a simple MOS silicon-gate transistor structure. The following is an explanation of each operation in the fabrication process.

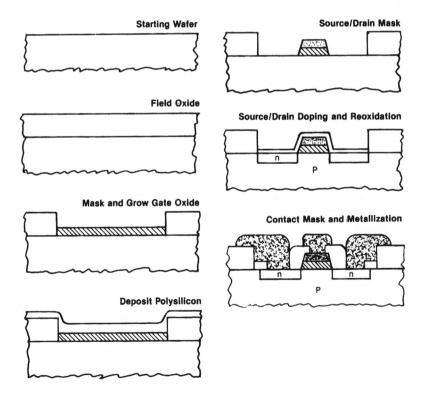

**Figure 4.15** Silicon gate MOS process steps.

**Step 1: Layering Operation.** The building starts with an oxidation of the wafer surface to form a thin protective layer and to serve as a doping barrier. This silicon dioxide layer is called the *field oxide*.

**Step 2: Patterning Operation.** The patterning process leaves a hole in the field oxide that defines the location of the source, gate, and drain areas of the transistor.

**Step 3: Layering Operation.** Next, the wafer goes to an silicon dioxide oxidation operation. A thin oxide is grown on the exposed silicon. It will service as the gate oxide.

**Step 4: Layering Operation.** In step 4, another layering operation is used to deposit a layer of polycrystalline (poly) silicon. This layer will also become part of the gate structure.

**Step 5: Patterning Operation.** Two openings are patterned in the oxide/polysilicon layer to define the source and drain areas of the transistor.

**Step 6: Doping Operation.** A doping operation is used to create an N-type pocket in the source and drain areas.

**Step 7: Layering Operation.** Another oxidation/layering process is used to grow a layer of silicon dioxide over the source/drain areas.

**Step 8: Patterning Operation.** Holes, called contact holes, are patterned in the source, gate, and drain areas.

**Step 9: Layering Operation.** A thin layer of conducting metal, usually an aluminum alloy, is deposited over the entire wafer.

**Step 10: Patterning Operation.** After deposition, the wafer goes back to the patterning area where portions of the metallization layer are removed from the chip area and the scribe lines. The remaining portions connect all the parts of the surface components to each other in the exact pattern required by the circuit design.

**Step 11: Heat Treatment Operation.** Following the metal patterning step, the wafer goes through a heating process in a nitrogen gas atmosphere. The purpose of the step is to "alloy" the metal to the exposed source and drain regions and the gate region to ensure good electrical contact.

**Step 12: Layering Operation.** The final layer of this device is a protective layer known variously as a *scratch* or *passivation layer*. Its purpose is to protect the components on the chip surface during the testing and packaging processes, and during use.

**Step 13: Patterning Operation.** The last step in the sequence is a patterning process that removes portions of the scratch protection layer over the metallization terminal pads on the periphery of the chip. This step is known as the *pad mask*.

The 13-step process illustrates how the four basic fabrication operations are used to build a particular transistor structure. The other components (diodes, resistors, and capacitors) required for the circuit are formed in other areas of the circuit as the transistors are being formed. For example, in this sequence, resistor patterns are put on the wafer at the same time as the source/drain pattern for the transistor. The subsequent doping operation creates the source/drain *and* the resistors. Other transistor types, such as bipolar and silicon gate MOS, are formed by the same basic four operations but using different materials and in different sequences.

In general, the circuit components are formed in the first series of fabrication operations and referred to as the *front end of the line* (FEOL). In the later series of operations, the various metallization layers that connect the circuit components are added to the wafer surface. These operations are called the *back end of the line* (BEOL).

Modern chip structures are many times more complicated than the simple process just described. They have multiple layers and pockets of dopants, numbers of layers added to the surface, including multiple layers of conductors interspersed with dielectrics (Fig. 4.16).

Achieving these complicated structures requires many processes. An each process, in turn, requires a number of steps and substeps. A speculative process for a 64-Gb CMOS device might require 180 major

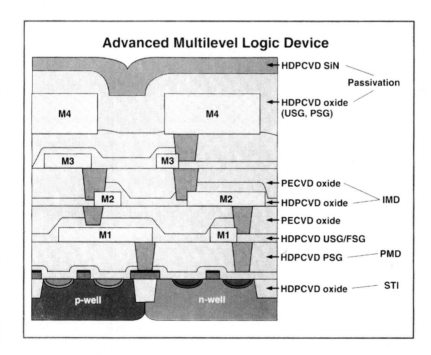

**Figure 4.16**  Modern chip structure.

steps, 52 clean/strips, and up to 28 masks.[1] Yet all of the major steps
are one of the four basic operations. Figure 4.17 lists the basic operations and the principle process options used for each. By the time the
industry reaches circuits with gate widths of several atoms and stacks
of metal on top of the circuit, the number of process steps will be in the
500, or higher, range.

| Basic Operation | Process | Options |
|---|---|---|
| Layering | Oxidation | Atmospheric |
| | | High Pressure |
| | | Rapid Thermal Oxidation (RTO) |
| | Chemical Vapor Deposition (CVD) | Atmospheric Pressure |
| | | Low Pressure (LPCVD) |
| | | Plasma Enhanced (PECVD) |
| | | Vapor Phase Epitaxy (VPE) |
| | | Metaloranic CVD (MOCVD) |
| | Molecular Beam Epitaxy (MBE) | |
| | Physical Vapor Deposition (PVD) | Vacuum Evaporation |
| | | Sputtering |
| Patterning | Resist | Positive |
| | | Negative |
| | Exposure Systems | Contact |
| | | Proximity |
| | | Scanning Projection |
| | | Stepper |
| | Exposure Sources | High Pressure Mercury |
| | | X-Rays |
| | | E-Beams |
| | Imaging Processes | Single Layer Resist |
| | | Multilayer Resist |
| | | Antireflecting layers |
| | | Off-Axis Illumination |
| | | Annular Ring Illumination |
| | | Planarization |
| | | Contrast Enhancement |
| | Etch | Wet Chemistry-Liquid/vapor |
| | | Dry (Plasma) |
| | | Lift-Off |
| | | Ion Milling |
| | | Reactive Ion Etch (RIE) |
| Doping | Diffusion | Open Tube-Horizontal/Vertical |
| | | Closed Tube |
| | | Rapid Thermal Processing (RTP) |
| | Ion Implantation | Medium/High Current |
| | | Low/High Voltage (energy) |
| Heating | Thermal | Hot Plates |
| | | Convection |
| | | RTP |
| | Radiation | Infrared (IR) |

**Figure 4.17**  Summary of wafer-fab operations/processes.

## Chip Terminology

Figure 4.18 depicts a photomicrograph of an MSI/bipolar integrated circuit. The level of integration was chosen so that some surface details could be seen. The components of higher-density circuits are so small that they cannot be distinguished on a photomicrograph of the entire chip. The chip features are:

1. A bipolar transistor
2. The circuit designation number
3. Bonding pads for connecting the chip into a package
4. A piece of contamination on a bonding pad
5. Metallization lines
6. Scribe (separation) lines
7. Unconnected component
8. Mask alignment marks
9. Resistor

## Wafer Sort

Following the wafer fabrication process comes a very important testing step: wafer sort. This test is the report card on the fabrication process. During the test, each chip is electrically tested for electrical performance and circuit functioning. Wafer sort is also known as *die sort* or *electrical sort.*

For the test, the wafer is mounted on a vacuum chuck and aligned to thin electrical probes that contact each of the bonding pads on the chip (Fig. 4.19). The probes are connected to power supplies that test the circuit and record the results. The number, sequence, and type of tests are directed by a computer program. Wafer probers are automated so that, after the probes are aligned to the first chip (manually or with an automatic vision system), subsequent testing proceeds without operator assistance.

The goal of the test is threefold. First is the identification of working chips before they go into the packaging operation. Second is characterization of the device/circuit electrical parameters. Engineers need to track the parameter distributions to maintain process quality levels. The third goal is an accounting of the working and nonworking chips to give fab personnel feedback on overall performance. The location of the working and nonworking chips is logged into a wafer map in the computer. Older technologies deposit a drop of ink on the *nonworking* chips.

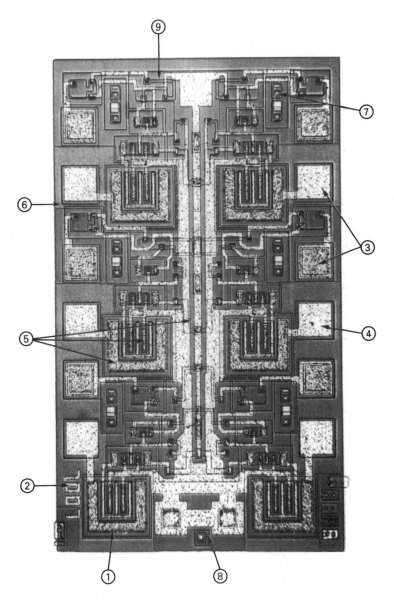

**Figure 4.18**   Chip terminology.

Wafer sort is one of the principal yield calculations in the chip production process. It also gets more expensive as the chips get larger and denser.[2] These chips require more time to probe, and power supplies, wafer handling mechanics, and computer systems have to be more sophisticated to perform the tests and track results. The vision systems

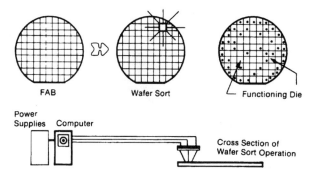

**Figure 4.19** Wafer sort.

must also evolve in sophistication (and cost) with expanding die size. Cutting the chip test time is also a challenge. Chip designers are being asked to include test modes for memory arrays. Test designers are exploring ways to streamline test sequences using stripped-down tests once the chip is fully characterized, performing scan tests of the circuit, and parallel testing different circuit parts. The details of wafer parameter yield are addressed in Chapter 6.

## Packaging

Most wafers continue on to stage four, called *packaging* (Fig. 4.20). The wafers are transferred to a packaging area on the same site or to a remote location. Many semiconductor producers package their chips in offshore facilities. (Chapter 18 details the packaging process.) In this stage, the wafers are separated into the individual chips, and the working chips are incorporated into a protective package. Some chips are directly incorporated into electrical systems without a package.

## Summary

The semiconductor microchip fabrication process is long, complicated, and includes many variations, depending on the type of product, level of integration, feature size, and other factors. Understanding a particular process is easier by separating it into the four stages. Wafer fabrication is further understood by identifying the four basic operations performed on the wafer. Several simple processes have been used to illustrate the basic wafer fabrication operations. Actual processes used are addressed in the process chapters and in Chapters 16 and 17. Industry drivers and manufacturing trends are explained in Chapter 15.

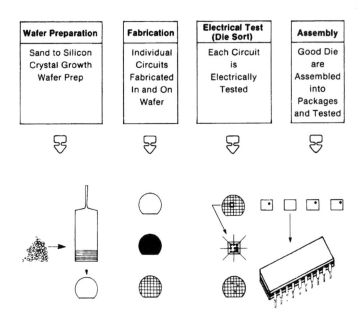

| Wafer Preparation | Fabrication | Electrical Test (Die Sort) | Assembly |
|---|---|---|---|
| Sand to Silicon Crystal Growth Wafer Prep | Individual Circuits Fabricated In and On Wafer | Each Circuit is Electrically Tested | Good Die are Assembled into Packages and Tested |

**Figure 4.20**    Integrated circuit manufacturing sequence.

## Review Questions

1. Name one layering technique.
2. Which basic operation uses the ion implant technique?
3. List the four basic wafer fabrication operations.
4. Draw and label cross sections for a oxidation/patterning/doping sequence.
5. Describe a composite drawing.
6. Which basic operation uses a photomask?
7. What parameter is tested at wafer sort (wafer thickness, defect density, circuit function)?
8. What step in the circuit design process uses a CAD system?
9. What is the purpose of a chip package?
10. What is the role of the holes created in the contact mask step?

## References

1. R. Kopp, *Kopp Semiconductor Engineering,* September 1996.
2. R. Iscoff, "VLSI Testing: The Stakes Get Higher," *Semiconductor International,* September 1993, p. 58.

# Contamination Control

## Overview

The effects of contamination on device processing, device performance, and device reliability are explained are identified along with the types and sources of contamination found in a fabrication area. Cleanroom layouts, major contamination-control procedures, and wafer surface cleaning techniques are explained.

## Objectives

Upon completion of this chapter, you should be able to:

1. Identify the three major effects of contamination on semiconductor devices and processing.

2. List the major sources of contamination in a fabrication area.

3. Define the "class number" of a cleanroom.

4. List the particle density of class 100, 10, and 1 fabrication areas.

5. Describe the role of positive pressure, air showers, and adhesive mats in maintaining cleanliness levels.

6. List at least three techniques used to minimize contamination from fabrication personnel.

7. Identify the three contaminants present in "normal" water, and their control in semiconductor plants.

8. Describe the differences between normal industrial chemicals and semiconductor-grade chemicals.

9. Name two problems associated with high static levels, and two methods of static control.

10. Describe a typical FEOL and BEOL wafer cleaning process.

11. List typical wafer rinsing techniques.

## Introduction

One of the first problems to plague the infant microchip fabrication efforts was contamination. The industry started with a cleanroom technology developed by the space industry. However, these techniques proved inadequate for large-scale manufacturing of chips. Cleanroom technology has had to keep pace with chip design and density evolution. The ability of the industry to grow has been dependent on keeping up solving contamination problems presented by each generation of chips. Yesterday's minor problems become the killer defects of today's chips.

### The problem

Semiconductor devices are very vulnerable to many types of contaminants. They fall into five major classes.

1. Particles

2. Metallic ions

3. Chemicals

4. Bacteria

5. Airborne molecular contaminants (AMCs)

**Particles.**  Semiconductor devices, especially dense integrated circuits, are vulnerable to all kinds of contamination. The sensitivity is due to the small feature sizes and the thinness of deposited layers on the wafer surface. These dimensions are down to the submicron range. A micron or micrometer (μm) is very small. A centimeter contains 10,000 μm. Another way to envision a micron is that a human hair is about 100 μm in diameter (Fig. 5.1). The small physical dimensions of the devices make them very vulnerable to particulate contamination in the air coming from workers, generated by the equipment, and present in processing chemicals (Fig. 5.2). As the feature size and films become smaller (Fig. 5.3), the allowable particle size must be controlled to smaller dimensions.

A rule of thumb is that the particle size must be ten times smaller than the minimum feature size.[1] A 0.30-μm feature size device is vulnerable to 0.03-μm diameter particles. Particles that locate in a critical part of the device and destroy its functioning are called *killer defects*. Killer defects also include crystal defects and other process-

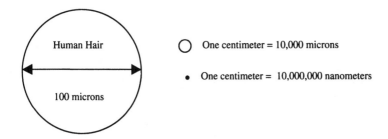

**Figure 5.1**  Relative sizes.

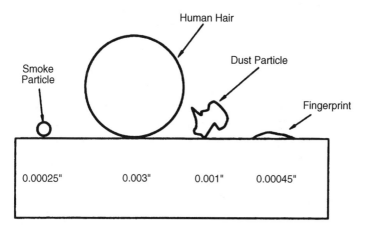

**Figure 5.2**  Relative size of contamination. (*Source:* Hybrid Microcircuit Technology Handbook.[12])

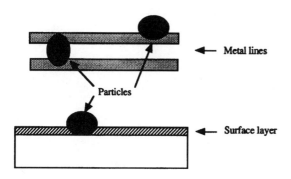

**Figure 5.3**  Relative size of airborne particles and wafer dimensions.

induced problems. On any wafer, there are a number of particles. Some are the killer variety, but others locate in less sensitive areas and do no harm. The 1994 *Semiconductor Industry Association (SIA) National Technology Roadmap for Semiconductors* (NTRS) calls for defect densities for pattering operations for 0.18-μm design rules to be 135 per layer/m$^2$ at a 0.06-μm defect size.

**Metallic ions.** In Chapter 2, it was established that semiconductor devices require controlled resistivity in the wafers and in the doped N- and P-regions, and precise N-P junctions. These three properties are achieved by the purposeful introduction of specific dopants into the crystal and into the wafer. These desired effects are achieved with very small amounts of the dopants. Unfortunately, it takes only a small amount of certain electrically active contaminants in the wafer to alter device electrical characteristics, changing performance and reliability parameters.

The contaminants causing these types of problems are known as *mobile ionic contaminants* (MICs). They are atoms of metals that exist in the material in an ionic form. Furthermore, these metallic ions are highly mobile in semiconductor materials. This mobility means that the metallic ions can move inside the device, even after passing electrical testing and shipping, causing the device to fail. Unfortunately, the metals (Fig. 5.4) that cause these problems in silicon devices are present in most chemicals. On a wafer, MIC contamination has to be in the 10$^{10}$ atoms/cm$^2$ range or less.[2]

Sodium is the most prevalent mobile ionic contaminant in most untreated chemicals and is the most mobile in silicon. Consequently, control of sodium is a prime goal in silicon processing. The MIC problem is most serious in MOS devices, a fact that has led some chemical suppliers to develop MOS or low-sodium-grade chemicals. These labels refer to low mobile ionic contaminant levels.

| Trace Metal Impurity | Parts per billion (ppb) |
|---|---|
| Sodium | 50 |
| Potassium | 50 |
| Iron | 50 |
| Copper | 60 |
| Nickel | 60 |
| Aluminum | 60 |
| Manganese | 60 |
| Lead | 60 |
| Zinc | 60 |
| Chlorides | 1000 |

**Figure 5.4** Example resist stripper trace metal contents *(EKC Technology–830 Photoresist stripper).*

**Chemicals.**  The third major contaminant in semiconductor process areas is unwanted chemicals. Process chemicals and process water can be contaminated with trace chemicals that interfere with the wafer processing. They may result in unwanted etching of the surface, create compounds that cannot be removed from the device, or cause nonuniform processes. Chlorine is such a contaminant and is rigorously controlled in process chemicals.

**Bacteria.**  Bacteria is the fourth major contaminant class. Bacteria are organisms that grow in water systems and on surfaces that are not cleaned regularly. Bacteria, once on the device, act as particulate contamination or may contribute unwanted metallic ions to the device surface.

**Airborne molecular contaminants.**  Airborne molecular contaminants (AMCs) are fugitive molecules that escape from the process tools or the chemical delivery systems, or are carried into the fabrication area on materials or personnel. Transfer of wafers from one process tool to another can carry hitchhiking molecules to the next tool. AMCs include all of the gasses, dopants, and process chemicals used in the fabrication area. These can be oxygen, moisture, organics, acids, bases, and others.[3]

Their harm is highest in processes that involve delicate chemical reactions, such the exposure of photoresist in the patterning operations. Other problems include the shifting of etch rates and unwanted dopants that shift device electrical parameters and change the wetting characteristics of etchants leading to incomplete etching.[4] Detecting and controlling AMCs is an emerging field that will need to reach maturity as device sizes continue to diminish as projected by the SIA's *International Roadmap for Semiconductor Technology* (IRST).

**Contamination-caused problems**

The five types of contaminants affect the processing and devices in three specific performance areas.

1. Device processing yield

2. Device performance

3. Device reliability

**Device processing yield.**  Device processing in a contaminated environment can cause a multitude of problems. Contamination may change the dimensions of device parts, change the cleanliness of the surfaces, and/or cause pitted layers. Within the fabrication process are a number of quality checks and inspections specifically designed to detect contaminated wafers. Contamination caused defects contribute the to rejection of wafers in the wafer fabrication process, reducing the overall yield (see Chapter 6).

**Device performance.**  A more serious problem is related to small pieces of contamination that may escape the in-process quality checks. Or unwanted chemicals and AMCs in the process steps may alter device dimensions or material quality. High levels of mobile ionic contaminants in the wafer can change the electrical performance of the devices. These problems usually show up at an electrical (called *wafer* or *die sort*) that checks each chip after the wafer fabrication process (see Chapter 6).

**Device reliability.**  Loss of device reliability is the most insidious of the contamination failures. Small amounts of metallic contaminants can get into the wafer during processing and not be detected during normal device testing. However, in the field, these contaminants can travel inside the device and end up in electrically sensitive areas, causing failure. This failure mode is a primary concern of the space and defense industries.

In the rest of this chapter, the sources, nature, and control of the types of contamination that affect semiconductor devices are identified. With the advent of LSI level circuits in the 1970s, the control of contamination became essential to the industry. Since that time, a great deal of knowledge about and control of contamination has been learned. Contamination control is now a discipline of its own and is one of the critical technologies that have to be mastered to profitably produce solid-state devices.

Contamination control issues addressed in this chapter apply to wafer fabrication areas, mask-making areas, some chip packaging areas, and areas in which semiconductor equipment and materials are manufactured.

## Contamination Sources

### General sources

Contamination in a cleanroom is defined as anything that interferes with the production of the product and/or its performance. The stringent requirements of solid-state devices define levels of cleanliness that far exceed those of almost any other industry. Literally everything that comes in contact with the product during manufacture is a potential source of contamination. The major contamination sources are:

1. Air

2. The production facility

3. Cleanroom personnel

4. Process water

5. Process chemicals

6. Process gases

7. Static charge

Each source produces specific types and levels of contamination and requires special controls to render it acceptable in the cleanroom.

**Air**

Normal air is so laden with contaminants that it must be treated before entering a cleanroom. A major problem is airborne particles, referred to as *particulates* or *aerosols*. Normal air contains copious amounts of small dust and particles, as illustrated in Fig. 5.5. A major problem with small particles (called *aerosols*) is that they "float" and remain in the air for long periods of time. Air cleanliness levels in cleanrooms are identified by the particulate diameters and their density in the air.

Air quality is designated by the *class number* of the air in the area as defined in Federal Standard 209E.[5] This standard designates air quality in the two categories of particle size and density. The class

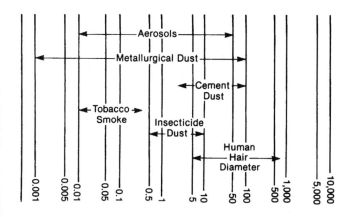

**Figure 5.5** Relative size of airborne particulates (in microns).

number of an area is defined as the number of particles 0.5 µm or larger in a cubic foot of air. The air in a typical city, filled with smoke, smog, and fumes, can contain up to 5 million particles per cubic foot, which is a class number of 5 million. Advancing chip sensitivity has identified smaller and smaller tolerable particle sizes for each generation of chip feature size.[6]

Figure 5.6 shows the relationship between particulate diameter and density as defined by Federal Standard 209E. Figure 5.7 lists the class numbers and associated particle size for various environments. Federal Standard 209E specifies cleanliness levels down to class 1 levels. While 209E defines class numbers at 0.5-µm particle size, successful wafer fabrication processing requires tighter controls. Engineers strive to achieve reduction of 0.3-µm particles in class 10 and class 1 environments. Specifications proposed by Semetech call for 64-Mbit memory process areas at class 0.1 and 256-Mbit air levels of class 0.01.[5]

### Clean air strategies

The design of a cleanroom is integral to its ability to produce contamination-free wafers. A major consideration in the design is the maintenance of clean air in the process areas. Automation is also an important cleanroom strategy for the reduction of contamination. This issue is explored in the equipment section and in Chapter 15. Four distinct room design strategies are used:

1. Clean workstations
2. Tunnel design
3. Total cleanroom
4. Mini-environments

### Cleanroom workstation strategy

The semiconductor industry adopted cleanroom techniques first developed by NASA to assemble space vehicles and satellites. However, the smallish cleanliness of rooms adequate for satellite assembly could not be maintained when expanded to larger fabrication areas with more production workers. The early semiconductor industry initially used ceiling and wall filters, which proved inadequate to keep the wafers clean. This problem was addressed with a cleanroom workstation strategy. This strategy focused on individual workstations with air filters and nonshedding materials. Outside the workstations, the wafers were stored and moved in covered boxes.

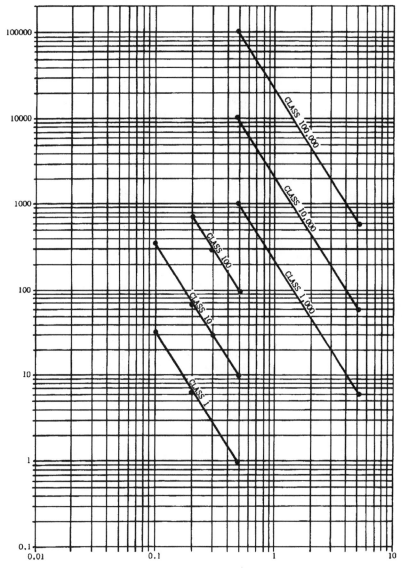

**Figure 5.6**  Air cleanliness class standard 209E.

| Environment | Class Number | Maximum particle size (micron) |
|---|---|---|
| Projected-256 Mbit | 0.01 | <<0.1 |
| Mini environment | 0.1 | <0.1 |
| ULSI Fab | 1 | 0.1 |
| VLSI Fab | 10 | 0.3 |
| VLF station | 100 | 0.5 |
| Assembly area | 1000-10,000 | 0.5 |
| House room | 100000 | |
| Outdoors | >500,000 | |

**Figure 5.7**    Typical class numbers for various environments.

The fabrication area consisted of a large room with the workstations (called *hoods*) arranged in rows so that the wafers could move sequentially through the process, never being exposed to dirty air. The filters in these clean hoods are either high-efficiency particulate attenuation (HEPA) filters or, for more demanding application, ultra-low-particle (ULPA) filters. These filters are constructed of fragile fibers with many small holes and are folded into the filter holder in an accordion design (Fig. 5.8). The high density of small holes and large area of the filter medium allow the passage of large volumes of air at low velocity. The low velocity contributes to the cleanliness of the hood by not causing air currents. The low velocity is also necessary for operator comfort. A typical airflow[6] is 90 to 100 ft/min. HEPA and ULPA filters achieve filtering efficiencies as high as 99.9999+ percent at 0.12-micron particle size.[7]

Typically, a clean hood (Fig. 5.9) has a HEPA/ULPA filter mounted in the top. Air is drawn from the room through a prefilter by a fan and forced through the HEPA filter. The air leaves the filter in a laminar pattern and, at the work surface, it turns and exits the hood. A shield directs the exiting air over the wafers in the hood. The formal name for the workstation is a *vertical laminar flow (VLF) station*. The term VLF is derived from the laminar nature of the airflow. Some worksta-

Dirty Air

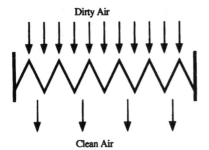

Clean Air

**Figure 5.8**    HEPA filter.

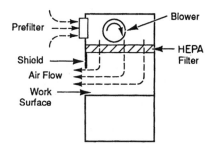

**Figure 5.9** Cross section of VLF process section.

tions are designed with the HEPA filter in the back of the work surface. These stations are called *horizontal laminar flow (HLF) hoods.*

Both types of stations keep the wafers clean in two ways. First is the filtered air inside the hood. The second cleaning action is the slight positive pressure built up in the station. This pressure prevents airborne dirt from operators and from the aisle areas from entering the hood.

A special design of VLF hood is required for wet chemical processing stations (Fig. 5.10). These stations must be connected to an exhaust system to remove the chemical fumes. Removal designs include an exit port in the rear and/or exhaust ports around the lip of the process tanks (lip exhaust). The fumes can pose both safety and contamination problems. In this design, care must be taken to balance the VLF air and the exhaust to maintain the required class number in the station.

The clean-station strategy also applies to equipment in modern fabs. Individual tools must be fitted with a VLF- or HLF-designed hood to keep the wafers clean during loading and unloading steps.

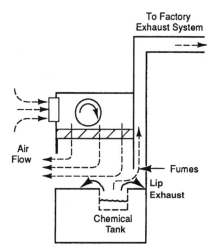

**Figure 5.10** Cross section of a VLF exhaust process section.

## Tunnel/bay concept

As more critical particulate control became necessary, it was noted that the VLF hood approach had several drawbacks. Chief among them was the vulnerability to contamination from the many personnel moving about in the room. People entering and exiting the fabrication area had the potential of contaminating all the process stations in the area.

This particular problem is solved by dividing the fabrication area into separate tunnels or bays (Fig. 5.11). Instead of the individual VLF hoods, filters are built into the ceilings and serve the same purpose. The wafers are kept clean by the filtered air from the ceiling filters and are less vulnerable to personnel-generated contamination because of fewer workers in the immediate vicinity. On the downside, tunnel arrangements cost more to construct and are less versatile when the process changes.

## Total cleanroom strategy

Developments in cleanroom design and filtering technology have allowed an evolution back to the open fabrication area (Fig. 5.12). In the latest version, air filtering is accomplished by HEPA filters in the ceiling with returns in the floor to give a continuous flow of clean air.

The workstations are tabletops with perforations to allow the filtered air to pass uninterrupted through the tabletop. This arrangement is often called a *ballroom* layout. Automation techniques used in wafer fabrication areas are presented in Chapter 15.

An important cleanroom parameter is recovery. This parameter is the time it takes for the filters to return the area to acceptable condi-

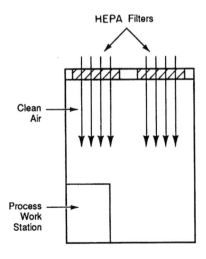

**Figure 5.11** Cross section of a cleanroom tunnel.

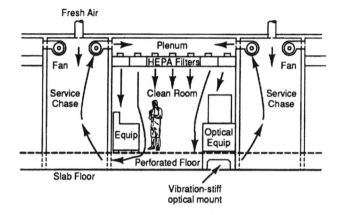

**Figure 5.12**  Cross section of a laminar flow cleanroom. *(Courtesy of Semiconductor International.)*

tions after a shift start, personnel break, or other disturbance. In a class 1 facility, the air may be turned over every six seconds.[9]

The trend of equipment and facility design has been to isolate the wafer from contamination sources. VLF hoods isolated the wafer from the room air, and tunnels isolate the wafer from excessive personnel exposure. The advent of CMOS ICs increased the number of process steps and the need to include more process stations in the cleanroom.

These larger rooms (and tunnels) bring with them the potential of contamination from the sheer volume of the air and the increased number of operators. These changes have driven the cost of cleanrooms up to several billion dollars.

### Micro- and mini-environments

Projections in the mid-1980s showed increasing cleanroom costs with diminishing returns on effectiveness. The concept of isolating the wafer in as small an environment as possible became the new direction. This concept was already in place with steppers and other process tools that had build in clean *micro-environments* for wafer loading and unloading (Fig. 5.13).

The challenge was to string together a series of mini-environments such that the wafer was never exposed to the room air. Hewlett Packard developed a critical link with the invention, in the mid 1980s, of the *standard mechanical interface* (SMIF).[10] With SMIF, the traditional wafer box is replaced with a wafer enclosure (*mini-environment*) that can be pressurized with air or nitrogen to keep out room air. This approach took on the general name of *wafer isolation technology* (WIT) or a mini-environment system. There are three parts to the system:

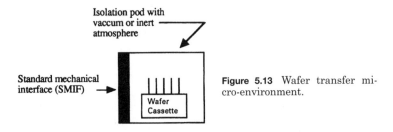

**Figure 5.13** Wafer transfer micro-environment.

the wafer box or pod for transporting the wafers, the isolated microenvironment at the tool, and a mechanism for extracting and loading the wafers. The wafer *pod* features a mechanical interface that allows direct connection to the micro-environment of the process tool (Fig. 5.13). Wafers may be loaded directly from and to the pod onto the tool wafer system by dedicated handlers. Another approach is to move the cassette from the pod to the tool wafer handling system with a robot. Mini-environments offer the advantage of greater temperature and humidity control.

The WIT/mini-environment strategy includes the benefit of upgrading existing fabs along with other benefits (Fig. 5.14). Yield losses from contamination are lowered. This critical benefit can be delivered at lower facility construction and operating costs. WIT allows keeping the aisle air cleanliness at a lower cleanliness level, which reduces construction and operating costs. With the wafer isolated, there is less

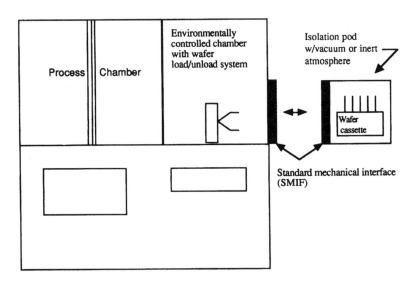

**Figure 5.14** Mini-environment system elements.

pressures on operator clothing, procedures, and constraints. However, the advent of larger-diameter wafers has driven up the weight of a pod of wafers to the point that they are too heavy for operators and too expensive to risk dropping. This situation requires robot handling, which increases cost and complexity. Mini-environment layouts must include storage for wafers (in pods) waiting for process tool availability. Current technology calls for storage units, called *stockers,* that hold the waiting pods or wafers. Layouts may include a central storage system with or without buffer storage at each tool.

### Temperature, humidity, and smog

In addition to control of particulates, the air parameters of temperature, humidity, and smog must be specified and controlled in a cleanroom. Temperature control is necessary for operator comfort and process control. Many of the wet processes of etching and cleaning take place in non-temperature-controlled baths, and rely on the room temperature for control. This control is important. A rule of thumb is that chemical reactions, such as an etch rate, change by a factor of 2 with a 10°C change in temperature. A typical temperature range is 72°F, ±2°F (22.2°C, ±1.1°C).

The relative humidity is also a critical process parameter, especially in patterning areas. In this area, thin films of a polymer are put on the wafer to act as an etch stencil. If the humidity is too high, the wafers collect moisture, preventing the polymer from sticking. The situation is the same as applying paint on a wet surface. On the other side, low humidity can foster the buildup of static charge on the wafer surface. The charge causes the wafer to attract particles out of the air. Relative humidity is controlled at between 15 and 50 percent.

Smog is another airborne contaminant in a cleanroom. Again the problem is most critical in the patterning areas. A step in the patterning process is similar to photographic film developing, which is a chemical process. Ozone, a major component of smog, interferes with the development process and must be controlled. Ozone is filtered out of the air by installing carbon filters in the incoming air ducts.

### Cleanroom layouts

The traditional cleanroom layout has individual process tunnels (or bays) opening onto a central hallway. Modular layouts have minifabs or major process areas clustered around a central area that serves the distribution of materials and personnel. These are two-level layouts, with the utility and some material services on the first level and the fab area on the second level. Three-level fabs, which have become the standard design, have the fabrication area on the second floor with gas, vacuum, deionizing (DI) water cleaners, and chemical supplies positioned on the first floor, from where they are piped up into the fab area. On the third floor are the air handlers and exhaust and scrubber systems.

## Cleanroom Construction

Selection of the clean air strategy is the first step in the design of a cleanroom. Every cleanroom is a trade-off between cleanliness and cost. Whatever the final design, every cleanroom is built on basic principles. The primary design is a sealed room that is supplied with clean air, is built with materials that are noncontaminating, and includes systems to prevent accidental contamination from the outside or from operators.

### Construction materials

The inside of a cleanroom is constructed entirely of materials that are nonshedding. This includes wall coverings, process station materials, and floor coverings. All piping holes are sealed, and even the light fixtures must have solid covers. Additionally, the design should minimize flat surfaces that can collect dust. Stainless-steel materials are favored for process stations and work surfaces.

### Cleanroom elements

Both the design of a cleanroom and its operation must be set up to keep dirt and contamination from getting into the room from the outside. Figure 5.15 shows a layout for a typical fab processing area. The following nine techniques are used to keep out and control dirt:

1. Adhesive floor mats
2. Gowning area
3. Air pressure
4. Air showers
5. Service bays
6. Double-door pass-throughs
7. Static control
8. Shoe cleaners
9. Glove cleaners

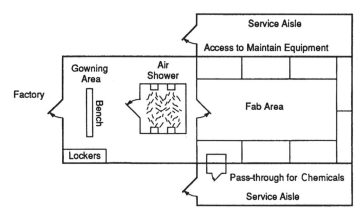

**Figure 5.15**   Fab area with gowning area, air showers, and service aisles.

**Adhesive floor mats.**   At the entrance to every cleanroom is a floor mat with an adhesive surface. The adhesive pulls off and holds dirt adhering to the bottoms of shoes. In some cleanrooms, the entire floor has a surface treated to hold dirt. Most mats have layers of the adhesive sheets. When the top sheet becomes dirty, it is removed to expose the next clean one.

**Gowning area.**   A major part of a cleanroom is the gowning area or anteroom. This area is a buffer between the cleanroom and the plant. It quite often is supplied with filtered air from ceiling HEPA filters. In this area, the operator's cleanroom apparel is stored. It is also the area where the cleanroom personnel change into their cleanroom garments. The management of this area varies with the degree of cleanliness required in the cleanroom. Quite often, it is managed to the same stringent requirements of the cleanroom itself. Often, the gowning area is divided into two sections by a bench. The operators don the garments on one side and put on their shoe coverings on the bench. The purpose is to keep the area between the bench and the cleanroom at a higher cleanliness level.

A good cleanroom procedure is to ensure that the doors between the factory and the cleanroom are never opened at the same time. This procedure ensures that the cleanroom is never exposed directly to the dirtier factory areas. Cleanroom management also includes lists of materials and garments that can and cannot come into the gowning room. Some areas will provide hallway lockers for coats and so on.

**Air pressure.**   A key design element is the air pressure balances between the cleanroom, the gowning room, and the factory. The well designed facility will have the air of these three sections balanced such that the highest pressure is in the cleanroom, the second highest in the gowning area, and the lowest in the factory hallways. The higher pressure in the cleanroom causes a low flow of air out of the doors, which, in turn, blows airborne particles back into the dirtier hall way.

**Air showers.**    The final design element protecting the cleanroom from outside contamination is the air shower located between the governing room and the cleanroom. Cleanroom personnel enter the air shower, where high-velocity air jets blow off particles from the outside of the garments. An air shower will have an interlocking system to prevent both doors from being opened at the same time.

**Service bays.**    A cleanroom is really a series of rooms (Fig. 5.15) within the factory, each contributing to the maintenance of the cleanroom. In the center is the processing cleanroom. Surrounding it is a bay area that is maintained at some designated class number that is generally higher than the cleanroom. Class 1000 or Class 10,000 are typical for service bays. In the bay are the process chemical pipes, electrical power lines, and cleanroom materials. Critical-process machines are backed up to the wall dividing the cleanroom and the bay. This arrangement allows technicians to service the equipment from the back without entering the cleanroom.

**Double-door pass-throughs.**    The bay also serves as a semi-clean area for the storage of materials and supplies. They are put into the cleanroom through double-door pass-through units that protect the cleanliness of the cleanroom. Pass-through units may be simple double-door boxes or may have a supply of positive-pressure filtered air with interlocking devices to prevent both doors from being opened at the same time. Often, the pass-throughs are fitted with HEPA filters. All materials and equipment brought into the cleanroom should be cleaned prior to entry.

**Static control.**    Higher-density circuits with submicron feature sizes are vulnerable to smaller particles of contamination attracted by static to the wafer. Static charges build up on the wafers, the storage boxes, work surfaces, and equipment. Each of these items can carry static charges as high as 50,000 V (volts) that attract aerosols out of the air and from personnel garments. The attracted particles end up contaminating the wafers. Statically held particulates are very difficult to remove with standard brush and wet cleaning techniques.

Most static charge is produced by *triboelectric* charging. This occurs when two materials initially in contact are separated. One surface becomes positively charged as it loses electrons. The other becomes negatively charged as it gains electrons. The triboelectric series table in Fig. 5.16 shows the charging potential for materials some materials found in a cleanroom.[11]

Static also represents a device operational problem. It occurs in devices with thin dielectric layers, as in MOS gate regions. An electrostatic discharge (ESD) of up to 10 A (amperes) is possible. This level of ESD can physically destroy a MOS device or circuit. ESD is a particular worry in device-packaging areas. This problem requires that sensitive devices, such as large-array memories, be handled and shipped in holders of antistatic materials.

Photomasks and reticles are particularly sensitive to ESD. A discharge can vaporize and destroy the chrome pattern.

Some equipment problems are static related—especially robots, wafer handlers, and measuring equipment. Wafers usually come to the equipment in carriers made of PFA-type materials. This carrier material is chosen for its chemical resistance, but it is not conductive. Charge builds up on the wafers but cannot dissipate to the carriers. When the carrier comes close to a piece of metal on the equipment, the wafer charge discharges to the equipment. The electromagnetic interference produced interferes with the machine operation.

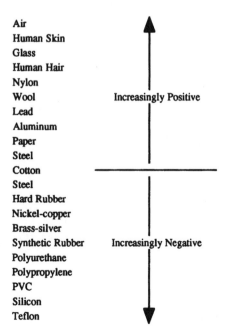

Figure 5.16 Triboelectric series. *(Source:* Hybrid Circuit Technology Handbook, *Noyes Publications.)*

Static is controlled by prevention of charge buildup and the use of discharge techniques (Fig. 5.17). Prevention techniques include use of antistatic materials in garments and in-process storage boxes. In some areas, a topical antistatic solution may be applied to the walls to prevent the buildup of static charge. These solutions work by leaving a neutralizing residue on the surface. Generally, they are not used in critical stations because of the possible contaminating effect of the residue.

Discharge techniques include the use of ionizers and grounded static-discharge straps. Ionizers are placed just underneath the HEPA filters, where they function to neutralize any charge buildup in the filtered air. Ionizers are also placed on nitrogen blow-off guns for the same effect. Some stations will have a portable ionizer blowing ionized air directly on the wafers being processed. Static discharge is also accomplished by grounding operators with wrist straps, having grounded mats at critical stations, and grounding work surfaces. Comprehensive static control programs, with prevention and discharge methods along with personnel training and third party monitoring, are features in most advanced fabs.

**Shoe cleaners.**    In any contamination-controlled area, the dirtiest region is the floor. Removal of dirt from the sides of shoes and shoe coverings is accomplished by shoe cleaners stationed at the cleanroom

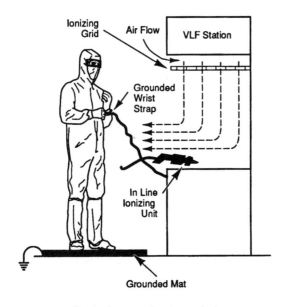

**Figure 5.17**   Static-charge reduction techniques.

door. Rotating brushes remove the dirt. Typical machines feature an internal vacuum to capture the loosened dirt, and bags to hold the dirt for removal from the area.[12]

**Glove cleaners.**   Maintaining clean gloves in the fab area is a challenge. Instructing operators to discard gloves when they are contaminated or dirty is one way. However, some contamination is hard to see, and the decision to discard becomes a judgment that can vary from operator to operator. Another approach is to discard the gloves after every shift. This can get very expensive. Some fab areas use glove cleaners that clean and dry the gloves in an enclosure.[13]

### Personnel-generated contamination

Cleanroom personnel are among the biggest sources of contamination. A cleanroom operator, even after showering and sitting, can give off between 100,000 and 1,000,000 particles per minute.[14] This number increases dramatically when a person is in motion. At 2 miles per hour, a human being gives off up to 5 million particles per minute. The particles come from flakes of dead hair and normal skin flaking. Additional particle sources are hair sprays, cosmetics, facial hair, and exposed clothing. The table in Fig. 5.18 lists the percentage increase in particulate generation over the background contamination level for different operator activities.[17]

Normal clothing can add more millions of particles to the area, even under a cleanroom garment. In cleanrooms with very high cleanliness levels, operators will be directed to wear street clothing that is made of tight-weave, nonshedding materials. Garments made of wools and cottons are to be avoided, as are ones with high collars.

A human's breath also contains high levels of contaminants. Every exhale puts numerous water droplets and particles into the air. The breath of smokers carries millions of particles for a long time after a cigarette is finished. Body fluids such as saliva contain sodium, a killer to many semiconductor devices. While healthy human beings are sources of many contaminants, sick individuals are even worse.

| | |
|---|---|
| Normal Breath | No Particles |
| Smoker's Breath after Smoking | 500 % |
| Sneezing | 2000% |
| Sitting Quietly | 20 % |
| Rubbing Hands On Face | 200% |
| Walking | 200% |
| Stamping Foot on Floor | 5000% |

**Figure   5.18** Activity-caused increase in particles. *(Source:* Hybrid Circuit Technology Handbook, *Noyes Publications.)*

Specifically, skin rashes and respiratory infections are additional sources of contaminants. Some fabrication areas reassign personnel with certain health problems.

Given the scope of the problem, the only feasible way to render humans acceptable in a cleanroom is to cover them up. The style and material of clothing selected for cleanroom personnel depends on the level of cleanliness required. For a typical area (Fig. 5.17), the clothing material will be nonshedding and may contain conductive fibers to draw off static charges. Materials used are polyester and GORETEX™. The trade-off is the filtering ability of the material and operator comfort. The reclean versus discard issue applies to cleanroom suits as well as gloves. Most ULSI fabs have found that reusable gowns, even with the cost of recleaning, provide a overall lowered cost of ownership.[18]

Every part of the body is covered up. The head will have an inner cap that keeps the hair in place. This is covered by an outer shell that is designed to fit close to the face and has snaps or a tail for securing the headgear under the body-covering smock. Covering the face will be a mask. Masks vary from surgical types to full ski-mask-style designs. In some cleanrooms, both an inner and outer face mask are required. The eyes, which are a major source of fluid particles, are covered by glasses (usually safety glasses) with side shields. In contamination-critical areas, the operators might wear a covering that totally encloses the head and face, connected to an air supply and filter. These are lower-tech models of a space suit helmet. The unit attaches to a belt filter, blower, and pump system. Fresh air is supplied by the pump, while the filter ensures that no breath-generated contaminates are discharged into the room.[1]

Body covers are oversuits (called *bunny suits*) that have closures for the legs, arms, and neck. Well designed suits will have covers over the zippers and no outside pockets.

The feet are covered with shoe coverings, some with attached leggings that come up the leg. In static-sensitive areas, straps are available to drain off static charge.

Hands are covered with at least one pair of gloves. Most favored are medical-type PVC gloves that permit good tactile feeling. Glove materials for chemical handling include orange latex (acid protection), green nitrile (solvent protection), and silver multilayered PVA solvent (special solvents).[16] In some areas, an inner pair of cotton gloves is permitted for comfort. Gloves should be pulled up over the sleeves to prevent contamination from traveling down the arm and into the cleanroom.

Skin flaking can be further controlled with the use of special lotions that moisten the skin. Any lotions used must be sodium- and chlorine-free.

In general, the order of gowning is from the head down. The theory is that dirt stirred up at each level is covered up by the next lower garment. Gloves are put on last. The garments and procedures needed to control contamination from cleanroom workers are well known. However, the primary level of defense is the dedication and training of the operators. It is easy for an area to become lax in maintaining cleanroom discipline and to suffer high levels of contamination.

## Process water

During the course of fabrication processing, a wafer will be chemically etched and cleaned many times. Each of the etching or cleaning steps is followed by a water rinse. Throughout the entire process, the wafer may spend a total of several hours in water rinse systems. A modern wafer fabrication facility may use up to several million gallons of water per day, representing a substantial investment in water processing, delivery to the process areas, and treatment and discharge of waste water.[19] Given the vulnerability of semiconductor devices to contamination, it is imperative that all process water be treated to meet very specific cleanliness requirements.

Water from a city system contains unacceptable amounts of the following contaminants:

1. Dissolved minerals

2. Particulates

3. Bacteria

4. Organics

5. Dissolved oxygen

6. Silica

The dissolved minerals come from salts in normal water. In the water, the salts separate into ions. For example, salt ($NaCl$) breaks up into $Na^+$ and $Cl^-$ ions. Each is a contaminant in semiconductor devices and circuits. They are removed from the water by reverse osmosis (RO) and ion-exchange systems.

The process of removing the electrically active ions changes the water from a conductive medium to a resistive one. This fact is used to improve the quality of deionized (DI) water. Deionized water has a resistivity of 18,000,000 $\Omega \cdot$ cm at 25°C. It is called 18-megohm (M$\Omega$) water. Figure 5.19 shows the effect on the resistivity of water when various amounts of dissolved minerals are present.

The resistivity of all process water is monitored at many points in the fabrication area. The goal and specification is 18 M$\Omega$ in VLSI ar-

| Resistivity Ohms-cm 25°C | Dissolved Solids (ppm) |
|---|---|
| 18,000,000 | 0.0277 |
| 15,000,000 | 0.0333 |
| 10,000,000 | 0.0500 |
| 1,000,000 | 0.500 |
| 100,000 | 5.00 |
| 10,000 | 50.00 |

**Figure 5.19** Resistivity of water versus concentration of dissolved solids.

eas, although some fabrication areas will run with 15-M$\Omega$ water levels. Solid particles (particulates) are removed from the water by sand filtration, earth filtration, and/or membranes to submicron levels. Bacteria and fungi find water a favorable host. They are removed by sterilizers that use ultraviolet radiation to kill the bacteria and filters to remove them from the stream of water.

Organic contaminates (plant and fecal materials) are removed with carbon bed filtration. Dissolved oxygen and carbon dioxide are removed with forced draft decarbonators and vacuum degasifiers.[20] Water specifications for a 4MB DRAM product are shown in Fig. 5.20.

The cost of cleaning process water to acceptable levels is a major operating expense of a fabrication area. In most fabrications, the process stations are fitted with water meters that monitor the used water. If the water falls within a certain range, it is recycled in the water system for cleanup. Excessively dirty water is treated as dictated by regulations and discharged from the plant. A typical fabrication system is shown in Fig. 5.21. Water stored in the system is blanketed with nitrogen to prevent the absorption of carbon dioxide. Carbon dioxide in the water interferes with resistivity measurements, causing false readings.

| Contaminant | 64Kb | 1Mb | 16Mb |
|---|---|---|---|
| Resistivity (M$\Omega$cm) | >17 | >18 | >18.1 |
| TOC (micro-g/L) | <200 | <50 | <10 |
| Bacteria (#/L) | <250 | <50 | <1 |
| Particles (#/mL) | <50 | <50 | <1 |
| Critical size (micron) | 0.2 | 0.2 | 0.05 |
| Dissolved oxygen (micro-g/L) | <200 | <100 | <20 |
| Sodium (micro-g/L) | <1 | <1 | <0.01 |
| Chlorine (micro-g/L) | <5 | <1 | <0.01 |
| Manganese (micro-g/L) | N/A | <1 | <0.005 |

**Figure 5.20** DRAM water specs. (*Source:* Semiconductor International, *July 1994, p. 178.*)

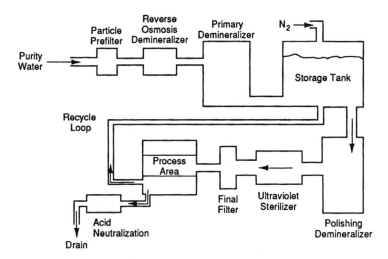

**Figure 5.21**  Typical deionized water system.

## Process chemicals

The acids, bases, and solvents used to etch and clean wafers and equipment have to be of the highest purity for use in a fabrication area. The contaminants of concern are metallics, particulates, and other chemicals. Unlike water, process chemicals are purchased and used as they come into the plant. Industrial chemicals are rated by grade. They are commercial, reagent, electronic, and semiconductor grade. The first two are generally too dirty for semiconductor use. Electronic grade and semiconductor grade are cleaner chemicals, but levels of cleanliness vary from manufacturer to manufacturer.

Trade organizations such as SEMI are establishing cleanliness specifications for the industry. However, most semiconductor plants purchase their chemicals to in-house generated specifications. Of primary concern are the metallic mobile ionic contaminants. These are usually limited to levels of 1 part per million (ppm) or less. Some suppliers are making available chemicals with MIC levels of only 1 part per billion (ppb). Particulate filtering levels are specified at 0.2 µm or lower.

Chemical purity is indicated by its assay number. The assay number indicates the percentage of the chemical in the container. For example, an assay of 99.9 percent on a bottle of sulfuric acid means that the bottle contains 99.9 percent sulfuric add and 0.01 percent other substances.

The delivery of a clean chemical to the processing area involves more than just making a clean chemical. Care must be taken to clean the inside of the containers, using containers that do not dissolve, using particulate-free labels, and placing the clean bottles in bags before shipping. Chemical bottle use is found only in older and lower technology fabs.

Most firms purchase clean process chemicals in bulk quantities. They are decanted into smaller vessels and are distributed to the process stations from a central location by pipes or distributed directly to the process station from a mini-unit. Bulk chemical distribution systems (BCDSs) offer cleaner chemicals and lower cost. Special care must be maintained to ensure that piping and transfer vessels are cleaned regularly to prevent contamination. A particular worry in secondary vessel filling is cross contamination. Vessels should be reserved for one type of chemical only.

The requirement for cleaner chemicals, tighter process control, and lower costs is being met by several techniques. One is point-of-use (POU) chemical mixers (another version of a BCDS). These units, connected to a wet bench or automatic machine, mix chemicals and deliver them directly to the process vessel or unit. Another technique is the use of chemical reprocessors. These units are located on the drain side of a wet process station. Discharged chemicals are refiltered and (in some cases) recharged for reuse. Critical recirculating etch baths are connected to filters to keep a clean supply of chemicals at the wafer. A newer process is point of use chemical generation (POUCG). Chemicals such as $NH_4OH$, HF, and $H_2O_2$ are made at the process station by combining the proper gases with DI water. Elimination of the contamination associated with chemical packaging and transfers gives this method the possibility of producing chemicals in the parts per trillion (ppt) range.[21]

**Process gases**

In addition to the many wet (liquid) chemical processes, a semiconductor wafer is processed with many gases. The gases are both the air-separation gases (oxygen, nitrogen, and hydrogen) and specialty gases such as arsine and carbon tetrafluoride.

Like the wet chemicals, they have to be delivered clean to the process stations and tools. Gas quality is measured in four categories:

1. Percentage of purity

2. Water vapor content

3. Particulates

4. Metallic ions

Extremely high purity is required for all process gases. Of particular concern are the gases used in oxidation, sputtering, plasma etch, chemical vapor deposition (CVD), reactive ion etch, ion implantation, and diffusion. All of these processes involve chemical reactions that are driven by an energy source. If the gases are contaminated with other gases, the anticipated reaction can be significantly altered or the result on the wafer changed. For example, chlorine contamination in a tank of argon used for sputtering can end up in the sputtered film, with disastrous results for the device. Gas purity is specified by the assay number, with typical values ranging from 99.99 to 99.999999 percent, depending on the gas and the use in the process. Purity is expressed as the number of 9s to the right of the decimal point. The highest purity is referred to as "six 9s pure."[22]

The challenge in fab areas is maintaining the gas purity from the manufacturer to the process station. From the source, the gas passes through a piping system, a gas panel containing valves and flowmeters, and a connection to the tool. Leaks in any part of the system are disastrous. Outside air (especially the oxygen) can enter into a chemical reaction with the process gas, changing its composition and the desired reaction in the process. Contamination of the gas can come from outgassing of the system materials. A typical system features stainless steel piping and valves along with some polymer components, such as connectors and seals. For ultra-clean systems, the stainless steel surfaces have electropolished and/or double vacuum melt inside surfaces to reduce outgassing.[23] Another technique is the growing of an iron oxide film, called *oxygen passivation* (OP), to further reduce outgassing. Polymer components are avoided. Gas panel design eliminates dead spaces that become repositories for contamination. Also critical is the use of clean welding processes to eliminate gases absorbed into the piping from the welding gases.

The control of water vapor is also critical. Water vapor is a gas and can enter into unwanted reactions just like other contaminating gases. In fabrication areas, processing silicon wafers with water vapor present is a particular problem. Silicon oxidizes easily wherever free oxygen or water is available. Control of unwanted water vapor is necessary to prevent accidental oxidation of silicon surfaces. Water-vapor limits are 3 to 5 ppm.

The presence of particulates and/or metallics in a gas has the same effect on the processing as in wet chemicals. Consequently, gases are filtered to the 0.2-μm level and metallic ions are controlled to parts per million or lower.

The air-separation gases are stored on the site in the liquid state. In this state, they are very cold, a situation that freezes some contaminants in the bottom of the tanks. Specialty gases are purchased in high-pressure cylinders. Since many of the specialty gases are toxic or flammable, they are stored in special cabinets outside the plant.

**Quartz.**   Wafers spend a lot of process time in quartz holders such as wafer holders, furnace tubes, and transfer holders. Quartz can be a significant source of contamination, both from outgassing and particulates. Evan high-purity quartz contains heavy metals that can outgas into diffusion and oxidation tube gas streams, especially during high-temperature processes. Particulates come from the abrasion of the wafers in the wafer boats and the scrapping of the boats against the furnace tubes. (Solutions to this problem are discussed in Chapter 11.) Both electric and flame fused quartz processes are used to produce quartz surfaces that are acceptable for semiconductor use.[24]

### Equipment

Successful contamination control is dependent on knowing the sources of contamination. Most analysis (Fig. 5.22) identifies process equipment as the largest source of particulates. By the 1990s, equipment-induced particulates rose to the level of 75 to 90 percent of all particle sources.[23] This does not mean that the equipment is getting dirtier. Advances in particulate control in air, chemicals, and from personnel has shifted the focus to equipment. Defect generation is part of equipment specifications. Generally, the number of particles added to the wafer per product pass (ppp) through the tool is specified. The term *particles per wafer pass* (PWP) is also used.

Reduction of particle generation starts with design and material selection. Other factors are the transport of the particles to the wafer and deposition mechanisms such as static. Most equipment is assembled in a cleanroom with the same class number as the customer's fab

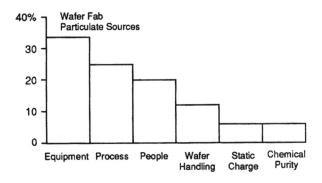

**Figure 5.22**   Sources of particulate contamination. This analysis, shown at SEMI Forecast by Dr. C. Rinn Cleavelin, Texas Instruments, revealed equipment-generated particles as the top enemy in 1985. *(Courtesy of Semiconductor Equipment and Materials Institute.)*

area. The clustering of several process tools with one clean micro-environment for loading and unloading minimizes the contamination generation associated with multiple loading stations. Cluster tools are discussed in Chapter 15. There is interest in having *in situ* particle monitors in the process chambers. Challenges to this level of automation and detection include correlation of in-chamber levels with on wafer levels and the lifetime of detectors in the hostile and corrosive environments of most process chambers.

## Cleanroom Materials and Supplies

In addition to the process chemicals, it takes a host of other materials and supplies to process the wafers. Each of these must meet cleanliness requirements. Logs, forms, and notebooks used will be either of nonshedding coated paper or of a polymer plastic. Pencils are not allowed, and pens are the nonretracting type.

Wafer storage boxes are made of specific non-particle-generating materials as are carts and tubing materials. Cart wheels and tools are used without greases or lubricants. In many areas, mechanics' tools and tool boxes are cleaned and left inside the cleanroom.

## Cleanroom Maintenance

Regular maintenance of the cleanroom is essential. The cleaning personnel must wear the same garments as the operators. Cleanroom cleaners and applicators, including mops, must be carefully specified. Normal household cleaners are far too dirty for use in cleanrooms. Special care must be taken when using vacuum cleaners. Special cleanroom vacuums with HEPA filtered exhausts are available. Many cleanrooms will have built-in vacuum systems to minimize dirt generation during cleaning.

The wipe-down of process stations is done with special wipes made from non-shedding polyester or nylon, prewashed to reduce contamination. Some are prewetted with an isopropyl alcohol and DI water solution, which provides convenience and eliminates secondary contamination from spraying cleaners in the cleanroom.[27]

The wiping procedure is critical. Wall surfaces should be wiped from top to bottom and deck surfaces from back to front. Cleaning chemicals in spray bottle, if used, should be sprayed into wipes, not onto the surfaces. This prevents overspray onto the wafers and equipment. Cleanroom cleaning has, itself, become a critical supporting operation. Many fabs use outside certification services[24] to identify and document cleanliness levels, practices, documentation, and control procedures. Certification standards for cleanroom maintenance schedules is identified in the ISO Global Cleanroon Standards (ISO 14644-2).

Copper metallization has become the metal of choice for advanced ULSI devices. While the copper has many advantages (see Chapter 13), there is a serious downside. Copper contamination inside a silicon wafer raises havoc with the electrical operation of the component devices. Isolating the copper processing areas and copper deposited wafers is essential. Separate areas and separate process tools are required with strict production control to ensure that no copper technology wafers get diverted into the other process areas.

### Wafer Surface Cleaning

Clean wafers are essential at all stages of the fabrication process but are especially necessary before any of the operations are performed at high temperature. Up to 20 percent of all process steps relate to wafer cleaning.[28] The cleaning techniques described here are used throughout the wafer fabrication process.

The story of semiconductor process development is in many respects the story of cleaning technology keeping pace with the increasing need for contamination-free wafers. Wafer surfaces have four general types of contamination. Each represents a different problem on the wafer, and each is removed by different processes. The four types are:

1. Particulates

2. Organic residues

3. Inorganic residues

4. Unwanted oxide layers

In general, a wafer cleaning process, or series of processes, must (a) remove all surface contaminants (listed above), (b) not etch or damage the wafer surface, (c) be safe and economical in a production setting, and (d) be ecologically acceptable. Cleaning processes are generally designed to accommodate two primary wafer conditions. One, called the *front end of the line* (FEOL), refers to the wafer fabrication steps used to form the active electrical components on the wafer surface. During these steps, the wafer surface, and particularly the gate areas of MOS transistors, are exposed and vulnerable. A critical parameter in these cleans is *surface roughness*. Excessive surface roughness can alter device performance and compromise the uniformity of layers deposited on top of the devices. Surface roughness is measured in nanometers as the root mean square of the vertical surface variation (nmRMS). Year 2000 requirements are on the order of 0.15 nm, dropping to less than 0.1 nm by the year 2010.[29] Other concerns in FEOL cleaning processes are the electrical condition of the bare surface.

Metal contaminants on the surface change the electrical characteristics of devices, with MOS transistors being particularly vulnerable. Sodium (Na) is a particular problem (see Chapter 4), along with Fe, Ni, Cu, and Zn. Cleaning processes will have to reduce concentrations to less than $2.5 \times 10^9$ atoms/cm$^2$ to meet 2010 device needs. Aluminum and calcium are also problems and need reduction on the surface to the $5 \times 10^9$ atoms/cm$^2$ level.[30]

A most critical factor is maintaining the integrity of gate oxides. Cleaning processes can attack and rough up gate oxides, with the thinner ones most vulnerable. A gate oxide is required to act as a dielectric in a MOS transistor, and as such must have a consistent structure, makeup, and thickness. *Gate oxide integrity* (GOI) is measured by testing the gate for electrical shorts. The NTRS indicates that, at 180-nm technology level, gates will have to exhibit less than 0.02 defects/cm$^2$ when tested at 5 MV/cm for 30 s.[33]

Specific concerns at the BEOL cleans, in addition to particles, metals, and general contamination, are anions, polysilicon gate integrity, contact resistance, via hole cleanliness, organics, and the overall numbers of shorts and opens in the metal system. These issues are explored in Chapter 13. Photoresist removal is also a cleaning process with both FEOL and BEOL consequences. These issues are explored in Chapter 9.

Different chemicals and cleaning methods are mixed and matched to accommodate the needs at particular steps in the process. A typical FEOL cleaning process (such as preoxidation clean) is listed in Fig. 5.23. The FEOL process listed is called a non-Hf-last process. Other variations have the HF removal step last. Non-Hf-last surfaces are hydrophilic that can be dried without water marks and have a thin oxide (grown in the cleaning steps) that can protect the surface. They also absorb more organic contamination. HF-last surfaces are hydrophobic, which can be difficult to dry without water marks if there are also hydrophilic (oxides) surface layers present. These surfaces are stable from hydrogen surface passivation.[29] A choice of either non-HF-last or HF-last depends on the sensitivity of the devices being fabricated in the wafer surface and general cleanliness requirements.

---

• Particle removal (mechanical)
• General chemical clean (such as sulfuric acid/H$_2$/O$_2$)
• Oxide removal (typically dilute HF)*
• Organic and metal removal (SC-1)
• Alkali metal and metal hydroxide removal (SC-2)
• Rinse steps
• Wafer drying

**Figure 5.23**  Typical FEOL cleaning process steps.

## Particulate removal

Particulates on the wafer surface vary from very large ones (50 μm in size) to very tiny ones (<1 μm in size). The larger ones can be cleaned off by conventional chemical baths and rinses. The smaller particulates are more difficult to remove, because they can be held to the surface by several strong forces. One is called the *van der Waals force*. It is a strong interatomic attraction between the electrons of one atom and the nucleus of another. A technique to minimize electrostatic attraction is manipulation of a factor called the *zeta potential*. Zeta potential arises from a charge zone around particles that is balanced by an oppositely charged zone in the cleaning liquid. This charge varies with velocity (the speed of movement of the cleaning liquid, as when the wafers are moved in a cleaning bath), the pH of the solution, and the concentration of electrolytes in the solution. It is also affected by additives to the cleaning solution, such as surfactants. These conditions can be set to create a large charge that has the same polarity of the wafer, thus creating a repulsion that serves to keep the particle in the solution and off the wafer surface.

Capillary force is another problem. It occurs when there is a liquid bridge between a particle and the surface (Fig. 5.24). Capillary forces can be higher than van der Waals forces.[30] Surfactants and mechanical assist, such as megasonics, are used to dislodge these particles from the surface. .

Cleaning processes are most often a series of steps designed to remove both the large and small particles. The simplest particulate removal process is to blow off the wafer surface using a spray of filtered high-pressure nitrogen from a hand-held gun located in the cleaning stations. In fabrication areas where small particles are a problem, the nitrogen guns are fitted with ionizers that strip static charges from the nitrogen stream and neutralize the wafer surface.

Nitrogen blow-off guns are effective in removing most large particles. Since the guns are hand held, the operators must use them in a manner that does not contaminate other wafers in the station or the station itself. Blow-off guns are not generally used in Class 1/10 cleanrooms.

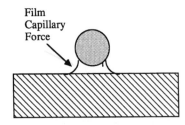

Film
Capillary
Force

**Figure 5.24** Capillary force from film.

## Wafer scrubbers

The stringent wafer cleanliness requirements for epitaxial growth led to the development of mechanical wafer surface scrubbers, which are used wherever particulate removal is critical.

The scrubbers hold the wafer on a rotating vacuum chuck (Fig. 5.25). A rotating brush is brought in near contact with the rotating wafer while a stream of deionized water is directed onto the wafer surface. The combination of the brush and wafer rotations creates a high-energy cleaning action at the wafer surface. The liquid is forced into the small space between the wafer surface and the brush ends where it achieves a high velocity, which aids the cleaning action. Caution must be exercised to keep the brushes and cleaning liquid lines clean to prevent secondary contamination. Also, the brush height above the wafer must be maintained to prevent scratching the wafer surface.

Surfactants may be added to the DI water to increase the cleaning effectiveness and prevent static buildup. In some applications, diluted ammonium hydroxide may be used as the cleaning liquid to prevent buildup of particles on the brush and to control the zeta potential in the system.[31]

Scrubbers are designed as stand-alone units with automatic loading capabilities or are built into other pieces of equipment to clean the wafers automatically before processing.

## High-pressure water cleaning

The removal of statically attached particles first became a necessity for the cleaning of glass and chrome photomasks. A high-pressure water spray process was developed, using a small stream of water pressurized 2000 to 4000 psi. The stream is swept across the mask or

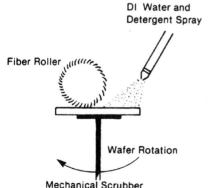

DI Water and Detergent Spray

Fiber Roller

Wafer Rotation

Mechanical Scrubber

**Figure 5.25** Mechanical scrubber.

wafer surface, dislodging both large and small particles. Often, a small amount of surfactant will be added to the water stream to act as a destatic agent.

### Organic residues

Organic residues are compounds that contain carbon, such as oils in fingerprints. These residues can be removed in solvent baths such as acetone, alcohol, or TCE. In general, solvent cleaning of wafers is avoided whenever possible due to the difficulty of drying the solvent completely off the wafer surface. Also, solvents always contain some impurities that themselves may represent a contamination source.

### Inorganic residues

Inorganic residues are those that do not contain carbon. Examples are inorganic acids such as hydrochloric or hydrofluoric acid, which may be introduced from other steps in wafer processing. The organic and inorganic residues are cleaned from the wafers in a variety of cleaning solutions described in the following section.

### Chemical cleaning solutions

A wide range of cleaning processes exist in the semiconductor industry. Each fabrication area has different cleanliness needs and different experiences with different solutions. The solutions described in this section are those in common use, although there are numerous variations and different combinations of solutions from one fabrication area to another. Described are processes used to clean wafers before doping, deposition, and metallization steps. (The special case of photoresist removal is addressed in Chapter 8.)

Liquid chemical cleaning processes are generally referred to as *wet processes* or *wet cleaning*. Immersion cleaning takes place in glass, quartz, or polypropylene tanks fitted into the deck of a cleaning station (see Chapter 4). Where heating of the solution is required, the tank may be sitting on a hot plate, be wrapped with heating elements, or have an immersion heater inside. Chemicals are also applied by spraying, either using direct impingement or in centrifugal tools (see spin rinse dryers).

### General chemical cleaning

**Sulfuric acid.**    A common general cleaning solution is hot sulfuric acid with an added oxidant. It is also a general photoresist stripper (see

Chapter 8). The sulfuric acid is an effective cleaner in the 90 to 125°C range. At these temperatures, it will remove most inorganic residues and particulates from the surface. Oxidants are added to the sulfuric acid to remove carbon residues. The chemical reaction converts the carbon to carbon dioxide, which leaves the bath as a vapor by the following reaction:

$$C + O_2 \rightarrow CO_2(gas)$$

The oxidants normally used are:

- Hydrogen peroxide ($H_2O_2$)
- Ammonium persulfate [$(NH_4)_2S_2O_8$]
- Nitric acid ($HNO_3$), and ozone ($O_2$)

**Sulfuric acid with hydrogen peroxide.**   Hydrogen peroxide with sulfuric acid is a common cleaner used to clean wafers at all stages of processing, especially before the tube processes. It is also used as a photoresist stripper in the patterning operation. Within the industry, this solution is known by a number of names, including *Carro's acid* and *piranha etch*. The latter term attests to the aggressiveness and effectiveness of the solution.

A manual method is to add about 30 percent (by volume) of hydrogen peroxide to a beaker of room-temperature sulfuric acid. In this ratio, an exothermic reaction takes place that quickly raises the temperature of the bath into the 110 to 130°C range. As time proceeds, the reaction slows, and the bath temperature falls below the effective range. At this point, the bath may be recharged with additional hydrogen peroxide or discarded. Recharging the bath eventually results in a lowered cleaning rate due to the conversion of the hydrogen peroxide to water, which dilutes the sulfuric acid.

In automated systems, the sulfuric acid is heated to the effective cleaning temperature range and small amounts (50 to 100 mL) of hydrogen peroxide are added before cleaning each batch of wafers. This method maintains the bath at the proper temperature, and the water created from the hydrogen peroxide evaporates out of the solution. The use of heated sulfuric acid is preferred for economic and process-control reasons. It is also easier to automate this approach to mixing the two chemicals.

**Ozone.**   The purpose of oxidant additives is the addition of oxygen to the solution. Some companies use a flow of gaseous ozone ($O_2$) directly in the container of sulfuric acid. Ozone and DI water constitute a

cleaning solution for light organic contamination.[32] A typical process is
1 to 2 ppm ozone in DI water for ten minutes at room temperature.[33]

### Oxide layer removal

The ease of silicon oxidation has been mentioned. The oxidation can
take place in air or in the presence of oxygen in the heated chemical
cleaning baths. Often, the oxide grown in the baths, while thin (100 to
200 Å), is thick enough to block the silicon surface from reacting properly during one of the other process operations. The thin surface oxide
can act as an insulator, preventing good electrical contact between the
silicon surface and a layer of conducting metal.

The removal of these thin oxides is a requirement in many processes. Silicon surfaces with an oxide are called *hydroscopic*. Surfaces
that are oxide free are termed *hydrophobic*. Hydrofluoric acid (HF) is
the acid favored for oxide removal. Prior to an initial oxidation, when
the surface is only silicon, the wafers are cleaned in a bath of full-
strength HF (49 percent). The HF etches away the oxide but does not
etch the silicon.

Later in the processing, when the surface is covered with previously
grown oxides, thin oxides in patterned holes are etched away with a
water and HF solution. These solutions vary in strength from 100:1 to
10:7 ($H_2O$ to HF). The strength is chosen depending on the amount of
oxide already on the wafer, since the water and HF solution will etch
both the oxide on the silicon surface in the hole and the oxide covering
the rest of the surface. A strength is chosen to ensure the removal of
the oxide in the holes while not excessively thinning the other oxide
layers. Typical dilutions are 1:50 to 1:100.

Management of the chemistry of the silicon surface is an ongoing
challenge for cleaning processes. Generally, pregate cleaning uses the
dilute HF as the last chemical step. It is called *HF-last*. HF-last surfaces are hydrophobic and passivated with low metallic contamination. However, hydrophobic surfaces are difficult to dry, often leaving
*watermarks*.[34] Another problem is increased particle adhesion and
copper plating out of the surface.[35]

**RCA clean.**   In the mid 1960s, Werner Kern, an RCA engineer, developed a two-step process to remove organic and inorganic residues from
silicon wafers. The process proved to be highly effective, and the formulas became known simply as the *RCA cleans*. Whenever an RCA
cleaning process is referred to, it means that hydrogen peroxide is
used along with some base or acid. The first step, Standard clean 1
(SC-1) uses a solution of water, hydrogen peroxide, and ammonium

hydroxide. Solutions vary in composition from 5:1:1 to 7:2:1 and are heated to the 75 to 85°C range. SC-1 removes organic residues and sets up a condition for the desorption of trace metals from the surface. During the process, an oxide film keeps forming and dissolving.

Standard clean-2 (SC-2) uses a solution of water, hydrogen peroxide, and hydrochloric acid mixed in ratios of 6:1:1 to 8:2:1 and is used at the 75 to 85°C temperature range. SC-2 removes alkali ions and hydroxides and complex residual metals. It leaves behind a protective layer of oxide. The original mixtures are shown in Fig. 5.26.

The RCA formulas have proven durable over the years and are still the base cleaning processes for most pre-furnace cleaning. Improvements in chemical purity have kept pace with industry cleaning needs. Depending on the application, the order of SC-1 and SC-2 steps may be reversed. Where an oxide-free surface is required, an HF step is used before, in between, or after the RCA cleans.

Many adaptations and changes have been made to the original cleaning solutions. One problem is the removal of metallic ions from the wafer surface. These ions exist in chemicals and are not dissolved (made soluble) in most cleaning and etching solutions. The addition of a *chelating agent*, such as ethylenediamine-tetra-acetic acid, serves to bind up the ions so they do not redeposit on the wafer.

Diluted RCA solutions are finding more use. SC-1 dilutions are 1:1:50 (instead of 1:1:5) and SC-2 dilutions are 1:1:60 (instead of 1:1:6). These solutions have been found to be as effective as the more concentrated versions. Additionally, they produce less microroughing and are cost effective and easier to remove.[36]

### Room-temperature and ozonated chemistries

The perfect cleaning process takes place at room temperature with totally safe chemicals that are easily and economically disposable. That

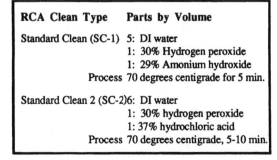

| RCA Clean Type | Parts by Volume |
|---|---|
| Standard Clean (SC-1) | 5: DI water |
| | 1: 30% Hydrogen peroxide |
| | 1: 29% Amonium hydroxide |
| | Process 70 degrees centigrade for 5 min. |
| Standard Clean 2 (SC-2) | 6: DI water |
| | 1: 30% hydrogen peroxide |
| | 1: 37% hydrochloric acid |
| | Process 70 degrees centigrade, 5-10 min. |

Figure 5.26   RCA clean formulas.

process does not exist. However, there is research into room-temperature chemistries. One process[37] combines room temperature baths of ultrapure water infused with ozone and two HF solutions (Fig. 5.27). Megasonic assist elevates the cleaning efficiency.

**Spray cleaning.** The standard cleaning technique has been immersion in chemical baths performed in wet benches or automated machines. While wet cleans are projected to carry the industry into the 0.35 to 0.50 μm era, there are growing considerations. The chemicals are becoming more expensive, immersion in a tank presents a redeposition of contaminants, and smaller and deeper patterns on the wafer surface constrain cleaning efficiency. Alternative cleaning methods have emerged. Spray cleaning offers several advantages. Chemical costs go down, since the chemicals expended are directed to the wafer rather than maintaining large reserves in tanks. Less chemical use reduces the cost of treating or removing hazardous chemical waste. Cleaning efficiency is improved. The pressure of the spray assists in cleaning small patterns on surfaces with deep holes. Also, there is less chance of recontamination, because the wafer receives only fresh chemicals. Spray methods allow immediate water rinsing after cleaning, eliminating transfer to a separate water rinse station.

**Dry cleaning.** The considerations concerning wet immersion methods have spurred interest in and development of *vapor or gas phase* cleaning. For cleaning, the wafers are exposed to vapors of the cleaning or etching chemical(s). HF/water vapor mixtures have found their way into processes for oxide removal. Available are systems for vapor phase replacement of peroxide-based cleaners.[38]

The ultimate industry dream is all dry cleaning and etching. Dry-etching methods (plasma; see Chapter 9) are well established. Dry cleaning is under development. Ultraviolet (UV) ozone can oxidize and photo-dissociate contaminants form the wafer surface.

**Cryogenic cleaning.** High-pressure carbon dioxide $CO_2$, or *snow cleaning,* is an emerging technique (Fig. 5.28). $CO_2$ is directed at the wafer

| Room Temperature Cleaning Process |
| --- |
| • Ozonated DI Water |
| • HF/$H_2O_2$/$H_2O$/Surfactant + Megasonic |
| • Ozonated Water + Megasonic |
| • Diluted HF (1%) |
| • DI Water Rinse + Megasonic |

**Figure 5.27** Experimental room-temperature cleaning process.

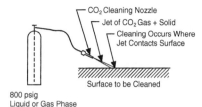

800 psig
Liquid or Gas Phase
Carbon Dioxide

**Figure 5.28** $CO_2$ "snow" cleaning. *(Courtesy of Walter Kern.)*

from a nozzle. As the gas leaves the nozzle, its pressure drops, causing a rapid cooling, which in turn forms $CO_2$ particles, or snow. The force of the impinging particle dislodges surface particles, and the flow carries them away. The physical bombardment of the surface supplies a cleaning action. Argon aerosol is another cryogenic technique. Argon is a fairly heavy and large atom that can dislodge particles when directed to the wafer under pressure.

A combination process, called *cryokinetic,* combines nitrogen and argon. Precooling of the gases under pressure forms a liquid/gas mixture that is flowed into a vacuum chamber. In the chamber, the liquid expands rapidly to form microscopic crystals that knock particles from the wafer surface.[39]

### Water rinsing

Every wet-cleaning sequence is followed by a rinse in deionized water. Rinsing performs the dual function of removing the cleaning chemicals from the surface and ending the etching action of an oxide etch. Great quantities of DI water, and great expense, are expended to achieve the goal of clean wafers. Rinsing is done by several different methods. Future focus will be on higher rinse efficiency and an order of magnitude reduction in quantity. The 1997 NTRS calls for reducing the present 30 gal/in$^2$ of silicon to 2 gal/in$^2$ of silicon in the 50-nm gate size sometime around 2012.

**Overflow rinsers.**    The need for atomically clean surfaces precludes just dunking the wafers in a tank of water. Thorough rinsing requires a continuous supply of clean water to the wafer surface. One such method is an overflow rinser (Fig. 5.29). It is a box sunk into the cleaning station deck. Deionized water is brought into the bottom of the box and flows through and around the wafers, exiting over a dam into a drain system. The rinsing action of the flowing water is enhanced by a stream of nitrogen bubbles introduced into the rinser from the bottom plate. As the nitrogen *bubbles* up through the water, it aids the mixing of the chemicals on the wafer surface with the water. This type of sys-

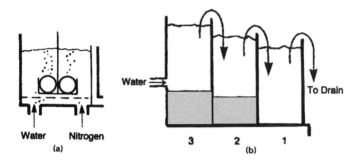

**Figure 5.29** Rinse systems. (*a*) Single overflow and (*b*) three-stage overflow.

tem is often called a *bubbler*. A variation is the *parallel downflow rinser*. In this configuration, the water is brought into the system outside the rinser and directed to flow down through the wafers (Fig. 5.30).

A rule of thumb is that adequate rinsing takes a minimum of five minutes (depending on wafer diameter) with a flow rate equivalent to five times the volume of the rinser per minute (the number of turnovers per minute). If the size of the bubbler is 3 L, the flow rate should be a minimum of 15 L/min.

The rinse time is determined by measuring the resistivity of the water as it exits the rinser. Cleaning chemicals act as charged molecules in the rinse water, and their presence can be inferred from the electrical resistivity of the water. If the water goes into the rinser at an 18-M$\Omega$ level, a reading of 15 to 18 M$\Omega$ on the exit side indicates that the wafers are cleaned and rinsed. The rinsing step is so critical that, generally, a minimum of two rinsers are used, and the total rinsing time is set at two to five times the minimum determined by resistivity

Parallel Down Flow

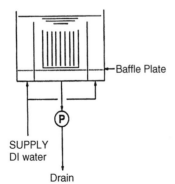

**Figure 5.30** Parallel downflow rinsing. (*Courtesy of Walter Kern.*)

measurements. Often, a water resistivity meter is mounted on the outlet to constantly measure the output resistivity and signal when rinsing is completed.

**Spray rinsing.**    Flowing water removes a water-soluble chemical from the wafer surface by a dilution mechanism. The top layer of the chemical dissolves into the water and is carried away by the flow. This action occurs over and over in a continuous manner. Rinsing is speeded up by a faster water flow rate, which can remove the dissolved chemical more quickly. The number of turnovers directly determines the rinse rate. This can be understood by considering a very fast water flow rate, but in a huge rinse tank. The chemical removed from the wafer surface would be evenly distributed throughout the tank, and therefore some would still cling to the surface. The chemical could be removed from the tank only when enough water has flowed in and out of it to carry away the chemical.

One way to have a faster rinse rate is by using a water spray. The spray removes the chemical with a physical force from its own momentum, and the many small droplets continually hitting the wafer surface have the effect of an extremely high turnover rinse. Along with more efficient rinsing, a spray rinser uses considerably less water than an overflow rinser. A problem with a spray rinser comes when the resistivity of the exiting water is measured with a resistivity monitor. Carbon dioxide from the air gets trapped in the water spray; the $CO_2$ molecules act as charged particles, and the resistivity meter reads them as contaminants, which they aren't.

**Dump rinsers.**    With respect to rinsing efficiency and water savings, a dump rinser (Fig. 5.31) is an attractive method. The system is like an overflow rinser but with a spray capability. The wafers are placed into the dry rinser and immediately sprayed with deionized water. While

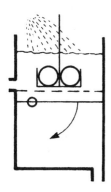

**Figure 5.31** Ultrasonic/megasonic wafer cleaning/etching bath.

they are being sprayed, the cavity of the rinser is rapidly filled with water. As the water overflows the top, a trap door in the bottom swings open, and the water is dumped instantly into the drain system. This fill-and-dump action is repeated several times until the wafers are entirely rinsed.

Dump rinsing is also favored, because all of the rinsing takes place in one cavity, which saves equipment and space. It is also a system that can be automated, so the operator only needs to load the wafers in (this can be done automatically) and push a button.

**Sonic-assisted cleaning and rinsing.**    The addition of sonic energy waves to a tank of cleaning chemicals or a rinse system aids and speeds wet processes. Use of sonic waves can increase efficiency, which in turn allows a lower bath temperature. Sonic waves are energy waves generated from transducers arranged on the outside of the tank (See Fig. 5.32). Two ranges are used. In the 20,000 to 50,000 hertz (1 hertz (Hz) = one cycle per second) range, they are called *ultrasonic* and, in the 850 kilohertz (kHz) range, they are called *megasonic* waves.[40] Ultrasonic assists rinsing through cavitation. The waves pass through the liquid, causing microscopic bubbles to form and collapse rapidly. This creates a microscopic scrubbing action that dislodges particles. The phenomenon is called *cavitation*. Megasonic assist offers a different mechanism. In fluid dynamics, there is a static or slow-moving boundary at surfaces, such as wafers. Small particles can be held in this layer, unexposed to the cleaning chemicals. Megasonic energy reduces this layer, exposing the particles to cleaning action. In addition, a phenomenon called *acoustic streaming* fosters an increase in the velocity of the rinse or cleaning solutions passing the wafer surface, increasing cleaning efficiency.[41]

**Spin-rinse dryers (SRDs).**    After rinsing, the wafers must be dried. This is not a trivial process. Any water that remains on the surface (even atoms) has the potential of interfering with any subsequent operation. There are three drying techniques (with variations) used.

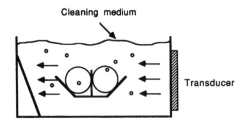

Cleaning medium

Transducer

Figure 5.32   Ultrasonic/megasonic wafer cleaning/etching bath.

## Drying techniques

- Spin rinse dryers
- Isopropyl alcohol (IPA) vapor dry
- Surface tension/Marangoni drying

Complete drying is accomplished in a centrifuge-like piece of equipment. In one version, the wafer boats are put in holders around the inside surface of a drum. In the center of the drum is a pipe with holes, and it is connected to a source of deionized water and hot nitrogen (Fig. 5.33).

The drying process actually starts with a rinse of the wafers as they rotate around the center pipe that sprays the water. Next, the SRD switches to a high-speed rotation as heated nitrogen comes out of the center pipe. The rotation literally throws the water off the wafer surfaces. The heated nitrogen assists in the removal of small droplets of water that may cling to the wafer.

SRDs are also built for drying single-wafer boats. The boat slips into a rotating holder in the center of a chamber. The water and nitrogen come into the chamber through its side rather than through a pipe in the center. The rinsing and drying take place as the boat spins about its own axis. This type of SRD is called an *axial dryer*. These two machines are used for automatic wafer cleaning and etching. As a wafer cleaner, the required chemicals are plumbed to the machine, and microprocessor-controlled valves direct the right chemicals into the chamber.

**Isopropyl alcohol (IPA) vapor drying.** A recently rediscovered drying technique is alcohol drying (Fig. 5.34). In the bottom of the dryer is a

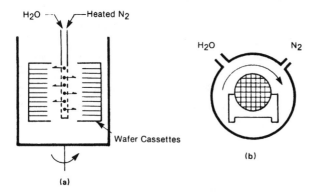

**Figure 5.33** Spin rinse dryer styles. (*a*) Multiboat and (*b*) single-boat axial.

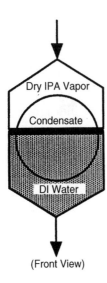

(Front View)

**Figure 5.34** Vapor dry. *(Courtesy of Walter Kern.)*

heated reserve of liquid IPA with a vapor cloud (vapor zone) above it. When a wafer with residual water on the surface is suspended in the vapor zone, the IPA replaces the water. Chilled coils around the vapor zone condense the water vapor out of the IPA vapors, leaving the wafers water free. A variation is the *direct displacement vapor dryer.* In this system, the wafers are pulled out of a DI water bath directly into an IPA vapor zone where the water displacement occurs.

**Surface tension/Marangoni drying.**  Water surface tension creates a unique condition when wafers are pulled slowly through a water surface. The tension draws the water away from the surface, leaving it dry. This effect is enhanced when a flow of an organic, such as IPA and nitrogen, is directed at the wafer water level interface. The IPA/N$_2$ flow creates a surface tension gradient, which in turns causes a water flow from the surface into the water. This internal flow further enhances the removal of water from the wafer. In practice, the wafers are either withdrawn from the water bath, or the water is allowed to slowly recede from the rinse tank.[42]

**Contamination detection/summary**

The detection of the various forms of contamination is detailed in Chapters 8 and 14. Figure 5.35 summarizes the specifications for cleanroom cleanliness.

| Year of Production | 2001 | 2006 | 2012 |
|---|---|---|---|
| Line width (nm) | 150 | 100 | 50 |
| Air Critical Particle Size (nm) | 8 | 2 | 1 |
| Water oxidizable carbon (ppb) | 1 | <1 | <1 |
| Water Dissolved Oxygen (ppb) | 1-10 | 1 | 1 |
| Water Particles>critical size | <0.2 per mL | <0.2 per mL | <0.2 per mL |
| Liquid Particles>critical size | <0.5 per mL | <0.5 per mL | <0.5 per mL |
| HCl, H2O impurities (ppt) | <1000 | <1000 | <1000 |
| BEOL solvents K,Li,Na (ppt each) | <1000 | <1000 | <1000 |
| POU Gases particles (per liter) | 2 | 2 | 2 |

Figure 5.35   SIA Roadmap Projections *(Source:* Micro, *October 1998, p. 54.)*

## Review Questions

1. State the class number required for VLSI processing and ULSI processing.
2. List three techniques used to keep contamination out of a cleanroom.
3. Draw a cross section of a "total" cleanroom, showing the filters and air patterns.
4. What does VLF mean?
5. What is the DI water specification?
6. Name three contaminants found in normal water.
7. What are the two ways that static is harmful in a fabrication area?
8. Explain the problem associated with high humidity in a fabrication area.
9. What is a mobile ionic contaminant, and why is it unwanted in a fabrication area?
10. How does a service bay contribute to the cleanliness of a cleanroom?
11. Describe a mini-environment.
12. Describe and sketch a SMIF/micro-environment unit.
13. List four types of contamination found on wafer surfaces.
14. List a cleaning chemical or chemical solution for each contaminant listed in question no. 13.
15. Describe the major techniques used to dry wafers.

# References

1. "New Challenges for robotics and automation," *Solid State Technology,* April 1994, p. 67.
2. *Ibid.*
3. Shiella Galatowitsch, Fabs Look to Minimize AMC, *Cleanrooms,* June 1999, p. 1.
4. Julia Sherry, Assessing Airborne Molecular Contaminant, *Future Fab 9,* International Issue, p. 135
5. Clean Room and Work Station Requirements, Federal Standard 209E, Sec. 1-5, Office of Technical Services, Dept. of Commerce, Washington, D.C.
6. 1997 Semiconductor Industry Association, *National Technology Roadmap for Semiconductors,* San Jose, CA.
7. A. Bonora, "Minienvironments and Their Place in the Fab of the Future," *Solid State Technology,* PennWell Publishing, September 1993.
8. Operator Training Course, Class-10 Technologies, Inc., San Jose, Calif., 1983, p. 13.
9. Hepa Corporation, Product Tour, www. Hepa.com/filters/tour.htm.
10. "SI Honors Top Fabs of 1994," *Semiconductor International,* April 1994, p. 78.
11. B. Newboe, "Minienvironments: Better Cleanrooms for Less," *Semiconductor International,* March 1993, p. 54.
12. J. Licari and L. Enlow, *Hybrid Microcircuit Technology Handbook,* Noyes Publications, Park Ridge, NJ, 1988, p. 281.
13. Dryden Engineering Product Description, Santa Clara, CA, 1995.
14. *Ibid.*
15. *The New American Revolution,* Araclean Services, La Grange, Ill., 1984, p. 9.
16. J. Licari and L. Enlow, *Hybrid Microcircuit Technology Handbook,* Noyes Publications, Park Ridge, NJ, 1988, p. 280.
17. R. Iscoff, "Cleanroom Apparel: A Question of Tradeoffs," *Semiconductor International,* Cahners Publishing, March 1994, p. 65.
18. Dryden Engineering Product Description, Santa Clara, CA, 1995.
19. R. Iscoff, "Cleanroom Apparel: A Question of Tradeoffs," *Semiconductor International,* Cahners Publishing, March 1994, p. 66.
20. M. Lancaster, "Ultrapure Water: The Real Cost," *Solid State Technology,* September 1996, p. 70.
21. R. Governal, "Ultrapure Water: A Battle Every Step of the Way," *Semiconductor International,* Cahners Publishing, July 1994, p. 177.
22. L. Peters, "Point-of-Use Generation: The Ultimate Solution for Chemical Purity," *Semiconductor International,* Cahners Publishing, January 1994, p. 62.
23. P. Carr, et al., "RTP Characterization Using In-situ Gas Analysis," *Semiconductor International,* Cahners Publishing, November 1993, p. 75.
24. H. Kobayashi, "How Gas Panels Affect Contamination," *Semiconductor International,* Cahners Publishing, September 1994, p. 86.
25. M. Hill, et al., "Quartzglass Components and Heavy-Metal Contamination," *Solid State Technology,* PennWell Publishing, March 1994, p. 49.
26. A. Busnaina, "Solving Process Tool Contamination Problems," *Semiconductor International,* Cahners Publishing, September 1993, p. 73.
27. Laureen Bellville, "Presaturated wipers optimize solvent use," *Cleanrooms,* April 2000, p. 30.
28. Dryden Engineering Service Description, Santa Clara, CA, 1995.
29. R. Allen, S. O'Brian, L. Loewenstein, M. Bennett, and B. Bohannon, "MMST Wafer Cleaning," *Solid State Technology,* PennWell Publishing, January 1996, p. 61.
30. *The National Technology Roadmap for Semiconductors,* The Semiconductor Industry Association, 1994, p. 116.
31. *Ibid.*
32. *Ibid,* p. 113.
33. W. Kern, "Silicon Wafer Cleaning: A Basic Review," 6th International SCP Surface Preparation Symposium, 1999.
34. J. Steigerwald, S. Murarka and R. Gutmann, *Chemical Mechanical Planarization of Microelectronic Materials,* John Wiley & Sons, 1997, p. 298.

35. D. Hymes and I. Malik, "Using Double-Sided Scrubbing Systems for Multiple General Fab Applications," *Micro,* October 1996, p. 55.
36. P. Burggraaf, "Keeping the 'RCA' in Wet Chemistry Cleaning," *Semiconductor International,* June 1994, p. 86.
37. W. Kern, "Silicon Wafer Cleaning: A Basic Review," 6th International SCP Surface Preparation Symposium, 1999.
38. *Ibid.*
39. F. Lin, et al., "Effects of Dilute Chemistries on Particle and Metal Removal Efficiency and on Gate Oxide Integrity," 5th International Symposium, SCP Global, 1998.
40. *Ibid.*
41. M. Wikol, et al., "Application of PTFE Membrane Contactors to the Bubble-Free Infusion of Ozone into Ultra-high Purity Water," 5th International Symposium, SCP Global, 1998.
42. R. Allen, S. O'Brian, L. Loewenstein, M. Bennett and B. Bohannon, "MMST Wafer Cleaning," *Solid State Technology,* PennWell Publishing, January 1996, p. 62.
43. J. Butterbaugh, "Enhancing Yield through Argon/Nitrogen Cryokinetic Aerosol Cleaning after Via Processing," *Micro,* June 1999, p. 33.
44. Wolf and Tauber, *Silicon Processing for the VLSI Era,* p. 519.
45. A. Busnaina and F. Dai, "Megasonic Cleaning," *Semiconductor International,* August, 1997, p. 85.
46. J. Wang, J. Hu, S. Puri, "Critical Drying Technology for Deep Submicron Processes," *Solid State Technology,* July 1998, p. 271.

# 6

# Productivity and Process Yields

## Overview

Wafer fabrication and packaging are incredibly long and complex processes involving hundreds of demanding steps. These steps are never performed perfectly every time, and contamination and material variations combine to wafer loss in the process. Additionally, some of the individual chips on the wafers fail to meet customer electrical and performance specifications. In this chapter, the major yield measurement points are identified along with the major process and material factors that effect yield. Typical yields for the different yield points and for different circuits are presented.

## Objectives

Upon the completion of this chapter, you should be able to:

1. Name the three major yield measurement points in the process.
2. Explain the effect of wafer diameter, die size, die density, number of edge die, and defect density on the wafer-sort yield.
3. From a list of individual process step yields, calculate the accumulative fabrication yield.
4. Be able to explain and calculate an overall process yield.
5. Explain the four major influences on fabrication yield.
6. Sketch a yield-versus-time curve for different process and circuit maturities.
7. Explain the relationship between high process yields and device reliability.

## Yield Measurement Points

Maintaining and improving process and product *yields* is the lifeblood of semiconductor manufacturing. To the casual observer, it would seem that the industry is fixated on production yields. It is. The demanding nature of the process and sheer number of processes required to produce a packaged chip result in product loss. These two factors result in a production process that typically ships only 20 to 80 percent of the chips it commits into the wafer-fabrication line.

These yields seem extraordinarily low compared to most manufacturing operations. Yet, when one considers the challenge of producing hundreds of circuits, composed of millions of micron-size patterns in layers that are equally thin, at very stringent cleanliness levels, all within the confines of a $0.5\text{-in}^2$ chip, it is a testament to the industry that any chips are produced at all.

Another factor that helps keep yields depressed is the nonrepairable nature of most production mistakes. While defective automobile parts can be replaced, few such options are available in semiconductor manufacturing. Defective chips or wafers generally cannot be recovered. In some cases, chips that fail performance tests can be downgraded for sale for a less demanding use. Scrapped wafers may find a new life as control wafers or "dummies" (see discussions of oxidation in Chapters 5 and 7).

Added to these process factors is the volume nature of the business. High capital costs and a higher-than-average percentage of engineering personnel translate to a high-overhead situation. This high overhead, coupled with competition that keeps downward pressure on selling prices, requires that most chip producers run a high-volume, high-yield process.

Given all of these factors, the preoccupation with yield is understandable. Most suppliers of equipment and materials tout the yield improvements possible with their products. Likewise, process engineering groups have as their prime responsibility the maintenance and improvement of process yields. Yield measurement starts at the individual process level and is tracked through the entire process sequence, from incoming blank wafer to shipment of the completed circuit.

Typically, a plant will monitor yields at three major points in the process. They are at the completion of the wafer fabrication processes, after wafer sort, and at the completion of the packaging and final test processes (Fig. 6.1).

## Accumulative Wafer-Fabrication Yield

The first major yield measurement point is at the completion of wafer fabrication. This yield is called by a variety of names, including fab yield, line yield, accumulative fab yield, or "cum" yield.

```
┌─────────────────────────────────────────────────────────┐
│         Major  Yield  Measurement  Points                 │
├─────────────────────────────────────────────────────────┤
│   Manufacturing Stage              Measured               │
│                                                           │
│  Wafer Fabrication- Yield =       # Wafers Out            │
│                                   # Wafers Started        │
│                                                           │
│   Wafer Sort-  Yield =        # Functioning (Good) Die    │
│                                   # Die On Wafer(s)       │
│                                                           │
│  Packaging-  Yield =    # Packages Die Passing Final Test │
│                         # Good Die Started Into Packaging │
└─────────────────────────────────────────────────────────┘
```

Figure 6.1   Major yield measurement points.

Whatever the name, it is expressed as a percentage of the wafers leaving the wafer fab divided by the number that entered the process. Since different product types have different components, feature sizes, and density factors, the wafer fab yield is calculated by product type rather than the entire fabrication line yield.

The cum fab yield starts by first calculating the number of wafers that leave each of the individual processes (called *station yields*) and dividing by the number that entered the station.

$$\text{Station yield} = \frac{\text{Number of wafers leaving station}}{\text{Number of wafers entering station}}$$

The station yields are in turn multiplied together to calculate the overall cum fab yield.

$$\text{Cum fab yield} = \text{Y(station 1)} \times \text{Y(station 2)} \times \ldots \times \text{Y(station } n)$$

Figure 6.2 lists an 11-step process such as the one illustrated in Chapter 5. Typical station yields are listed in column 3 and the accumulated yield in column 5. For a single product, the cum fab yield calculated from the station yields is the same as the yield calculated by dividing the number of wafers *out* by the number of wafers *in*. The accumulated yield equals the simple cum fab yield calculation for this individual circuit. Note that, even with very high individual station yields, the cum fab yield will continue to fall as the wafers come through the process. A modern integrated circuit will require 300 to 500 individual process steps, which represents a huge challenge to maintain profitable productivity. Successful wafer fabrication operations must achieve accumulative fabrication yields over 90 percent to stay profitable and competitive.

| Step | Wafers in | Yield* | Wafers out | Accumulated yield |
|------|-----------|--------|------------|-------------------|
| 1. Field oxidation | 1000 | 99.5 | 995 | 99.5 |
| 2. Source/drain mask | 995 | 99.0 | 965 | 96.5 |
| 3. Source/drain doping | 965 | 99.3 | 978 | 97.8 |
| 4. Gate region mask | 978 | 99.0 | 968 | 96.8 |
| 5. Gate oxidation | 968 | 99.5 | 964 | 96.4 |
| 6. Contact hole mask | 964 | 94.0 | 906 | 90.6 |
| 7. Deposit metal layer | 906 | 99.2 | 899 | 89.9 |
| 8. Metal layer mask | 899 | 97.5 | 876 | 87.6 |
| 9. Alloy metal layer | 876 | 100 | 876 | 87.6 |
| 10. Passivation layer deposition | 876 | 99.5 | 872 | 87.2 |
| 11. Passivation layer mask | 872 | 98.5 | 859 | 85.9 |
| *Yield values are typical for the particular steps. | | | | |

**Figure 6.2**  Accumulated (wafer fab) yield calculation.

Wafer-fabrication cum yields vary from 50 to 95 percent, depending on a number of factors. The calculated cum yield is used for production planning and by engineering and management and as a measure of the process effectiveness.

## Wafer-Fabrication Yield Limiters

Wafer-fabrication yield is limited by a number of factors. The five listed below are fundamental factors that must be controlled in any wafer-fabrication facility. These basic factors, in combination with device or circuit specific factors, result in the overall yield of good chips out of a given facility.

1. Number of process steps
2. Wafer breakage and warping
3. Process variation
4. Process defects
5. Mask defects

### Number of process steps

Note in the calculation in Fig. 6.2 that each individual process operation yield must be in the high 90 percent range to produce the 85.9

percent cum fab yield. Illustrated is a fairly simple 11-step process. ULSI circuits require hundreds of major process operations. Into the year 2012, the total number of major steps will reach 600.[1] Each operation requires several steps, each of which in turn involves a number of substeps. It is easy to appreciate the continual pressures on fabrication areas to maintain high cum yields generated by the number of process steps. The more complicated the circuit, with a high number of steps, the lower the expected cum yield.

More process steps also increase the probability that one of the other four yield limiters will affect the wafer during the process. This factor is a tyranny of numbers. For example, to achieve a 75 percent accumulated fabrication yield with a 50-step process, each of the individual steps would have to be 99.4 percent! A further tyranny of this type of calculation is that the cum yield can never exceed the lowest individual step yield. If one process step can achieve only a 50 percent yield, the overall cum yield can never be higher than 50 percent.

More manufacturing pressure is present in the fact that, for each major process operation, there are a number of steps and substeps. In the illustrated 11-step process, the first operation is oxidation. A simple oxidation process requires several steps: cleaning, oxidation, and evaluation. Each of the steps requires substeps. Figure 6.3 lists the eight substeps for a typical oxidation process. Each of the substeps represents an opportunity to contaminate, break, or damage the wafer.

| Substep | Number of Wafer Handlings |
| --- | --- |
| 1. Wafers are removed from carrier and placed in cleaning boat. | 2 |
| 2. Wafers are cleaned, rinsed, and dried. | 1 |
| 3. Wafers are removed from cleaning boat, inspected, and placed on oxidation boat. | 2 |
| 4. Boat is removed from furnace. | 0 |
| 5. Wafers are removed from boat and placed back in carrier. | 1 |
| 6. Test wafers are removed from carrier and measured. | 2 |
| TOTAL NUMBER OF HANDLINGS | 8 |

Figure 6.3  Substeps of oxidation process.

## Wafer breakage and warping

During the course of the fabrication process, the wafers are handled many times by a combination of manual and automatic techniques. Each handling presents an opportunity to break the relatively fragile wafers. A typical 300-mm (12-in) diameter wafer is only about 800 μm thick. Careful wafer handling is a required, and automatic handlers must be maintained to minimize breakage.

Heat treatments add to the susceptibility of the wafers to breaking. Strains are induced in the crystalline material that make them vulnerable to breaking in subsequent steps. Automatic processing machines accommodate only full-diameter wafers. Therefore, any breakage, however small, is a cause for rejecting the wafer from the process.

Silicon wafers are relatively easy to handle with good practices, and automatic equipment has reduced wafer breakage to a low level. Gallium arsenide wafers, however, are not that resilient, and breakage is a major wafer-yield limiter. In gallium arsenide fabrication lines, where the circuits command a high selling price, partial wafers are often processed.

Along with minimizing breakage, the wafer surfaces must remain flat throughout the processing. This is especially true on fabrication lines that use patterning techniques that project the pattern onto the wafer surface. If the surface is warped or wavy, the projected image will become distorted and change the required image dimensions. This is similar to projecting a slide onto a distorted screen. Warping comes about from rapid heating and/or cooling of the wafers in tube furnaces. (The solution for this problem is addressed in Chapter 7.)

## Process variation

As the wafer comes through the fabrication process, it receives a number of doping, layering, and patterning processes, each of which must meet incredibly stringent physical and cleanliness requirements. But even the most sophisticated processes vary from wafer to wafer, lot to lot, and day to day. When a process exceeds its process limits (goes out of spec), it will cause some unallowed result on the wafer or within the chips on the wafer.

A goal of process engineering and process control programs is not only to keep each process operating within its control specifications but to maintain a constant distribution of the process parameters, such as time, temperature, pressure, and others. These process parameters are monitored with statistical *process control techniques* explained in Chapter 15.

Throughout the process, there are a number of inspections and tests designed to detect unwanted variations as well as frequent calibration of the equipment parameters to process specifications. Some of these tests are performed by production personnel and some by quality control organizations. However, even the best maintained and monitored process exhibits some variations. One of the challenges of process engineering and circuit design is to accommodate the variations and still have a functioning device.

**Process defects.**    Process defects are defined as isolated regions (or spots) of contamination or irregularities on the wafer surface. These defects are often called *spot defects* or *point defects*. They occur randomly on the wafer surface. Some are non-fatal, and some will render the circuit inoperable. These are called *killer defects* (see Fig. 6.11, p. 151). Unfortunately, smaller defects are sometimes not detectable during the fabrication process. They become evident at wafer sort as rejected chips.

The major sources of these defects are the various liquids, gases, room air, personnel, process machines, and water used in the fabrication area. Particulates and other small contaminants become lodged in or on the wafer surface. Many of these defects occur in the patterning process. Recall that the patterning process requires using a thin, fragile layer of photoresist to protect the wafer surface during the etch steps. Any holes or tears in the photoresist layer from particulates will end up as tiny etched holes in the wafer surface layer. These holes are called *pinholes* and are a major concern of photomasking engineers. Consequently, the wafers are inspected often for contamination, usually after each major step for contamination. Wafers that exceed the established allowable density are rejected. The SIA *International Technology Roadmap for Semiconductors* (ITRS) calls for maximum defect densities on 300-mm wafer surfaces of 0.68 per square centimeter ($cm^2$).

### Mask defects

A photomask or reticle is the source of the pattern that is transferred to the wafer surface in the patterning process. Defects on the mask/reticle end up on the wafer as defects or pattern distortions. There are three common mask/reticle originated defects. First is contamination, such as dirt or stains on the clear part of the mask/reticle. In optical lithography, they can block the light and print onto the wafer as though they were an opaque part of the pattern. Second are cracks in the quartz. They, too, can block the patterning light and/or scatter the light, causing unwanted images and/or distorted images. Third are pattern distortions that occur in the mask/reticle making process. These include pinholes or chrome spots, pattern extensions or missing parts, breaks in the pattern, or *bridges* between adjacent patterns (Fig. 6.4).

Control of mask generated defects is more critical for device/circuits with smaller feature sizes, higher densities, and larger die sizes.

### Wafer-Sort Yield Factors

After fabrication, the wafers go to the wafer sort tester. During the test, each chip will be tested electrically for device specifications and

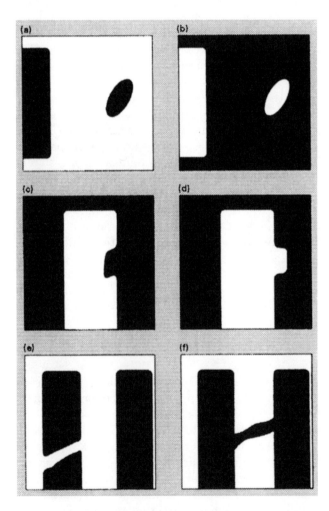

**Figure 6.4**  Mask defects. (*a*) Spot, (*b*) hold, (*c*) inclusion, (*d*) protrusion, (*e*) break, (*f*) bridge. *(Source: Solid State Technology, July 1993, p. 95.)*

functionality. Up to several hundred individual electrical tests may be performed on each circuit. While these tests measure the electrical performance of the device(s), they indirectly measure the precision and cleanliness of the fabrication processes. Because of natural process variations and undetected defects, the wafer may have passed all the in-process checks and still have many chips that do not function.

Since wafer sort is a comprehensive test, many factors influence the yield. They are:

1. Wafer diameter
2. Die size (area)
3. Number of processing steps
4. Circuit density
5. Defect density
6. Crystal defect density
7. Process cycle time

### Wafer diameter and edge die

The semiconductor industry went to round wafers with the introduction of silicon. The first wafers were less than one inch in diameter. Since that time, there has been a regular progression to larger-diameter wafers, with 150 mm (6 in) being the standard of VLSI fabrication lines in the late 1980s, and 200-mm wafers being developed for production use in the 1990s. Projections call for 300- and 450-mm diameter wafers early in this century.[3]

The move to larger-diameter wafers has been driven by production efficiency, increasing die sizes, and the influence on the wafer-sort yield. Production efficiency is easily understood when one considers that there is an incremental increase in the cost of processing a larger-diameter wafer, while the number of available whole chips on the wafer can increase substantially as illustrated in Fig. 6.5.

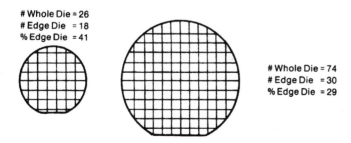

# Whole Die = 26
# Edge Die  = 18
% Edge Die  = 41

# Whole Die = 74
# Edge Die  = 30
% Edge Die  = 29

**Figure 6.5**   Effect of larger wafer diameter on percentage of partial die.

The effect of increasing the wafer diameter also has positive effects on the wafer-sort yield. Figure 6.6 shows two wafers of the same diameter but with different die sizes. Note that the smaller-diameter wafer has a very large proportion of its surface covered with partial die—die that cannot function. The larger-diameter wafer, with its greater number and percentage of whole die, if all other factors are equal, will have a higher wafer-sort yield.

**Wafer diameter and die size**

Another driving force for larger-diameter wafers is the trend to larger die sizes. As shown in Fig. 6.6 increasing the die size *without* increasing the wafer diameter also results in a wafer surface with a smaller percentage of whole die. Maintaining a decent wafer-sort yield as the die size increases requires increasing the wafer diameter. Figure 6.7 lists the number of various size chips that will fit on different size wafers. The bottom line is that larger-diameter wafers are more cost effective.

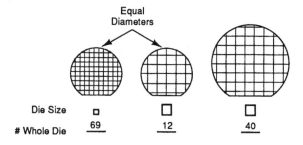

**Figure 6.6**   Effect of larger wafer diameter on percentage of partial die.

|  | Wafer diameter (mm) | | |
|---|---|---|---|
| Die size (mils) | 150 | 200 | 300 |
| 300 × 300 | 293 | 531 | 820 |
| 400 × 400 | 113 | 165 | 550 |
| 500 × 500 | 108 | 191 | 410 |

**Figure 6.7**   Die size versus number of die on wafer.

## Wafer diameter and crystal defects

In Chapter 3, the concept of a crystal dislocation was introduced. A crystal dislocation is a point defect *in the wafer* that comes from a local discontinuity of the crystal structure. Dislocations exist throughout the crystal structure and, like contamination and process defect density, affect the wafer-sort yield.

Dislocations also are generated during the fabrication process. They generate (or nucleate) at sites where there are chips and abrasions of the edge of the wafer. These chips and abrasions come from poor handling techniques and automatic handling equipment. The abraded area causes a crystal dislocation. Unfortunately, the dislocation is propagated into the center of the wafer (Fig. 6.8) during subsequent heat treatments, such as oxidations and diffusions. The length of the dislocation line into the interior of the wafer is a function of the thermal history of the wafer. Consequently, wafers receiving more process steps and/or more heating steps will have more and longer dislocation lines affecting a greater number of chips. One obvious solution to the problem is larger-diameter wafers, which leaves a larger number of unaffected die in the center of the wafer.

## Wafer diameter and process variations

The process variations discussed in the section of wafer fabrication yields affect the wafer-sort yield. In the fabrication area, process variations are detected by sampling inspection and measurement techniques. The nature of inspection sampling is that not all of the variations and defects are detected, so that wafers are passed on with some number of problems. These problems show up at wafer sort as failed devices.

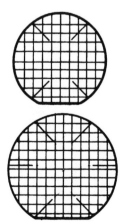

**Figure 6.8** Effect of dislocations on wafer-sort yield for different wafer dimensions.

Process variations occur at a higher rate around the edge of the wafer. In the high-temperature processes performed in tube furnaces, there is always some temperature nonuniformity across the wafers. The change in temperature results in uniformity differences on the wafer. Variations occur more at the wafer outer edges where heating and cooling occur at a faster rate. Another contributor to this wafer edge phenomenon is contamination and physical abuse of the wafer layers that emanates from handling and touching the wafers on their edges. In the patterning process, there can be feature size uniformity problems in the mask-driven processes (full mask projection, proximity and contact exposure). The nature of the light systems is such that the center will be of higher uniformity than the outside edges. In the reticle-driven masking processes (steppers), there is a smaller area of exposure (one or several die), which reduces the image variations across the wafer.

All of these problems result in a lower wafer-sort yield around the edge of the wafer, as illustrated in Fig. 6.9. Larger-diameter wafers help to maintain wafer-sort yields by having a larger area of unaffected die in the center of the wafer.

### Die area and defect density

The die size also affects wafer-sort yield relative to the defect density on the wafer surface. The relationship is illustrated in Fig. 6.10. In Fig. 6.10$a$, a wafer is shown with five defects and no die pattern. This situation illustrates a background defect density produced by all of the fab area factors, regardless of the die size, device type, process control requirements, and so forth. The wafers in Fig. 6.10$b$ and $c$ illustrate the effect of this background defect density on the wafer-sort yield for two different die sizes. The larger the die size for a given defect density, the lower the yield.

### Circuit density and defect density

The defects on the wafer surface result in die failures by causing a malfunction of some part of the die. Some of the defects are located in

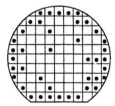

**Figure 6.9** Typical location of functioning die after wafer sort.

• **Nonfunctioning Die**

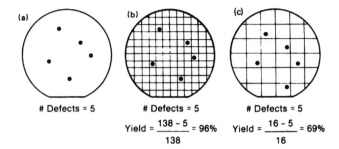

$$\text{Yield} = \frac{138 - 5}{138} = 96\% \qquad \text{Yield} = \frac{16 - 5}{16} = 69\%$$

**Figure 6.10**  Effect of background defects on wafer-sort yield for different die sizes.

nonsensitive parts of the die and do not cause a failure. However, the trend is to higher levels of circuit integration, which came about because of smaller feature size and a higher density of die components. The result of these trends is a higher probability that any given defect will be in an active part of the circuit, thus lowering the wafer-sort yield as illustrated in Fig. 6.11.

### Number of process steps

The number of process steps was indicated as a limiter of the fab cum yield. The more steps, the greater the opportunities to break or misprocess a wafer. The effect also influences the wafer-sort yield. As the number of process steps increases, the background defect density increases unless procedures are implemented to lower it. A higher background defect density affects more chips, lowering the wafer-sort yield.

### Feature size and defect size

Smaller feature sizes make maintaining an acceptable sort yield difficult from two major factors. First, the smaller images are more difficult to print (see "Mask defects" section, p. 145, and Chapter 8). Second, the

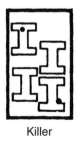

**Figure 6.11**  Killer defects (failed die) and nonfatal (passed die) defects.

Killer          Nonfatal

smaller images are vulnerable to ever-smaller defect sizes as well as the over all defect density. The 10:1 rule of minimum feature size to allowable defect size has been discussed. One assessment is that, at a defect density of 1 defect per $cm^2$, a circuit with 0.35-μm feature size will have a wafer-sort yield 10 percent less than that of a 0.5-μm circuit processed under the same conditions.[4]

### Process cycle time

The time that the wafers are actually being processed can be measured in days. But due to queuing at the process stations and temporary slowdowns due to process problems, the wafer often stays in the fab area for several weeks. The longer the wafer is sitting around, the more opportunity for contamination that lowers wafer-sort yield. The move to just-in-time manufacturing (see Chapter 15) is one attempt to increase yields and decrease the manufacturing costs associated with increased in-line inventories.

### Wafer-sort yield formulas

The ability to understand and predict wafer-sort yields with some accuracy is essential to the operation of a profitable and reliable chip supplier. Over the years, a number of models have been developed that relate process, defect densities, and chip size parameters to the wafer-sort yield. Five yield model formulas are shown in Fig. 6.12. Each relates different parameters to the wafer-sort yield. As the chips get larger, the number of process steps increases, the feature size decreases the sensitivity to smaller defect sizes increases, and more of the background defects become killer defects.

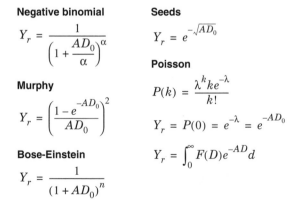

**Negative binomial**

$$Y_r = \frac{1}{\left(1 + \frac{AD_0}{\alpha}\right)^{\alpha}}$$

**Murphy**

$$Y_r = \left(\frac{1 - e^{-AD_0}}{AD_0}\right)^2$$

**Bose-Einstein**

$$Y_r = \frac{1}{(1 + AD_0)^n}$$

**Seeds**

$$Y_r = e^{-\sqrt{AD_0}}$$

**Poisson**

$$P(k) = \frac{\lambda^k k e^{-\lambda}}{k!}$$

$$Y_r = P(0) = e^{-\lambda} = e^{-AD_0}$$

$$Y_r = \int_0^{\infty} F(D)e^{-AD}d$$

**Figure 6.12**  Wafer-sort yield models.

**Exponential model.**    The exponential relationship (Fig. 6.12a) or Poisson model is the simplest and one of the first yield models[5] developed. It is applicable to individual process steps and assumes a random distribution of defects ($D_0$) across the wafer. For multistep analysis, the factor (n) equating to the number of process steps is used (Fig 6.12b). This model generally is used for products that contain over 300 die and MSI circuits of lower densities. Die sizes that are smaller are predicted by the Seeds model.

The exponential, Poisson, and Seeds models all illustrate the primary relationship between die area, defect density, and wafer-sort yield. In these, e is a constant with a value of 2.718.

B. T. Murphy proposed a model using a more sophisticated distribution of defects. The Bose-Einstein model adds the number of process steps (n), In the *negative binomial* model, there is a cluster factor. It accounts for defect distributions that tend to be "clustered" on the wafer surface rather than simply exhibiting a random distribution. Adopted by the SIA in the *ITRS*, the cluster factor is assigned a value of 2.[5]

In most yield models, the factor for processing steps (n) is actually the number of patterning steps. Experience has proved that the patterning steps contribute the greatest number of point defects and therefore have a direct bearing on sort yield.

Figure 6.13 illustrates the different predictions of the various yield models.[6] No two complex circuits have comparable designs or processes. Processes vary from company to company, as does the basic background defect density. These factors make the development of an accurate universal yield model difficult. Most chip companies have developed their own models that reflect their particular manufacturing process and product designs. The models are all defect driven. That is, they assume that all of the fab processes are under control and that the defect levels are those built into the process. They do not include major process problems, such as a contaminated tank of process gas.

The defect density used in all the models is not the same as a defect density determined by optical inspection of the wafer surface. The defect density that shows up in the yield models is all-inclusive; it includes contaminants and surface and crystal defects. Further, it predicts only the defects that destroy die: the "killer defects." Defects that fall in noncritical areas of the chip are not part of the models, nor are situations where two or more defects fall in the same sensitive area.

It is also important to keep in mind that the yield numbers predicted by the formulas are those expected from a process that is basically under control. In reality, the wafer-sort yield will vary from

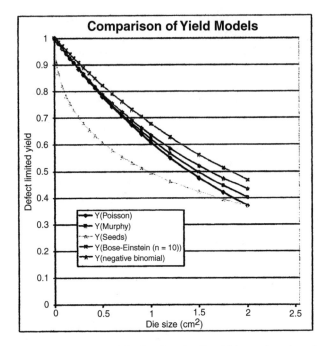

**Figure 6.13**   Yield models showing the die yield as a function of die size defect density.

wafer to wafer because of the normal process variations in the fabrication process. A typical wafer-sort yield plot is shown in Fig. 6.14.

Note that wafer 13 falls far below the normal range of sort yields. In a situation like this, the process engineer would look for some catastrophic process failure such as an out-of-spec layer thickness or a doping layer that is too deep or too shallow.

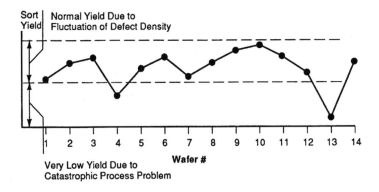

**Figure 6.14**   Plot of wafer-sort yields.

## Assembly and Final Test Yields

After wafer sort, the wafers go to the packaging process, also called *assembly and test*. There, the wafers are separated into the individual die and packaged into a protective enclosure. During this series of steps, there are a number of visual inspections and quality checks of the assembly process.

The last steps of the packaging process is a series of physical, environmental, and electrical tests, known collectively as the *final tests*. (The details of the processes, inspections, and final tests are described in Chapter 18.) After the final tests, the third major yield is calculated, which is the ratio of die passing the final tests compared with the number of *good die* that entered packaging after passing the wafer-sort test.

## Overall Process Yields

The overall process yield is the mathematical product of the three major yield points (Fig. 6.15). This number, expressed in percent, gives the percentage of shipped die as compared with the number of whole die on the starting wafer. It is an inclusive measurement of the success of the entire process.

Overall yields vary with several major factors. In Fig. 6.16 is a list of typical process yields and their calculated overall yield. In the first two columns are major process factors that influence the individual and overall yields.

First is the integration level of the particular circuit. The more highly integrated the circuit, the lower the expected yield in all categories. Higher integration levels assume a corresponding decrease in feature size. Column 2 lists the maturity of the manufacturing process. Process yields almost always follow an *S* curve pattern (Fig. 6.17) through the lifetime of the product in manufacturing. In the beginning, the yield rises rather slowly as the initial bugs are worked out of the process. This is followed by a period when the yields rise rapidly, eventually leveling off as the limits imposed by the process maturity

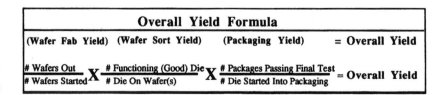

Figure 6.15   Overall yield formula.

| Integration Level | Product Maturity | Fab Yield % | Sort Yield % | Assembly Final Test % | Overall Yield % |
|---|---|---|---|---|---|
| ULSI | Mature | 95 | 85 | 97 | 78 |
| ULSI | Mid | 88 | 65 | 92 | 53 |
| ULSI | Introduction | 65 | 35 | 70 | 16 |
| LSI | Mature | 98 | 95 | 98 | 91 |
| Discrete | Mature | 99 | 97 | 98 | 94 |

**Figure 6.16**  Typical yields for various products.

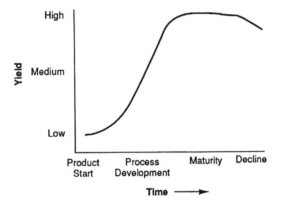

**Figure 6.17**  Yield changes with process maturity levels.

die size, integration level, circuit density, and defect density. As the table in Fig. 6.16 shows, overall yields can vary from very low (maybe even zero for new or poorly designed products) to the 90 percent range for simpler and mature products. Semiconductor producers consider their yield performance very proprietary, since profit and production control are a direct function of the process yields.

An examination of the yield values in the table reveals that wafer-sort is the lowest of the three yield points. This fact illustrates why yield-improvement programs are directed at the many factors that influence the wafer-sort yield. At one time, the improvement of wafer-sort yields had the biggest impact factor on productivity. The advent of larger and more complex chips (such as the megabit memories) has shifted productivity improvements to include other factors, including the cost of ownership of equipment (see Chapter 15). Successful competition in the mega-chip era will require wafer-sort yields 90 percent or higher.[7]

## Review Questions

1. Name the three yield measurement points in the manufacturing process.

2. Of the three yield points, which is typically the lowest?

3. Indicate if the wafer-sort yield would go up or down with the following changes:

   a. Change to larger-diameter wafers

   b. Change to smaller die size

   c. More process steps

   d. Decrease in defect density

   e. Increase in die density

4. If 1000 wafers are started into the process and 875 are passed to wafer sort, what is the accumulated fabrication yield?

5. Calculate the overall yield for a process that has an 82 percent fabrication yield, a 47 percent wafer-sort yield, and a 92 percent assembly and final test yield.

6. Which situation would you expect to have the highest overall yield, a 22-step LSI process or a 33-step VLSI process?

## References

1. J. Baliga, "Yield Management," *Semiconductor International,* January 1998, p. 74.
2. L. Peters, "Speeding the Transition to 0.18 μm," *Semiconductor International,* January 1998, p. 66.
3. APT Presentation "Overall Road Map Technology Characteristics," Industry Strategy Symposium sponsored by The Semiconductor Equipment and Materials Institute, January, 1995.
4. Bill Walker, "Motorola VP Defines Sub-Micron Manufacturing Challenges," *Semiconductor International,* Cahners Publishing, October 1994, p. 21.
5. S. M. Sze, *VLSI Technology,* McGraw-Hill Publishing Company, New York, 1983, p. 605.
6. D. Horton, "Modeling the yield of mixed-technology die," *Solid State Technology,* September 1998, p. 109.
7. Bill George, COO of Sematech, and Susan Billatin, quoted in "Process Control: Covering all of the Bases," *Semiconductor International,* Cahners Publishing, September 1993, p. 80.

# 7

# Oxidation

## Overview

The ability of a silicon surface to form a silicon dioxide passivation layer is one of the key factors in silicon technology. In this chapter, the uses, formation, and processes of silicon dioxide growth are explained. Detailed is the all-important tube furnace, which is a mainstay of oxidation, diffusion, heat treatment, and chemical vapor deposition processes. Other oxidation methods, including rapid thermal processing, are also explained.

## Objectives

Upon completion of this chapter, you should be able to:

1. List the three principal uses of a silicon dioxide layer in silicon devices.
2. Describe the mechanism of thermal oxidation.
3. Sketch and identify the principal sections of a tube furnace.
4. List the two oxidants used in thermal oxidation.
5. Sketch a diagram of a dryox oxidation system.
6. Draw a flow diagram of a typical oxidation process.
7. Explain the relationship of process time, pressure, and temperature on the thickness of a thermally grown silicon dioxide layer.
8. Describe the principles and uses of rapid thermal, high-pressure, and anodic oxidation.

## Silicon Dioxide Layer Uses

Of all the advantages of silicon for the formation of semiconductor devices, the ease of growing a silicon dioxide layer is perhaps the most useful. Whenever a silicon surface is exposed to oxygen, it is converted to silicon dioxide (Fig. 7.1). Silicon dioxide is composed of one silicon atom and two oxygen atoms ($SiO_2$). We encounter silicon dioxide daily. It is the chemical composition of ordinary window glass. It's semiconductor version, however, is purer and formed in a specific way. Silicon dioxide layers are formed on bare silicon surfaces at elevated temperatures in the presence of an oxidant. The process is called *thermal oxidation*.

Although silicon is a semiconducting material, silicon dioxide is a dielectric material. This combination, a dielectric layer formed on a semiconductor, along with other properties of silicon dioxide, makes it one of the most commonly used layers in silicon devices. Silicon dioxide layers find use in devices to pacify the silicon surface, to act as doping barriers and surface dielectrics, and to serve as dielectric parts of device structures.

### Surface passivation

In Chapter 4, the extreme sensitivity of semiconductor devices to contamination was examined. While a major focus of a semiconductor facility is the control and elimination of contamination, the techniques are not always 100 percent effective. Silicon dioxide layers play an important role in protecting semiconductor devices from contamination.

Silicon dioxide performs this role in two ways. First is the physical protection of the surface and underlying devices. Silicon dioxide layers formed on the surface are very dense (nonporous) and very hard. Thus, a silicon dioxide layer (Fig. 7.1) acts as a contamination barrier by physically preventing dirt in the processing environment from getting to the sensitive wafer surface. The hardness of the layer protects the wafer surface from scratches and abuse endured by the wafer in the fabrication processes.

The second way silicon dioxide protects devices is chemical in nature. Regardless of the cleanliness of the processing environment, some electrically active contaminants (mobile ionic contaminants) end up in or on the wafer surface. During the oxidation process, the top layer of silicon is converted to silicon dioxide. Contaminants on the

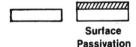

**Surface
Passivation**

**Figure 7.1** Surface passivation with silicon dioxide layers.

surface end up in the new layer of oxide, away from the electrically active surface. Other contaminants are drawn up into the silicon dioxide film where they are less harmful to the devices. In the early days of MOS device processing, it was common to oxidize the wafers and then remove the oxide before further processing to rid the surface of unwanted mobile ionic contamination.

### Doping barrier

In Chapter 5, doping was identified as one of the four basic fabrication operations. Doping requires creating holes in a surface layer through which specific dopants are introduced into the exposed wafer surface through diffusion or ion implantation. In silicon technology, the surface layer is most often silicon dioxide (Fig. 7.2). The silicon dioxide left on the wafer acts to block the dopant from reaching the silicon surface. All of the dopants used in silicon technology have a very slow rate of movement in silicon dioxide as compared to silicon. While the dopants penetrate to the required depth in the exposed silicon, they penetrate only a short distance into the silicon dioxide surface. It takes only a relatively thin silicon dioxide layer to block the dopants from reaching the silicon surface.

Another factor favoring the use of silicon dioxide is a coefficient of thermal expansion similar to that of silicon. In the high-temperature processes of oxidation, diffusion doping, and others, the wafer expands and contracts as it is heated and cooled. The silicon dioxide expands and contracts at close to the same rate as silicon, which means that the wafer will not warp during the heating and cooling.

### Surface dielectric

Silicon dioxide is classified as a dielectric. This means that, under normal circumstances, it does not conduct electricity. When dielectrics are used in electrical circuits or devices, they are referred to as *insulators*. Acting as an insulator is an important role of silicon dioxide layers. Figure 7.3 shows a cross section of a wafer with a conductive layer of metal on top of a layer of silicon dioxide. The oxide prevents shorting of the metal layer to the underlying metal just as the insulation on an

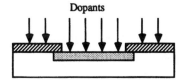

Dopants

**Figure 7.2** Silicon dioxide layer as dopant barrier.

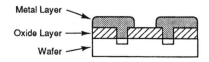

Metal Layer

Oxide Layer

Wafer

**Figure 7.3** Oxide layer used as dielectric layer between wafer and metal.

electric cord prevents the wires from shorting. In this capacity, the oxide must be continuous; that is, have no holes or voids.

The oxide must also be thick enough to prevent a phenomenon known as *induction*. Induction can occur when the separating layer of oxide is thin enough to allow an electrical charge in a metal layer to cause a buildup of charge in the wafer surface. The surface charge can cause shorting and other unwanted electrical effects. A thick enough layer will prevent an induced charge in the wafer surface. Most of the wafer surface is covered with an oxide layer thick enough to prevent induction from the metal layers. This is called the *field oxide*.

**Device dielectric**

At the other end of the induction phenomenon is MOS technology. In an MOS transistor, a thin layer of silicon dioxide is grown in the gate region (Fig. 7.4). The oxide functions as a dielectric whose thickness is chosen *specifically* to allow induction of a charge in the gate region under the oxide. The gate is the part that controls the flow of current through the device (see Chapter 16). The dominance of MOS technology for ultra-large-scale integrated (ULSI) circuits has made the formation of gate regions a prime focus of process development and concern. Thermally grown oxides are also used as the dielectric layer in *capacitors* formed between the silicon wafer and a surface conduction layer (Fig. 7.5).

Silicon dioxide dielectric layers are also used in structures with two or more metallization layers. In this application, the silicon dioxide layers are deposited with *chemical vapor deposition* (CVD) techniques rather than thermal oxidation (see Chapter 12).

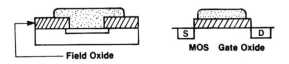

Field Oxide

MOS    Gate Oxide

S    D

**Figure 7.4** Silicon dioxide as field oxide and in MOS gate.

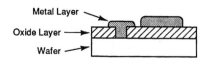

Metal Layer

Oxide Layer

Wafer

**Figure 7.5** Silicon dioxide layer in solid-state capacitor.

## Device oxide thicknesses

The silicon dioxide layers used in silicon-based devices vary in thickness. At the thin end of the scale are advanced MOS gate oxides, in the 35 to 80 A range.[1] At the thick end are field oxides. Figure 7.6 lists the thickness ranges for the major uses.

## Thermal Oxidation Mechanisms

Thermal oxide growth is a simple chemical reaction as shown in Fig. 7.7. This reaction takes place even at room temperature. However, an elevated temperature is required to achieve quality oxides in reasonable process times for practical use in circuits and devices. Oxidation temperatures are between 900 and 1200°C.

Although the formula shows the reaction of silicon with oxygen, it does not illustrate the growth mechanism of the oxide. To understand the growth mechanism, consider a wafer placed in a heated chamber and exposed to oxygen gas (Fig. 7.8a). Initially, the oxygen atoms combine readily with the silicon atoms. This stage is called *linear* because the oxide grows in equal amounts for each unit of time (Fig. 7.8b). After approximately 1000 angstroms (Å) of oxide is grown, a limit is imposed on the linear growth rate. [An angstrom is one ten-thousandth of a micron (μm); in other words there are 10,000 Å in 1 μm.]

For the oxide layer to keep growing, the oxygen and silicon atoms must come in contact. However the initially grown layer of silicon dioxide separates the oxygen in the chamber from the silicon atoms of the wafer surface. For oxide growth to continue, either the silicon in the wafer must migrate through the all-ready grown oxide layer to the oxygen in the vapor, or the oxygen must migrate to the wafer surface. In the thermal growth of silicon dioxide, the oxygen migrates (the techni-

| Silicon Dioxide Thickness, Å | Application |
|---|---|
| 60–100 | Tunneling Gates |
| 150–500 | Gate Oxides, Capacitor Dielectrics |
| 200–500 | LOCOS Pad Oxide |
| 2,000–5,000 | Masking Oxides, Surface Passivation |
| 3,000–10,000 | Field Oxides |

Figure 7.6   Silicon dioxide thickness chart.

$$Si \text{ (solid)} + O_2 \text{ (gas)} \xrightarrow{Heat} SiO_2 \text{(solid)}$$

Figure 7.7   Reaction of silicon and oxygen to form silicon dioxide.

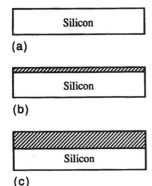

(a)

(b)

Figure 7.8  Silicon dioxide growth
states. (a) Initial, (b) linear, and
(c) parabolic.

(c)

cal term is *diffuses*) through the existing oxide layer to the silicon wafer
surface. Thus, the layer of silicon dioxide consumes silicon atoms from
the wafer surface—the oxide layer grows *into* the silicon surface.

With each succeeding new growth layer, the diffusing oxygen must
move farther to reach the wafer. The effect is a slowing of the oxide
growth *rate* with time. This stage of oxidation is called the *parabolic*
stage. When graphed, the mathematical relationship of the oxide
thickness, growth rate, and time takes the shape of a parabola. Other
terms used for this second stage of growth is a *transport-limited reaction* or *diffusion limited reaction,* which means that the growth rate is
limited by the transportation or diffusion of the oxygen through the
oxide layer already grown. The linear and parabolic stages of growth
are illustrated in Fig. 7.9. The formula in Fig. 7.10 expresses the fundamental parabolic relationship for oxide layers above approximately
1200 Å.

Thus, a growing oxide goes through two stages: the linear stage and
the parabolic stage. The change from linear to parabolic is dependent
on the oxidizing temperature and other factors (see the following section, "Influences on the oxidation rate"). In general oxides less than
1000 Å (0.1 μm) are controlled by the linear mechanism. This is the
range of most MOS gate oxides.[2]

$$X = B/A \; t$$
Linear oxidation of silicon

$$X = \sqrt{B \; t}$$
Parabolic oxidation of silicon

X = oxide thickness
B = parabolic rate constant
B/A = linear rate constant
t = oxidation time

Figure 7.9  Linear and parabolic
growth of silicon dioxide.

$$R = \frac{X^2}{t}$$

where $R$ = silicon dioxide growth rate
    $X$ = oxide thickness
    $t$ = oxidation time

**Figure 7.10** Parabolic relationship of $SiO_2$ growth parameters.

The major implication of this parabolic relationship is that thicker oxides require much more time to grow than thinner oxides. For example, growth of a 2000-Å (0.20-μm) film at 1200°C in dry oxygen requires 6 min (Fig. 7.11).[3] To double the oxide thickness to 4000 Å requires some 220 min—over 36 times as long. This longer oxidation time presents a problem for semiconductor processing. When pure dry oxygen is used as the oxidizing gas, the growth of thick oxide layers requires even longer oxidation times, especially at the lower temperatures. Generally, process engineers want to have the shortest process times possible as is consistent with quality control. The 220 min in the example given is excessive, i.e., only one oxidation would be possible in one shift of operation.

One way to achieve faster oxidations is to use water vapor ($H_2O$) instead of oxygen as the oxidizing gas (oxidant). The growth of silicon dioxide in water vapor proceeds by the reaction shown in Fig. 7.12. In the vapor state, the water is in the form $H-OH^-$. It is composed of one atom of hydrogen (H) and a molecule of oxygen and hydrogen with a negative charge ($OH^-$). This molecule is called the *hydroxyl ion.* The hydroxyl ion diffuses through the oxide layers already on the wafer faster than oxygen. The net effect is a faster oxidation of the silicon, as shown in the growth curves in Fig. 7.11.

Water vapor at the oxidation temperatures is in the form of steam, and the process is called either *steam oxidation, wet oxidation,* or *pyrogenic steam.* The term *wet oxidation* comes from the time when liquid water was the primary water vapor source. An oxygen-only oxidation process is called *dry oxidation.* If oxygen only is used, it must be free of any water vapor (dry) or the oxide growth would be that of water vapor.

Notice in the reaction of water vapor and silicon that there are two hydrogen molecules ($2H_2$) on the right side of the equation. Initially, these hydrogen molecules are trapped in the solid silicon dioxide layer, making the layer less dense than an oxide grown in dry oxygen. However, after a heating of the oxide in an inert atmosphere, such as nitrogen (see "Oxidation Processes," p. 191), the two oxides become similar in structure and properties.

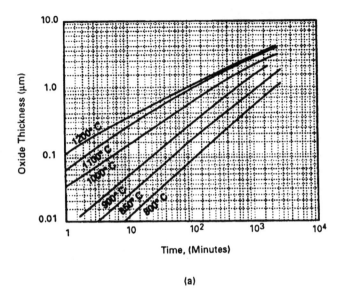

(a)

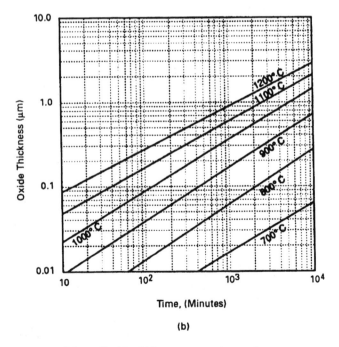

(b)

**Figure 7.11**   Silicon dioxide thickness versus time and temperature in (a) dry oxygen and (b) steam.

$$\text{Si (solid)} + H_2O \text{ (gas)} \xrightarrow[\text{Heat}]{} SiO_2\text{(solid)} + 2H_2 \text{ (gas)} \uparrow$$

**Figure 7.12**   Reaction of silicon and water vapor to form silicon dioxide and hydrogen gas.

## Influences on the oxidation rate

The original oxide thickness vs. time curves were determined on $\langle 111 \rangle$-oriented, undoped wafers.[4] MOS devices are fabricated in $\langle 100 \rangle$-oriented wafers and wafers surfaces are doped. Both of these factors influence the oxidation rate for a particular temperature and oxidant environment. Other factors influencing the oxidation growth are impurities intentionally included in the oxide (such as HCl) and oxidation of polysilicon layers.

**Wafer orientation.**   The orientation of the wafer has an effect on the oxidation growth rate. $\langle 111 \rangle$ planes have more silicon atoms than $\langle 100 \rangle$ planes. The larger number of atoms allows for a faster oxide growth on $\langle 111 \rangle$-oriented wafers than for $\langle 100 \rangle$-oriented wafers. Figure 7.13 shows the growth rates for the two orientations in steam. This difference is seen more in the linear growth stage and at lower temperatures.

**Wafer dopant redistribution.**   The silicon surface being oxidized always has dopants. A production silicon wafer starts into the line doped as

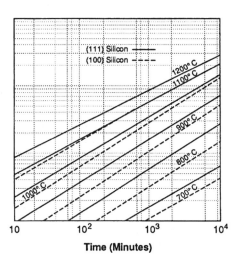

**Figure 7.13** Oxidation of $\langle 111 \rangle$ and $\langle 111 \rangle$ silicon in steam.

Time (Minutes)

either an N-type or P-type. Later on in the process, wafers have dopant(s) in the surface from diffusion or ion implant operations. The dopant elements used and their concentration both have effects on the oxidation growth rate. For example, oxides grown over a highly doped phosphorus layer are less dense than those grown over the other silicon dopants. These phosphorus-doped oxides also etch faster and present an etching challenge in the patterning operation due to resist lifting and rapid undercutting.

Another effect on oxidation growth rate is the distribution of the dopant atoms in the silicon after the oxidation is completed.[5] Recall that during thermal oxidation, the oxide layer grows *down* into the wafer. A question is "What happens to the dopant atoms that were in the layer of silicon converted to silicon dioxide?" The answer depends on the conductivity type of the dopant. The N-type dopants of phosphorus, arsenic, and antimony have a higher solubility in silicon than in silicon dioxide. When the advancing oxide layer reaches them, they move down into the wafer. The silicon silicon-dioxide interface acts like a snowplow pushing ahead an ever-greater pile of snow. The effect is that there is a higher concentration (called *pile-up*) of N-type dopants at the silicon dioxide-silicon interface than was originally in the wafer.

When the dopant is the P-type boron, the opposite effect happens. The boron is drawn up into the silicon dioxide layer, causing the silicon at the interface to be depleted of the original boron atoms. Both of these effects, pile-up and depletion, have significant impact on electrical performance of devices. The exact effects of pile-up and depletion on the dopant concentration profile are illustrated in Chapter 17.

Doping concentration effects on the oxidation rate vary with the dopant type and concentration level. In general, higher doped regions oxidize faster than more lightly doped regions. Heavily doped phosphorus regions can oxidize 2 to 5 times the undoped oxidation rate.6

However doping-induced oxidation effects are more pronounced in the linear stage (thin oxides) of oxidation.

**Oxide impurities.**    Certain impurities, particularly chlorine from hydrochloric acid (HCl), are included in the oxidizing atmosphere for inclusion in the growing oxide (see "Oxidation Processes," p. 191). These impurities have an influence on the growth rate. In the case of HCl, the growth rate can increase from 1 to 5 percent.[7]

**Oxidation of polysilicon.**    Polysilicon conductors and gates are a feature of most MOS devices/circuits. The device/circuit processes require oxidation of the polysilicon. Compared to the oxidation of single-crystal silicon, polysilicon can be faster, lower, or similar. A number of factors related to the formation of the polysilicon structure influence a subsequent oxidation. They are the polysilicon deposition method, deposition temperature, deposition pressure, the type and concentration of doping, and the grain structure of the polysilicon.[8]

**Differential oxidation rates and oxide steps.**    After the initial oxidation step in a device/circuit fabrication process, the wafer surface has a variety of conditions. Some areas have the field oxide, some are doped, and some are polysilicon regions, and so on. Each of these areas has a different oxidation rate and will increase in oxide thickness depending on the condition. This oxidation thickness difference is called *differential oxidation*. For example, an oxidation of a MOS wafer after a polysilicon gate has been formed next to lightly doped source/drain areas (Fig. 7.14a) results in a thicker oxide growth on gate because silicon dioxide grows faster on polysilicon.

Differential oxidation rates cause the formation of steps in the wafer surface (Fig. 7.14b). Illustrated is a step created by the oxidation of an exposed area next to a relatively thick field oxide. The oxide will grow faster in the exposed area, since additional oxide growth in the field oxide is limited by the parabolic rate limitation. In the exposed area, the faster growing oxide will use up more silicon than under the field oxide. The step is shown in Fig. 7.14b.

**Thermal Oxidation Methods**

The oxide formation reaction formulas include a triangle under the reaction direction arrows. These triangles indicate that the reaction requires energy to proceed. In silicon technology, that energy is usually supplied by heating the wafers and is called *thermal oxidation*. Silicon

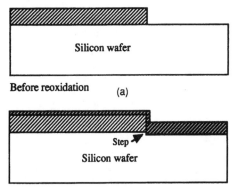

Before reoxidation    (a)

Step

Silicon wafer

After reoxidation    (b)

**Figure 7.14** Differential oxidation of silicon.

dioxide layers are grown either at atmospheric pressure or at high pressure. An atmospheric pressure oxidation takes place in a system without intentional pressure control—the pressure is simply that of the atmosphere for the location. There are two atmospheric techniques: tube furnaces and rapid thermal systems (Fig. 7.15).

### Horizontal Tube Furnaces

Horizontal tube furnaces have been used in the industry since the early 1960s for oxidation, diffusion, heat treating, and various deposition processes. They were first developed for diffusion processes in germanium technology and to this day are often called simply *diffusion furnaces*. The more correct, generic, term is a *tube furnace*. Horizontal designs evolved into vertical designs with a number of advantages detailed in a following section. The basic tube furnace principles apply to both systems.

A cross section of a basic single horizontal three-zone tube furnace is shown in Fig. 7.16. It consists of a long ceramic tube made of mullite (a ceramic), with coils of copper tubing on the inside surface. Each of the coiled tubes defines a zone and is connected to a separate power supply operated by a proportional band controller. Furnaces may have up to seven separate zones. Inside the furnace tube is a quartz reaction tube that serves as the reaction chamber for the oxidation (or other processes). The reaction tube may itself be inside a ceramic liner called a *muffle*. The muffle acts as a heat sink fostering a more even heat distribution along the quartz tube.

Thermocouples are positioned against the quartz tube and send temperature information to the proportional band controllers. The controllers proportion power to the coils, which in turn heat the reac-

| Thermal Oxidation | | |
|---|---|---|
| • Atmospheric Pressure | Tube furnace | Dry Oxygen<br>Wet Oxygen<br>Bubbler<br>Flash System<br>Dry Oxidation |
| | Rapid Thermal | Dry Oxygen |
| • High Pressure | Tube Furnace | Dry or Wet Oxygen |
| Chemical Oxidation | | |
| • Anodic Oxidation | Electrolytic Cell | Chemical |

**Figure 7.15**  Oxidation methods.

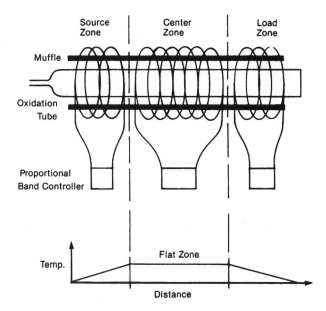

**Figure 7.16**  Cross section of single horizontal tube furnace with three heating zones.

tion tube by radiation and conduction. Radiation heating comes from the energy given off by the coils and impinging on the tube. Conduction takes place where the coils touch the tube. These controllers are very sophisticated and can control temperatures in the center zone (flat zone) to ±0.5°. For a process that operates at 1000°C, this variation is only ±0.05 percent. For the oxidation, the wafers are placed on a holder and positioned in the flat zone. The oxidant gas is passed into the tube, where the oxidation takes place. The details and the options used in actual practice are discussed in the following sections.

A production tube furnace is an integrated system of seven various sections (Fig. 7.17):

1. Reaction chamber(s)

2. Temperature control system

3. Furnace section

4. Source cabinet

5. Wafer cleaning station

6. Wafer load station

7. Process automation

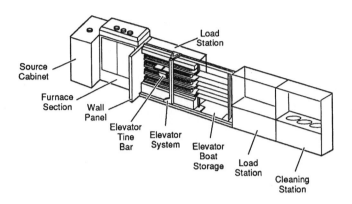

**Figure 7.17**    Tube furnace.

### Reaction chamber

The basic tube furnace operation was previously described. An important part of the system is the reaction chamber. This part protects the wafers from outside contamination and serves as a heat sink to even out the temperature inside the tube. Tube furnace reaction tubes are round, with a gas inlet end and a wafer load end. The gas inlet end, also known as the source end, tapers down to a ground fitting to provide a leak-free connection to the gas sources. The load end also has a ground fitting to receive either an end cap that keeps dirt out of the furnace or a wafer transfer unit (see sections on "Automatic wafer loading," p. 188, and "Manual wafer loading," p. 189).

The traditional reaction chamber material is high-purity quartz. Quartz is a highly purified glass favored for its inherent stability at high temperatures and its basic cleanliness. Drawbacks to quartz are its fragility and the presence of some metallic ions. Quartz also allows sodium ions from the heating coils to pass into the reaction chamber.

Another drawback is the tendency of quartz to break up and sag at temperatures above 1200°C. The breakup is called *devitrification* and results in small flakes of the quartz tube surface falling onto the wafers. Sagging impedes the easy placement of the wafer holders in and out of the tube. Quartz tubes are manufactured by two methods: electric fusion and flame fusion. As the names imply, two different heat sources are used to fuse the quartz. Both materials have about the same levels of trace metals but, in one evaluation, flame-fused tubes produced devices with better characteristics.[9]

Quartz tubes require periodic cleaning. One method is to place them in a tank of hydrofluoric (HF) acid or a solution of HF and water. This cleaning process takes place outside of the fabrication area, necessitating the removal and cooling of the tube. The cooling and heating hasten the devitrification process. The HF cleans by removing a thin layer of the quartz. This continual etching eventually weakens the tube wall, thus limiting its lifetime.

Some firms do an *in situ* tube cleaning with a portable plasma generator. The plasma generator is positioned inside the tube, where it activates etching gases passed into the tube. Both nitrogen trifluoride ($NH_3$) and the gases etch away contaminants. A second *in situ* cleaning method uses etching gases in the tube, which are activated by a plasma field created in the tube. This type of cleaning is more applicable to tube furnaces used for chemical vapor deposition (CVD) processes where the buildup of reaction-created particles is greater than in oxidation processes.

An alternative material to quartz process tubes and wafer holders is silicon carbide. Silicon carbide is structurally stronger and does not break down with repeated heating and cooling. This resistance to temperature cycling also makes the material a better metallic ion barrier over a longer period of time. During processes where oxygen is in the tube, the inside surface grows a thin layer of silicon dioxide. When this oxide is removed in HF, there is no attack of the tube material, which also contributes to the extended lifetime. The widespread use of silicon carbide tubes and wafer holders has been slowed by their high cost and weight.

### Temperature control system

The temperature control system connects thermocouples touching the reaction tube to proportional band controllers that feed the power to the heating coils. Proportional band controllers maintain even temperatures in the tube by feeding in or turning off the current to the coils in proportion to the deviation of the tube temperature from the set point. The closer the tube is to the set-point temperature, the smaller the amount of power that is fed to the coils. This system allows fast recovery of the tube to a cold load without overshoot. Adjustments are made to the controllers until the desired temperatures in the processing section of the tube (flat zone) are achieved.

*Overshoot* is the raising of the tube temperature too high above the desired process temperature as a result of applying too much power to the coils (Fig. 7.18).

Advanced systems have thermocouples positioned against the outside of the tube wall. These feed a microprocessor that, in turn, gives the information to the controllers. This system is called *autoprofiling*.[10] Good process control requires that the temperature profile inside the furnace be checked periodically with thermocouples

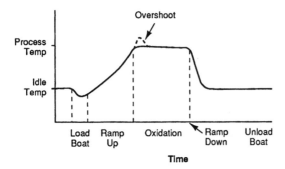

**Figure 7.18**  Temperature levels during oxidation.

inserted in the tube and the temperature measured by independent recorders.

The processing of larger-diameter wafers has brought with it the concern of wafer warping (Fig. 7.19). Silicon expands faster than silicon dioxide. When heated, the silicon dioxide pulls the wafer combination into a concave shape. Wafers that are heated or cooled rapidly will warp to the point of being useless. The degree of warping increases with higher process temperatures, that is, those above 1150°C.

Two methods are employed to minimize warping of wafers in tube furnaces. One is called *ramping* (or temperature ramping). Ramping is the procedure of maintaining the furnace at a temperature several hundred degrees below the process temperature. The wafers are slowly inserted into the furnace at this lower temperature and, after a short stabilization period, the controllers automatically take the furnace up to the process temperature. At the end of the process cycles, the furnace is cooled to the lower temperature before the wafers are removed. During the ramping process, the controllers must maintain the temperature control in the flat zone.

Tube furnaces are maintained at close to the process temperature 24 hours a day due to the devitrification of quartzware and the length of time it takes to stabilize the flat zone. In the interest of economy, some fabrication areas will keep the furnaces at the lower temperature. This is called an *idle* condition.

The second anti-warping procedure is the slow loading of the wafer boat into the tube. At loading rates of about 1 in/min, warping is mini-

**Figure 7.19**  Wafer warping.

mized. For large-diameter wafers and large batch sizes, both methods are used.

Another requirement of the heating system is a fast recovery time after the wafers are loaded in the tube. A full load of wafers can drop the tube temperature as much as 50°C or more.[11] The heating system works to bring the flat zone to temperature as fast as possible without introducing warping conditions or over shoot. Figure 7.18 illustrates a typical temperature-time recovery curve for a five-zone tube furnace.

**Furnace section**

A production-level tube furnace will contain three or four tubes (reaction chambers) and a separate temperature control system for each of the tubes. The tubes are arranged vertically above each other in a stack. The tubes open into an exhaust chamber that draws away the spent and heated gases as they exit the tube. This section of the furnace is called a *scavenger*. The scavenger is connected to the facility's exhaust system, which contains a scrubber to remove toxic gases from the withdrawn gases.

The tubes within a bank may be all used for the same purpose, such as oxidations, or be designated for different operations, such as oxidation, diffusion, alloy, or CVD, depending on the process requirements and the volume of wafers being processed. To change the use requires a change of tubeware and contamination monitoring to ensure that contaminants from a previous use are not present in the individual tube section.

**Source cabinet**

Each tube requires a number of gases to accomplish the desired chemical reaction. In the case of oxidation, the gas oxidants of oxygen or water vapor have been detailed. In addition, almost every tube process has nitrogen-flow capability. The nitrogen is used during the loading and unloading stage of the process to prevent accidental oxidation. In the idle condition, nitrogen is kept constantly flowing through the tube. The flow serves to keep dirt out of the system and maintain the pre-established flat zone.

Each of the tube processes requires that the gases be delivered to the tube in a specified sequence, at a specified pressure, at a specific flow rate, and for a specific time. The equipment used to regulate the gases is located in a cabinet attached to the furnace section of the system and is known as the *source cabinet*. There is a separate unit, called a *gas control panel* or *gas flow controller,* connected to each tube. The panel consists of solenoids, pressure gauges, mass flow controllers or flow meters, filters, and timers. In its simplest version, the gas flow controller consists of manually operated valves and timers. In production systems, the sequencing and timing of the various gases into the tube is controlled by a microprocessor. Required gases are plumbed to the panel. During operation, a timer opens a solenoid to admit the required gas to the panel. Its pressure is controlled by a pressure gauge. Flow amount into the tube is controlled by a flow meter or mass flow controller.

A *mass flow meter* is preferred in place of a flow meter for its inherent superior control. Stoichiometric considerations require that the same amount of material, as measured by its mass, be delivered into the reaction chamber. Standard flow meters measure the volume of material, and equal volumes of the same material may contain different amounts of material due to pressure and temperature differences. Semiconductor processes use a thermal-type of mass flow meter. The system consists of a heated gas passage tube with two temperature sensors. When no gas is flowing, the temperature sensors are at the same temperature. With the introduction of a gas flow, the downstream sensor reads higher. The difference between the two sensors is related to the amount of heat mass (not volume) that has moved downstream. The meter has a feedback mechanism to control the gas flow such that a steady amount of material flows through the meter. The source section also contains the microprocessor-controlled valves that meter the gas into the reaction chamber in the right sequence for the right amount of time. A general schematic of a mass flow meter is shown in Fig. 7.20. Mass flow meters can be set for a specific amount and on-board sensors measure and control the output with a feedback control system.[12] The piping material used in gas flow controllers is stainless steel to maintain high levels of cleanliness and to minimize chemical reactions between the gas and tube material.

Often, the gas flow controller is called a *jungle*. This term came about when the gas controllers were built in-house and had the look of a "jungle" of tubing and valves. Gases are supplied to the gas flow controller through piping from the liquid gas supplies in the pad section of the facility, or by smaller lecture bottles of gas located at the process tool.

Some processes require a chemical that is difficult to deliver in gas form. In this situation, a bubbler and liquid source is used. A bubbler consists of a quartz vial designed to admit a gas into the liquid. As the gas bubbles through the liquid and mixes with the source vapors in the top of the bubbler, it picks up the source chemicals and carriers them into the tube. Bubblers are used in oxidation, diffusion, and CVD processes. An oxidation bubbler is shown in Fig. 7.21. Diffusion and CVD bubblers generally have a cylindrical shape.

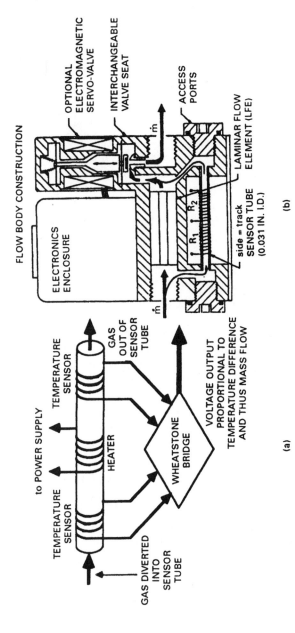

**Figure 7.20** (a) Operating principle of a mass flow meter and (b) cutaway.

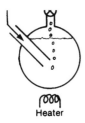

Heater

Figure 7.21   Bubbler water vapor source.

## Vertical Tube Furnaces

Horizontal tube furnaces have enjoyed great popularity over the years. A lot of process development has taken place in the 40 years of their use. However, the need for greater contamination control and the move to ever increasing wafer diameters as well as the need for more productive processing has pushed horizontal tube furnaces to a limit.

While contamination control has led to the development of cantilever systems, they have had to get larger and stronger as the wafers have grown in diameter and weight. Also, the larger wafer diameters require larger tube diameters, which puts additional pressures on the maintenance of extended temperature flat zones. Few firms are reporting the use of horizontal tube furnaces for 200-mm and larger wafers.[14]

The larger wafer loads have caused a lengthening of the furnaces and their associated load stations. In cleanroom terms, the "footprint" of the equipment is increasing. The problem is that larger footprints require larger and more expensive cleanrooms.

There are also process problems associated with larger-diameter horizontal tubes. One is keeping the gas streams in a laminar flow pattern in the tube. *Laminar gas flow* is uniform with no separation of the gases into layers and without turbulence that causes uneven reactions within the tube.

These considerations have resulted in the development of vertical tube furnaces, which are the configuration of choice for high-production, large-diameter processes.[15] In this configuration, the tube is held in a vertical position (Fig. 7.22) with loading taking place from the top or bottom. Tube materials and heating systems are the same as for horizontal systems.

The wafers are loaded in standard cassettes and lowered or raised into the flat zone. This action is accomplished without the particulates generated by the cassettes scraping the sides of the tubes. In this configuration, the wafers are in the most dense loading for a tube furnace. An added plus for vertical tube furnaces is the ease of rotating the wafers in the tube, which produces a more uniform temperature across

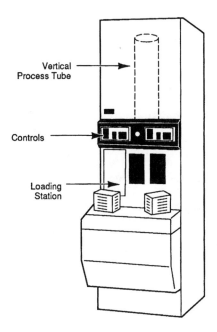

Figure 7.22 Vertical tube furnace.

the wafer. These furnaces have the same subsystems as horizontal furnaces.

Process uniformity is also enhanced by a more uniform (laminar) gas flow in a vertical tube. In a horizontal system, gravity tends to separate mixed gases as they flow down the tube. In a vertical system, the gas moves parallel to gravity minimizing the gas separation problem and the boat rotation minimizes gas turbulence. Vertical furnaces are capable of producing process variations 60 percent of those in horizontal furnaces.[16]

Particle generation associated with boat scrapping in horizontal systems is virtually eliminated in vertical systems, and the smaller area required for loading results in a cleaner system. Particle densities can reach into the $0.01/cm^2$ range.[17]

Perhaps one of the most appealing cost aspects of vertical tube furnaces is the small footprint. The system is smaller than a conventional four-stack system. Vertical systems offer the possibility of locating the furnaces outside the cleanroom with only a load station door opening into the cleanroom. In this arrangement, the cleanroom footprint of the furnace is practically zero, and maintenance can take place from the service chase. Another possible arrangement of vertical furnaces is in an island/cluster configuration. The furnaces are arranged around a central robot that alternately load several furnaces. A simpler design translates into a more reliable furnace with lower maintenance costs and longer periods of up-time. Vertical furnaces can be configured to perform any of the oxidation, diffusion, annealing, and deposition processes required in wafer fabrication.

### Fast Ramp Furnaces

As wafer sizes (and mass) become larger, ramp-up and ramp-down times become longer, creating more cost and bottlenecks in the fab flow. One way to offset the time increase is make sure that the batch sizes are maximized. But this can slow up the flow if the furnace is sitting idle waiting for enough wafers to form a large batch. Into this breech has come the fast-ramp, low-batch furnaces. They are mini-horizontal furnaces with high-powered heating systems. Where conventional tube systems operate at a few degrees per minute of ramping, fast-ramp furnaces can achieve rates of tens of degrees per minute.[18] Their lower capacity can be offset by the faster process times.

### Rapid Thermal Processing (RTP)

Ion implantation has replaced thermal diffusion due to its inherent doping control. However, ion implantation requires a follow-on heating operation, called *annealing,* to cure out crystal damage induced by the implant process. The annealing step has been traditionally done in a tube furnace. Although the heating anneals out the crystal damage, it also causes the dopant atoms to spread out in the wafer, an undesirable result. This problem led to the investigation of alternate energy sources to achieve the annealing without the spreading of the dopants. The investigations led to the development of rapid thermal process (RTP) technology.

RTP technology is based on the principle of radiation heating (Fig. 7.23). The wafer is automatically placed in a chamber fitted with gas inlets and exhaust outlets. Inside a heat source above (and possibly below), the wafer provides the rapid heating. Heat sources include graphite heaters, microwave, plasma arc, and tungsten halogen lamps.[19] Tungsten halogen lamps are the most popular.[20] The radiation from the heat source couples into the wafer surface and brings it up to the process temperature at rates of 75 to 125°C per second. The same temperature would take minutes to reach in a conventional tube furnace. A typical time-temperature cycle is shown in Fig. 7.24. Likewise, cooling takes place in seconds. With radiation heating, because of its very short heating times, the body of the wafer never comes up to temperature. For the ion implant annealing step, this means that the crystal damage is annealed while the implanted atoms stay in their original location.

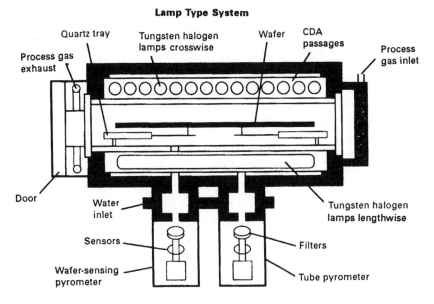

**Figure 7.23** RTP design. *(Source:* Semiconductor International, *May 1993.)*

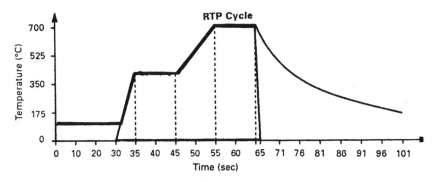

**Figure 7.24** Example RTP time/temperature curve. *(Source:* Semiconductor International, *May 1993.)*

Use of RTP reduces the *thermal budget* required for a process. Every time a wafer is heated near diffusion temperatures, the doped regions in the wafer continue to spread down and sideways (see Chapter 11). Every time a wafer is heated and cooled, more crystal dislocations form (see Chapter 3). Thus, minimizing the total time a wafer is heated allows more dense designs and fewer failures from dislocations.

Another advantage is single-wafer processing. The move to larger-diameter wafers has introduced uniformity requirements that, in many processes, are best met in a single-wafer process tool.

RTP technology is a natural choice for the growth of thin oxides used in MOS gates. The trend to smaller feature sizes on the wafer surface has brought along with it a decrease in the thickness of layers added to the wafer. Layers undergoing dramatic reduction in thickness are thermally grown gate oxides. Advanced production devices are requiring gate oxides less than 100 Å thick. Oxides this thin are sometimes hard to control in conventional tube furnaces due to the problem of quickly supplying and removing the oxygen from the system. RTP systems can offer the needed control by their ability to heat and cool the wafer temperature very rapidly. RTP systems used for oxidation, called *rapid thermal oxidation* (RTO) systems, are similar to the annealing systems but have an oxygen atmosphere instead of an inert gas. A typical time/temperature/thickness relationship for RTO is shown in Fig. 7.25.

Other processes using RTP technology include wet oxide (steam) growth, localized oxide growth, source/drain activation after ion implant, LPCVD polysilicon, amorphous silicon, tungsten, salicide contacts, LPCVD nitride, and LPCVD oxide.[21] RTP systems come in atmospheric, low-pressure, and ultra-high-vacuum designs.

Temperature control across a wafer is different in a radiation chamber as opposed to in a furnace tube. In an RTP system, the wafer never comes to thermal stability. The problem is particularly acute at the wafer edges. Another problem comes from the number and different layers already on an in-process wafer. These different layers each absorb the heating radiation in a different way, resulting in temperature differences across the wafer, which in turn contribute to temperature nonuniformity. This phenomenon is called *emissivity* and is a property of the particular material and the wavelength of the heating radiation. Temperature nonuniformity creates nonuniform process results in and on the wafer surface and, if the temperature differential is high enough, crystal slip at the wafer's edge.

Solutions to the problem include lamp placement and control of individual lamps in the system along with top and bottom lamps. Some systems have a heated annular ring to keep the edge of the wafer within the required temperature range. Process temperatures are usually measured by thermocouples; however, they require back contact with the heated wafer, which is impractical in a single-wafer system, and thermocouples have a response time that is longer than some RTP heating cycles. Optical pyrometers, which gauge temperatures by measuring characteristic energies given off by the heated object, are preferred. However, they too are prone to errors, especially on wafers with a number of layers. The difficulty is relating the emission given off by the wafer to the actual temperature on the surface. Solutions to this problem include a backside seal layer of silicon nitride to mini-

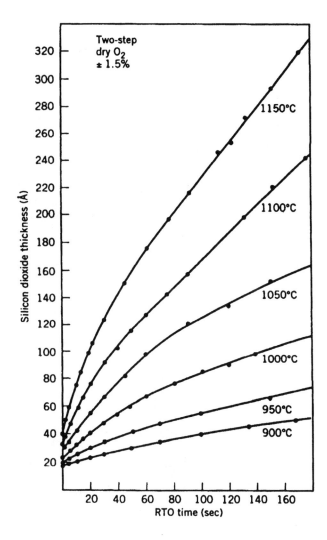

**Figure 7.25** Oxidation of silicon by RTO. *(Source: Ghandhi, VLSI Fabrication Principles.)*

mize backside emissivity variations[22] and open-loop lamp control. *Open-loop control* is based on converting lamp control to direct current (dc) to get away from voltage variations to the lamps. Other approaches involve elaborate sampling of and/or filtering of the radiation coming off the wafer to more closely relate the measurement to the wafer surface temperature. It has also been suggested that measurement of the wafer expansion, which is directly due to temperature increases, may be a more reliable and direct measurement technique.[23]

Given the benefits of RTP, especially in a high-production, single-wafer processing environment, it can be expected that the development will continue.

### High-Pressure Oxidation

The thermal budget problem also was an impetus (along with others) for high-pressure oxidation. The growth of dislocations in the bulk of the wafer and the growth of hydrogen-induced dislocations along the edge of openings in layers on the wafer surface[24] are two high-temperature oxidation problems. In the first case, the dislocations cause various device performance problems. In the latter case, surface dislocations cause electrical leakage along the surface or the degradation of silicon layers grown on the wafer for bipolar circuits.

The growth of dislocations is a function of the temperature at which the wafer is processed and the time it spends at that temperature. A solution to this problem is to perform thermal oxidation processes at a lower temperature. This solution by itself causes the production problem of longer oxidation times. The solution that addresses both problems is high-pressure oxidation (Fig. 7.26). These systems are configured like conventional horizontal tube furnaces but with one major exception: the tube is sealed, and the oxidant is pumped into the tube at pressures of 10 to 25 atm (10 to 25 times the pressure of the atmosphere). The containment of the high pressure requires encasing the quartz tube in a stainless steel jacket to prevent it from cracking.

At these pressures, the oxidation proceeds at a faster rate than in atmospheric systems. A rule of thumb is that a 1-atm increase in pressure allows a 30°C drop in the temperature. In a high-pressure system, that increase relates to a drop of 300 to 750°C in temperature. This reduction is sufficient to minimize the growth of dislocations in and on the wafers.

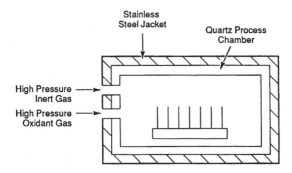

**Figure 7.26**  High-pressure oxidation.

Another option using high-pressure systems is to maintain the regular process temperature and reduce the time of the oxidation. Other considerations concerning high-pressure systems focus on the safe operation of the system and possible contamination from the additional pumps and piping needed to create the high pressures inside the tube.

Very thin MOS gate oxide growth is a candidate for high-pressure oxidation. The thin oxide must have structural integrity (no holes, and so forth) and have a dielectric strength high enough to prevent charge induction in the gate region. Gate oxides grown in high-pressure processes have higher dielectric strength than similar oxides grown at atmospheric pressure.[25] High-pressure oxidation is also a solution for the *bird's beak* problem that occurs during local oxidation of silicon (LOCOS). See Chapter 16, "LOCOS process." An unwanted bird's beak of oxide grows into the active region of an MOS device as in Fig. 7.27. High-pressure oxidation can minimize the bird beak encroachment into the device area and minimize field oxide thinning during LOCOS processing.[26]

In addition to oxidation, high-pressure systems are finding some use in CVD epitaxial depositions and for flowing glass layers on the wafer surface.[27] Both of these processes are of higher quality when performed at lower temperatures.

## Oxidant sources

**Dry oxygen.**  When oxygen is used as the oxidant, it is supplied from the facility source or from tanks of compressed oxygen located in or near the source cabinet. It is imperative that the gas be dry, that is, not contaminated with water vapor. The presence of water vapor in the oxygen would increase the oxidation rate and cause the oxide layer to be out of specification. Dry oxygen oxidation is the preferred method for growing the very thin ($\approx$1000 Å) gate oxides required for MOS devices.

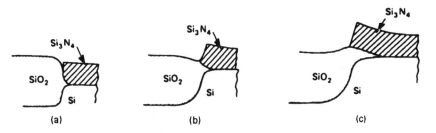

**Figure 7.27**   Bird's beak growth. (*a*) No pre-etch, (*b*) 1000 Å pre-etch, and (*c*) 2000 Å pre-etch. (*From Ghandhi,* VLSI Fabrication Principles.)

**Water vapor sources.**    Several methods are used to supply water vapor (steam) into the oxidation tube. The choice of method depends on the level of thickness and cleanliness control required of the oxide layer in the device.

**Bubblers.**    The historic method of creating a steam vapor in the tube is by a bubbler (see "Source Cabinet"). For oxidation, the bubbler liquid is deionized (DI) water (refer back to Fig. 7.21) heated close to the boiling point (98 to 99°C), which creates a water vapor in the space above the liquid.

As the carrier gas is bubbled through the water and passes through the vapor, it becomes saturated with water. Under the influence of the elevated temperature inside the tube, the water vapor becomes steam and causes the oxidation of the silicon surfaces.

A primary drawback with a bubbler system is that control of the amount of water vapor entering the tube as the water level in the bubbler changes and fluctuations in the water temperature. With bubblers, there is always concern about contamination of the tube and oxide layer from dirty water or dirty flasks. This contamination potential is heightened by the need to open the system periodically to replenish the water.

**Dry oxidation (dryox).**    Bubbler and flash new levels of thickness control and cleanliness came with the introduction of MOS devices. The heart of an MOS transistor is the gate structure, and the critical layer in the gate is a thin, thermally grown oxide. Liquid-water-steam systems are unreliable for growing thin, clean gate oxides. The answer was found in dry oxidation (or dry steam) process (Fig. 7.28).

In the dry oxidation system, gaseous oxygen and hydrogen are introduced directly into the oxidation tube. Inside the tube, the two gases mix and, under the influence of the high temperature, form steam. The result is a wet oxidation in steam. Dryox systems offer improved control and cleanliness over liquid systems. First, gases can be purchased in a very clean and dry state. Second, the amounts going into the tube can be very precisely controlled by the mass flow controllers. Dryox is the preferred general oxidation method for production for all advanced devices.

A drawback to dryox systems is the explosive property of hydrogen. At oxidation temperatures, hydrogen is very explosive. Precautions

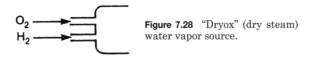

$O_2$ ⟶
$H_2$ ⟶

**Figure 7.28**  "Dryox" (dry steam) water vapor source.

used to reduce the explosion potential include separate oxygen and hydrogen lines to the tube and flowing excess oxygen into the tube. The excess oxygen ensures that every hydrogen molecule ($H_2$) will combine with an oxygen atom to form the nonexplosive water molecule, $H_2O$. Other precautions used are hydrogen alarms and a hot filament in the source cabinet and in the scavenger end of the furnace to immediately burn off any free hydrogen before it can explode.

**Chlorine-added oxidation.** The thinner MOS gate oxides require very clean layers. Improvements in cleanliness and device performance are achieved when chlorine is incorporated into the oxide. The chlorine tends to reduce mobile ionic charges in the oxide layer, reduce structural defects in the oxide and silicon surface, and reduce charges at the oxide-silicon interface. The chlorine comes from the inclusion of anhydrous chlorine ($Cl_2$), anhydrous hydrogen chloride (HCl), trichloroethylene (TCE), or trichloroethane (TCA) in the dry oxygen gas stream. The gas sources, chlorine and hydrogen chloride, are metered into the tube along with the oxygen from separate flow meters in gas flow controller. The liquid sources, TCE and TCA, are carried into the tube as vapors from liquid bubblers. For safety and ease of delivery, TCA is the preferred source of chlorine. The oxidation-chlorine cycle may take place in one step or be preceded or followed by a dry oxidation cycle.

**Wafer-cleaning station**

When the wafers come to a tube furnace, they first go to the attached wafer-cleaning station. The need for stringent cleanliness control has been stressed throughout this text. It is especially important before the tube operations because of the heating of the wafers. Contamination left on the wafer surfaces can pit the surface, diffuse into the surface, or interfere with the quality of the layer being grown.

The cleaning station may be a VLF "wet" bench that has built-in tanks for cleaning chemicals, units for rinsing the wafers in deionized water, and drying units. The station may be, or include, an automatic cleaning machine The station also has a wafer rinse and drying machine. Wafer cleaning is detailed in the section on "Oxidation Processes."

**Wafer-load station.** After cleaning, the wafers are passed to the in-line load station. Here, the wafers are inspected for cleanliness and loaded into holders for insertion into the tube. The station is located under a ceiling-mounted HEPA filter or in a VLF hood. The loading station and the cleaning station are located next to each other in such a manner that the wafers stay in a clean environment during the transfer. A quick surface inspection of the wafers is normally done with the aid of an ultraviolet light. These high-intensity light sources allow the operator to see small particles and stains that are not visible to the naked eye. Sometimes, a microscope inspection of the surface is performed.

**Automatic wafer loading.**    Production efficiency and VLSI cleanliness requirements have led to the development of a variety of automatic boat- and tube-loading mechanisms. Pick-and-place machines (sometimes called *robots*) pick each of the wafers out of one cassette and place it into an empty one. Some versions of pick-and-place machines pick up the entire load of wafers and transfer it in one operation to the empty cassette. A challenge to any wafer boat-loading system is the correct placement of test wafers within the load of device wafers as well as "dummy wafers" often placed at the ends (or top and bottom) of a boat load of wafers (see p. 190). These wafers must be *picked* from other boats.

An additional challenge is the loading two wafers back-to-back in the same slot. This procedure is used to increase the productivity of the operation.

The loaded cassettes are put into the furnace by a number of techniques. One is to place the loaded boat into a transfer tube called an *elephant*. The elephant has a ground joint that mates with the joint on the load end of the tube. It also has a hole in the back end to admit a quartz rod that hooks into the back end of the wafer boat. The tube is used to push the boat into the flat zone of the tube furnace (Fig. 7.29). Other versions automate the push-pull action with a push rod connected to a motor. A number of cassettes are placed by the operator on a platform called a *paddle*. The paddle moves the cassettes into the tube under the control of a motor (Fig. 7.30).

Boat loading can be a source of severe wafer contamination from particles scraped from the inside of the tube as the cassettes are shoved down the tube. One technique for solving this problem is to use

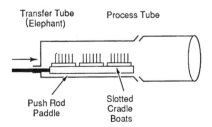

Transfer Tube
(Elephant)    Process Tube

Push Rod
Paddle    Slotted
Cradle
Boats

**Figure 7.29** Transfer tube loading of wafers.

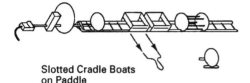

Slotted Cradle Boats
on Paddle

Figure 7.30   Wafer boat and cradle.

cassettes or paddles with small wheels. The rolling wheels kick up fewer particles than does scraping. Another solution is a cantilever system, which features a rigid rod with the cassettes positioned on the end. The rod moves the cassettes down the center of the tube without ever touching its sides. Some systems leave the cassettes suspended in the center of the tube, while others give the cassettes a soft landing on the tube bottom.

Automated push-pull machines are necessary to achieve the control required for slow entry and exit of the wafers to prevent warping. A typical push-pull rate is 1 in/min. A standard four-stack furnace system requires an automatic push-puller for each of the tubes. Although these systems load and unload automatically, they require an operator to place the loaded boats on the paddle.

An overriding goal of process equipment design is "hands-off" operation. That need is addressed in tube furnace systems built on the elevator design.[13] The concept employs a storage buffer of loaded tube cassettes whose tube destination is known by a computer. When it is time for the cassettes to be loaded in the tube, a robot selects the proper cassettes and moves to the correct tube position and loads them into the tube. The advantages of these systems are the computer production control and the need for only one loading mechanism per furnace stack. For a properly designed system, robotic activity is less contaminating to the wafers than human operators.

**Manual wafer loading.**   Wafers are processed through the cleaning steps in Teflon® or Teflon-derivative wafer holders, also called *boats* or *cassettes*. They are transferred to quartz or silicon carbide holders for the furnace process. The boats come in various designs (Fig. 7.30), depending on the degree of control required of the operation and the loading density. The most productive designs have the wafers standing up in machined slots, positioned crosswise to the gas flow. In tube operations where uniformity is critical, the wafers may be loaded parallel to the gas flow.

The weight and value of larger diameter wafers dictates wafer handling by automatic systems. However there is always a need to select single wafers for examination or testing. For these operation wafer handling is by vacuum wands or limited-grasp tweezers (Fig. 7.31).

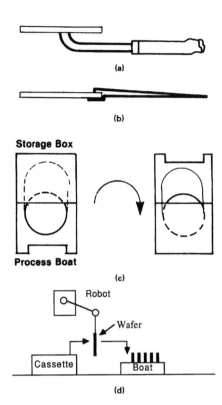

**Figure 7.31** Manual wafer handling devices. (*a*) Vacuum pickup, (*b*) limited-grasp tweezer, (*c*) flip transfer boats, and (*d*) auto pick-and-place.

Vacuum wands are attached to a vacuum source and are designed to allow grasping of the wafers from the backside. This arrangement minimizes damage and contamination of the sensitive front side of the wafer.

**Oxidation Process Automation**

Tube furnaces have evolved from operator-controlled manual systems to various levels of automation. A modern production tube furnace is automated from wafer load to wafer unload in a one-button operation. A goal is to automate the process from wafer clean to unload. The operator places the cassettes of cleaned wafers on the load mechanism and pushes the start button. The computer takes over and directs the tube furnace through the predetermined process, which is called a *recipe*. The times, temperatures, gas sequences, and push-pull rates are programmed into an on-board computer or a host computer. Full automation of a tube operation involves delivery of the proper wafers to the station with robots or automatic equipment taking the wafers entirely through the process and passing them on to the next process.

The various components of a furnace process may be organized in a *cluster* arrangement (see Chapter 15).

### Preoxidation Wafer Cleaning

Removal of surface contamination and unwanted native oxides is essential to a successful oxidation process. Contamination can diffuse into the wafer causing electrical problems in devices and structural integrity problems in the silicon dioxide film. Thin layers of native oxides can alter the thickness and integrity of the grown oxide layer.

Preoxidation processes (see Chapter 5) typically start with a mechanical scrubber, followed by an RCA wet cleaning sequence to remove organic and inorganic contamination. Finally, an HF or diluted HF etch is performed to rid the surface of native or chemically grown oxides. This is an HF-last process.

The processing of the wafers through the oxidation process is divided into several distinct steps, as shown in the flow diagram in Fig. 7.32. After the wafers are logged into the station, they receive a thorough cleaning. Clean wafers are essential at all stages of the fabrication process but are especially necessary before any of the operations performed at high temperature.

### Oxidation Processes

After the wafers are precleaned, cleaned and etched, and loaded on the push-pull mechanism, they are ready for surface oxidation. The actual oxidation proceeds in different gas cycles (Fig. 7.33) within the furnace tube. The first gas cycle occurs as the wafers are being loaded into the tube. Since the wafers are at room temperature and precise oxide thickness is a goal of the operation, the gas metered into the

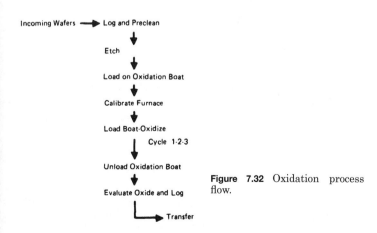

Incoming Wafers ➡ Log and Preclean

Etch

Load on Oxidation Boat

Calibrate Furnace

Load Boat-Oxidize

Cycle 1-2-3

Unload Oxidation Boat

Evaluate Oxide and Log

Transfer

**Figure 7.32** Oxidation process flow.

| Cycle        Gas | Purpose |
|---|---|
| 1. Nitrogen | Temperature Stabilization in an Inert Atmosphere |
| 2. Oxygen or Water Vapor | Oxide Growth |
| 3. Nitrogen | Stop Oxidation and Removal of Wafers in an Inert Atmosphere |

**Figure 7.33**  Oxidation process cycles.

tube during loading is dry nitrogen. The nitrogen is necessary to prevent any oxidation while the wafers are coming up to the required oxidation temperature.

Once the wafers are stabilized at the correct temperature, the flow gas controller switches the gas flow to the selected oxidant. For oxides greater than 1200 Å, the oxidant is usually steam from one of the sources previously discussed. For oxides less than 1200 Å, pure oxygen is usually used because of its greater process control and the cleaner, denser oxide it produces. Thin MOS transistor gate oxides are usually grown at lower temperatures (900°C) in oxygen. However, oxidation process times can require hours in the furnace. One alternative is to grow thin gates oxides in wet oxygen environments to reduce process time, but in a reduced pressure. Lowering the pressure maintains oxide density and structural integrity.[28] The thinner MOS gate oxides require very clean layers. Improvements in cleanliness and device performance are achieved when chlorine is incorporated into the oxide.[29] The chlorine tends to reduce mobile ionic charges in the oxide layer, reduce structural defects in the oxide and silicon surface, and reduce charges at the oxide-silicon interface.

The chlorine comes from the inclusion of anhydrous chlorine ($Cl_2$), anhydrous hydrogen chloride (HCl), trichloroethylene (TCE), or trichloroethane (TCA) in the dry oxygen gas stream. The gas sources, chlorine and hydrogen chloride, are metered into the tube along with the oxygen from separate flow meters in gas flow controller. The liquid sources, TCE and TCA, are carried into the tube as vapors from liquid bubblers. For safety and ease of delivery, TCA is the preferred source of chlorine.

The oxidation-chlorine cycle may take place in one step or be preceded or followed by a dry oxidation cycle. After the oxidation cycle, the furnace gas is switched back to dry nitrogen. The nitrogen terminates the oxidation of the silicon by diluting and removing the oxidant used. It also prevents any oxidation during the wafer exit step.

## Postoxidation Evaluation

After the wafers are removed from the oxidation boats, they will receive an inspection and evaluation(s). The nature and number of the evaluations depend on the oxide layer and the precision and cleanliness required of the particular circuit being fabricated. (The details of the evaluations performed are explained in Chapter 14.)

A requirement of the oxidation process is a uniform noncontaminated layer of silicon dioxide on the wafer. As the wafers proceed through the wafer fabrication operations, there is a buildup of both thermally grown oxides and other deposited layers on the wafer surface. These other layers interfere with the determination of the quality of a particular oxide. For this reason, each batch of oxidized wafers going into the tube includes a number of test wafers, with bare surfaces placed at strategic locations on the wafer boat. Test wafers are necessary for the evaluations that are destructive or require large undisturbed areas of oxide. At the conclusion of the oxidation operation, they are used for the evaluation of the process. Some of the evaluations are performed by the oxidation operator and some are performed off line in quality control (QC) labs.

### Surface inspection

A quick check of the cleanliness of the oxide is performed by the operator as the wafers are unloaded from the oxidation boat. Each wafer is viewed under a high-intensity ultraviolet (UV) light. Surface particulates, irregularities, and stains are readily apparent in UV light.

### Oxide thickness

The thickness of the oxide is of major importance. It is measured on test wafers by a number of techniques (see Chapter 14). The techniques are color comparison, fringe counting, interference, ellipsometers, stylus apparatus, and scanning electron microscopes (SEMs).

### Oxide and furnace cleanliness

In addition to the physical contaminants of particles and stains, the oxide should have a minimum number of mobile ionic contaminants. These are detected by the sophisticated capacitance-voltage (C/V) technique, which detects the total number of mobile ionic contaminants present in the oxide. It cannot identify the origin of those contaminants, which may come from the tubes, the gases, the wafers, or the cleaning process. Therefore, C/V evaluation is a go-no-go assessment of the wafers and serves as a check of the total furnace operation.

In most fabrication lines, C/V analysis is also used to certify the cleanliness of the furnace and its associated parts. A oxide with a low-mobile ionic contamination level certifies that the entire system is clean. When the oxide fails the test, more investigation is necessary to identify the source.

A second oxide-cleanliness-related parameter is dielectric strength. This parameter measures the dielectric (nonconducting) property of the oxide by the destructive oxide rupture test.

A third cleanliness factor is the index of refraction of the oxide. Refraction is the property of a transparent substance that causes light to bend as it travels through it. The apparent versus actual location of an object on the bottom of a body of water is an example of refraction. The index of refraction of a pure oxide is 1.46. Variations of this value come about from impurities in the oxide. A constant index of refraction is the starting point for several of the interference thickness-measurement techniques. Variations in the index can lead to erroneous thickness measurements. The index of refraction is measured by interference and ellipsometry techniques (see Chapter 14).

**Thermal nitridation**

An important factor in the production of small high-performance MOS transistors is a thin gate oxide. However, in the 100-Å (or less) range, silicon dioxide films tend to be of poor quality and difficult to control. An alternative to silicon dioxide films is a thermally grown silicon nitride ($Si_3N_4$) film. $Si_3N_4$ is denser than silicon oxide and has fewer pinholes in these thin ranges. It also is a good diffusion barrier. Growth control of thin films is enhanced by a flat growth mechanism (after an initial rapid growth). This characteristic is shown (Fig. 7.34) in the

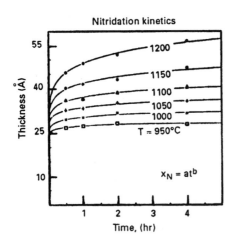

Figure 7.34  Nitridation of ⟨100⟩ silicon. *(Source: Wolf, Silicon Process.)*

growth of silicon nitride formed by the exposure of the silicon surface to ammonia ($NH_3$) between 950 and 1200°C.[30]

Some advanced devices use silicon oxynitride ($SiO_xN_y$) films. They are also called *nitrided-oxide* or *nitroixide* films. These are formed from the nitridation of silicon oxide films. Unlike silicon dioxide films, oxynitride films vary in composition depending on the growth process.[31] Another MOS gate structure is a sandwich of oxide/nitrite/oxide (ONO).[32]

## Review Questions

1. Describe the role and movement of oxygen during thermal oxidation.

2. Draw a sketch of a horizontal tube furnace and identify all the sections.

3. Explain why dryox is favored for VLSI oxidation processes.

4. Why are temperature ramping techniques required?

5. What is the advantage of rapid thermal oxidation?

6. Name the two oxidants added to sulfuric acid for wafer cleaning.

7. Make a flow diagram of the oxidation steps, from incoming wafer to evaluation.

8. Name the three factors that determine the oxide thickness.

9. What advantages are offered by high-pressure oxidation?

10. List three uses of silicon dioxide layers in semiconductor devices.

## References

1. C. Hu, "MOSFET Scaling in the Next Decade and Beyond," *Semiconductor International,* Cahners Publishing, June 1994, p. 105.
2. S. Wolf and R. Tauber, *Silicon Processing for the VLSI Era,* Lattice Press, Sunset Beach, CA, 1986.
3. P. Gise and R. Blanchard, *Modern Semiconductor Fabrication Technology,* Reston Books, Reston, VA, p. 43.
4. S. M. Sze, *VLSI Technology,* McGraw-Hill Book Company, 1983, p. 137.
5. P. Gise and R. Blanchard, *Modern Semiconductor Fabrication Technology,* Reston Books, Reston, VA, p. 46.
6. S. K. Ghandhi, *VLSI Fabrication Principles,* John Wiley & Sons, Inc., 1994, p. 464.
7. S. M. Sze, *VLSI Technology,* McGraw-Hill Book Company, 1983, p. 147.
8. S. M. Sze, *VLSI Technology,* McGraw-Hill Book Company, 1983, p. 159.
9. M. Hill, D. Helman, and M. Rother, "Quartzglass Components and Heavy Metal Contamination," *Solid State Technology,* PennWell Publishing Corp., March 1994, p. 49.
10. J. Maliakal, D. Fisher, Jr., and A. Waugh, "Trends in Automated Diffusion Furnace Systems for Large Wafers," *Solid State Technology,* December 1984, p. 107.
11. *Ibid.*

12. C. Murray, "Mass Flow Controllers: Assuring Precise Process Gas Flows," *Semiconductor International,* Cahners Publishing, October 1985, p. 72.
13. P. Burggraaf, "'Hands-off' Furnace Systems," *Semiconductor International,* September 1987, p. 78.
14. P. Burggraaf, "Verticals: Leading Edge Furnace Technology," *Semiconductor International,* Cahners Publishing, September 1993, p. 46.
15. P. Singer, "Trends in Vertical Diffusion Furnaces," *Semiconductor International,* April 1986, p. 56.
16. P. Burggraaf, Verticals: "Leading Edge Furnace Technology," *Semiconductor International,* Cahners Publishing, September 1993, p. 46.
17. *Ibid.*
18. P. Singer, "Furnaces Evolving To Meet Diverse Thermal Processing Needs," *Semiconductor International,* March, 1997, p. 86.
19. P. Singer, "Rapid Thermal Processing: A Progress Report," *Semiconductor International,* Cahners Publishing, May 1993, p. 68.
20. S. Leavitt, "RTP: On the Edge of Acceptance," *Semiconductor International,* March 1987.
21. M. Moslehi, A. Paranjpe, L. Velo, and J. Kuehne, RTP: "Key to Future Semiconductor Fabrication," *Solid State Technology,* PennWell Publishing, May 1994, p. 37.
22. *Ibid,* p. 38.
23. P. Singer, "Rapid Thermal Processing: A Progress Report," *Semiconductor International,* Cahners Publishing, May 1993, p. 69.
24. D. Toole and P. Crabtree, "Trends in High-Pressure Oxidation," *Microelectronic Manufacturing and Test,* October 1988, p. 1.
25. S. K. Ghandhi, *VLSO Fabrication Principles,* John Wiley and Sons, 1994, p. 466.
26. S. Kim, A. Emami, and S. Deleonibus, "High-Pressure and High-Temperature Furnace Oxidation for Advanced Poly-Buffered LOCOS," *Semiconductor International,* May 1994, p. 64.
27. D. Toole and P. Crabtree, "Trends in High-Pressure Oxidation," *Microelectronic Manufacturing and Test,* October 1988, p. 8.
28. Brian Dance, Growth of Ultrathin Silicon Oxides by Wet Oxidation, Semiconductor International, February 2002, p. 44.
29. Wolf and Tauber, *Silicon Processing for the VLSI Era,* p. 226.
30. Wolf and Tauber, *Silicon Processing for the VLSI Era,* p. 210.
31. S. K. Ghandhi, *VLSI Fabrication Principles,* John Wiley & Sons, Inc., New York, 1994, p. 484.
32. P. Singer, "Directions in Dielectrics in CMOS and DRAMs," *Semiconductor International,* Cahners Publishing, April 1994, p. 57.

# The Ten-Step Patterning Process—Surface Preparation to Exposure

## Overview

Patterning is the series of processes that establishes the shapes, dimensions, and placement of the required physical "parts" (components) of the IC in and on the wafer surface layers. This chapter presents the first four steps of a basic ten-step photo process and a discussion of photoresist chemistry.

## Objectives

Upon completion of this chapter, you should be able to:

1. Sketch wafer cross sections showing the basic ten-step photomasking process.

2. Explain the reaction of negative and positive photoresists to light.

3. Describe the correct resist and mask polarities required to produce holes and islands in wafer surface layers.

4. Make a list of the major process options for each of the ten basic steps.

5. Select from the list in objective 4 the processes used to pattern features in micron and submicron sizes.

6. Describe the need for, and process steps used in, double masking, multilayer resist processing, and planarization techniques.

7. Explain the use of antireflective coatings and contrast enhancement in the patterning of "small" feature sizes.

8. List the optical and nonoptical methods used for alignment and exposure.

9. Compare the equipment and advantages of each alignment and exposure method.

## Introduction

Patterning is one of the basic operations. At the end of the operation, a surface layer is left with either a hole or an island. (See Fig. 8.1.) *Patterning* is also called *photolithography, photomasking, masking, oxide removal (OR)), metal removal* (MR), and *microlithography.*

Patterning is one of the most critical operations in semiconductor processing. It is the process that sets the surface (horizontal) dimensions on the various parts of the devices and circuits. The goal of the operation is twofold. First is to create, in and on the wafer surface, a pattern with the dimensions established in the design phase of the IC or device. This goal is referred to as the *resolution* of the images on the wafer.

The second goal is the correct placement of the circuit pattern on the wafer. The entire circuit pattern must be correctly placed on the wafer surface relative to the crystal pattern of the wafer substrate, and the individual parts of the circuit must line up relative to each other (Fig. 8.2). This is called *alignment* or *registration* of the various circuit patterns. A typical IC requires 20 to 40 individual patterning (or masking) steps. This registration requirement is similar to the correct alignment of the different floors of a building. It is easy to visualize that misalignment of elevator shafts and stair wells would render the building useless. In a circuit, the effects of misaligned mask layers can cause the entire circuit to fail.

Control of the dimensions and defect levels is difficult, because each step in the patterning process contributes variations. A patterning

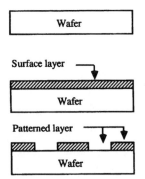

**Figure 8.1** Basic patterning process.

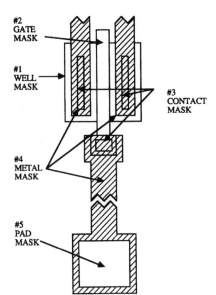

#2
GATE
MASK

#1
WELL
MASK

#3
CONTACT
MASK

#4
METAL
MASK

#5
PAD
MASK

**Figure 8.2**  Five mask set silicon gate transistor.

process is one of trade-offs and balancing (see sections on individual patterning processes). In addition to dimensional control and pattern alignment, defect control during the process steps is critical. Given the number of steps in each patterning operation and the number of mask layers, the masking process is the chief source of defects.

## Overview of the Photomasking Process

Photolithography is a multistep pattern transfer process similar to photography and stenciling. The required pattern is first formed in reticles or photomasks and transferred into the surface layer(s) of the wafer through the photomasking steps.

The transfer takes place in two steps. First, the pattern on the reticle or mask is transferred into a layer of photoresist (Fig. 8.3). Photoresist is a light-sensitive material similar to the coating on a regular photographic film. Exposure to light causes changes in its structure and properties. In the example in Fig. 8.3, the photoresist in the region exposed to the light was changed from a soluble condition to an insoluble one. Resists of this type are called *negatively acting,* and the chemical change is called *polymerization.* Removing the soluble portion with chemical solvents (developers) leaves a hole in the resist layer that corresponds to the opaque pattern on the reticle.

The second transfer takes place from the photoresist layer into the wafer surface layer (Fig. 8.4). The transfer occurs when etchants re-

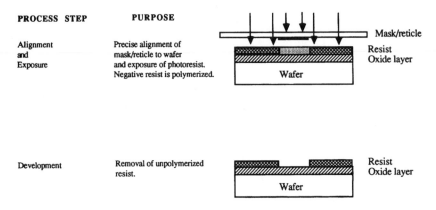

| PROCESS STEP | PURPOSE | |
| --- | --- | --- |
| Alignment and Exposure | Precise alignment of mask/reticle to wafer and exposure of photoresist. Negative resist is polymerized. | Mask/reticle<br>Resist<br>Oxide layer<br>Wafer |
| Development | Removal of unpolymerized resist. | Resist<br>Oxide layer<br>Wafer |

**Figure 8.3**   First pattern transfer—mask/reticle to resist layer.

| PROCESS STEP | PURPOSE | |
| --- | --- | --- |
| Etch | Top layer of wafer is removed through opening in resist layer. | Resist<br>Oxide layer<br>Wafer |
| Photoresist removal (strip) | Remove photoresist layer from wafer. | Oxide layer<br>Wafer |

**Figure 8.4**   Second pattern transfer—resist layer to surface layer.

move the portion of the wafer's top layer that is not covered by the photoresist. The chemistry of photoresists is such that they do not dissolve (or dissolve slowly) in the chemical etching solutions; they are *etch-resistant,* hence the name *resists* or *photoresists.*

In the examples shown in Figs. 8.3 and 8.4, the result is a hole etched in the wafer layer. The hole came about because the pattern in the mask was opaque to the exposing light. A mask whose pattern exists in the opaque regions is called a *clear-field mask* (Fig. 8.5). The pattern could also be coded in the mask in the reverse, in a dark-field mask. If the same steps were followed, the result of the process would be an island of material left on the wafer surface (Fig. 8.6).

The resist reaction to light just described is a character of negative-acting photo resists. There are also positive-acting photo resists. Within these resists, the light changes the chemical structure from relatively nonsoluble to much more soluble. The term describing this

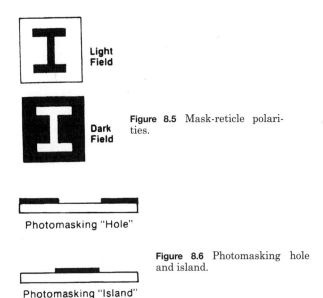

Figure 8.5  Mask-reticle polarities.

Figure 8.6  Photomasking hole and island.

change is *photosolubilization*. Figure 8.7 shows that an island is produced when a light-field mask is used with a positive photoresist.

The result obtained from the photomasking process from different combinations of mask and resist polarities is shown in Fig. 8.8. The choice of mask and resist polarity is a function of the level of dimensional control and defect protection required to make the circuit work. These issues are discussed in the process sections of the chapter.

## Ten-Step Process

Transferring the image from the reticle or mask onto the wafer surface layer is a multistep procedure (Fig. 8.9). Feature size, alignment tolerance, the wafer surface, and the masking layer number all influence the difficulty and steps for a particular masking process. Many photo processes are customized to the particular conditions. However, most are variations or options of a basic ten-step process. The process illustrated is shown with a light-field mask and a negative photoresist.

The first image transfer takes place in steps 1 through 7. In steps 8, 9, and 10, the image is transferred (second image transfer) into the wafer surface layer. The reader is challenged to list the steps and draw the corresponding cross sections using combinations of a dark field mask and a positive photoresist. It is strongly recommended that the reader master this ten-step process before proceeding to the advanced photolithography processes.

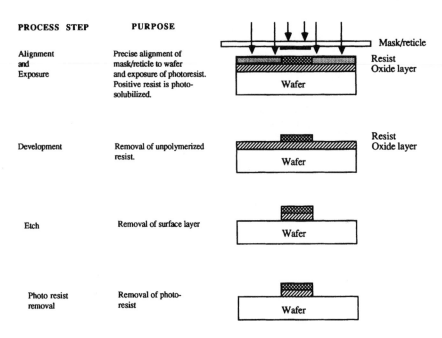

| PROCESS STEP | PURPOSE |
| --- | --- |
| Alignment and Exposure | Precise alignment of mask/reticle to wafer and exposure of photoresist. Positive resist is photo-solubilized. |
| Development | Removal of unpolymerized resist. |
| Etch | Removal of surface layer |
| Photo resist removal | Removal of photo-resist |

**Figure 8.7** Image transfer from a light-field mask with a positive photoresist to create an island.

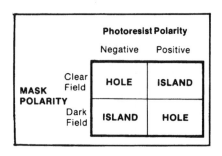

**Figure 8.8** Mask and photoresist polarity results.

## Basic Photoresist Chemistry

Photoresists have been used in the printing industry for over a century. In the 1920s, they found wide application in the printed circuit board industry. The semiconductor industry adapted this technology to wafer fabrication in the 1950s. Negative and positive photoresists designed for semiconductor use were introduced by Eastman Kodak and the Shipley Company, respectively, in the late 1950s.

The photoresist is the heart of the masking process. The preparation, bake, exposure, etch, and removal processes are fine-tuned to ac-

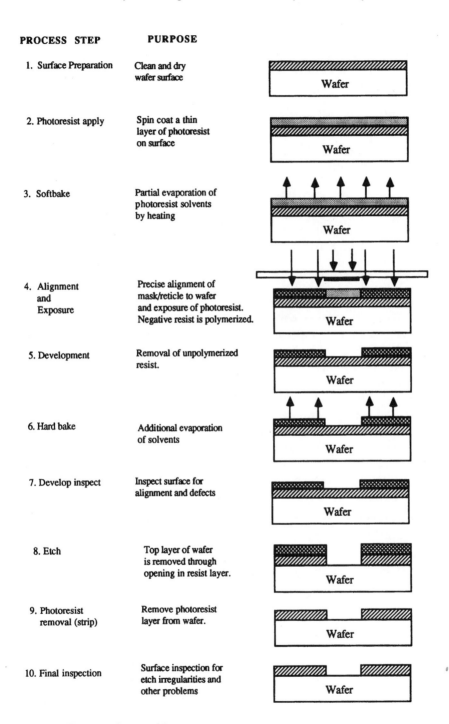

| PROCESS STEP | PURPOSE |
|---|---|
| 1. Surface Preparation | Clean and dry wafer surface |
| 2. Photoresist apply | Spin coat a thin layer of photoresist on surface |
| 3. Softbake | Partial evaporation of photoresist solvents by heating |
| 4. Alignment and Exposure | Precise alignment of mask/reticle to wafer and exposure of photoresist. Negative resist is polymerized. |
| 5. Development | Removal of unpolymerized resist. |
| 6. Hard bake | Additional evaporation of solvents |
| 7. Develop inspect | Inspect surface for alignment and defects |
| 8. Etch | Top layer of wafer is removed through opening in resist layer. |
| 9. Photoresist removal (strip) | Remove photoresist layer from wafer. |
| 10. Final inspection | Surface inspection for etch irregularities and other problems |

**Figure 8.9**  Ten-step photomasking process.

commodate the particular resist used and the desired results. The selection of a resist and development of a resist process is a detailed and lengthy procedure. Once a resist process is established, it is changed very reluctantly.

### Photoresist composition

Photoresists are manufactured for both general and specific applications. They are tuned to respond to specific wavelengths of light and different exposing sources. They are given specific thermal flow characteristics and formulated to adhere to specific surfaces. These properties come about from the type, quantity, and mixing procedures of the particular chemical components in the resist. There are four basic ingredients (Fig. 8.10) in photoresists: polymers, solvents, sensitizes, and additives.

**Light-sensitive and energy-sensitive polymers.** The ingredients that contribute the photosensitive properties to the photoresist are special light- and energy-sensitive polymers. Polymers are groups of large, heavy molecules containing carbon, hydrogen, and oxygen that are formed into a repeated pattern. Plastics are a form of polymers.

The most commonly used resists are designed to react to ultraviolet or laser sources. They are called *optical resists*. Others respond to X-rays or electron beams.

In a negative resist, the polymers change from unpolymerized to polymerized after exposure to a light or energy source. Physically, the polymers form a cross-linked material that is etch resistant (see Fig. 8.11). In most negative resists, the polymers are a polyisopreme,

| Component | Function |
|---|---|
| Polymer | Polymer structure changes from soluble to polymerized (or vice versa) when exposed by the exposure source in an aligner. |
| Solvent | Thins resist to allow application of thin layers by spinning. |
| Sensitizers | Controls and/or modifies chemical reaction of resist during exposure. |
| Additives | Various added chemical to achieve process results, such as dyes. |

**Figure 8.10** Photoresist components.

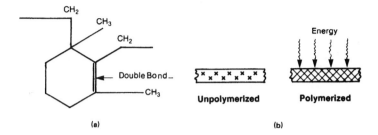

**Figure 8.11**  Negative resist chemistry.

which is a synthetic form of natural rubber. Polymerization also happens when the resist is exposed to heat and/or normal light. To prevent accidental exposure, photomasking areas processing negative resist use yellow filters or yellow lighting.

The basic positive photoresist polymer is the phenol-formaldehyde polymer, also called the phenol-formaldehyde novolak resin (Fig. 8.12). Within the resist, the polymer is relatively insoluble. After exposure to the proper light energy, the resist converts to a more soluble state. This reaction is called *photosolubilization*. The photosolubilized part of the resist can be removed by a solvent in the development process.

Photoresists respond to many forms of energy. The forms are often referred to by their general category (light, heat radiation, and so on), or by a specific portion of the electromagnetic spectrum [ultraviolet light (UV), deep ultraviolet (DUV), I line, and so forth] (see "Exposure sources," p. 233). The exposing energies used are detailed in the section on alignment and exposure. A number of strategies are used to resolve small images (see "Comparison of Positive and Negative Resists," p. 212). One is to use a narrower (or single) wavelength exposing source. The traditional novolak-based positive resist has been

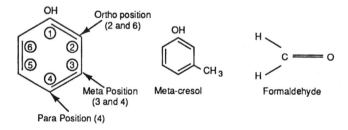

**Figure 8.12**  Phenol-formaldehyde novolak resin structure. *(After: W. S. DeForest,* Photoresist: Materials and Processes, *McGraw-Hill, New York, 1975.)*

fine tuned for use with I-line exposure sources. However, it does not work as well with DUV sources. Resist manufacturers have developed *chemically amplified resists* for this exposure source. Chemically amplified means that the chemical reactions of the polymer are increased by chemical additives. Resists for X-ray and electron beam (e-beam) are based on polymers different from conventional positive and negative resist chemistry.

Figure 8.13 contains a list of resist polymers used in production resists. The relative merits of negative and positive resists are given at the end of this section, and uses of these two types of resists and others are discussed at the end of the process sections.

**Solvents.**  The largest ingredient by volume in a photoresist is the solvent. It is the solvent that makes the resist a liquid and allows the resist to be applied to the wafer surface as a thin layer by spinning. Photoresist is analogous to paint, which is composed of the coloring pigment dissolved in an appropriate solvent. It is the solvent that allows the application of the paint onto a surface in a thin layer. For negative photoresist, the solvent is an aromatic type, xylene. In positive resist, the solvent is either ethoxyethyl acetate or 2-methoxyethyl.

**Sensitizers.**  Chemical sensitizers are added to the resists to cause or control certain reactions of the polymer. In negative resists, the untreated polymer responds (polymerizes) to a certain range of the ultra-

| Resist | Polymer | Polarity | Sensitivity (Coul/cm$^2$) | Exposure Radiation |
|--------|---------|----------|---------------------------|--------------------|
| Positive | Novolak (M–Cresol-formaldehyde) | + | $3-5 \times 10^{-5}$ | UV |
| Negative | Poly Isoprene | – | | UV |
| PMMA | Poly-(Methyl Methacrylate) | + | $5 \times 10^{-5}$ | E-Beam |
| PMIPK | Poly-(Methyl Isopropenyl Ketone | + | $1 \times 10^{-5}$ | E-Beam/ Deep UV |
| PBS | Poly-(Butene-1-Sulfone) | + | $2 \times 10^{-6}$ | E-Beam |
| TFECA | Poly-(Trifluoroethyl Chloroacrylate | + | $8 \times 10^{-7}$ | E-Beam |
| COP (PCA) | Copolymer-($\alpha$–Cyano Ethyl Acrylate–$\alpha$–Amido Ethyl Acrylate) | – | $5 \times 10^{-7}$ | E-Beam X-Ray |
| PMPS | Poly-(2-Methyl Pentene-1-Sulfone) | + | $2 \times 10^{-7}$ | E-Beam |

**Figure 8.13**  Resist comparison table.

violet spectrum. Sensitizers are added to either broaden the response range or narrow it to a specific wavelength. In negative resists, a compound called bis-aryldiazide is added to the polymer to provide the light sensitivity.[1] In positive resists, the sensitizer is o-naphthaquino-nediazide.

**Additives.**    Various additives are mixed with resists to achieve particular results. Some negative resists have dyes intended to absorb and control light rays in the resist film. Positive resists may have chemical *dissolution inhibitor systems.* These are additives that inhibit the dissolution of nonexposed portions of the resist during the development step.

## Photoresist Performance Factors

The selection of a photoresist is a complicated procedure. The primary driving force is the dimensions required on the wafer surface. The resist must first have the capability of producing those dimensions. Beyond that, it must also function as an etch barrier during the etching step, a function that requires a certain thickness for mechanical strength. In the role of etch barrier, it must be free of pinholes, which also requires a certain thickness. In addition, it must adhere to the top wafer surface, or the etched pattern will be distorted, just as a paint stencil will give a sloppy image if it is not taped tightly to the surface. These, along with process latitude and step coverage capabilities, are resist performance factors. In the selection of a resist, the process engineer often must make trade-off decisions between the various performance factors. The photoresist is one part of a complicated system of chemical processes and equipment that must work together to produce the image results and be productive; that is, it must enable an acceptable cost of ownership for the whole patterning process.

### Resolution capability

The smallest opening or space that can be produced in a particular photoresist is generally referred to as its *resolution capability.* The smaller the opening or space produced, the better the resolution capability. Resolution capability for a particular resist is referenced to a particular process, including the exposing source and developing process. Changing the other process parameters will alter the inherent resolution capability of the resist. Generally, smaller line openings are produced with thinner resist film thicknesses. However, a resist layer must be thick enough to function as an etch barrier and be pinhole free. The selection of a resist thickness is a trade-off between these two goals.

The capability of a particular resist relative to resolution and thickness is measured by its *aspect ratio* (Fig. 8.14). The aspect ratio is calculated as the ratio of the resist thickness to the image opening. As the industry requires smaller patterns, the factor of pattern density and shape become an influence on photoresist design. For a given resist, isolated patterns, small contact holes, and high-density pattern areas will all expose and develop differently as a result of reflection and chemical reaction factors. Consequently, there are available resists specifically designed for use in these situations.

Positive resists have a higher aspect ratio as compared to negative resists, which means that, for a given image-size opening, the resist layer can be thicker. The ability of positive resist to resolve a smaller opening is a result of the smaller size of the polymer. It is a little like using a smaller brush to paint a thinner line.

### Adhesion capability

In its role as an etch barrier, a photoresist layer must adhere well to the surface layer to faithfully transfer the resist opening into the layer. Lack of adhesion results in distorted images. Resists differ in their ability to adhere to the various surfaces used in chip fabrication. Within the photomasking process, a number of steps are specifically included to promote the natural adhesion of the resist to the wafer surface. Negative resists generally have a higher adhesion capability than positive resists.

### Photoresist exposure speed, sensitivity, and exposure source

The primary action of a photoresist is a change in structure in response to an exposing light or radiation. An important process factor is the speed at which that reaction takes place. The faster the speed, the faster the wafers can be processed through the masking area. Negative resists typically require 5 to 15 sec of exposure time, whereas positive resists take three to four times longer.

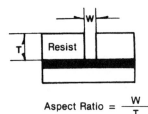

**Figure 8.14**   Aspect ratio.

$$\text{Aspect Ratio} = \frac{W}{T}$$

The sensitivity of a resist relates to the amount of energy required to cause the polymerization or photosolubilization to occur. Furthermore, sensitivity relates to the energy associated with specific wavelength of the exposing source. Understanding this property requires a familiarization with the nature of the electromagnetic spectrum (Fig. 8.15). Within nature, we identify a number of different types of energy: light, short and long radio waves, X-rays, and so on. In reality, they are all electromagnetic energy (or radiation) and are differentiated from each other by their wavelengths, with the shorter wavelength radiation having higher energies.

Common positive and negative photoresists respond to energies in the ultraviolet and deep ultraviolet (DUV) portion of the spectrum (Fig. 8.16). Some are designed to respond to particular wavelength peaks (g, h, i lines in Fig. 8.18) within those ranges. Some resists are designed to work with X-rays or electron beams (e-beams). Resist sensitivity, as a parameter, is measured as the amount of energy required to initiate the basic reaction. The units are millijoules per square centimeter $(mJ/cm^2)$.[2] The specific wavelengths the resist reacts to are called the *spectral response characteristic* of the resist. Figure 8.17 shows the spectral response characteristic of a typical production resist. The peaks in the spectrum are regions (wavelengths) that carry higher amounts of energy (Fig. 8.18). The different light sources used

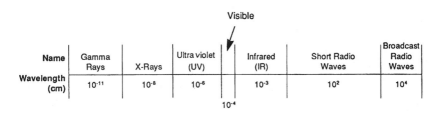

**Figure 8.15**   Electromagnetic spectrum.

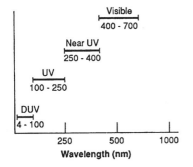

**Figure 8.16**   Ultraviolet and visible spectrum. (*After Elliott.*[1])

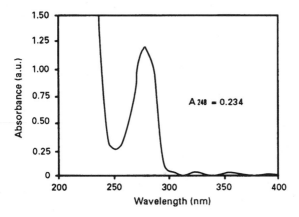

**Figure 8.17** Exposure response curve. *(Source: Shipley Megaposit XP-89131 Photo Resist.)*

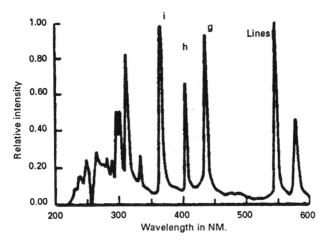

**Figure 8.18** Mercury (Hg) spectrum. *(From* Silicon Processing for the VLSI, *by Wolf and Tauber.)*

in masking areas are covered in the "Alignment and Exposure" section, p. 231.

## Process latitude

While reading the sections on the individual masking process steps, the reader should keep in mind the fact that the goal of the overall process is a faithful reproduction of the required image size in the wafer layer(s). Every step has an influence on the final image size, and each of the steps has inherent process variations. Some resists are more tolerant of these variations; that is, they have a wider process latitude. The wider the process latitude, the higher the probability that the images on the wafer will meet the required dimensional specifications.

### Pinholes

Pinholes are microscopically small voids in the resist layer. They are detrimental, because they allow etchants to seep through the resist layer and etch small holes in the surface layer. Pinholes come from particulate contamination in the environment, the spin process, and from structural voids in the resist layer.

The thinner the resist layer, the more pinholes. Therefore, thicker films have fewer pinholes, but they also make the resolution of small openings more difficult. These two factors present one of the classic trade-offs in determining a process resist thickness. One of the principal advantages of positive resists is their higher aspect ratio, which allows a thicker resist film and a lower pinhole count for a given image size.

### Particle and contamination levels

Resists, like other process chemicals, must meet stringent standards for particle content, sodium and trace metal contaminants, and water content.

### Step coverage

By the time the wafer is ready for the second masking process, the surface has a number of steps. As the wafer proceeds through the fabrication process, the surface gains more layers. For the resist to perform its etch barrier role, it must maintain an adequate thickness over these earlier layer steps. The ability of a resist to cover surface steps with adequate resist is an important parameter.

### Thermal flow

During the masking process, there are two heating steps. The first, called *soft bake,* evaporates solvents from the resist. The second one, *hard bake,* takes place after the image has been developed in the resist layer. The purpose of the hard bake is to increase the adhesion of the resist to the wafer surface. However, the resist, being a plastic-like material, will soften and flow during the hard-bake step. The amount of flow has an important effect on the final image size. The resist has to maintain its shape and structure during the bake, or the process design must account for dimensional changes due to thermal flow.

The goal of the process engineer is to achieve as high a bake temperature as possible to maximize adhesion. This temperature is limited
by the flow characteristics of the resist. In general, the more stable the
thermal flow of the resist, the better it is in the process.

### Comparison of Positive and Negative Resists

Up to the mid-1970s, negative resist was dominant in the masking
process. The advent of VLSI circuits and image sizes in the 2- to 5-μm
range strained the resolution capability of negative resists. Positive
resists had been around for over 20 years, but their poorer adhesion
properties were a drawback, and their superior resolution capability
and pinhole protection were not needed.

In the 1980s, positive resist became the resist of choice. The transition was not easy. Switching to a positive resist requires changing the
polarity of the masks or reticles. Unfortunately, it is not a matter of
simply reversing the fields in the mask-making process. Reticle/mask
dimensions print differently with the two resists (Fig. 8.19). With negative resist and a light-field masks, the dimension in the resist is
smaller than the mask/reticle dimension as a result of light wrapping
(diffraction) around the image. With a positive resist and a dark-field
mask, the diffraction tends to widen the image. These changes must
be considered in making the mask/reticle and designing the other
masking processes. In other words, an entirely new process is required
to switch resist types.

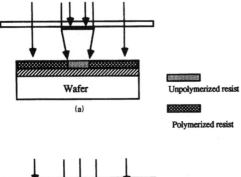

Wafer

Unpolymerized resist

(a)

Polymerized resist

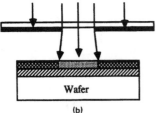

Wafer

(b)

**Figure 8.19** Changes in image
size with (a) image size reduction with light-field mask and
negative resist, and (b) image
size enlargement with dark-field
mask and positive photoresist.

Most of the images on most of the mask layers are holes. With positive resist, the mask polarity is dark field, which results in additional pinhole protection for the wafer (Fig. 8.20). Clear-field masks are prone to small cracks in the glass surface. These cracks, called *glass damage,* block the exposing light, creating unwanted holes in the photoresist layer, which in turn etch into the wafer surface as holes. The same is true for dirt particles that locate on the clear area of the mask/ reticle. On dark-field masks, the majority of the surface is covered by chrome, which is hard and less likely to have pinholes. Thus, the wafer has fewer unwanted pinholes.

Another problem with negative resists is oxygenation. This is a reaction of the resist to oxygen in the atmosphere, and it can cause a thinning of the resist film by as much as 20 percent. Positive resists do not have this property. Positive resists are more expensive than negative resists but, for a demanding process, the higher cost is offset by higher yield. For very small image sizes, positive resist is the only choice.

Developing characteristics differ between the two types of resists. Negative resists develop in readily available solvents and feature a high solubility differential between the polymerized and unpolymerized areas. The image dimensions remain relatively constant during the development step. Positive resists have a lower solubility between the polymerized and unpolymerized areas requiring carefully prepared developer solutions and require temperature control of the pro-

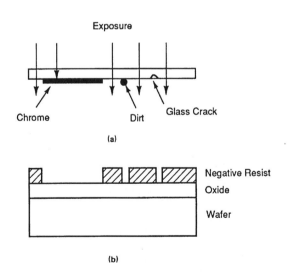

**Figure 8.20** (*a*) Clear-field mask with dirt particle and glass crack, and (*b*) result in negative resist after develop.

cess. Dissolution inhibitor systems are added specifically to control dimensions during development.

Photoresist removal is also a factor. Generally, the removal of positive resists is easier and takes place in chemicals that are more environmentally sound. Fabrication areas producing devices and circuits with image sizes greater than 2 μm may still use negative resists. Figure 8.21 shows a comparison of properties of the two resists.

## Physical Properties of Photoresists

The performance factors just detailed and all of the ten basic process steps are related to a number of physical and chemical properties of the resist. These properties are rigorously controlled by resist manufacturers.

### Solids content

A photoresist is a liquid that is applied to the wafer by a spinning technique. The thickness of resist left on the wafer is a function of the spin step parameters and several resist properties: *solids content* and *viscosity*.

Recall that the photoresist is a suspension of polymers, sensitizers, and additives in a solvent. Different resists will contain different amounts of these solids. The amount is referred to as the solids content of the resist and is expressed as the weight percent in the resist. Solids contents are in the 20 to 40 percent range.[3]

### Viscosity

Viscosity is the quantitative measure of liquid flow. High-viscosity liquids, such as tractor oils, flow in a sluggish manner. Low-viscosity liq-

| Parameter | Negative | Positive |
|---|---|---|
| Aspect Ratio (Resolution) | | Higher |
| Adhesion | Better | |
| Exposure Speed | Faster | |
| Pinhole Count | | Lower |
| Step Coverage | | Better |
| Cost | | Higher |
| Developers | Organic Solvents | Aqueous |
| Strippers | | |
|    Oxide Steps | Acid | Acid |
|    Metal Steps | Chlorinated Solvent Compounds | Simple Solvents |

Figure 8.21   Comparison of negative and positive results.

uids, such as water, flow more readily. In both cases, the mechanism of flow is the same. The molecules in the liquid roll over each other as the liquid is being poured. As the molecules roll about, there exists an attraction between them that acts as an internal friction. Viscosity is the measurement of that friction.

Viscosity is measured by several techniques. Most photoresist manufacturers measure viscosity with a rotating vane in the resist. The higher the viscosity, the more force required to move the vane through the liquid at a constant speed. The rotating-vane apparatus illustrates the force-related character of viscosity.

The unit of viscosity is the *centipoise* (one-hundredth of a poise). It is named after the French scientist Poisseulle, who investigated the viscous flow of liquids. One poise is equal to one dyne second per centimeter. The viscosity unit of centipoise is more correctly named the absolute viscosity.

Another unit, called the *kinematic viscosity*, is the centistoke. This value is calculated from the absolute viscosity (centipoise) divided by the density of the resist. Viscosity varies with temperature; therefore, its specified value is stated at a particular temperature, usually 25°C. Viscosity is a major parameter determining the resist thickness during the spin process. Viscosity is closely related to the solids content. The higher the solids content, the higher the viscosity.

## Surface tension

The surface tension of a resist also influences the outcome at spin. Surface tension is a measure of the attractive forces in the surface of the liquid (Fig. 8.22). Liquids with high surface tension flow less readily on a flat surface. It is the surface tension that draws a liquid into a spherical shape on a surface or in a tube.

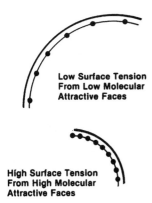

Low Surface Tension
From Low Molecular
Attractive Faces

High Surface Tension
From High Molecular
Attractive Faces

Figure 8.22  Surface tension.

### Index of refraction

The optical properties of the resist play a role in the exposure mechanism. One property is *refraction* or the bending of light as it passes through a transparent or semitransparent medium. Refraction is the same phenomenon that makes an object in a pool of water appear at a different location. This comes about as the light ray is slowed up by the material. The *index of refraction* is a measurement of the speed of light in a material compared to its speed in air, as shown in Fig. 8.23. It is calculated as the ratio of the reflecting angle to the impinging angle. For photoresists, the index of refraction is close to that of glass, approximately 1.45.

### Storage and control of photoresists

Photoresists are delicate high-technology mixtures. Great care and precision go into their manufacture. Once a photomasking process is developed, its continuing success depends on the day-in, day-out control of the process parameters and a consistent photoresist product. Delivered batch-to-batch consistency is a responsibility of the manufacturer. Maintaining that consistency is the responsibility of the user. Several properties of resists dictate the required storage and control conditions.

### Light and heat sensitivity

Both light and heat can activate the sensitive mechanisms in the resist intended to be activated in the closely controlled masking process

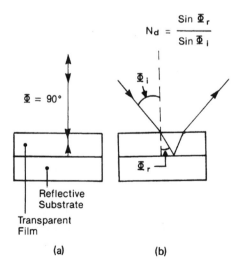

$$N_d = \frac{\text{Sin } \Phi_r}{\text{Sin } \Phi_i}$$

$\Phi = 90°$

$\Phi_i$

$\Phi_r$

Reflective
Substrate

Transparent
Film

(a)                    (b)

**Figure 8.23**  Index of refraction. (*a*) 90° incident light and (*b*) angled light refracted in the transparent film.

steps. It is imperative that the resist be protected during storage and handling to prevent unwanted reactions that would interfere with the process results. This is the reason why masking areas use yellow light. It is also the reason why resist bottles are brown. The colored glass protects the resist from stray light. Proper transportation and storage of resist require temperature control within the limits specified by the manufacturer.

### Viscosity sensitivity

Viscosity control is essential for good film thickness control. To maintain the photoresist's viscosity, resist bottles must remain capped prior to use. Opened bottles will allow evaporation of the solvent, which results in a higher solids content, which in turn results in a higher viscosity. If photoresist is dispensed from plastic tubing, the material should be tested to ensure that the resist is not leaching plasticizers out of the material. The plasticizers will increase the viscosity of the resist.

Resist is also available in sealed vacuum pouches. The resist is protected during shipping and storage. During use, the pouch continues to collapse as the resist is dispensed, preventing air from reaching the resist surface.

### Shelf life

A container of photoresist comes with a recommended shelf life. The problem again has to do with the self-polymerization or photosolubilization of the resist. In time, changes to the polymer will take place, altering the resist performance when it reaches the production line.

### Cleanliness

Needless to say, any and all equipment used to dispense photoresist must be maintained in the cleanest condition possible. Besides the effects of particular contamination from the system, the resist tubing must be cleaned regularly because of the possible buildup of dried photoresist. Cleaning agents should be checked for their compatibility with the resist. For example, TCE should not be used with negative resist, because it can cause bubbles in the resist.

### Photomasking Processes

In the following sections, each of the ten basic masking steps is examined. Presented is the purpose of the step, technical considerations and challenges, options, and process control methods.

## Surface Preparation

Throughout this text, various analogies will be used to aid the reader in understanding the complicated processes. A good analogy to the photomasking process is painting. Even the amateur painter soon learns that, to end up with a smooth film of paint that adheres well, the surface must be dry and clean. The same is true in photomasking technology. Ensuring that the resist will stick to the wafer surface requires surface preparation. This step is performed in three stages: particle removal, dehydration, and priming.

### Particle removal

The wafers coming to the photomasking area almost always come in a clean condition from another process such as oxidation, doping, or CVD. However, during storage, loading, and unloading into carriers, they may pick up some particulate contamination that must be removed from the surface. Depending on the level of contamination and/or the process requirements, several particulate removal techniques may be used (Fig. 8.24). In extreme cases, the wafers may be put through a wet chemical cleaning similar to the preoxidation cleaning processes, including an acid cleaning, water rinsing, and drying. The particular acid used must be compatible with the top layer on the wafer surface. Particle removal methods are the same as described in Chapter 7: manual blow-off, mechanical scrubbers, and/or high-pressure water spray.

### Dehydration baking

It has been mentioned that the wafer surface has to be dry to promote adhesion. A dry surface is called *hydrophobic,* a chemical condition. Liquids form into small droplets on a hydrophobic surface, such as water beads on a newly waxed car. A hydrophobic surface is conducive to good photoresist adhesion (Fig. 8.25). Wafers coming to the masking operation usually have a hydrophobic surface.

Unfortunately, when the wafer is exposed to moisture, either from the air or from post-cleaning rinses, the surface condition changes to a *hydrophilic* one. This condition is evidenced by a liquid on the surface spreading out in a wide puddle, such as water on a nonwaxed car sur-

- High-pressure nitrogen blowoff
- Wet chemical cleaning
- Rotating brush scrubber
- High-pressure water stream

**Figure 8.24** Prespin wafer cleaning methods.

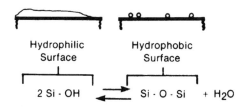

Hydrophilic Surface

Hydrophobic Surface

Figure 8.25  Hydrophilic versus hydrophobic surfaces.

$$2\ Si \cdot OH \rightleftharpoons Si \cdot O \cdot Si\ +\ H_2O$$

face. A hydrophilic surface is also said to be *hydrated*. Resist does not adhere very well to hydrated surfaces (Fig. 8.25).

Two important ways to maintain a hydrophobic surface are to keep the room humidity below 50 percent and to coat the wafers with photoresist as quickly as possible after being received from a previous process. Storage of the wafers in desiccators purged with dry, filtered nitrogen or in the dry mini-environment of a SMIF box (see Chapter 4). Additional steps may be taken to establish a wafer surface with acceptable adhesion properties. These steps include a dehydration bake and priming with a chemical.

A heating operation may be used to reset the wafer surface to a dehydrated condition. Dehydration bakes take place in three temperature ranges to address three different dehydration mechanisms. In the range of 150 to 200°C (low temperature), surface water is evaporated. At 400°C (medium temperature), water molecules loosely attached to the surface will leave. At temperatures above 750°C (high temperature), the surface is chemically restored to a hydrated condition.

In most masking processes, only low-temperature dehydration bake temperatures are used. This is because this temperature range is easily obtainable with hot plates and chest-type convection and vacuum ovens. Another advantage of low-temperature dehydration is that the wafers do not have to wait for a cool-down before the spin process. Systems to perform this step can easily be integrated into a spin-bake system, making them dehydration-spin-bake systems. An explanation of these heating systems is provided in "Soft Bake," p. 227.

High-temperature dehydration bakes are rare. One reason is that reaching a temperature of 750°C usually requires the use of a tube furnace, and tube furnaces are large and cannot be integrated into the spin-bake processes. A second reason is the temperature level itself. At 750°C, doped junctions in the wafer can move (which is undesirable), and mobile ionic contaminants on the surface can move into the wafer causing reliability and performance problems.

**Wafer priming**

In addition to dehydration baking, the wafers may go through a chemical priming step to ensure good adhesion of the resist. In painting, primers are a subcoat selected for their ability to adhere to the surface and provide a good surface to which the paint will stick. In semiconductor photomasking, the primer effect is similar. The primer chemically ties up molecular water on the wafer surface, thereby increasing its adhesion property.[4]

A number of chemicals provide priming capabilities, but one, hexamethyldisilazane (HMDS), is in universal use. The use of HMDS is described in U.S. Patent 3,549,368 by R. H. Collins and F. T. Deverse (1970) of IBM. The HMDS is mixed with xylene in solutions of from 10 to 100 percent. The exact mixture is determined by the particular surfaces and environmental factors in the cleanroom. Unlike a painting primer, a thickness of only several molecules is sufficient to provide the necessary adhesion promotion.

### Spin priming

In most photomasking areas, liquid primers are applied to the wafer while it is on the resist spinner chuck (Fig. 8.26). The dispensing of the HMDS can be manual from a syringe or automatic. Automatic spinners (see next section) have a separate system to dispense HMDS onto the wafer surface just prior to the application of the resist. After the spinner dispenses the primer onto the rotating wafer, the chuck is ramped to a higher speed to dry the HMDS layer. A major production advantage of spin priming is that it takes place in-line with the spin step.

### Vapor priming

Both immersion and spin priming require the direct contact of the HMDS liquid with the wafer surface. Whenever a liquid is in contact with the wafer, there is the danger of contamination from the liquid. Another consideration is that the HMDS must be dry before the resist is applied. Wet HMDS can dissolve the bottom layer of the resist and interfere with the exposure, development, and etching. Lastly, HMDS

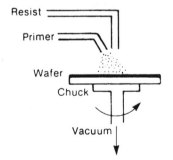

**Figure 8.26** Spin dispense of primer.

is relatively expensive, and in spin priming, an excess of HMDS is sprayed on the wafer to ensure adequate coverage. The excess is thrown off the wafer and discarded.

The preceding considerations are overcome by vapor priming techniques. Vapor priming is practiced in three forms. Two are performed at atmospheric pressure and one in a vacuum (Fig. 8.27). One atmospheric system employs a bubbler chamber connected to a desiccator-type chamber. Nitrogen is bubbled through the HMDS and carried into the chamber, where it coats the wafers. Another method employs a vapor degreaser with a reserve of liquid HMDS. The HMDS is heated to the vapor point and the wafers are suspended in the vapors for coating.

The third vapor technique is vacuum vapor priming, which uses a sealed flask of HMDS connected to a vacuum oven or single-wafer chamber. The wafers are first heated in the oven in a nitrogen atmosphere. After a temperature of about 150°C is reached, the atmosphere is switched to a vacuum. Once the vacuum level is reached, a valve is opened, and HMDS vapors are drawn into the chamber by the low pressure. Within the chamber, the wafers become completely coated as the vapors fill the entire chamber. This method has shown good adhesion longevity, even in the presence of high humidity.

Vacuum vapor priming offers the additional advantage of a combined dehydration bake and prime step and a significant reduction in HMDS usage. Vacuum vapor priming practiced in a chest-type oven adds an additional step to the process. Many automatic spinner systems incorporate in-line vacuum vapor primers.

## Photoresist Application (Spinning)

The purpose of the photoresist application step is the establishment of a thin, uniform, defect-free film of photoresist on the wafer surface.

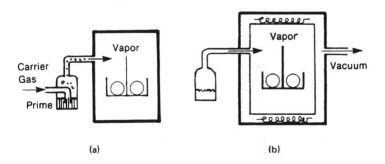

**Figure 8.27**   Vapor prime methods. (*a*) Atmospheric and (*b*) vacuum bake-vapor prime.

These goals are easy to state, but they require sophisticated equipment and stringent controls to achieve. A typical resist layer varies from 0.5 to 1.5 µm in thickness and has to have a uniformity of only ±0.01 µm (100 Å). This variation is 1 percent of a 1.0-µm thickness.

The usual methods of applying thin layers of liquids to surfaces are brushing, rolling, and dipping. None of these methods is adequate to achieve the quality resist film necessary for photomasking. The method used is spinning, which was briefly described in the section on priming. Spinners are built in manual, semiautomatic, and automatic designs. The systems differ in the degree of automation and are described in the following text. However, the deposit of the film on the wafer is common to each of the systems.

Spin processes are designed to prevent or minimize the buildup of a bead of resist around the outer edge of the wafer. Called an *edge bead,* the buildup causes image distortions during the exposure and etch processes.

### The static dispense spin process

Application of the resist occurs immediately after the priming process. The wafer is placed on a vacuum chuck and several cubic centimeters ($cm^3$) of the photoresist is deposited in the center of the wafer (Fig. 8.28) and allowed to spread out into a puddle. The puddle is allowed to spread until it covers the majority of the wafer surface. The amount of resist deposited in the puddle is critical only in the extremes. Too small an amount will result in incomplete resist coverage, and too much will cause a buildup of a resist rim or result in resist on the back of the wafer (Fig. 8.29).

When the puddle reaches its specified diameter, the chuck is rapidly accelerated to a predetermined speed. During the acceleration, centrifugal forces spread the resist to the wafer edge and throw off excess resist, leaving a thin uniform layer on the wafer. The high-speed spin

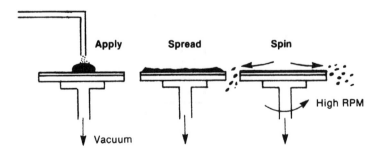

**Figure 8.28**  Static spin process.

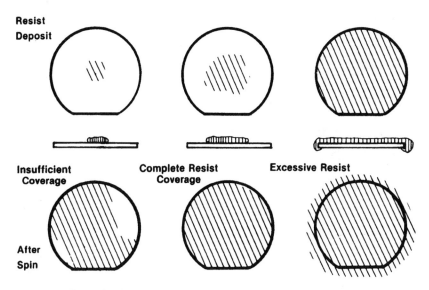

**Figure 8.29**   Example of resist coverage.

continues for some time after the resist is spread to allow drying of the resist.

The final thickness of the film is established as the result of the resist viscosity, the spin speed, the surface tension, and the drying characteristics of the resist. In practice, surface tension and the drying characteristics are properties of the resist, and the viscosity-spin speed relationship is determined from curves supplied by the resist manufacturer or established for the particular spin system used (Fig. 8.30).

Although spin speed is specified to control resist thickness, it is actually the acceleration rate that establishes the final resist thickness. The acceleration characteristic of the spinner must be specified and maintained as a constant in the spin process.

### Dynamic dispense

The need for uniform resist films on larger-diameter wafers led to the development of the dynamic spin dispensing technique in the 1970s (Fig. 8.31). For this technique, the wafer is rotated at a low speed of approximately 500 rpm. While the wafer is rotating, the resist is dispensed onto the surface. The action of the rotation assists in the initial spreading of the resist. Less resist is used, and a more uniform layer is achieved. After spreading of the resist, the spinner is accelerated to a high speed to complete the spread and thin the resist into a uniform film.

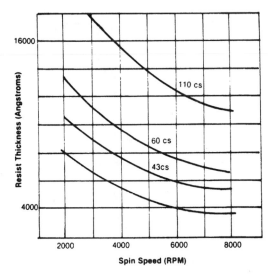

Figure 8.30  Resist speed thickness versus spin speed. *(Courtesy of KTI Chemicals.)*

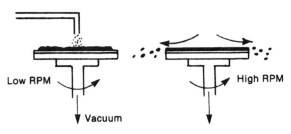

Figure 8.31  Dynamic spin dispense.

## Moving-arm dispensing

An improvement on the dynamic dispense technique is the addition of a moving-arm resist dispenser (Fig. 8.32). The arm moves in a slow motion from the center of the wafer toward its edge. This action creates more uniform initial and final layers. A moving-arm dispenser also saves resist material, especially for larger-diameter wafers.

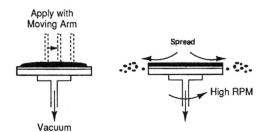

Figure 8.32  Moving-arm dispensing.

## Manual spinners

A manual spinner is a simple machine consisting of from one to four vacuum chucks (called *heads*), a motor, a tachometer, and a connection for a vacuum source. Each head is surrounded by a catch cup, which serves to collect the excess resist and direct it to a collection vessel. The catch cup also prevents "balls" of resist thrown off the wafers during the acceleration from landing on adjacent wafers. The process starts with the removal of particles from the wafer with a filtered nitrogen gun. The wafer(s) are mounted on the head with tweezers or a vacuum wand, and the chuck vacuum is turned on. Next, the HMDS is dispensed onto the surface from a syringe or squeeze bottle, and the wafer is spun and dried. Finally, the resist puddle is dispensed from another syringe or squeeze bottle. Most spin processes performed on manual spinners use the static dispense method.

## Edge bead removal

A consequence of the high-speed spin can be a buildup of resist around the edge of the wafer, called an *edge bead*. It is removed by a solvent spray directed on the front and back of the wafer, near the edge.

## Automatic spinners

A semiautomatic spinner adds automation to the resist blow-off, resist dispense, and spinning cycles. The nitrogen blow-off is accomplished from a separate tube over the vacuum chuck that is connected to a pressurized nitrogen source (Fig. 8.33). Also in the dispense chamber are a primer resist tube and a resist dispense tube. The resist tube is fed resist from either a nitrogen pressurized vessel or through a diaphragm-type pump. In general, the industry has moved away from pressurized, resist dispense systems because of problems that arise from the absorption of nitrogen into the resist. The nitrogen comes out of the resist after the dispense cycle, causing voids in the film. Diaphragm pumps eliminate this problem.

Automatic resist dispensers have a negative-pressure capability that automatically draws the resist back up into the dispenser tube after each dispensing operation. This *drawback* (or *suckback*) minimizes the exposed surface of the resist in the tube from drying into a hard ball that can be deposited on the wafer (Fig. 8.34). In fully automatic systems, all of the events of the spin process are controlled by microprocessors. The systems have mechanisms to extract the wafers from the carriers, place them on the chucks, perform the priming step, dispense the photoresist, remove the edge bead, perform a soft bake, and place the wafers back in their carriers. The standard system configu-

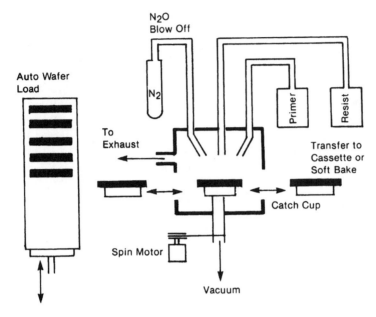

**Figure 8.33** Automated spinner diagram.

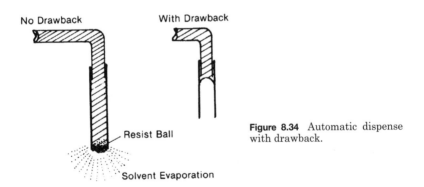

**Figure 8.34** Automatic dispense with drawback.

ration is in a line, referred to as a *track*. Production-level spinners will have from two to four side-by-side tracks.

### Backside coating

In some device processes, it is required that the oxide on the back of the wafer remain in place through the masking process. One way this is accomplished is by coating the back of the wafer with photoresist. The requirement for this backside coating is simply a thick enough layer to survive the etch process.

## Soft Bake

The soft bake step is a heating operation with the purpose of evaporating a portion of the solvents in the photoresist. After the bake, the resist film is still "soft," as opposed to being baked to a varnish-like finish. The solvents are evaporated for two reasons. Remember that the principal role of the solvents is to allow the application of the resist in a thin layer on the wafer. After that role is fulfilled, the presence of the solvent can interfere with the rest of the processing. The first interference occurs during the exposure step. The solvent(s) in the resist can absorb exposing radiation, thus interfering with the proper chemical change in the photosensitive polymers. The second problem is with the resist adhesion. Using the painting analogy, we know that complete drying (evaporation of the solvent) is necessary for good adhesion.

The principle soft-bake parameters are time and temperature. Two major goals of the patterning process are correct image definition and the adhesion of the resist to the wafer during the etch step. Both of these goals are influenced by soft bake. In the extreme, an underbaking will result in incomplete image formation at exposure and excessive lifting (poor adhesion) at the etch step. Overbaking will cause the polymers in the resist to polymerize and not react to exposing radiation.

Temperature and time ranges for the soft bake are provided by the resist manufacturer and are fine-tuned by the masking engineer. Negative photoresists have to be baked in a nitrogen atmosphere, whereas positive resists can be baked in air. A number of different methods are used to accomplish the soft-bake step, and various pieces of equipment incorporating all three heat transfer methods are used.

The three methods of heat transfer are conduction, convection, and radiation. Conduction is the transfer of heat by direct physical contact of the object with a heated surface. Hot plates heat by conduction. In the conduction process, the vibrating atoms of the hotter surface cause the object atoms to vibrate also. As they vibrate and collide, the atoms become hotter.

Some dehydration baking is in convection ovens. Systems using convection heating include home forced-air furnaces, hair dryers, air- and nitrogen-fed ovens, and oxidation furnaces. In these systems, a unit heats the gas and a blower or pressure pushes the gas to a space where it, in turn, heats the object.

The third method is radiation. The term *radiation* describes the travel of electromagnetic energy waves through space. Radiation waves travel in vacuums as well as through gases. The sun transfers heat to the Earth by radiation. Heating lamps also transfer heat by radiation. Radiation is the heating method used by RTP systems. When the radiation strikes an object, the energy carried by the wave is transferred directly to the atoms of the object.

### Convection ovens

The mainstay baking oven for nonautomated fabrication lines is the convection oven (Fig. 8.35). It is a stainless steel chamber in an insulated enclosure. Either nitrogen or air is supplied by ducts surrounding the chamber and passed through a heater before being directed into the chamber. The inside of the chamber is fitted with racks for the wafer carriers. The carriers stay inside the oven for a predetermined time while the heated gas brings them up to temperature. Convection ovens used for VLSI applications have proportional band controllers and HEPA filters to maintain a clean baking environment.

There are several drawbacks to convection ovens for soft baking. One is batch-to-batch temperature variation, which arises from the amount of time the door is open for loading, the size of the load, and the variable time for all parts of the oven to reach a constant temperature. A process problem associated with convection heating is the tendency of the top layer of resist to "crust," trapping solvents in the resist (Fig. 8.36).

### Manual hot plates

In manual and laboratory operations, a simple hot plate can be used for the soft bake. The wafers are placed on an aluminum holder, which

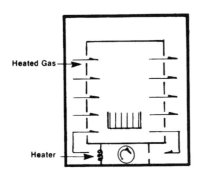

Figure 8.35  Convection oven.

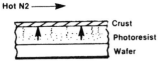

Figure 8.36  Crusting effect of ovens.

has a dial thermometer set in a hole drilled into it. The wafers are put on the holder, which is set on the hot plate (Fig. 8.37). The operator monitors the rise in temperature on the thermometer and removes the holder when the proper temperature is reached. With a well controlled hot plate, an effective soft bake can be achieved. A process advantage of hot-plate heating is that the bottom of the wafer is heated first. This allows the solvents to escape through the top surface and also minimizes "crusting" of the surface. A hot-plate process is operator dependent and has low productivity.

### In-line, single-wafer hot plates

The backside advantage of hot-plate heating can be gained in track systems when a single-wafer hot plate is built into the system. Wafers leaving the spinner are positioned on the hot plate and clamped to it with a vacuum. The wafer and resist are heated for a predetermined time, the vacuum released, and the wafer transferred to a carrier. These in-line systems are connected to the facility exhaust system for the removal of the solvent vapors.

### Moving-belt hot plates

A constraint on the productivity of single-wafer hot plates in an integrated system is the total time of the spin step. A typical spin time is

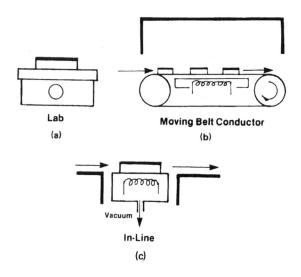

**Figure 8.37**  (*a*) Manual hot plate, (*b*) in-line continuous hot plate, and (*c*) in-line single-wafer hot plate.

25 to 40 sec, which means the soft bake would have to be completed in that amount of time to keep the wafers moving in a continuous flow. For some resists and for some processes, this is too short a time. A way around the problem is a moving-belt hot plate. The wafers are placed on a heated, moving steel belt, and the temperature and belt speed are set to meet the soft bake requirements and process the wafers in a continuous flow.

### Moving-belt infrared ovens

The desire for fast, uniform, and noncrusting soft bake methods led to the development of infrared (IR) radiation sources (Fig. 8.38). Infrared baking is much faster than conduction baking and heats from the "inside out." Inside-out baking is the principle of conduction hot plate baking. The infrared waves pass through the resist layer without heating it, much like sunlight will pass through a window without heating it. The wafer, however, absorbs the energy, gets hot, and in turn heats the resist layer from the bottom.

### Microwave baking

Microwaves as a soft bake heating source have the advantage of infrared heating but at a much faster rate because of the higher energy carried in a microwave. Soft-bake temperatures can be well under 1 min. This brief time lends itself to on-chuck soft baking. Immediately after spinning, a microwave source is directed at the wafer, completing the soft bake (Fig. 8.39).

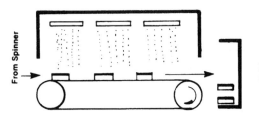

**Figure 8.38**  Moving-belt infrared (IR) heating.

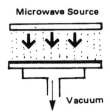

**Figure 8.39**  Microwave heating.

## Vacuum baking

Vacuum ovens offer several advantages for a number of process steps. A vacuum oven is configured similarly to a convection oven, but is fitted to a vacuum source. Vacuum is particularly efficient for evaporation processes, because the reduced pressure aids the evaporation of the solvents, reducing the reliance on the temperature. However, for soft baking, the wafers must be heated to a uniform temperature. A problem arises because heating in a vacuum oven is by radiation from the heated chamber walls to the wafers. This heat transfer method is sometimes called *line-of-sight* because, for uniformity, each wafer must have a clear line of sight to the heat source. In a chest-type vacuum oven packed with carriers of vertically held wafers, this condition cannot be met. The result is poor temperature uniformity in most vacuum ovens.

The benefits of vacuum and uniform hot-plate heating can be achieved with in-line, single-wafer systems. Figure 8.40 is a table summarizing the different soft bake methods.

## Alignment and Exposure

Alignment and exposure (A&E) is, as the name implies, is one of the basic ten patterning steps, with two separate actions. The first part of the A&E step is the positioning or alignment of the required image on the correct location on the wafer surface. The second part is the encoding of the image in the photoresist layer from an exposing light or other radiation source. Correct alignment of the image patterns and establishment of the precise image dimensions in the resist are absolute requirements for the functioning of the devices and circuits. Furthermore, the wafers spend 60 percent of the process time in the lithography area.

Consider the challenge of building a typical ULSI megabit DRAM circuit. The SIA ITRS projects that by 2007 the functions in this device will be 17 Gb, in a chip size of 568 mm². The individual parts of the devices will have feature sizes in the submicron range and have to "fit" or overlay within or next to each other within tolerances of one-third the nominal feature size. This tolerance has to be maintained as each of the level patterns is exposed on the wafer. The SIA *International Technology Roadmap for Semiconductors* projects that, by 2007, advanced ICs will require 24 to 27 patterning steps. The allowable variation in wafer placement must be maintained over all of the pattern alignments.

### Aligner system performance capabilities

An aligner system is composed of two major subsystems. One correctly positions the pattern(s) on the wafer surface. The different aligner

| Method | Bake Time Min. | Temperature Control | Productivity Type | Rate Waf/Hr. | Queuing |
|---|---|---|---|---|---|
| Hot Plate | 5-15 | Good | single to small batch | 60 | Yes |
| Convection Oven | 30 | Average - Good | Batch | 400 | Yes |
| Vacuum Oven | 30 | Poor - Average | Batch | 200 | Yes |
| I.R. Moving Belt | 5-7 | Poor - Average | Single | 90 | No |
| Conductive Moving Belt | 5-7 | Average | Single | 90 | No |
| Microwave | 0.25 | Poor Average | Single | 60 | No |

Figure 8.40   Soft-bake chart.

types have different alignment subsystems, which are explained in the equipment sections. The second part is the exposure subsystem. This subsystem includes an exposure source and a mechanism for directing the exposing radiation rays to the wafer surface.

Aligners are selected and compared by several criteria (Fig. 8.41) that relate to their ability to produce the required images in a consistent and productive manner. Perhaps the most important parameter is the *resolution capability,* or the ability of the machine to produce a particular size image. The higher the resolution capability, the better the machine. In addition to the resolution of the required image size, the aligner must be capable of placing the images in their correct position relative to each other. This performance parameter is called the *registration capability* of the aligner. These two factors must be performed over the entire wafer, a factor called *dimensional control.* The final performance factor is cost of ownership, which includes initial purchase cost, *wafer throughput* (the time required to load, align, expose, and unload the wafer), maintenance cost, and the uptime of the machine. These factors are discussed in Chapter 15.

### Alignment and exposure systems

Up to the mid-1970s, the photoresist engineer had the choice of only two A&E systems: contact and proximity aligners. Today, the choice has expanded to include both optical and nonoptical aligners (Fig. 8.42), such as projection and steppers. Optical aligners use an ultraviolet light source, whereas nonoptical systems use exposure sources from other parts of the electromagnetic spectrum. The systems in use today were developed to keep pace with the reduction of feature size, increased circuit density, and required defect reductions of the ULSI era.

### Exposure sources

While aligners are very sophisticated machines, they operate on several basic optical principles. Consider producing a shadow image of a

| Alignment Selection Criteria |
| --- |
| • Resolution Capability/ Limit |
| • Alignment Accuracy |
| • Contamination Level |
| • Reliability |
| • Productivity |
| • Overall cost of ownership (COO) |

**Figure 8.41**  Aligner selection criteria.

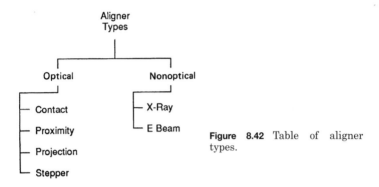

**Figure 8.42** Table of aligner types.

fork on a wall by shining a flashlight through a real fork held near the wall. By semiconductor standards, the image of the fork is quite imprecise. But there are ways to improve the image. One way would be to replace the flashlight with a narrower-wavelength light. The white light that comes from a flashlight contains many different wavelengths (colors), all mixed. At the edge of the fork, a phenomenon called *diffraction* takes place. Diffraction is the bending of the light rays at an opaque edge (or going through a narrow slit). The amount of bending depends on the wavelength, and, with many wavelengths (white light), there are many rays scattering from the edge and fuzzing up the image. Using a shorter-wavelength light, or a single-wavelength source, minimizes the diffraction. Another way to improve the image is to get all of the rays traveling along the same path. In normal white light, the rays are coming from the light bulb in many directions and leave the edge of the fork with many directions, again fuzzing the image. With mirrors and lenses, the light rays can be *collimated* into a band of parallel rays, thus improving image quality. The sharpness of the image and its dimensions are also influenced by the distance of the light behind the fork and the distance of the fork to the wall. Closing up both of these distances sharpens the image. These techniques (narrower or single wavelength exposure sources, collimated light, and strict control of the distances) are all used in aligners to produce the required images.

Exposure sources are chosen to create the required image size in conjunction with a specific photoresist (see "Photoresist exposure speed, sensitivity, and exposure source," p. 208). The workhorse exposure source has been the high-pressure mercury lamp. It produces light in the ultraviolet (UV) range. The reduction of feature size has driven the development of improvements to the basic lamp and photoresists. To achieve greater definition, resists are tailored to respond to only narrow bands (lines) in the mercury lamp spectrum. This need also has resulted in the development of lamps and resists that operate in the shorter wavelengths of the spectrum. This region of the spectrum is known as the *deep ultraviolet* or DUV.

Other approaches to lower wavelength and higher-energy exposure sources are excimer lasers, X-rays, and electron beams. A more detailed discussion of the various exposure sources is provided in Chapter 10.

### Alignment criteria

The first mask is aligned by positioning the $y$ axis of the mask at a 90° angle to the major wafer flat on the wafer (Fig. 8.43). Subsequent masks are aligned to a previously patterned mask with the use of alignment marks (also called *targets*). These are special patterns (Fig. 8.44) located in an easily found position on the edge of each chip pattern or in the separation lines surrounding each chip on the wafer.

Alignment is accomplished by the operator positioning a mark on the mask to a corresponding mark in the wafer pattern. For automatic systems (see "Steppers," p. 238), the alignment marks serve the same purpose. Alignment marks become a permanent part of the chip surface through the etching process. They are then in place for the alignment of the next layer.

Alignment errors, called *misalignment,* fall into several categories (Fig. 8.45). A common one is a simple misplacement in the $x$-$y$ directions. Another common misalignment is rotational, where one side of

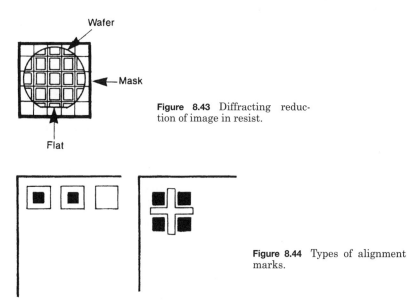

Figure 8.43  Diffracting reduction of image in resist.

Figure 8.44  Types of alignment marks.

(a)

(b)

(c)

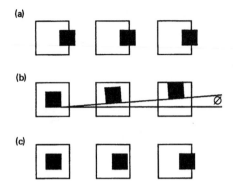

**Figure 8.45** Misalignment types. (*a*) *x* direction, (*b*) rotational, and (*c*) run-out.

the wafer is aligned, but the patterns become increasingly misaligned across the wafer. A third rotational misalignment comes about when the die pattern is rotated on the mask or reticle.

Other misalignment problems associated with masks and stepper aligners are run-out and run-in. These problems arise when the chip patterns are not formed on the mask on constant centers or are placed on the chip off center. The result is that only a portion of the mask chip patterns can be properly aligned to the wafer patterns. The pattern becomes progressively misaligned across the wafer.

A rule of thumb is that circuits with micron or submicron feature sizes must meet registration tolerances of one-third the minimum feature. An *overlay budget* is calculated for the total circuit. It is the allowable accumulated alignment error for the entire mask set (see Fig. 8.2). For a 0.35-μm product, the allowable overlay budget is about 0.1 μm.[5]

### Aligner types

**Contact aligners.** Until the mid-1970s, the contact aligner was the workhorse aligner of the semiconductor industry. The alignment part of the system uses a full wafer-size photomask positioned over a vacuum wafer chuck. The wafer is mounted on the chuck and viewed through a split-field objective microscope (Fig. 8.46). The microscope presents the operator with a simultaneous view of each side of the mask and wafer. The chuck is moved left, right, and/or rotated (*x, y,* and *z* movement) by manual controls until the wafer is aligned to the mask pattern.

Once the mask and wafer are aligned properly, the wafer chuck moves up on a piston, pushing the wafer into contact with the mask. Next, the collimated ultraviolet rays coming from a reflection and lens system pass through the mask and into the photoresist.

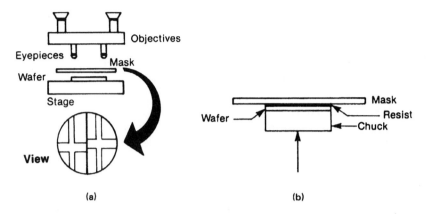

Figure 8.46    Contact aligner system. (*a*) Alignment stage and (*b*) contact stage.

Contact aligners are used in production for discrete devices and circuits with SSI and MSI densities and feature sizes of approximately 5 μm and above. They also are used for flat panel displays, infrared sensors, device packages, and multichip modules (MCMs).[6] A contact aligner is capable of submicron imaging with the proper resist and a well tuned process. Its replacement by other systems is more related to yield losses associated from the contact of the mask and wafer.

Contact can damage the soft resist layer, the mask, or both. Dirt adhering to the clear portions of the mask blocks light during exposure. Epitaxial layer spikes on bipolar wafers can degrade the mask. Mask damage is so prevalent that masks have to be removed and discarded or cleaned every 15 to 25 exposures. Dirt between the mask and wafer will cause resolution problems in the immediate area of the piece of dirt. Alignment of larger diameter wafers presents a light uniformity problem that causes image size variations and alignment problems.

**Proximity aligners.**    Proximity aligners were a natural evolution of contact aligners. The systems are essentially contact aligners but with mechanisms that hold the wafer either in near or soft contact with the mask. Sometimes proximity aligners are called *soft-contact machines.*

The performance of a proximity aligner is a trade-off between resolution capability and defect density. With the wafer in soft contact with the mask, there is always some scattering of the light, which fuzzes the definition of the image in the resist. On the other hand, the soft contact also greatly reduces the number of defects associated with mask and resist damage. Even with the improved defect density, proximity aligners do not find much use in VLSI photomasking processes.

**Scanning projection aligners.**   The end of contact aligners was foreseen for years, and development work was ongoing in the search for an alternative. The search centered on the concept of projecting (Fig. 8.47) the mask image onto the wafer surface, much as a slide (the mask) is projected onto a screen (the wafer). While simple in concept, the technique requires an excellent optical system to expose an entire wafer surface in one exposure. The problem was solved with the introduction of the Perkin Elmer scanning projection aligner. This system avoided the problems of a full mask projection exposure in favor of a scanning technique that used a mirror system with a slit blocking part of the light coming from the light source. The slit allows a more uniform portion of the light to shine on the mirror system, which is in turn projected onto the wafer (Fig. 8.48). Since the size of the slit is smaller than the wafer, the light beam is scanned across the wafer. With this system, a new parameter, scan speed, became a parameter requiring control. They are called 1:1 aligners, since the image dimensions on the mask are the same size as the intended image dimensions on the wafer surface.

**Steppers.**   Scanning projection aligners were a great leap forward over contact aligners for production work, but they still had some limitations, such as alignment and overlay (registration) problems associated with full-size masks, image distortion, and mask-induced defects from dust and glass damage.

By the mid-1970s there were thoughts of stepping images (Fig. 8.49) from a reticle directly onto the wafer surface—the same technique used to make masks. A reticle, carrying the pattern of one or several

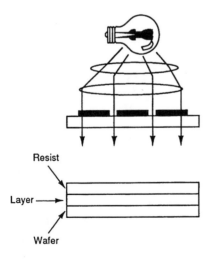

**Figure 8.47** Concept of projection exposure.

Resist

Layer

Wafer

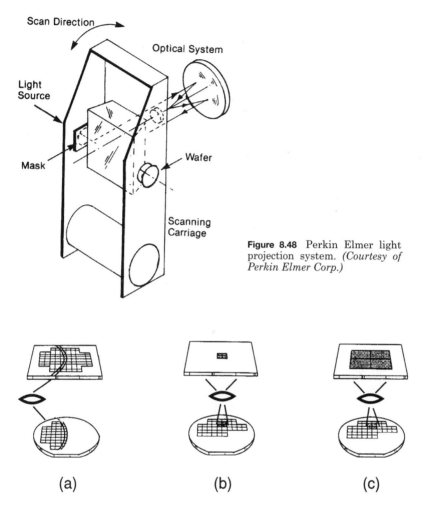

Figure 8.48  Perkin Elmer light projection system. *(Courtesy of Perkin Elmer Corp.)*

(a)          (b)          (c)

Figure 8.49  Projection imaging techniques. (*a*) Scan, (b) 1:1 step/repeat, and (c) reduction step/repeat.

chips, is aligned and exposed, then is *stepped* to the next site and the process *repeated*. A reticle is of a higher quality than a full-size mask so fewer defects occur. There is better overlay and alignment, because each chip is individually aligned. The procedure of stepping allows precise matching of larger-diameter wafers. Other advantages are resolution improvements, because a smaller area is being exposed each time and a lessened vulnerability to dust and dirt. Some steppers are 1:1; i.e., the image on the reticle is the same dimensions required on the wafer. Others use a reticle 5 to 10 times the final dimensions. These are called *reduction steppers*. Making an oversize reticle is easier, and any dirt and small glass distortions are reduced out of existence during the exposure (Fig. 8.50). Generally, 5× reduction is preferred.

A 5× reticle is smaller and easier to make than a 10× one. Also, a 5× reticle projects a larger (up to 20 × 20 mm) field onto the wafer, resulting in faster wafer throughput.[7] Field size is projected in the SIA NTRS to grow beyond the 25 × 32 mm capability of 6-in reticles. Under development are 9-in reticles with the capability of 25 × 50 mm field sizes.[8]

The key to production use of steppers was automatic alignment systems. There is no way that an operator could individually align several hundred die on a wafer at a productive rate. Automatic alignment is accomplished by passing low-energy laser beams through alignment marks on the reticle and reflecting them off corresponding alignment marks on the wafer surface. The signal is analyzed, and information is fed to the $x$-$y$-$z$ wafer chuck controls by a computer, which moves the wafer around until the wafer and reticle are aligned. The images are placed in the photoresist by sequentially exposing each die pattern across and down the wafer (Fig. 8.51).

An alternative to laser signal control is a vision system. These systems use a camera to capture a vision of the die and compare it to a data base. The wafer is moved until it and the mask/reticle image match the data base.

Most production steppers have UV exposure sources with G- or I-line capabilities. Steppers intended for small geometries are fitted with laser sources operating in the DUV range.[9] The maintenance of

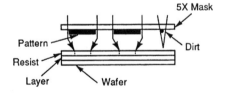

**Figure 8.50** 5× mask pattern transfer.

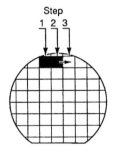

**Figure 8.51** Step-and-repeat die alignment and exposure.

the correct image size during the exposure part of the process requires tight humidity and temperature control. Most steppers are enclosed in an environmental chamber that controls these important parameters and keeps the wafers clean.

**Step and scan aligners.**    Larger die sizes would normally require larger lens systems with larger fields of vision. Increasing the field of vision shortens the time of alignment and exposure. However, larger lenses become expensive. An alternative is a stepper with a smaller lens and the capability to scan the smaller field over the required area (see Fig. 8.52).

**Post-exposure bake (PEB)**

Standing waves are a problem that occurs with optical exposure and positive resists (see Chapter 10). One technique for minimizing the effect of standing waves is to bake the wafers after exposure. The bake method could be any one of the ones described earlier. The time/temperature specifications for the PEB are a function of the baking method, exposure conditions, and resist chemistry.

**X-ray aligners.**    X-ray systems are similar to UV and DUV systems in function. However, the exposure sources are X-rays. These are high-energy beams with small wavelengths capable of patterning very small patterns. One drawback is the need for masks made of gold and other materials capable of blocking the high-energy rays. Also, the development of high-performance X-ray-capable resists has been slow. The challenge is to balance high sensitivity to X-rays while maintaining a good etch barrier.

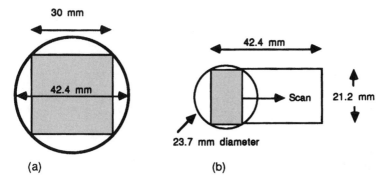

**Figure 8.52**   Step and scan comparison. (a) Step-and-repeat requires 42.4-mm diameter lens field for 9-cm$^2$ die; (b) step-and-scan for same die requires 23.7-mm lens field.

X-ray systems are similar to other mask-using systems with the mask positioned between the source and wafer and with mechanisms for viewing and alignment.

**Electron beam aligners.**    Electron beam lithography is a mature technology used in the production of high-quality masks and reticles. The system consists of an electron source that produces a small-diameter spot and a "blanker" capable of turning the beam on and off. The exposure must take place in a vacuum to prevent air molecules from interfering with the electron beam. The beam passes through electrostatic plates capable of directing (or steering) the beam in the x-y direction on the mask, reticle, or wafer. This system is functionally similar to the beam-steering mechanisms of television sets. Precise direction of the beam requires that the beam travel in a vacuum chamber in which there is the electron beam source, support mechanisms, and the substrate being exposed.

Since the pattern required generates from the computer, there is no mask. The beam is directed to specific positions on the surface by the deflection subsystem and the beam turned on where the resist is to be exposed. Larger substrates are mounted on an x-y stage and are moved under the beam to achieve full surface exposure. This alignment and exposure technique is called *direct writing*.

The pattern is exposed in the resist by either raster or vector scanning (Fig. 8.53). Raster scanning is the movement of the electron beam side to side and down the wafer. The computer directs the movement and activates the blanker in the regions where the resist is to be exposed. One drawback to raster scanning is the time required for the beam to scan, since it must travel over the entire surface. In vector scanning, the beam is moved directly to the regions that have to be ex-

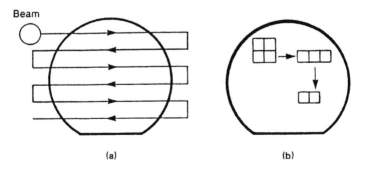

Beam

(a)                    (b)

**Figure 8.53**   Electron beam scanning. (a) Raster scan and (b) vector scanning.

posed. At each location, small square or rectangular shaped areas are exposed, building up the desired shape of the exposed area.

**Mix and match aligners.**    Small-geometry imaging is expensive. Fortunately, only certain layers of a product mask set are critical enough to require the advanced imaging techniques. For advanced circuits, fully 50 percent of the layers may be noncritical.[10] The other, less critical layers can be imaged with more established techniques such as projection scanners or less expensive steppers. Mix and match will probably be a feature of fab operations that use X-ray or e-beam technologies.

## Advanced Lithography

The industry is careening along Moore's law toward a 50-nm mode in 2016. The basic processes described in Chapters 8 and 9 would not suffice to produce feature sizes much below the 200-nm mode. Advancing beyond this milestone requires a whole host of improvements on the basic processes. They include new resists, new exposure sources, improved masks, and more.

## Review Questions

1. Draw cross-sectional diagrams of the ten-step process with a positive resist and dark-field mask.

2. Make a list of the major components in photoresist and explain the role of each.

3. What type of polymers are in negative resist?

4. What type of polymers are in positive resist?

5. Describe the changes in positive and negative resist during exposure to light.

6. Name the exposure and alignment systems that require a mask or reticle.

7. Which exposure system(s) are used for ULSI processes? Explain why.

8. Name ten factors that effect the image size.

9. What is the result of too low a soft-bake temperature?

10. What process steps are required to complete the first image transfer?

# References

1. D. J. Elliott, *Integrated Circuit Fabrication Technology,* McGraw-Hill, New York, 1976, p. 168.
2. "Photoresists for Microlithography," *Solid State Technology,* PennWell Publishing Company, June 1993, p. 42.
3. *Ibid.,* p. 71.
4. *Ibid,* p. 116.
5. K. Simon, et al., "Abstract Alignment Accuracy Improvement by Consideration of Wafer Processing Impacts," SPIE Symposium on Microlithography, 1994, p. 35.
6. E. Cromer, "Mask Aligners and Steppers for Precision Microlithography," *Solid State Technology,* PennWell Publishing Company, April 1993, p. 24.
7. S. Wolf and R. N. Tauber, *Silicon Processing,* vol. 1, Lattice Press, Sunset Beach, CA, p. 473.
8. R. Singh, S. Vu, and J. Sousa, "Nine-Inch Reticles: An Analysis," *Solid State Technology,* October 1998, p. 83.
9. E. Cromer, "Mask Aligners and Steppers for Precision Microlithography," *Solid State Technology,* PennWell Publishing Company, April 1993, p. 26.
10. J. Greeneich, "Mixing Critical and Noncritical Steppers, *Solid State Technology,* PennWell Publishing Company, October 1994, p. 79.

# The Ten-Step Patterning Process—Developing to Final Inspection

## Overview

In this chapter, the basic patterning process step used for photoresist developing through final inspection (steps 5 through 10 of the basic process) are explained. The end of the chapter examines the processes used for mask making and a discussion of alignment error budgets.

## Objectives

Upon completion of this chapter, you should be able to:

1. Draw a cross section of a wafer before and after developing.
2. Make a list of the developing methods.
3. Explain the purpose and methods of hard bake.
4. List at least five reasons why a wafer can be rejected at develop inspect.
5. Draw a diagram of the develop-inspect-rework loop.
6. Explain the methods and relative merits of wet and dry etch.
7. Make a list of the resist strippers used to strip photoresist from oxide and metal films.
8. Explain the purpose and methods of final inspection.

## Development

After the wafer completes the alignment and exposure step, the device or circuit pattern is coded (*latent image*) in the photoresist as regions of exposed and unexposed resist (Fig. 9.1). The pattern is "developed" in the resist by the chemical dissolution of the unpolymerized resist regions. Development processes are designed to form in the resist layer a pattern with the exact dimensions designated during the circuit design process. A problem resulting from a poor developing process (Fig. 9.1) is underdevelopment, which leaves the hole incompletely developed to the correct dimensions, or a coved sidewall. In some cases, the development will not be long enough (incomplete) and will leave a layer of resist in the hole. The third problem is overdevelopment, which removes too much resist from the image edge or top surface. High-aspect plug holes are a particular challenge with uniform hole diameter and cleaning being difficult due to the lack of fluid circulation in the deep hole.

Negative and positive resists have different developing characteristics and require different chemicals and processes (Fig. 9.2).

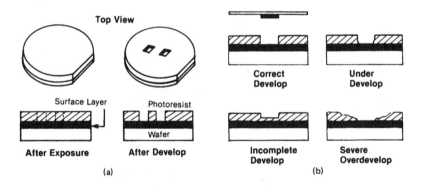

**Figure 9.1**  Photoresist development. (*a*) Process; (*b*) problems.

|  | **Positive Resist** | **Negative Resist** |
|---|---|---|
| **Developer** | Sodium hydroxide (NaOH) | Xylene |
|  | Tetramethyl amonium hydroxide (TMAH) | Stoddard solvent |
| **Rinse** | Water ($H_2O$) | n-Butylacetate |

**Figure 9.2**  Resist developer and rinse chemicals.

## Negative resist development

The successful development of the image coded in the resist is dependent on the nature of the resist's exposure mechanisms. Negative resist, upon exposure to light, goes through a process of polymerization that renders the resist resistant to dissolution in the developer chemical. The dissolving rate between the two regions is high enough that little of the resist layer is lost from the polymerized regions.

The chemical preferred for most negative-resist developing situations is xylene, which is also used as the solvent in negative-resist formulas.

The development step is done with a chemical developer followed by a rinse. For negative resists, the rinse chemical is usually $n$-butyl acetate, because it neither swells nor contracts the resist. Either of these actions could change pattern dimensions. For wafers that have been patterned with a stepper, a milder-acting Stoddart solvent may be used.

The action of the rinse is twofold. First, it rapidly dilutes the developer chemical to stop the developing action. Although the polymerized resist is developer resistant, there is always a transition region at the exposed edge (Fig. 9.3) that contains partially polymerized molecules.

If the developer is allowed to stay in full strength on the wafer, it dissolves into this region and changes the image dimensions. The second action of the rinse is to remove partially polymerized pieces of resist from the open regions in the resist film.

Use of an all-aqueous negative-resist development system is an industry goal. Eliminating the organic solvents from the fab area would serve to reduce the safety systems and lower the cost of treating or removing the spent chemicals. To date, little success with an acceptable replacement has been reported.

## Positive resist development

Positive resists present a different developing condition. The two regions, polymerized and unpolymerized, have a dissolving rate differ-

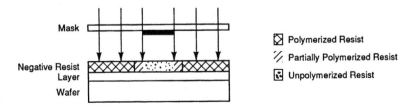

**Figure 9.3**  Transition region at resist image edge.

ence of about 1:4. This means that, during the developing step, some resist is always lost from the polymerized region. The use of developers that are too aggressive or that have overly long developing times may result in an unacceptable thinning of the resist film, which in turn may cause it to lift or break down during the etch step.

Two types of chemical developers are used with positive resist, alkaline-water solutions and nonionic solutions. The alkaline-water solutions can be sodium hydroxide or potassium hydroxide. Since both of these solutions contain mobile ionic contaminants, they are not desirable in processing sensitive circuits. Most positive-resist fabrication lines use a nonionic solution of tetramethylammonium hydroxide (TMAH). Sometimes a surfactant is added to break down the surface tension and make the solution more wettable to the wafer surface. The aqueous nature of positive developers makes them more environmentally attractive than the solvent developers required for negative resists.

The positive-resist developing process is more sensitive than negative processes.[1] Factors influencing the outcome are the soft bake time and temperature; degree of exposure; developer concentration; and time, temperature, and method of developing. The development process parameters are determined by matrix testing of all the variables. The effect on line width for a particular process is shown in Fig. 9.4.

Tight control of the development and rinse process is critical for dimensional control when using a positive resist. The rinse chemical for positive-resist developers is water. It serves the same role as negative-resist rinsers but is cheaper, safer to use, and allows easier disposal.

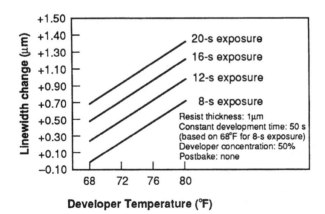

**Figure 9.4** Developer temperature and exposure relationship versus line-width change.

## Wet development methods

- Immersion
- Spray
- Puddle

Figure 9.5 Development methods.

Several methods are used to develop resist films (Fig. 9.5). The selection of a method is dependent on the resist polarity, the feature size, defect density considerations, the thickness of the layer to be etched, and productivity.

**Immersion.** Immersion is the oldest development method. In its simplest form, the wafers, in a chemically resistant carrier, are immersed in a tank of the developer solution for a specific time before being transferred to a second tank of the rinse chemical (Fig. 9.6). The problems associated with such a simple wet procedure are the following:

1. The surface tension of the liquids can prevent the chemicals from penetrating into small openings.

2. Partially dissolved pieces of resist can cling to the wafer surface.

3. The tanks can become contaminated as hundreds of wafers are processed through them.

4. The wafers can become contaminated as they are drawn through the liquid surface.

5. Developer chemicals (especially positive developers) can become diluted through use.

6. Frequent changing of solutions to eliminate problems 1, 2, and 3 raises the cost of the process.

7. Fluctuations in room temperature can cause changes in the developing rate of the solution.

8. The wafers have to be quickly transferred to a drying process step, which introduces a third step.

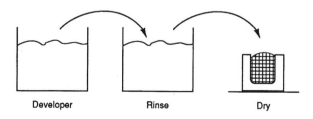

Developer          Rinse          Dry

Figure 9.6 Immersion developer steps.

Additions are often made to the immersion tanks to improve the development process. Uniformity and penetration of small openings are aided by mechanical agitation of the bath by stirring or rocking mechanisms. A popular stirring system is a Teflon®-encapsulated magnet that is coupled to a rotating magnetic field outside the tank.

Agitation is also achieved by passing ultrasonic or megasonic waves through the liquid. The ultrasonic waves cause a phenomenon called *cavitation*. The energy in the waves causes the liquid to separate into tiny cavities that immediately collapse. The rapid generation and collapse of the millions of microscopic cavities create a uniformity of development and help the liquid penetrate into small openings. Sonic energy in the megasonic range (1 MHz) reduces a stagnant boundary layer that naturally clings to the wafer surface.[2] Uniform development rates are also enhanced by the addition of heaters and temperature controllers to the bath.

**Spray development.**     The preferred method of chemical development is by spray. In fact, spray processing is generally preferred over an immersion system for any wet process (clean, develop, etch) for a number of reasons. For instance, there is a major reduction in chemical use with spray systems. Process improvements include better image definition due to the mechanical action of the spray pressure in defining the resist edge and removing partially polymerized pieces of resist. Spray systems are always cleaner than immersion systems, because each wafer is developed (or etched or cleaned) only with fresh chemicals.

Spray processing is done in either single or batch systems. In the single-wafer configuration (Fig. 9.7), the wafer is clamped on a vacuum chuck and rotated while the developer and rinse are sequentially sprayed onto the surface. Drying takes place immediately after the rinse cycle by increasing the rotational speed of the wafer chuck. In

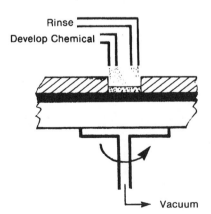

**Figure 9.7** Spray development and rinse.

appearance and design, a spray developer system for single wafers is the same as a resist spinner but plumbed for different chemicals. Single-wafer spray developers offer the advantage of track automation by integrating the developing and hard bake processes. A major process advantage of these systems is the increased uniformity from the direct impingement of the chemical spray on the wafer.

Spray development had been a standard process for negative resist for years, because negative resist is fairly insensitive to the temperature of the developer. For the temperature-sensitive positive resists, spray development is a more demanding process. The problem is the phenomenon of rapid cooling of a fluid dispensed through an orifice under pressure. Called *adiabatic cooling*, it is the same phenomenon that causes a can of pressurized household cleaner to cool during dispensing. Spray developers used for positive resist often have a heated wafer chuck or a heated spray nozzle to control the develop temperature. Other problems encountered with the spray development of positive resist are machine deterioration when alkaline developers are used and foaming as the water-based developer comes out of the nozzle under pressure. Batch developers come in two versions, single boat and multiple boat. These machines are the spin-rinse-dryers described in Chapter 7. As developers, they require additional plumbing to accommodate the developing chemicals. Batch developing systems, in general, are less uniform than direct-spray, single-wafer systems, because the spray does not impinge directly on each wafer surface, and temperature control for positive resist processing is more complicated.

**Puddle development.**    Spray development is very attractive for its uniformity and productivity. A process variation used to gain the advantages of spray for development of positive resist is the *puddle procedure*. This system uses a standard single-wafer spray unit. The difference between regular spray development and the puddle procedure is in the application of the developer chemical to the wafer. The process starts with the deposit of enough developer on the static wafer to cover the surface (Fig. 9.8). Surface tension holds the developer in a puddle on the wafer. The puddle sits there for some required time, usually on a chuck-heated wafer, causing the majority of the development to take place. Puddle development is, in effect, a single-wafer, front-side-only immersion process. After the required puddle time, the wafer surface is sprayed with more developer and rinsed, dried, and passed on to the next step.

**Plasma descum.**    A particularly difficult form of incomplete development is a condition called *scumming*. The scum may be undissolved

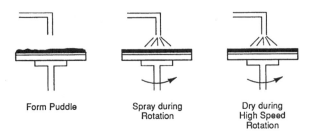

Form Puddle          Spray during          Dry during
                     Rotation              High Speed
                                           Rotation

**Figure 9.8**  Puddle-spray development.

pieces of resist or dried developer[3] left on the surface. The film is very thin and is difficult to detect with visual inspection. In reaction to this problem, advanced ULSI lines with micron and submicron openings will remove (descum) the film from the wafers in a oxygen-rich plasma chamber after a chemical develop.

### Dry (or plasma) development

The elimination of liquid processes has been a long-term goal. They are difficult to integrate into automated lines, and the chemicals are a substantial expense to purchase, store, control, and remove/dispose. One approach to replacing liquid chemical developers is to use a plasma etch process. Dry plasma etch is a well established process for etching wafer surface layers (see "Dry Etch"). In a plasma etcher, ions, energized by a plasma field, chemically dissolve (etch away) exposed layer surfaces. Dry resist development requires a photoresist chemistry that leaves either the exposed or unexposed portions of the resist layer readily removable by plasma energized oxygen. In other words, one part of the pattern is oxidized off the wafer surface. A dry development process, called DESIRE, uses silylation and plasma $O_2$ and is described in Chapter 10.

## Hard Bake

Hard bake is the second heat treatment operation in the masking process. Its purpose is essentially the same as the soft bake step: the evaporation of solvents to harden the resist. For hard bake, however, the goal is exclusively to achieve good adhesion of the resist to the wafer surface. As such, this step is sometimes called *pre-etch bake* or *prebake*.

### Hard bake methods

Hard bake is similar to soft bake in the equipment and methods used. Convection ovens, in-line and manual hot plates, infrared tunnel ovens, moving-belt conduction ovens, and vacuum ovens are all used for hard baking. The track systems are preferred for automated lines. See the "Soft Bake" section in Chapter 8 (p. 227).

### Hard bake process

The exact time and temperature of the hard bake are determined much the same as in the soft bake process. The starting point is the process recommended by the resist manufacturer. After that, the process is fine-tuned to achieve the adhesion and dimensional control required. Nominal hard bake temperatures are from 130 to 200°C for 30 min in a convection oven. Temperatures and times vary in the other methods. The minimum temperature is set to achieve good adhesion of the resist image edge to the surface. The heat-caused adhesion mechanism is dehydration and polymerization. The heat drives water out of the resist, at the same time further polymerizing it, thereby increasing its etch-resistant properties.

The upper temperature limit of the hard bake is set by the flow point of the resist. Resist is a plastic-like material that softens and flows when heated (Fig. 9.9). When the resist flows, the image dimensions are changed. Resist flow is first evident as a thickening of the resist edge when viewed with a microscope. Extreme flow exhibits itself as a series of fringe lines around the image. The fringes are an optical effect from the slope left in the resist after the flow.

Hard bake takes place either immediately after the developing step or just before the etching step, as shown in Fig. 9.10. In most produc-

Normal temperature        High temperature

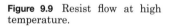

**Figure 9.9** Resist flow at high temperature.

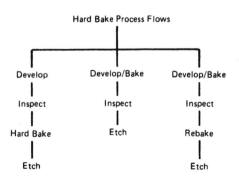

Hard Bake Process Flows

| Develop | Develop/Bake | Develop/Bake |
| Inspect | Inspect | Inspect |
| Hard Bake | Etch | Rebake |
| Etch | | Etch |

**Figure 9.10** Hard bake process flow options.

tion situations, the hard bake is performed in a tunnel oven that is in-line with the developer. When this procedure is used, it is important that the wafers be stored in a nitrogen atmosphere and/or be pro-cessed through the develop inspection step as quickly as possible to prevent the reabsorption of water into the resist film.

A goal of process engineering is to have as many common processes as possible. For hard baking, that is sometimes difficult due to the dif-ferent adhesion characteristics of various wafer surfaces. The more difficult surfaces, such as aluminum- and phosphorus-doped oxides, are sometimes given a higher-temperature hard bake or a second hard bake in a convection oven just prior to being etched.

## Integrated Image Processing

Often, the exact image dimensions cannot be obtained at the align-ment and expose process. In some cases, the image can be brought to the correct dimension with a controlled overetch.

### Develop inspect

The first quality check in the photomasking process is performed after developing and baking. It is appropriately called *develop inspect* or simply DI. The purpose of the inspection is to identify wafers that have a low probability passing the final masking inspection, provide process performance and process control data, and identify wafers for rework.

This inspection yield (that is, the number of wafers that pass this first quality check) is not factored into the overall yield formula, but it is a much-watched yield for two principal reasons. The critical nature of the photomasking process to the functioning of the circuit has been emphasized throughout this text. It is at develop inspect that the pro-cess engineer has the first chance to judge the performance of the pro-cess. The second importance of the develop inspect step is related to the two types of rejects made at this inspection. First, some of the wa-fers will have problems from previous steps that prevent their contin-uation in the process. These wafers are rejected at develop inspect and discarded. Other wafers have problems associated specifically with the quality of the pattern in the resist film. These wafers can be reworked (Fig. 9.11) by removal of the photoresist and reinsertion into the pat-terning process. This is one of the few places in the entire fabrication process where a general rework of mistakes is possible, since no per-manent changes have been made to the wafer.

Wafers sent back into the masking process are called *reworks* or *re-dos*. The goal is to keep the rework rate as low as possible—certainly

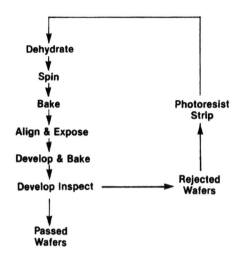

Figure 9.11   Rework process loop.

under 10 percent and preferably less than 5 percent. Experience has shown that wafers that have gone through a masking rework have a lower wafer sort yield at the end of the fab process. Reworking causes adhesion problems, and the additional handling can result in contamination and breakage. If too many wafers are in the rework loop, the overall wafer sort yield will suffer, and the production line will be clogged. A second reason for keeping the rework rate under control is related to the additional accounting and identification required to process the rework wafers.

The develop inspect yield and rework rate vary from mask level to mask level. In general, the first levels in the masking sequence have wider feature sizes, flatter surfaces, and lower density, all of which make for a higher yield out of the mask step. By the time the wafers are at the critical contact and metal masks, the rework rate tends to rise.

### Develop inspect methods

Descriptions of the inspection equipment are presented in Chapter 14. In general, there are four primary categories of wafer problems that occur at both develop inspect and final inspect. There are deviations in the pattern dimensions (critical dimension measurements). There are misaligned patterns. There are surface problems such as contamination, holes or scratches in the photoresist, and stains or other surface irregularities. Finally, there are patterns that are distorted in shape.

**Manual inspection.**   The flow diagram in Fig. 9.12 shows a typical manual develop inspection sequence. The first step is a naked-eye in-

| Step | Looking For: | Methods |
|------|-------------|---------|
| 1. | Stains/ gross contamination | Naked eye/UV light |
| 2. | Stains/contamination<br><br>Pattern irregularities<br><br>Missalignment | 100-400X microscope/<br><br>SEM/ AFM/ Automatic<br><br>Inspection Systems |
| 3. | Critical dimensions | microscope/SEM/ AFM |

**Figure 9.12**  Order of develop inspection.

spection of the wafer surface. Since no magnification is used, this inspection is sometimes called a 1× inspection (a magnification of 1 is equal to naked-eye viewing). It may take place in normal room light, but it is more likely that the wafer will be viewed under a collimated white light or a high-intensity ultraviolet light. The wafer is viewed at an angle in the light beam. This method is surprisingly effective in showing film thickness irregularities, gross developing problems, scratches, and contamination, especially stains.

Wafers that pass the 1× visual inspection are inspected with a microscope. Most industrial inspection techniques use a system called *random sampling* based on probability theory. The theory determines the size of an area that must be inspected to accurately detect the level of defects on the complete wafer. Such an inspection system is based on the operator selecting sites for inspection in a random fashion.

Random sampling is not used in manual inspection. As the defect density goes down, as it must for VLSI circuits, the area of inspection must go up if the defects are to be statistically sampled. However, as the die density goes up, the individual parts also get smaller, which in turn requires a higher magnification to see them. The increasing magnification narrows the field of view, which in turn increases the time for an operator to inspect a wafer. The time required to statistically sample a large-diameter, low-defect wafer is prohibitive.

The usual procedure is to select a number of specific locations for inspection. The chosen die are assumed to geographically represent the wafer surface. This method also assumes that gross problems will show up on the selected die. This method is adequate for gross wafer-wide problems and for alerting the process engineer to problems in the masking process. Often, the microscopes will have motorized/programmable stages that automatically go to the inspection areas on the wafer.

In addition to the issue of the die sampling plan, there is the issue of the wafer sampling plan. The number of wafers inspected varies with the sensitivity and difficulty of the particular mask level. The more sensitive and difficult the layer, the more wafers are likely to be inspected. The number of wafers inspected per batch varies from a few to 100 percent. In addition to process defects, dimensional measurement and control is part of the develop inspection. On each mask level, there is a region or set of patterns whose dimensions are critical to the functioning of the whole circuit. A representative pattern is chosen for each level as the *critical dimension* (CD). During the develop inspection, the dimensions of these patterns are measured using manual or automatic microscopic techniques.

**Automatic inspection.**   With die getting larger, dimensions getting smaller, and processes getting more numerous and more sophisticated, the relatively slow manual inspection becomes nonproductive. Automatic inspection systems designed to detect surface and pattern distortion problems are the inspection system of choice for both off-line and on-line inspections. The systems are described in Chapter 14. Automatic systems offer the prospect of higher amounts of data, and this in turn enables the process engineers to characterize and control the processes. They also are consistent, whereas humans vary in abilities and fatigue doing repetitive tasks. However, some process defects such as very small particles and problems from previous patterning steps can escape detection with automated systems. Optical microscopes are still useful for analysis.

**Causes of develop inspection reject**

There are many reasons why a wafer can be rejected or sent for rework at the develop inspection step. Generally, the only defects looked for are those added to the wafer during the current photomasking step. Defects from previous masking steps are generally overlooked under the theory that every wafer is passed on with some defects or problems and that the wafer arrived at the current step with acceptable quality. If there is a serious problem with a wafer that somehow escaped previous inspections, it is pulled from the batch.

The inspection is generally on a *first-fail basis*. The operator or machine inspects until a rejectable level is reached on the wafer and the wafer is pulled. The information for each wafer is logged (Fig. 9.13) for data accumulation and analysis. Automatic and semiautomatic optical inspection stations have electronic memories for accumulating and correlating this reject data.

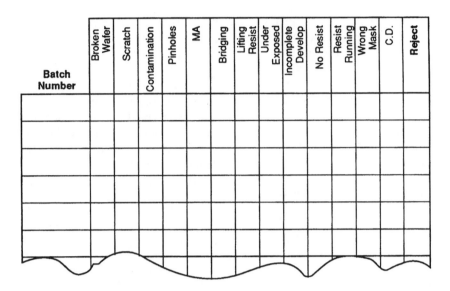

| Batch Number | Broken Wafer | Scratch | Contamination | Pinholes | MA | Bridging | Lifting Resist | Under Exposed | Incomplete Develop | No Resist | Resist Running | Wrong Mask | C.D. | Reject |
|---|---|---|---|---|---|---|---|---|---|---|---|---|---|---|
| | | | | | | | | | | | | | | |
| | | | | | | | | | | | | | | |
| | | | | | | | | | | | | | | |
| | | | | | | | | | | | | | | |
| | | | | | | | | | | | | | | |
| | | | | | | | | | | | | | | |

**Figure 9.13**  Typical develop inspect log.

Most of the causes for rejects have been discussed. One problem not discussed is bridging (Fig. 9.14). It is a condition in which two patterns are connected (bridged) by a thin layer of photoresist, usually at the metal layer. If passed on to the etch step, the photoresist bridge results in an electrical short between the patterns. Bridging comes from an overexposure, poor mask definition, or a resist film that is too thick. Bridging is a particularly vexing problem as patterns get closer together.

## Etch

At the completion of the develop inspect step, the mask (or reticle) pattern is defined in the photoresist layer and is ready for etch. It is in

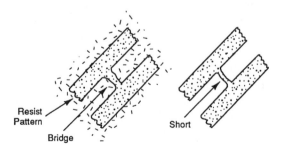

**Figure 9.14**  Bridged conduction lines.

the etch step that the image is permanently transferred into the surface layer on the wafer. Etching is the process of removing the top layer(s) from the wafer surface through the openings in the resist pattern.

Etching processes fall into two main categories: wet and dry (see Fig. 9.21, p. 265). The primary goal of each is an exact transfer of the image from the mask/reticle into the wafer surface. Other etch process goals include uniformity, edge profile control, selectivity, cleanliness, and minimizing cost of ownership (COO).

## Wet Etching

The historic method of etching has been by immersion techniques using wet etchants. The procedure is similar to the preoxidation clean-rinse-dry process (Chapter 7) and immersion development. Wafers are immersed in a tank of an etchant for a specific time, transferred to a rinse station for acid removal, and transferred to a station for final rinse and a spin-dry step. Wet etching is used for products with feature sizes greater than 3 μm. Below that level, the control and precision needed requires dry-etching techniques.

Etching uniformity and process control are enhanced by the addition of heaters and agitation devices, such as stirrers or ultrasonic and megasonic waves, to the immersion tanks.

Wet bench methods are giving way to spray systems similar to those described in Chapter 7. For etching, the control of the chemical composition and timing becomes critical.

Contamination problems is addressed by *point-of-use filters*. These are special filters fitted to automatic chemical dispensing systems to filter-clean the chemicals just prior to filling the immersion tank. This placement catches particulate contamination from the chemicals, the pumps, and the tubing systems.

Wet etchants are selected for their ability to uniformly remove the top wafer layer without attacking the underlying material (good selectivity).

Etch time variability is a process parameter influenced by temperature variations as the boat and wafers come to temperature equilibrium in the tank and the continued etching action as the wafers are transferred to a rinse tank. Generally, the process is set at the shortest time compatible with uniform etching and high productivity. The maximum time is limited to the amount of time the resist will continue to adhere to the wafer surface.

The exactness of the image transfer is dependent on several process factors. They include incomplete etch, overetching, undercutting, selectivity, and anisotropic/isotropic etching of the sidewalls.

## Incomplete etch

Incomplete etch is a situation in which a portion of the surface layer remains in the pattern hole or on the surface (Fig. 9.15). The causes of incomplete etch are too short an etch time, the presence of a surface layer that slows the etching, or an uneven surface layer that results in incomplete etch in the more thickly coated portions of the wafer. If wet-chemical etching is used, a lowered temperature or weak etch solution will cause incomplete etch. If dry-plasma etching is used, a wrong gas mixture or an improperly operated system can cause the same effect.

## Overetch and undercutting

The opposite condition to incomplete etch is overetch. In any etch process, there is always some degree of overetch planned. This allows for thickness variations in the surface layer. Planned overetch also allows for the etch to break through any slow-etching layers on the top surface.

The ideal etch leaves vertical sidewalls in the layer (Fig. 9.16). Etch techniques that produce this ideal result are said to be *anisotropic*. However, the etchant removes material in all directions. This phenomenon is called *isotropic* etching. Etching takes place at the top of the layer for the entire time it takes to reach the bottom of the layer. The result is a sloped side. This action is also called *undercutting* (Fig. 9.17), because the surface layer is undercut below the resist edge. An ongoing goal of the etch step is the control of undercutting to an acceptable level. Circuit layout designers take undercutting into account when planning the circuit. Adjacent patterns must be separated a certain distance to prevent shorting. The amount of undercutting must be calculated when the pattern is designed. The anisotropic etching available with plasma etching is preferred for advanced circuits. Reducing undercutting allows denser circuits.

Severe undercutting (or overetch) takes place when the etch time is excessive, the etching temperature is too high, or the etch mixture is too strong. Undercutting is also present when the adhesion bond be-

Resist
Layer
Wafer

**Figure 9.15**  Incomplete etch.

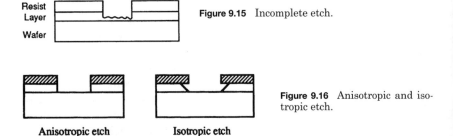

Anisotropic etch          Isotropic etch

**Figure 9.16**  Anisotropic and isotropic etch.

**Normal**

**Over Etch**

**Over Etch and Resist Lifting**

**Figure 9.17**  Degrees of undercutting.

tween the photoresist and the wafer surface is weak. This is a constant worry, and the purpose of the dehydration, prime, soft bake, and hard bake steps is to prevent this type of failure. Failure of the resist bond at the edge of the etch hole can result in severe undercutting. If the bond is very poor, the resist can lift from the wafer surface, causing catastrophic undercutting.

### Selectivity

Another goal in the etch step is the preservation of the surface underlying the etched layer. If the underlying surface of the wafer is partially etched away, the physical dimensions and electrical performance of the devices are changed. The property of the etch process that relates to preservation of the surface is *selectivity*. It is expressed as the ratio of the etching rate of the layer material to the etch rate of the underlying surface. Oxide/silicon selectivity varies from 20 to 40,[4] depending on the etching method. High selectivity implies little or no attack of the underlying surface. Good selectivity becomes a problem in etching small contacts with aspect ratios greater than 3:1.[5] Selectivity also applies to the removal of the photoresist. This is more of a consideration in dry etch processes. While the top layer(s) is being removed by the etchant, some photoresist is also being removed. The selectivity factor must be high enough to ensure that the photoresist is not removed before the etched layer.

### Silicon wet etching

Silicon layers are typically etched with a solution of nitric and hydrofluoric (HF) acids mixed in water. The formula becomes an important factor in control of the etch. In some ratios, the etch has an exothermic reaction with the silicon. Exothermic reactions are those that produce heat, which in turn speeds up the etch reaction, which in turn creates more heat, and so on, resulting in an uncontrollable process. Sometimes, acetic acid is mixed in with the other ingredients to control the exothermic reaction.

Some devices require the etching of a trough or trench into the silicon surface. The etch formula is adjusted to make the etch rate dependent on the orientation of the wafer. $\langle 111 \rangle$-oriented wafers etch at a 45° angle, whereas $\langle 100 \rangle$-oriented wafers etch with a "flat" bottom.[6]

Other orientations result in different-shaped trenches. Polysilicon films are also etched with the same basic formula.

### Silicon dioxide wet etching

The most common etched layer is a thermally grown silicon dioxide. The basic etchant is hydrofluoric acid, which has the advantage of dissolving silicon dioxide without attacking silicon. However, full-strength HF has an etch rate of about 300 Å/s at room temperature.[7] This rate is too fast for a controllable process (a 3000-Å layer would etch in only 10 s).

In practice, the HF (assay of 49 percent) is mixed with water or ammonium fluoride and water. The ammonium fluoride ($NH\langle INF/4 \rangle F$) acts as a buffer to the unwanted generation of hydrogen ions which accelerate the etch rate. These solutions are known as *buffered oxide etches* or BOEs. They are mixed in different strengths to create reasonable etch times for the particular oxide thickness (Fig. 9.18). Some BOE formulas include a wetting agent (a surfactant such as Triton X-

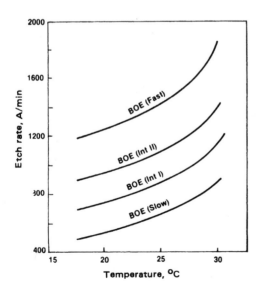

**Figure 9.18** Etch rate versus temperature for BOEs.

100 or equivalent) to reduce the surface tension of the etch, allowing it to uniformly penetrate into smaller openings.

### Aluminum film wet etching

Selective etching solutions for aluminum and aluminum alloy layers are based on phosphoric acid. An unfortunate by-product of the reaction of aluminum and phosphoric acid is tiny bubbles of hydrogen, as shown in the reaction in Fig. 9.19. These bubbles cling to the wafer surface and block the etch action. The result is either bridges of aluminum that can cause electrical shorts between adjacent leads or spots of unwanted aluminum, called *snowballs*, left on the surface.

Neutralization of this problem is accomplished by use of an aluminum etching solution that contains phosphoric acid, nitric acid, acetic acid, water, and wetting agents. A typical solution of the active ingredients (less wetting agent) is 16:1:1:2.

In addition to the special formulas, a typical aluminum etch process will include wafer agitation by stirring or moving the wafer boat up and down in the solution. Sometimes, ultrasonic or megasonic waves are used to collapse and move the bubbles around.

### Deposited oxide wet etching

One of the final layers on a wafer is a silicon dioxide passivation film deposited over the aluminum metallization pattern. These films are known as *vapox* or *silox* films. While the chemical composition of the films is that of silicon dioxide, the same as thermally grown silicon dioxide, they require a different etch solution. The difference is in the selectivity required of the etchant.

The usual etchant for silicon dioxide is a BOE solution. Unfortunately, the BOE attacks the underlying aluminum pads, causing bond-

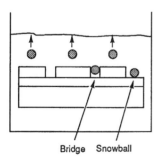

Bridge   Snowball

⊛  Hydrogen Bubble

**Figure 9.19** Hydrogen bubble blockage etchant.

ing problems in the packaging process. This condition is called *brown*, or *stained, pads*. The preferred etchant for this layer is a solution of ammonium fluoride and acetic acid mixed in a ratio of 1:2.

### Silicon nitride wet etching

Another compound favored for the passivation layer is silicon nitride. It is possible to etch this layer by wet chemical means, but it is not as easy as for the other layers. The chemical used is hot (180°C) phosphoric acid. Because the acid evaporates rapidly at this temperature, the etch must be done in a closed reflux container equipped with a cooled lid to condense the vapors. The major problem is that photoresist layers do not stand up to the etchant temperature and aggressive etch rate. Consequently, a layer of silicon dioxide or some other material is required to block the etchant. These two factors have led to the use of dry-etching techniques for silicon nitride.

### Wet spray etching

Wet spray etching offers several advantages over immersion etching. Primary is the added definition gained from the mechanical pressure of the spray.[8] Spray etching also minimizes contamination from the etchants. From a process control point of view, spray etching is more controllable, since the etchant can be instantly removed from the surface by switching the system to a water rinse. Single-wafer spinning-chuck spray systems offer considerable process uniformity advantages.

Disadvantages to spray etching are system cost, safety considerations associated with caustic etchants in a pressurized system, and the requirement of etch-resistant materials to prevent the deterioration of the machine. On the plus side, spray systems are usually enclosed, which adds to worker safety. Figure 9.20 is a table of common semiconductor films and their common etchants.

### Vapor etching

Vapor etching is the exposure of the wafer to etchant vapors. HF is the most common. Advantages are continued replenishment of the etchant at the surface with fresh vapors and instant etch termination. Keeping the toxic vapors contained in the system is a safety concern.

### Dry Etch

The limits of wet etching for small dimensions have been mentioned in the previous section. For review, they include the following:

| | COMMON ETCHANT | ETCH TEMP | RATE $\overset{o}{A}$ /MIN | METHOD |
|---|---|---|---|---|
| SiO₂ | HF & NH₄F (1 : 8) | Room | 700 | Dip & wetting agent predip |
| SiO₂ | HF & NH₄F (1 : 8) | Room | 700 | Dip & wetting agent predip |
| SiO₂ (Vapox) | Acetic Acid & NH₄F(2 : 1) | Room | 1000 | Dip |
| Aluminum | H₃PO₄ : 16 HNO₃ : 1 Acetic : 1 H₂O : 2 Wetting Agent | 40 – 50°C | 2000 | a) Dip & agitation b) Spray |
| Si₃N₄ | H₃PO₄ | 150 – 180°C | 80 | Dip |
| POLYSi | HNO₃ : 50 H₂O : 20 HF : 3 | Room | 1000 | Dip |

**Figure 9.20**   Summary of wet-etching process.

1. Wet etching is limited to pattern sizes of 3 µm.

2. Wet etching is isotropic, resulting in sloped sidewalls.

3. A wet-etch process requires rinse and dry steps.

4. The wet chemicals are hazardous and/or toxic.

5. Wet processes represent a contamination potential.

6. Failure of the resist-wafer bond causes undercutting.

These considerations have led to the use of dry-etch processes for the definition of small feature sizes on advanced circuits. Figure 9.21 provides an overview of the dry etching techniques used.

*Dry etching* is a generic term that refers to the etching techniques in which gases are the primary etch medium, and the wafers are etched without wet chemicals or rinsing. The wafers enter and exit the system in a dry state. There are three dry-etching techniques: plasma, ion beam milling (etching), and reactive ion etch (RIE).

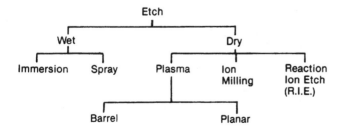

**Figure 9.21**   Guide to etch methods.

## Plasma etching

Plasma etching, like wet etching, is a chemical process but uses gases and plasma energy to cause the chemical reaction. Comparison of silicon dioxide etching in the two systems illustrates the differences. In wet etching of silicon dioxide, the fluorine in the BOE etchant is the ingredient that dissolves the silicon dioxide, converting it to water-rinsable components. The energy required to drive the reaction comes from the internal energy in the BOE solution or from an external heater.

A plasma etcher requires the same elements: a chemical etchant and an energy source. Physically, a plasma etcher consists of a chamber, vacuum system, gas supply, an end-point detector, and a power supply (Fig. 9.22). The wafers are loaded into the chamber, and the pressure inside is reduced by the vacuum system. After the vacuum is established, the chamber is filled with the reactive gas. For the etching of silicon dioxide, the gas is usually $CF_4$ mixed with oxygen. A power supply creates a radio frequency (RF) field through electrodes in the chamber. The field energizes the gas mixture to a plasma state. In the energized state, the fluorine attacks the silicon dioxide, converting it into volatile components that are removed from the system by the vacuum system.

Earlier plasma systems were designed with circular chambers and are called *barrel etchers*. In this design, ions are energized by the plasma in a nondirectional pattern. The etching ions attack the surface layer from all directions (isotropic etching), creating a tapered sidewall. Etch uniformity is a problem. It is hard to supply a constant amount of etchant to all the wafers in the system because of the nondirectionality of the etching ions and the close packing of the wafers.

Another consideration of barrel plasma etching is radiation damage resulting from the high-energy plasma field. The high energy causes charges to build up in the wafer surface that compromise the electrical functioning of the circuit. Protection of the wafers from the radiation is provided by perforated metal cylinders that isolate the wafers from the plasma field (Fig. 9.23).

Reactive Gas

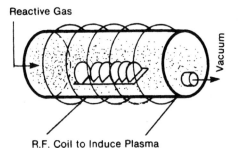

Vacuum

R.F. Coil to Induce Plasma

**Figure 9.22** Barrel plasma etch.

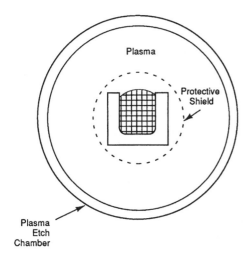

**Figure 9.23**  Barrel plasma stripper with protective shield.

**Planar plasma etching.**    For more precise etching, plasma planar systems are preferred. These systems contain the basic components of the barrel system, but the wafers are placed on a grounded pallet under the RF electrode (Fig. 9.24). Etching takes place with the wafers actually in the plasma field. The etching ions are more directional than those in a barrel system, resulting in a more anisotropic etch. Almost vertical sidewalls are possible with plasma etch. Etching uniformity is increased with the rotation of the wafer pallet in the system.

Planar plasma etch systems are designed in both batch and single-wafer chamber configurations. Single-wafer systems are popular for their ability to have the etch parameters tightly controlled for uniform etching. Also, with load-lock chambers, single-wafer systems can maintain high production rates and are amenable to in-line automation.

RF-generated parallel plate plasma sources are giving way to new sources for 0.35-μm processing.[9] High-density, low-pressure plasma sources under consideration are electron cyclotron resonance (ECR),

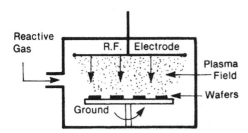

**Figure 9.24**  Planar plasma etch.

high-density reflected electron, helicon wave, inductively coupled plasma (ICP), and transformer-coupled plasma (TCP).

The figures of merit for a dry etch process are: etch rate, radiation damage, selectivity, particulate generation, post-etch corrosion, and cost of ownership.

### Etch rate

The etch rate of a plasma system is determined by a number of factors. System design and chemistry (Fig. 9.25) are two of them. Others are the ion density and system pressure. Ion density (no. ions/cm$^3$) is a function of the power supplied to the electrodes. (Power supply configurations are described in Chapter 12.) Increasing power creates more ions, which in turn increase the etch rate. Ion density is similar to increasing the strength of a liquid chemical etch solution. Ion densities are in the $3 \times 10^{10}$ to $3 \times 10^{12}$ range.[10] System pressure influences etch rate and uniformity through a phenomenon called *mean free path*. This is the average distance a gas atom or molecule will travel before a

| Film | Etchant | Typical Gas Compounds |
|------|---------|----------------------|
| Al | Chlorine | $BCl_3$, $CCl_4$, $Cl_2$, $SiCl_4$ |
| Mo | Fluorine | $CF_4$, $SF_4$, $SF_8$ |
| Polymers | Oxygen $DF_4$, $SF_4$, $SF_6$ | |
| Si | Chlorine, fluorine $CF_4$, $SF_4$, $SF_6$ | $BCl_3$, $CCl_4$, $Cl_2$, $SiCl_4$ |
| $SiO_2$ | Chlorine, fluorine | $CF_4$, $CHF_3$, $C_2F_6$, $C_3F_8$ |
| Ta | Fluorine | " |
| Ti | Chlorine, fluorine | " |
| W | Fluorine | " |

**Figure 9.25** Plasma etch chemicals.

collision with another particle. At higher pressures, there are many collisions that give the particles many directions, which in turn causes loss of edge profile control. Low pressures are preferred, but there is a trade-off with plasma damage as explained below. System pressures typically run in the 0.4 to 50 m torr range.[11] Etch rates vary from 600 to 2000 Å/min.[12]

### Radiation damage

It would seem that higher-density sources with low pressure is the preferred system design. However, there is a countervailing process of *radiation* or *plasma* damage to the wafers. Within the plasma field are energetic atoms, radicals, ions, electrons, and photons.[13] These species, depending on their concentration and energy levels, cause various damage in semiconductors. The damage includes surface leakage, changes in electrical parameters, degradation of films (especially oxides), and damage to silicon among others. There are two damage mechanisms. One is simple overexposure to the high-energy species in the plasma. The other is *dielectric wearout* from currents flowing across dielectrics during the etch cycle.[14] Higher-density sources also cause a problem for photoresist removal. The combination of the energy and low pressure tend to harden the resist to a level that is difficult to remove with conventional processes (see "Resist Stripping," p. 273). System designers are looking to plasma sources that feature high-density, low-energy ions (to reduce damage) and low-pressure operations.

Besides balancing the ion density/pressure parameters, *downstream plasma* processing is an option to reduce plasma damage. The damaging species come from the high energy applied to the gas by the plasma source. Downstream systems create the plasma field in one chamber and transport it downstream to the wafer(s). The wafers are separated from the damaging plasma. To minimize the damage, the system must allow for distinguishing the plasma discharge, ionic recombination, and reduction of the electron density.[15] Downstream plasma systems were developed to minimize damage during plasma resist removal. They are attracting interest for etch applications even though they add more complexity to an etch system.

### Selectivity

Selectivity is a major consideration in plasma etching processes, especially when balanced against the need for overetch. Ideally, an etch time could be calculated to remove the anticipated layer thickness with just a little overetch time for safety. Unfortunately, the accumulated thickness and composition variations on multiple layer *stacks* on high-density devices present etch uniformity problems. Also on high-density devices, a phenomenon called *microloading* introduces etch rate variations. Microloading is a change in the local etch rate relative to the area of material being etched. A large area will *load* the etching process with removed material, slowing it down in that area while a smaller etch area proceeds at a faster rate. Topography issues also drive overetch considerations. A typical situation is the opening of contact holes in both thin regions and thick regions on the device/circuit (see Fig. 10.13). These factors can lead to an overetch of 50 to 80 percent[16] for metal etches and up to 200 percent for oxides and polysilicon etches.[17]

Overetch makes the issue of selectivity critical. There are two factors to consider: the photoresist and the underlying layer (usually silicon or silicon nitride). Dry-etch process have a higher resist removal rate than wet processes. Given the thinner photoresist layers used to define small geometries and the increasing use of stacks of layers, the photoresist selectivity becomes critical. Compounding the selectivity issue are high-aspect-ratio patterns. Advanced devices have patterns with up to 4:1 aspect ratios. The holes are so narrow compared to their height that etching can slow up or stop near the bottom.[18]

Four methods used to control selectivity are the selection of the etching gas formula, the etch rate, the dilution of the gas near the end of the process to slow down the attack of the underlying layer, and end-point detectors in the system.

Terminating the etch when the top layers have been removed requires an in-system end-point detector. Typically, a laser interferometer is used. A laser beam is reflected off of the wafer surface as the etch proceeds. It returns to the detector in an oscillating mode that varies with the type of material being etched.[*] An endpoint detector senses the presence of the etching layer material in the exhaust stream and automatically signals an end to the etch when no more material is detected.

**Contamination, residues, corrosion, and cost of ownership.**    Other process problems of concern, especially in the submicron range, are particulate generation, residues, post-etch corrosion, and all the cost-of-ownership (COO) factors. One approach to reduction of particles is an electrostatic wafer holder to replace mechanical holders. Mechanical

---

[*]If the material(s) is transparent or semitransparent. When the characteristic signal from the last layer etched changes, the detector terminates the etch process. A typical reflectance pattern for a dual layer of salicide/polysilicon is shown in Fig. 9.26.

holders generate particles and cause wafer breakage, and the clamps shadow part of the wafer surface. The electrostatic clamp holds the wafer with a direct current (dc) potential between the wafer and the chuck.[19]

The plasma etch environment is highly reactive, and many chemical reactions take place. Hydroxyl groups in the photoresist react with metal halide gases to form stable metal halide (such as $AlF_3$, $WF_5$, $WF_6$) and oxides such as $TiO_3$, $TiO$, and/or $WO_2$.[20] These residues create contamination problems and can interfere with the selective deposition of tungsten.[21]

Post-etch corrosion results from some etch residues left on metal patterns after the etch process. The addition of copper to aluminum metal and the use of titanium/tungsten metallization increase the corrosion problem from residual chlorine after plasma etch. Minimizing this problem includes substituting fluorine-based etchants for chlorine etchants, passivating the sidewalls, and post-etch processes such as removing the residual chlorine or using a native oxide to passivate the surface.[22] Other solutions include an oxygen plasma treatment, fuming nitric acid, and a wet photoresist stripper step.[23] Cost-of-ownership factors are detailed in Chapter 15. For etchers, a primary COO parameter is the etch rate.

Figure 9.25 lists the common gas etchants for various materials. Silicon and silicon dioxide processes favor fluorine-based etchants such as $CF_4$. Aluminum etching generally takes place with chlorine-based gases such as $BCl_3$.

### Ion beam etching

A second type of dry-etch system is the ion beam system (Fig. 9.26). Unlike the chemical plasma systems, ion beam etching is a physical process. The wafers are placed on a holder in a vacuum chamber, and a stream of argon is introduced into the chamber. Upon entering the chamber, the argon is subjected to a stream of high-energy electrons from a set of cathode (–)-anode (+) electrodes. The electrons ionize the argon atoms to a high-energy state with a positive charge. The wafers are held on a negatively grounded holder, which attracts the ionized argon atoms. As the argon atoms travel to the wafer holder, they accelerate, picking up energy. At the wafer surface, they crash into the exposed wafer layer and literally blast small amounts from the wafer surface. Scientists call this physical process *momentum transfer*. No chemical reaction takes place between the argon atoms and the wafer material. Ion beam etching is also called *sputter etching* or *ion milling*.

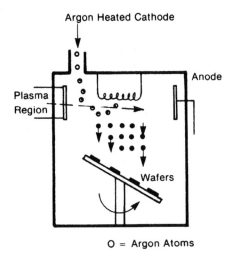

O = Argon Atoms

• = Ionized Argon Atoms

**Figure 9.26**  Ion beam milling.

The material removal (etching) is highly directional (anisotropic), resulting in good definition of small openings. Being a physical process, ion milling has poor selectivity, especially with photoresist layers.

### Reactive ion etching

Reactive ion etching (RIE) systems combine plasma etching and ion beam etching principles. The systems are similar in construction to the plasma systems but have a capability of ion milling. The combination brings the benefits of chemical plasma etching along with the benefits of directional ion milling. A major advantage of RIE systems is in the etching of silicon dioxide over silicon layers. The combination etch results in a selectivity ratio of 35:1,[24] whereas ratios of only 10:1 are available with plasma-only etching. RIE systems have become the etching system of choice for most advanced product lines.

### Resist effects in dry etching

For both wet- and dry-etching processes, a patterned photoresist layer is the preferred etch barrier. In wet etching, there is almost no attack of the resist by the etchants. However, in dry etching, residual oxygen in the system attacks the resist layer. The resist must remain thick enough to stand up to the etchants without becoming so thin that pinholes are present. Some structures use deposited layers as etch barriers to avoid the problem of resist lost (see Chapter 10).

Another resist-related dry etch problem is resist baking. Within the dry-etch chamber, the temperature can rise as high as 200°C, a temperature that can bake the resist to a condition that makes it difficult to remove from the wafer. Another temperature-related problem is the tendency of resist patterns to flow and distort the images.

One unwanted effect of plasma etching is the deposition of *sidewall polymer* strings on the sides of etch patterns. The polymer comes from the photoresist. During a subsequent oxygen-plasma resist strip step, the strings become metal oxides[25] that are difficult to remove.

## Resist Stripping

After etching, the pattern is a permanent part of the top layer of the wafer. The resist layer that has acted as an etch barrier is no longer needed and is removed (or stripped) from the surface. Traditionally, the resist layer has been removed by wet chemical processing. Despite the issues, wet chemistry is the preferred method in the *front end of the line* (FEOL), where the surface and sensitive MOS gates are exposed and vulnerable to plasma damage.[26] There is a growing use of plasma $O_2$ stripping, primarily in the *back end of the line* (BEOL), where the sensitive devices parts are covered by surface layers of dielectrics and metals.

A number of different chemicals are used for stripping. Choices depend on the wafer surface (under the photoresist), production considerations, the polarity of the resist, and the condition of the resist (Fig. 9.27). Wafers are stripped of photoresist after a number of processes: wet etch, dry etch, and ion implantation. There are different degrees of difficulty depending on the prior process. High-temperature hard bakes, plasma etch residues and sidewall polymers, and ion implantation *crusting* all present challenges for the resist removal process.

| Stripper Chemistry | Strip Temperature (Centigrade) | Surface Oxide | Metallized | Resist Polarity |
|---|---|---|---|---|
| **Acids:** | | | | |
| Sulfuric acid+oxidant | 125 | X | | +/- |
| Organic Acids | 90-110 | X | X | +/- |
| Chromic/Sulfuric | 20 | X | | +/- |
| **Solvents:** | | | | |
| NMP/Alkanolamine | 95 | | X | + |
| DMSO/Monothanolamine | 95 | | X | + |
| DMAC/Diethanolamine | 100 | | X | + |
| Hydroxylamine (HDA) | 65 | | X | + |

**Figure 9.27** Wet photoresist stripper chart.

Generally, the strippers are divided into the categories of universal strippers, positive-resist-only, and negative-resist-only strippers. They are also divided by the type of wafer surface: metallized or nonmetallized.

Wet stripping is used for the following reasons:

1. It has a long process history.

2. It is cost effective.

3. It is effective in the removal of metallic ions.

4. It is a low-temperature process and does not expose the wafers to potentially damaging radiation.

### Wet chemical stripping of nonmetallized surfaces

**Sulfuric acid and oxidant solutions.**    Solutions of sulfuric acid and an oxidant (hydrogen peroxide or ammonium persulfate[27]) are the most common wet strippers used for the removal of resist from *nonmetallic* surfaces. Nonmetallic surfaces are either silicon dioxide, silicon nitride, or polysilicon. This solution strips both negative and positive resists. These are the same chemical solutions and processes used for pre-tube-cleaning wafers described in Chapter 7.

Nitric acid is sometimes used as an additive oxidant in a sulfuric acid bath. A mixing ratio of about 10:1 is typical. A drawback to nitric acid is that it turns the bath a light orange color, which can mask the buildup of carbon in the bath. Fuming nitric acid is used as a stripper in Europe and Japan. All of these solutions dissolve the resist by an oxidation mechanism.

### Wet chemical stripping of metallized surfaces

Stripping from metallized surfaces is a more difficult task, because the metals are subject to attack or oxidation. Four types of wet chemicals are used for stripping metallized surfaces. They are:

1. Organic strippers

2. Solvent strippers

3. Solvent/amine strippers

4. Specialty strippers

**Phenolic organic strippers.**    Organic strippers contain a combination of sulfonic acid (an organic acid) and chlorinated hydrocarbon solvents such as duodexabenzene. The formula requires phenol to create a rinsable solution. In the 1970s, concern over the toxic ingredients in these formulas led to the development of sulfonic acid, nonphenolic, nonchlorinated[28] resist strippers. Stripping the photoresist requires heating the solutions into the 90 to 120°C range. Often, the process uses two or three heated strip baths. Rinsing is in two steps, the first being a solvent followed by a water rinse and drying step.

**Solvent/amine strippers.**    One of the advantages of positive resists is their ease of removal from the wafer surface. A positive resist layer that has not been hard baked is easily removed from the wafer with a simple acetone soak. In fact, acetone has been the traditional positive-resist stripper. Unfortunately, acetone represents a fire hazard, and its use is discouraged.

Several manufacturers supply positive-only strippers based on solvent and organic amine solutions. *N*-methyl pyrrolidine (NMP)[29] is the most used solvent. Others are dimethylsulfoxide (DMSO), sulfolane, dimethylforamide (DMF), and dimethylacetamide (DMAC). These strippers are effective, water-rinsable, and drain-dumpable. The strippers may be heated to increase the removal rate and/or to remove resist films that have been through a high-temperature hard bake. Solvent and solvent/amine strippers remove the resist by a chemical dissolution mechanism.

**Specialty wet strippers.**    A number of wet chemical strippers have been developed to solve specific problems. One is a positive-resist stripper based on hydroxylamine (HDA) chemistry.[30] Another chemistry relies on chelating agents to, in effect, chemically bind metal contaminants in the solution.[31] The stripper removes a host of plasma etch residues and polyimide layers not removed by solvent/amine strippers. Other strippers include corrosion inhibitors.[32]

Figure 9.27 is a table of the most common wet resist strippers and their uses. The advent of multilayer metallization systems with transition metal connecting plugs requires wet strippers that do not attack these metals.

**Dry stripping**

Like etching, the dry-plasma process can also be applied to resist stripping. The wafers are placed in a chamber, and oxygen is introduced (Fig. 9.28). The plasma field energizes the oxygen to a high-energy state, which in turn oxidizes the resist components to gases that are removed from the chamber by the vacuum pump. The term *ashing* is used to disignate plasma processes designed to remove only organic

$$C_xH_y \text{ (resist)} + O_2 \text{ (plasma energized)} \rightarrow CO \text{ (gas)} + CO_2 \text{ (gas)} + H_2O$$

**Figure 9.28**  Resist removal by plasma oxygen.

residues. Plasma stripping indicates a process designed to remove both organic and inorganic residues. In dry strippers, the plasma is generated by microwave, RF, and UV-ozone sources.[33]

The major advantage of plasma resist stripping is the elimination of wet hoods and the handling of chemicals. The principal disadvantage is its ineffectiveness in the removal of metallic ions. There is not enough energy in the plasma field to volatilize the metallic ions. Another consideration of plasma stripping is radiation damage to the circuits from the high-energy plasma field. This problem is reduced with system designs that have the plasma chamber removed from the stripper chamber. They are called *downstream strippers,* because the plasma is created downstream from the wafers. MOS wafers are more sensitive to radiation effects during stripping.

Replacement of wet stripping by dry/plasma techniques has been a long-term industry desire. However, the inability of oxygen plasma to remove mobile ionic metal contamination and certain metal residues, and radiation damage concerns, have maintained wet stripping and wet/dry combinations as the mainstream photoresist removal processes. Plasma stripping is used to remove hardened resist layers. A following wet strip step is used to remove the residuals not removed by the plasma. Specialty wet strippers are available to attack these hardened layers.

**Post-ion implant and plasma etch stripping**

Two problem areas are photoresist removal after ion implant and after a plasma strip. Ion implant causes extreme polymerization of the resist and crusting of the top. Generally, the resist is removed or reduced with a dry process followed by a wet process. Post-plasma etch resist layers are similarly difficult to remove. In addition, the etch process can leave residues, such as $AlCl_3$ and/or $AlBr_3$, that react with water or air to form compounds that corrode the metal system.[34] Low-temperature plasma can remove the offending compounds before they take on a corrosive chemistry. Another approach is to add halogens to the plasma atmosphere to minimize the formation of the insoluble metal oxides. This is another instance of setting process parameters to achieve efficient processing (resist removal) without inducing wafer surface damage or metal corrosion.

## Final Inspection

The final step in the basic photomasking process is a visual inspection. It is essentially the same procedure as the develop inspect, with the exception that the majority of the rejects are fatal (no rework is possible).

The one exception is contaminated wafers that may be recleaned and reinspected. Final inspection certifies the quality of the outgoing wafers and serves as a check on the effectiveness of the develop inspection. Wafers that should have been identified and pulled from the batch at develop inspect are called *develop inspect escapes.*

The wafers receive a first surface inspection in incident white or ultraviolet light for stains and large particulate contamination. This inspection is followed by a microscopic or automatic inspection for defects and pattern distortions. Measurement of the critical dimensions for the particular mask level is also part of the final inspection. Of primary interest is the quality of the etched pattern, with underetching and undercutting being two parameters of concern. The table shown in Fig. 9.29 is a list of typical causes of wafer rejection found in the final inspection.

| Possible Process Cause | Contamination | Misalign | Undercut | Incomplete Etch | Wrong Mask | Pin Holes | C.D.'s | Visual Reject |
|---|---|---|---|---|---|---|---|---|
| Contaminated Etch | X | | X | X | | | | |
| Contaminated Stripper | X | | | | | | | |
| Contaminated $H_2O$ | X | | | | | | | |
| Insufficient Rinse | X | | X | | | | | |
| No Wet Agent | | | | X | | | | |
| Under Etch | | | | X | | | X | |
| Over Etch | | | X | | | | X | |
| Wrong Etch | | | X | X | | | X | |
| Hard Bake Too High | | | X | X | | | X | |
| Poor Develop | | | | X | | | X | |
| $P_2O_5$ & $SiO_2$ | | | X | | | | X | |
| $B_2O_3$ & $SiO_2$ | | | | X | | | X | |
| Low Hard Bake | | | X | | | | | |
| Develop Inspect Escapes | | X | X | X | X | X | | |

**Figure 9.29**  Final inspect rejects and process causes.

## Mask Making

In Chapter 5, the steps of circuit design were detailed. In this section, the process used to construct a photomask or reticle is examined. Originally, the masks were made from emulsion-coated glass plates. The emulsions are similar to those found on camera film. These masks were vulnerable to scratches, deteriorated during use, and were not capable of resolving images in the sub-3-micron range. Masks for most modern work use a chrome-on-glass technology. This mask-making technology is almost identical to the basic wafer-patterning operation Figure 9.30. In fact, the goal is the same—the creation of a pattern in the thin chrome layer on the glass reticle surface. The preferred materials for mask/reticles are borosilicate glass or quartz, which have good dimensional stability and transmission properties for the wavelengths of the exposing sources. Chrome layers are in the 1000-Å range and deposited on the glass by sputtering (see Chapter 12). Advanced mask/reticles use layers of chromium, chromium oxide, and chromium nitride.[35]

Mask/reticle making follows a number of different paths depending on the starting exposure method (pattern generation, laser, e-beam) and the end result (reticle or mask) (Fig. 9.31). Flow A shows the process for making a reticle using a pattern generator, which is an older

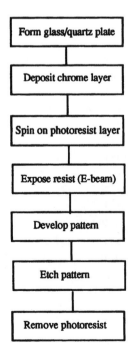

**Figure 9.30** Major steps in mask/reticle plate processing.

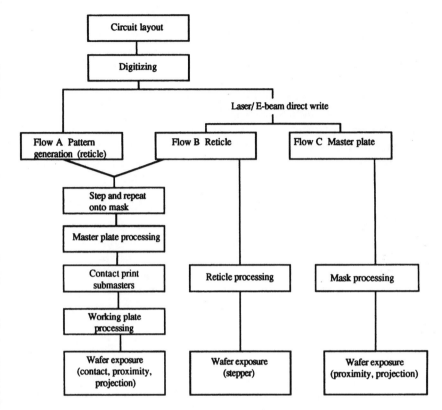

**Figure 9.31**  Mask/reticle-making processing flows.

technology. A pattern generator consists of a light source and a series of motor-driven shutters. The chrome-covered mask/reticle, with a layer of photoresist, is moved under the light source as the shutters are moved and opened to allow precisely shaped patterns of light to shine onto the resist, creating the desired pattern. The reticle pattern is transferred to the resist-covered mask blank by a step-and-repeat process to create a master plate. The master plate is used to create multiple working mask plates in a contact printer. This tool brings the master into contact with a resist-covered mask blank and has a UV light source for transferring the image. After each of the exposure steps (pattern generation, laser, e-beam, master plate expose, and contact print), the reticle/mask is processed through development, inspection, etch, strip, and inspection steps that transfer the pattern permanently into the chrome layer. Inspections are very critical, since any undetected mistake or defect has the potential of creating thousands of scrap wafers. Reticles for this use are generally 5 to 20 times the final image size on the mask.[36]

Advanced products with very small geometries and tight alignment budgets require high-quality reticle and/or masks. The reticles and masks for these processes are made with lasers or e-beam direct write exposure (flows A & B). Laser exposure uses a wavelength of 364 nm, making it an I-line system. It allows using standard optical resists and is faster than e-beam. Direct-write laser sources are turned on and off with an acousto-optical modulator (AOM).[37] In all cases, the reticle or mask is processed to etch the pattern in the chrome.

Other mask/reticle process flows may be employed. The reticle in flow A may be laser/e-beam generated, or the master plate may be laser/e-beam generated.

VLSI and ULSI-level circuits require virtually defect-free and dimensionally perfect masks and reticles. Critical dimension (CD) budgets from all sources are 10 percent or better, leaving the reticles with a 4 percent error margin.[38] There are procedures to eliminate unwanted chrome spots and pattern protrusions with laser "zapping" techniques. Focused ion beams (FIB) is the preferred repair technology for small image masks and reticles. Clear or missing pattern parts are "patched" with a carbon deposit. Opaque or unwanted chrome areas are removed by sputtering from the beam.

## Summary

For VLSI and ULSI work, the resolution and registration requirements are very stringent. In 1977, the minimum feature size was 3 μm. By the mid-1980s, it had passed the 1-micron barrier. By the 1990s, 0.5-micron sizes were common with 0.35-μm technology planned for production circuits. Circuit design projections call for minimum gate sizes of 10 to 15 nm in 2016.[39]

Chip manufacturers calculate several *budgets* for each circuit product. A *critical dimension (CD) budget* calculates the allowable variation in the image dimensions on the wafer surface. For products with submicron minimum feature sizes, the CD tolerances are 10 to 15 percent.[40] Also of concern is the critical defect size relative to the minimum feature size. These two parameters are brought together in an *error budget* calculated for the product. An *overlay budget* is the allowable accumulated alignment error for the entire mask set. A rule of thumb is that circuits with micron or submicron feature sizes must meet registration tolerances of one-third the minimum feature. For a 0.35-μm product, the allowable overlay budget is about 0.1μm.[41]

## Review Questions

1. Name the major methods of resist development.

2. What are the chemicals used to develop negative and positive resist?

3. What is the purpose of the hard bake step?

4. Name three methods used for hard bake.

5. What problems arise if the hard bake temperature is too low? Too high?

6. Name the preferred wet etchants for etching: silicon dioxide layers, silicon nitride layers, and aluminum layers.

7. Name two advantages of dry etching over wet etching.

8. List three principal dry etching techniques.

9. Write a wet stripping flow diagram.

10. Name the purpose and methods of final inspect.

11. List at least five problems and solutions for resolving submicron patterns.

## References

1. D. Elliott, *Integrated Circuit Fabrication Technology*, McGraw-Hill, New York, NY, 1976, p. 216.
2. A. Busnaina and F. Dai, "Megasonic Cleaning," *Semiconductor International*, August 1997.
3. S. Wolf, R. Tauber, *Silicon Processing for the VLSI Era*, Lattice Press, Newport Beach, CA, 1986, p. 530.
4. P. Singer, "Meeting Oxide, Poly and Metal Etch Requirements," *Semiconductor International*, Cahners Publishing, April 1993, p. 51.
5. *Ibid*, p. 51.
6. S. Wolf and R. Tauber, *Silicon Processing for the VLSI Era*, Lattice Press, Newport Beach, CA, 1986, p. 532.
7. *Ibid*.
8. C. Murray, "Wet etching update," *Semiconductor International*, May 1986, p. 82.
9. P. Burggraaf, "Advanced Plasma Sources: What's Working?" *Semiconductor International*, Cahners Publishing, May 1994, p. 57.
10. P. Singer, "Meeting Oxide, Poly and Metal Etch Requirements," *Semiconductor International*, Cahners Publishing, April 1993, p. 53.
11. *Ibid*.
12. D. Elliott, *Integrated Circuit Fabrication Technology*, McGraw-Hill, New York, NY, 1976, p. 275.
13. S. Fonsh, C. Viswanathan and Y. Chan, "A Survey of Damage Effects in Plasma Etching," *Solid State Technology*, PennWell Publishing Company, July 1994, p. 99.
14. *Ibid*.
15. C. Boitnott, "Downstream Plasma Processing: Considerations for Selective Etch and Other Processes," *Solid State Technology*, PennWell Publishing Company, October 1994, p. 51.
16. P. Riley, S. Pengm and L. Fang, "Plasma Etching of Aluminum for ULFI Circuits," *Solid State Technology*, PennWell Publishing Company, February 1993, p. 4.
17. M. Engelhardt, "Advanced Polysilicon Etching in a Magnetically Confined Reactor," *Solid State Technology*, PennWell Publishing Company, June 1993, p. 57.
18. P. Singer, "Meeting Oxide, Poly and Metal Etch Requirements, *Semiconductor International*," Cahners Publishing, April 1993, p. 51.

19. B. Newboe, "Wafer Chucks Now Have an Electrostatic Hold," *Semiconductor International*, Cahners Publishing, February 1993, p. 30.
20. C. Cardinaud, M. Peignon and G. Turban, "Surface Modification of Positive Photoresist Mask During Reactive Ion Etching of Si and W in SF6 Plasm," *J. Electrochemical Soc.*, vol. 198, 1991, p. 284.
21. W. M. Lee, *A Proven Sub-Micron Photoresist Stripper Solution for Post Metal and Via Hole Processes*, EKC Technology, Inc., Hayward, CA, 1993.
22. F. Clayton and S. Beeson, "High-Rate Anisotropic Etching of Aluminum on a Single-Wafer Reactive Ion Etcher," *Solid State Technology*, PennWell Publishing Company, July 1993, p. 93.
23. W. M. Lee, *A Proven Sub-Micron Photoresist Stripper Solution for Post Metal and Via Hole Processes*, EKC Technology, Inc., Hayward, CA, 1993.
24. D. Elliott, *Integrated Circuit Fabrication Technology*, McGraw-Hill, New York, NY, 1976 p. 282.
25. W. M. Lee, *A Proven Sub-Micron Photoresist Stripper Solution for Post Metal and Via Hole Processes*, EKC Technology, Inc., Hayward, CA, 1993.
26. R. Dejule, "Managing Etch and Implant Residue," *Semiconductor International*, August 1997, p. 62.
27. EKC Technology Inc., Technical Bulletin SA-80.
28. EKC Technology Inc., Technical Bulletin—Nophenol 922.
29. EKC Technology Inc., Technical Bulletin—Posistrip Series.
30. W. M. Lee, *A Proven Sub-Micron Photoresist Stripper Solution for Post Metal and Via Hole Processes*, EKC Technology, Inc., Hayward, CA, 1993.
31. R. Dejule, "Managing Etch and Implant Residue," *Semiconductor International*, August 1997, p. 57.
32. M. D. Levenson, "Wet Stripper Companies Clean Up," *Solid State Technology*, PennWell Publishing Company, April 1994, p. 31.
33. P. Burggraaf, "What's Driving Resist Dry Stripping?," *Solid State Technology*, PennWell Publishing Company, November 1994, p. 61.
34. R. Dejule, "Managing Etch and Implant Residue," *Semiconductor International*, August 1997, p. 58.
35. B. Grenon, et al., "A Comparison of Commercially Available Chromium-Coated Quartz Mask Substrates," OCG Microlithography Seminar, Interface 94, p. 37.
36. S. Wolf and R. N. Tauber, *Silicon Processing*, vol. 1, Lattice Press, Sunset Beach, CA, p. 477.
37. J. Reynolds, "Mask Making Tour Video Course," *Semiconductor Services*, Redwood City, CA, August 1991, Segment 5.
38. J. Reynolds, "Elusive Mask Defects: Random Reticle CD Variation," *Solid State Technology*, PennWell Publishing Company, September 1994, p. 99.
39. Semiconductor Industry Association, International Technology Roadmap for Semiconductors, 2001, 2002, web site.
40. J. Wiley and J. Reynolds, "Device Yield and Reliability by Specification of Mask Defects," *Solid State Technology*, PennWell Publishing Company, July 1993, p. 65.
41. K. Simon, et al., *Abstract-Alignment Accuracy Improvement by Consideration of Wafer Processing Impacts*, SPIE Symposium on Microlithography, 1994, p. 35.

# Advanced Photolithography Processes

## Overview

Feature sizes decreasing to the nanometer range, the increasing need for low defect densities, increasing chip density and size, along with larger diameter wafers have challenged the industry to squeeze every capability out of traditional processes and develop new ones. The problems encountered and current solutions to reaching nanometer circuit dimensions are explored in this chapter. Included is a discussion of post-optical lithography techniques, collectively known as *next-generation lithography* (NGL).

## Objectives

Upon completion of this chapter, you should be able to:

1. Describe four exposure-related effects that cause image distortion.
2. Draw a cross-sectional flow diagram of a dual-layer resist process.
3. Draw a cross-section flow diagram of a dual damascene process.
4. List two planarization techniques.
5. State an advantage of an image-reversal process.
6. Describe how antireflective layers, contrast enhancement layers, and resist dye additives improve resolution.
7. Identify the parts of a pellicle and the advantages it offers to a resist process.

## Issues of VLSI/ULSI Patterning

The ten-step patterning process detailed in Chapters 8 and 9 is a one-photoresist-layer basic process. It would be sufficient for the production of MSI and some simple LSI and VLSI circuits. However, the VLSI/ULSI demands of decreasing feature sizes and defect densities is beyond the capabilities of the basic process. The limits showed up at the sub-three (3) micron level and became critical in the submicron era. Problems include physical limitations associated with the optical exposure equipment, resolution limits of photoresists, and a host of surface problems, including reflective surfaces and multilevel topographies.

"The report of my demise has been greatly exaggerated."—Mark Twain.

In the mid-1970s, it was widely accepted that optical photoresist processes had a lower resolution limit of about 1.5 μm. This projection gave rise to the interest in X-ray and e-beam exposure systems.

However, manipulation and improvements on the basic processes have successively lowered the usable range of optical lithography to the 0.2-μm range.[1] In the first edition of *Microchip Fabrication* (1984), I reported that "industry futurists that either e-beam or X-ray exposure will replace the UV and DUV sources by the mid-1990s. It did not happen. Optical lithography has been walking the plank for decades. And every generation of engineers has tweaked and improved patterning based on optical exposure systems to bring the industry to the 100-nm node.[2] The crystal ball of the past has been replaced with the SIA's *International Technology Roadmap for Semiconductors* (ITRS). Figure 10.1 shows the various "nodes" of future devices and their year of introduction.[3] A node is essentially the gate

| Year of Production | 2001 | 2006 | 2012 |
|---|---|---|---|
| Line width (nm) | 150 | 100 | 50 |
| Memory size | 1 Gb | 16Gb | 64 Gb |
| Logic Bits/cm2 | 380M | 2.2B | 17B |
| Chip Size-DRAM (mm2) | 445 | 790 | 1580 |
| Max wiring levels | 7 | 7-8 | 9 |
| Mask layers | 23 | 24/26 | 28 |
| Defect density-DRAM (D/m | 875 | 490 | 250 |
| Chip conections-I/O's | 1195 | 1970 | 3585 |
| Wafer diameter (mm) | 300 | 300 | 450 |

**Figure 10.1** SAI roadmap projections. *(Source:* Semiconductor International, *January 1998.)*

width. The lithography requirement is characterized as the 1/2 pitch length of adjacent lines (Fig. 10.2). Another feature of future lithography processes is that the wavelength of the exposing radiation is larger than shown in Fig. 10.1. There are some future requirements that will drive lithography circa 2016. In most cases, the techniques required have yet to be defined.

In this chapter, some of the limits of optical imaging and advanced process solutions are presented. Also, industry developments extending optical lithography and development of NGL [called *next generation lithography* (NGL)] have proceeded on almost all of the individual elements of basic patterning processes. They include resist development, mask materials and designs, exposure sources (Fig. 10.2), alignment and exposure schemes, reflection control, and process schemes. In the present and future eras, lithography advances along the technology node scale will involve combinations of the factors listed. No single lithography process step is expected to provide a comprehensive breakthrough. Figure 10.3 illustrates major exposure systems identified for the future.

**Resist.** The general resist resolution problem is the wavefront that arrives at the resist layer. Simple cross-sectional drawings show the arriving rays as uniform arrows. The actual radiation in the wavefront has a mixture of directions and energies that vary across surface and in the vertical direction. This is called the *aerial image*.[4] Resolving one-half and one-third micron images with optical lithography requires management and manipulation of the aerial image. Control methods fall into three general areas: optical resolution, resist resolution limits, and surface problems. A fourth problem area is etch definition. Resist resolution is intimately mated to the exposure source and exposure system used. Basic resist formulation has improved with manufacturing control, matching the chemistry to the wavelength.

| | | DRAM level | | |
|---|---|---|---|---|
| | 4 kb | 64 kb | 256 Mb | 64 Mb → 1 GB |
| Aligner type | Contact | Projection | Stepper | Stepper-scanner |
| Exposure | Hg | Hg | g, I line | KrF, Arf |
| Image | Mask | Mask | Reticle | Reticle |

**Figure 10.2**  Patterning techniques versus DRAM level.

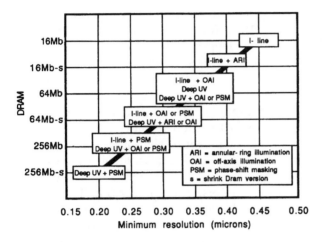

DRAM

| | 0.15 | 0.20 | 0.25 | 0.30 | 0.35 | 0.40 | 0.45 | 0.50 |

Minimum resolution (microns)

**Figure 10.3** Lithography resolution process roadmap. *(Source: Semiconductor International, February 1993.)*

Chemically amplified (CA) resist moved the processes into 0.25-micron range. CA resists are considered the basic platform (with future improvements) to carry lithography to the 90-nm node and beyond. These resists, like older resists, are based on the photosensitive polymers, photoacid generators (PAGs), dissolution inhibitors, etch barriers, and an acid labile, base soluble group.[5] New resists must operate in the tougher etch environments of plasma etch. They not only must image the smallest features, usually gates, but also deal with imaging dense patterns and small metallization contacts. Additionally, etch *line edge roughness* (LER) becomes a factor as line size dimensions approach the size of the molecules in the resist.

### Contrast effects

Good resolution becomes difficult in regions of the mask where an opaque line is surrounded by a large clear area. The large amount of radiation coming around the opaque line tends to shrink the dimension of the line in the resist layer (Fig. 10.4) as the exposing rays diffract around the edge of the pattern. This problem is called a *proximity effect.*

Another contrast effect is called *subject contrast* (Fig. 10.5). This situation comes about when some exposure radiation penetrates the opaque region of a mask or reflects off the wafer surface into the resist.

The result is a partially exposed region that leaves a distorted image after the development step. This is more a problem with negative resist than with positive resist.

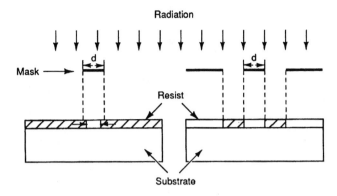

Figure 10.4   Proximity effects.

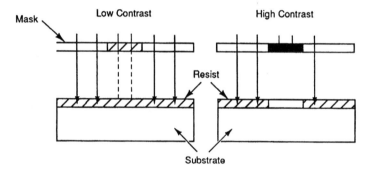

Figure 10.5   Subject contrast.

**Exposure sources.** Exposure sources are chosen to match the spectral response characteristic of the resist and the feature size of the images (see section on "Photoresist exposure speed, sensitivity, and exposure source"). Most optical aligner systems use a high-pressure mercury lamp, which produces light as a current is passed through a tube of mercury. A high-pressure atmosphere allows a higher level of stimulation of the mercury without evaporation.

The energy from the mercury comes out in bundles of waves grouped in ranges. Some resists are designed to react to the entire range of the wavelengths and some to specific wavelengths. Other resists are designed to respond to the specific high-energy peaks of a mercury lamp. The three peaks are at the 365-, 405-, and 436-nm wavelengths.

They are referred to as the I, H, and G lines, respectively. Steppers used for advanced imaging often have filters to expose the resist to the G- or I-lines. I-line exposure has been the source of choice in the submicron era. The I-line has a wavelength of 365 nm (or 0.365 μm), which is near the image size for 0.350 products. The problem is the physical difficulty of resolving images smaller than the wavelength of the light.

The advantage of confining the exposure source to a narrow wavelength is to foster complete and faster exposure of the resist with little scattering. Narrowband exposure limits resolution problems associated with partial exposure of the resist at the clear/opaque edge on the mask. The exposure time is faster for a specific spectral resist, since the peaks represent higher-energy portions of the spectrum.

Resolution capabilities of the aligner and the resist are primarily a function of the wavelength of the exposing light. The smaller or narrower the exposing wavelength, the higher the resolution capability. This is due to the diffraction effect (Fig. 10.6). The longer the wavelength, the more the diffraction effect, which ultimately acts as a limit on the resolution capability of the system. Shorter wavelengths of light also carry more energy, allowing shorter exposure times and in turn limiting poor resolution coming from scattering of the light in the resist and from the wafer surface (Fig. 10.7).

Shorter ultraviolet wavelength peaks are in a region of the mercury lamp output spectrum called the mid- or deep ultraviolet (DUV). The peaks occur at 313 (mid UV) and at 254 nm (DUV). Deep- and mid-ultraviolet exposing sources are created by the use of filters with a standard ultraviolet source such as mercury-xenon, xenon, or deuterium lamps.[6]

**Figure 10.6** Diffracting reduction of image in resist.

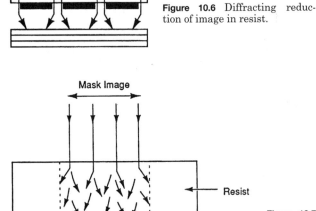

**Figure 10.7** Light scattering in resist film.

**Excimer lasers.** There are also DUV sources available from excimer lasers. The following gas lasers and their associated wavelength are used: XeF (351 nm), XeCl (308 nm), KrF (248 nm), ArF (193 nm).[7] Both KrF and ArF have been developed for above the 130-nm node processes, with ArF considered the source of choice for below 130-nm patterning.[8] $F_2$ is also considered a candidate to follow ArF.[9] Figure 10.8 shows the progression of exposure sources to pattern ever-smaller images.

**Focused ion beams.** The literature always includes the possibility of focused ion beams for photoresist exposure. The lure is less scattering of the beam and lower energy scattering in the resist, which should allow smaller and sharper images. Also, the energy of ion beams is higher than electrons, which could lead to faster exposure times. Unfortunately, development has been hampered by a host of production considerations, including vibration and ion-optical systems required for submicron imaging.[10]

**X-ray.** The desire for higher-resolution exposure sources inevitably led to the consideration of two nonoptical sources X-ray sources (Fig. 10.9) and electron beams (e-beams). X rays are high-energy photons with wavelengths of 4 to 50 Å.[11] This range of wavelengths is capable of very small image sizes (down to the 0.1-μm level) due to the lack of diffraction effects. X-ray aligners are projection systems using a full-size mask (1:1 mask to wafer image). They generally have higher

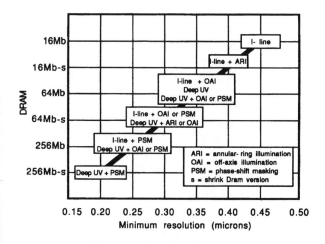

**Figure 10.8** Lithography resolution process roadmap. (*Source: Semiconductor International, February 1993.*)

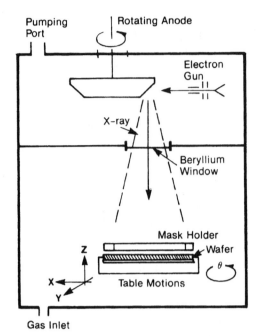

**Figure 10.9** X-ray exposure system.

output through shorter exposure times. Reflection and scattering in the resist is minimum, and there are few depth-of-focus problems. X-ray-exposed wafers show a lower level of defects from dust and organic matter on the mask, because the X-rays pass through the spots.

A number of difficulties have surfaced with production-level X-ray aligners. A major problem has been with the development of X-ray blocking masks. Because X-rays pass through conventional chrome and glass masks, a process that requires gold as the blocking layer and other materials that stand up to the high energy of the X-rays needs to be developed (see "Maskmaking").

While development work goes on in defining X-ray equipment, it also goes on in developing X-ray resists. This work is complicated, because there is no standard X-ray source, and the resists must show high sensitivity to X-rays as well as being good etch barriers. These last two factors have proven difficult to balance in resist chemistries. Another barrier limits X-ray aligners to a 1:1 printing. The high-energy X-rays destroy conventional optics needed for reduction systems. And the limitation of reticle sizes means that only steppers are practical for production machines.

X-ray sources include standard X-ray tubes or laser-driven sources as point sources or synchrotron generators. Point sources are similar to conventional systems with one exposure source per machine. Synchrotrons are large, expensive machines that can accelerate electrons in an orbit. The orbiting electrons give off X-rays in a process called *synchrotron radiation*. As the X-rays rotate, they can be directed through ports to a number of individual aligners.

**Electron beam.**    Electron beam lithography is a mature technology used in the production of high-quality masks and reticles. The system (Fig. 10.6) consists of an electron source that produces a small-diameter spot and a "blanker" capable of turning the beam on and off.

The exposure must take place in a vacuum to prevent air molecules from interfering with the electron beam. The beam passes through electrostatic plates capable of directing (or steering) the beam in the *x*-*y* direction on the mask, reticle, or wafer. This system is functionally similar to the beam-steering mechanisms of a television set. Precise direction of the beam requires that the beam travel in a vacuum chamber in which there is the electron beam source, support mechanisms, and the substrate being exposed.

There is no mask or reticle used to generate the pattern. With no mask, a source of defects and errors is eliminated along with the ex-

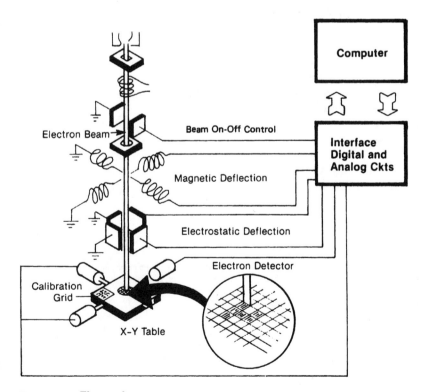

**Figure 10.10**  Electron beam exposure system.

pense of the mask or reticle. The blanking and steering functions are controlled by a computer that has in its memory the wafer pattern taken directly from the CAD design stage. The beam is directed to specific positions on the surface by the deflection subsystem and the beam turned on where the resist is to be exposed. Larger substrates are mounted on an *x-y* stage and are moved under the beam to achieve full surface exposure. This alignment and exposure technique is called *direct writing*.

The pattern is exposed in the resist by either raster or vector scanning (Fig. 10.11). Raster scanning is the movement of the electron beam side to side and down the wafer. The computer directs the movement and activates the blanker in the regions where the resist is to be exposed. One drawback to raster scanning is the time required for the beam to scan, since it must travel over the entire surface. In vector scanning, the beam is moved directly to the regions that have to be exposed. At each location, small square or rectangular shaped areas are exposed, building up the desired shape of the exposed area.

Alignment and overlay parameters are very good with electron beam systems, because no distortions are introduced from masks or from optical effects such as diffraction. Resolution is also good, with current machines capable of 0.25-μm feature sizes.[9] Drawbacks to full use of electron-beam systems in wafer production are speed and cost. Current electron-beam aligners can expose fewer than ten 6-in diameter wafers per shift and cost several million dollars per system. A factor in the slowness of the system is the time required to create the vacuum and release it in the exposing chamber.

## Other Exposure Issues

In Chapter 8, the issue of image resolution and exposure wavelength was explored. In general, the way to expose smaller images is to use a

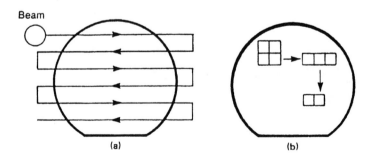

**Figure 10.11**   Electron beam scanning. (*a*) Raster scan and (*b*) vector scanning.

smaller-wavelength exposure source. However, this leads to a smaller depth of field. In the sub-0.5-μm range, exposure processes extend from I-line to deep UV. The DOF problem requires other refinements, including variable numerical aperture lens, annular-ring illumination, off-axis illumination, and phase shift masks. In addition, there are other optical effects that come into play as the image size gets smaller and the pattern density gets larger. These issues and solutions are examined next.

**Numerical aperture (NA) of a lens.**    In the example of the shadow image of the fork, distance was introduced as a resolution factor. In semiconductor imaging, there are contact/proximity exposure systems where the wafer and mask are touching or in close proximity (see "Contact, Proximity Aligners"). But most VLSI and ULSI products are exposed with projection systems where the mask/reticle and wafer are separated. Projection optics present particular problems. The challenge is to project the image from the mask/reticle to the wafer surface with as little loss of resolution or dimensional control as possible. Small image sizes require the use of short wavelengths. The minimum image that can be resolved on the wafer is constrained by the physical attributes of the projection optical system. One attribute is the numerical aperture (NA) of the projection lens. The NA is the ability of the lens to gather light. The relationship is:

$$\sigma = k\frac{\lambda}{NA}$$

where    $\sigma$ = the minimum feature size
   $k$ = a constant (sometimes called the Rayleigh constant)
   $\lambda$ = the wavelength of the exposing light
   NA = is the numerical aperture of the lens

The term $k$ (or $k_1$ in some formulas) relates to the ability of the lens (or total lithography system) to differentiate two adjacent images. Diffraction effects will cause even the most perfect lens to blur images as the get closer to each other. Values of $k$ are about 0.5.

   The formula suggests that decreasing the wavelength and/or increasing the NA are ways to print a smaller image size. Using deep UV, X-rays, or e-beams are ways to reduce the wavelength. However, there are limits to increasing the NA. That is because of a trade-off parameter called *depth of focus* or *depth of field*. In regular photography, we run into depth-of-field problems when the foreground is in focus and the background is out of focus, or vice versa. At the wavelengths

used in patterning, even the micron-size topography on a wafer is
enough to cause depth-of-field problems. Images, both on the top surface and at the bottom of valleys, must have good resolution and the
correct dimensions. Another trade-off with increasing the *NA* is a decrease in the field of view. This is the same phenomenon experienced
when going to a higher magnification with a zoom lens. At the higher
powers, the breadth of the view is narrowed. Field of view becomes a
production limit with steppers. A narrower field requires more time to
complete exposure of the entire wafer.

All of these technical problems are complicated by the increasing
amount of information that must be printed on advanced circuits. The
technology driver product is the dynamic random access memory
(DRAM). This product evolves about every three years to a new generation with four times more information. With feature size shrinkage,
chip size increases, stacking of components, and better designs, three
times more information (per square centimeter) is required of the
masking process.[8] This trend pressures alignment makers, especially
steppers, to develop ever more sophisticated lens systems, exposure
sources, resists, and other image enhancement techniques such as
light source improvements, phase shift masks, and so forth (see Chapter 10). These techniques effectively lower the constant $k$ in the resolution limit formula above.

**Variable numerical aperture (NA) lenses.**   The NA factor is an overall
measurement of a lens's ability to gather light. The trade off is depth
of field (DOF). A lens with a better ability to gather light (higher NA)
suffers from a decreased depth of field (see Chapter 8). Unfortunately,
advanced circuits have many layers and high and low points (*surface
topography*) that require a large DOF. A stepper with one lens is limited to exposing wafers within its associated depth of field, which may
or may not include the levels on the wafer surface. Newer steppers
come with a variable NA lens, allowing their use over a wider range of
DOF requirements.

**Off-axis illumination.**   Shifting the direction of the exposure beam from
the perpendicular (off-axis) interrupts the interference pattern that
causes standing waves in the resist.

**Lens issues and reflection systems.**   At the extremes of lithographic
patterning shaping, the exposing beam through a lens system becomes an issue. The issue is absorption. An exposure system should
deliver a specific wavelength (or bundle of controlled wavelengths) to
the resist surface. The materials used for lenses can absorb radiation

in the required ranges, and this becomes a serious problem below 193 nm. Calcium fluoride ($CaF_2$) is one material that is transparent in this radiation range, and it is expected to be used in the 157-nm node.[13]

Moving beyond optical lithography and into NGL, there are two primary candidates: extreme ultraviolet (EUV) and electron-beam projection lithography (EPL). EUV systems use a deep ultraviolet (DUV) source from the shorter end of that spectrum with a wavelength in the 10 to 15 nm range (Fig. 10.12). A projection system can more faithfully deliver the radiation to the wafer without the distortions that occur with lens systems. At the wafer surface, the e-beam is steered in a step-and-scan process.

The basics of an electron beam system have been described. Unfortunately maskless (raster or vector scanning) systems are too slow for patterning onto wafer surfaces. An advanced system using e-beam exposure is electron-beam projection lithography (EPL). This system uses an e-beam exposure source, a mask, and a scanning projection method (Fig. 10.13). A system developed by Lucent Technologies, called SCALPEL, uses a scattering mask. Electrons are highly energetic and pass through most materials. While conventional masks block portions of the exposing beam and allow other portions to pass through, a scattering mask allows passage of the e-beam through both segments. However, one part of the mask scatters the e-beam as shown. A reduction lens focuses the beam down toward the wafer. In between is an aperture that essentially allows the unimpeded beam to pass onto the wafer surface and blocks the scattered beams.

Current mask design for this system uses a multiplayer structure. The "pattern" is defined in a layer of silicon nitride that was deposited on a silicon wafer. Most of the wafer is etched away, except for silicon "struts" left to keep the mask rigid. This arrangement is mounted on a thin membrane. All of the parts of this mask are transparent to the e-beam. However, the pattern is defined because the incoming beam is broken into two components—scattered and nonscattered. Because

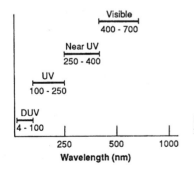

**Figure 10.12**  Ultraviolet and visible spectrum. *(After Elliott.[36])*

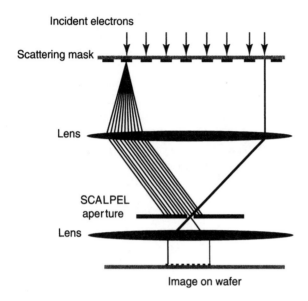

**Figure 10.13** Basic SCALPEL principle of operation showing contrast generation by differentiating more or less scattered electrons. *(Bell Laboratories, Lucent Technologies web site.)*

this pattern is scanned onto the wafer surface, the struts do not come into the scanned pattern.

### Phase shift masks (PSMs)

For conventional optical patterning, several techniques are used to improve image fidelity from mask to wafer. A diffraction problem occurs as two mask patterns get closer together. At some point, the normal diffraction of the exposure rays start touching, leaving the patterns unresolved in the resist. The blending of the two diffraction patterns into one is because all the rays are in the same phase. *Phase* is a wave term that relates to the relative positions of the peaks and valleys of a wave (Fig. 10.14). The waves in (*a*) are in phase, those in (*b*) out of phase. One way to prevent the diffraction patterns from wiping out two adjacent mask patterns is to cover one of the openings (Fig. 10.15) with a transparent layer that shifts one of the sets of exposing rays *out of phase*, which in turn nulls the blending. This approach to an *alternating phase shift mask* (also called *alternating aperture phase shift mask—AAPSM*) requires the deposition of a layer of silicon dioxide on the mask/reticle and a photomasking process to remove the oxide layer from alternate patterns. Covering every other

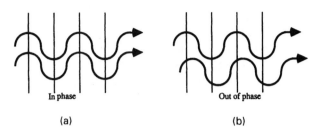

In phase          Out of phase

(a)                    (b)

**Figure 10.14**  *Wave phases.*

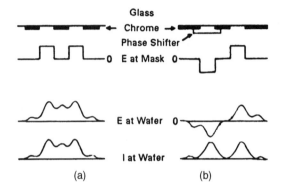

Glass
Chrome
Phase Shifter
E at Mask
E at Wafer
I at Wafer

(a)                    (b)

**Figure 10.15**  Light intensity patterns. (*a*) Without phase shifting and (*b*) with phase shifting. (*Source: VLSI Fabrication Principles, by Ghandhi.*)

clear opening works well for repeated array patterns such as those found in memory products.

Another solution is the addition of phase shifting layers to the edges of the mask/reticle patterns. This process also requires a oxide deposition and full masking process. There are several variations for this approach. They are subresolution (or outrigger) and rim phase shift masks.[14]

### Optical proximity corrected or optical process correction (OPC)

It is known that, in the sub-0.5-µm range, a perfect image on the mask/reticle can, from proximity effects, result in a distorted pattern in the resist. OPC masks attempt to reverse that situation by having a distorted image on the mask/reticle that is designed to produce a perfect image in the resist. Situations that are particularly vulnerable are dense patterns of adjacent open/opaque regions and the shortening and/or rounding of pattern corners such as small contact holes. A computer is used to analyze exposure process conditions (contrast effects) and design the on-mask pattern shapes.[15] An example of the correction of an end rounded pattern is shown in Fig. 10.16. Another technique is *double masking*. The first mask, a phase mask, creates part of the pattern in the resist. A follow-on mask, a trim mask, is somewhat oversized and completes the desired pattern.[16]

### Annular ring illumination

*Annular ring illumination* is a technique that was first introduced in Perkin Elmer scanning projection aligners. One of the guns in the resolution arsenal is a more uniform exposing light. Unfortunately, conventional optical exposure sources produce a light spot that is too nonuniform for small image exposure. However, within the spot, there are areas (rings) of more uniform energy. An annular ring illuminator blocks off all but a *ring* portion of the spot, directing a more uniform wave of exposing radiation to the wafer.

### Pellicles

The development of projection exposure systems (projection aligners and steppers) brought with them an increased mask and reticle lifetime. With the increased lifetime came the incentive to make higher quality masks and reticles. In a production line where masks are used for a long time, wafer-sort yield loss comes from dirt and scratches picked up during handling and use. One source of damage comes from mask and reticle cleaning steps. This situation is a "Catch-22." The

(a)

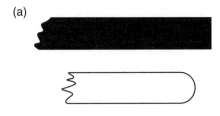

(b)

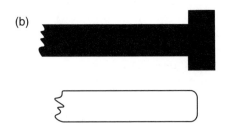

Figure 10.16 (*a*) Conventional image formation and (*b*) image enhancement with use of mask image "hammerhead."

masks and reticles become dirty during the process and require clean-ing. The cleaning procedure then itself becomes a source of contamina-tion, scratches, and breakage.

A solution to these problems is a pellicle (Fig. 10.17). A *pellicle* is a thin layer of an optically neutral polymer stretched onto a frame. The frame is designed to fit onto the mask or reticle. The pellicle is fitted to the mask or reticle after the mask is made and cleaned. Once in place, the pellicle membrane is the surface collecting any dirt or dust in the environment. The height of the membrane above the mask surface holds the dirt particles out of the focal plane of the mask. In effect, the particles are transparent to the exposing rays.

Another benefit of a pellicle is the elimination of scratches from the mask surface, since the surface is covered. A third benefit is that, once pelliclized, a mask or reticle does not need in-process cleaning. For some applications, the pellicle membrane is given an antireflective coating, which assists in the imaging of small geometries, especially on reflective wafer surfaces. These benefits translate into wafer-sort yield increases of 5 to 30 percent.[17]

Pellicle membranes are made from either nitrocellulose (NC) or cel-lulose acetate (AC). NC films are used in exposure systems with broadband exposure sources (340 to 460 nm)[18] while AC films are used in mid-ultraviolet applications. The membranes are thin (0.80 to 2.5 μm) and must show a high transmission rate for the rate exposure wavelengths. A typical pellicle will exhibit over a 99 percent transmis-sion rate for the peaks of the exposing wavelengths. Pellicle effective-ness requires stringent thickness control, on the order of ±800 Å, and control of particles to less than 25 μm in diameter.

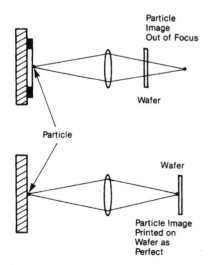

**Figure 10.17**  Pellicle.

A pellicle membrane is made by a spin casting technique. The pellicle material is dissolved in a solvent and spun onto a rigid substrate, such as a glass plate. This is the same technique used to spin photoresist onto wafers. Thickness of the membrane is controlled by the viscosity of the solution and the spin speed of the spin coater. The membrane is removed from the substrate and fixed onto the frame. Frame shapes are determined by the size and shape of the mask or reticle. Cleanliness control requires class 10 or better cleanrooms and antistatic packaging.

## Surface Problems

The resolution of small images is affected by several conditions on the wafer surface. Reflections off the surface layers, increasing variation of the topography, and the etching of multilayer stacks all require special process steps or "tweaking" of the process.

### Resist light scattering

In addition to light radiation reflecting off the wafer surface, the radiation tends to diffuse into the resist causing poor image definition. The amount of diffusion is in proportion to the resist thickness. Some additives put in the photoresist to increase radiation absorption also increase the amount of radiation diffusion, thus reducing image resolution.

### Subsurface reflectivity

The high-intensity exposing radiation ideally is directed at a 90° angle to the wafer surface. When this ideal situation exists, exposing waves reflect directly up and down in the resist, leaving a well defined exposed image (Fig. 10.18). In reality, some of the exposing waves are traveling at angles other than 90° and expose unwanted portions of the resist.

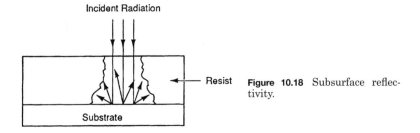

**Figure 10.18** Subsurface reflectivity.

This subsurface reflectivity varies with the surface layer material and the surface smoothness. Metal layers, especially aluminum and aluminum alloys, have higher reflectivity properties. A goal of the deposition processes is a consistent and smooth surface to control this form of reflection.

Reflection problems are intensified on wafers with many steps, also called a varied topography. The sidewalls of the steps reflect radiation at angles into the resist, causing poor image resolution. A particular problem is light interference at the step that causes a "notching" of the pattern as it crosses the step (Fig. 10.19).

## Antireflective Coatings

Antireflective coatings (ARCs) spun onto the wafer surface before the resist (Fig. 10.20) can aid the patterning of small images. The ARC layer brings several advantages to the masking process. First is a planarizing of the surface, which makes for a more planarized resist layer. Second, an ARC cuts down on light scattering from the surface into the resist, which helps in the definition of small images. An ARC can also minimize standing wave effects and improve the image contrast. The latter benefit comes from increased exposure latitude with a proper ARC.

An ARC is spun onto the wafer and baked. After the resist is spun on top of the ARC, the wafer is aligned and exposed. The pattern is developed in both the resist and the ARC. During the etch, the ARC acts as an etch barrier. To be effective, an ARC material must transmit light in the same range as the resist. It must also have good adhesion properties with the wafer surface and the resist. Two other require-

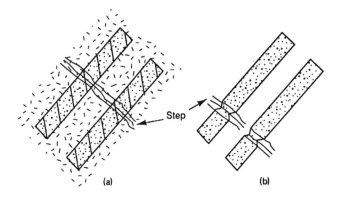

**Figure 10.19** Metal line "notching" over step. (*a*) Before etch and (*b*) after etch.

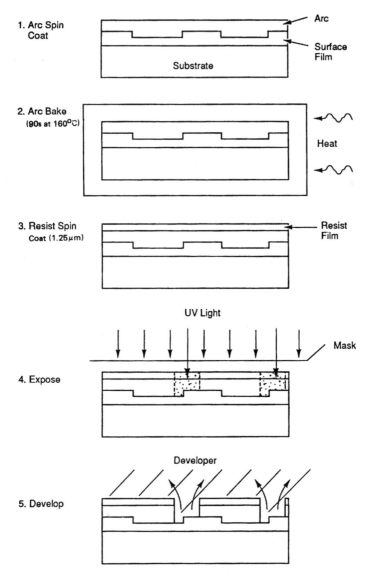

**Figure 10.20**  Antireflective process sequence.

ments are that the ARC must have a refractive index that matches the resist, and the ARC must develop and be stripped in the same chemicals as the resist.

There are several penalties associated with the use of an ARC. One is an additional layer requiring a separate spin and bake. The resolution gains offered by an ARC can be offset with poor thickness control and/or an ill-controlled developing step. The time of exposure can increase 30 to 50 percent increasing the wafer throughput time. ARC layers may also be used as the intermediate layer in a trilayer resist process or used on the top of the photoresist (top antireflective coating, or TAR).

## Standing waves

In the "Subsurface reflectivity" section, it was mentioned that the ideal exposure situation is when the radiation waves are directed to the wafer surface at 90°. This is true when only reflection problems are under consideration. However, 90° reflection causes another problem in positive photoresists—the creation of standing waves. As the radiation wave reflects off the surface and travels back up through the resist, it interferes constructively and destructively with the incoming wave, creating regions of varying energy (Fig. 10.21). The result, after development, is a rippled sidewall and a loss of resolution. A number of solutions are used to moderate standing waves, including dyes in the resist and separate antireflective coatings directly on the wafer surface. Most positive resist processes include a post-exposure bake (PEB) step before development of the resist layer. The bake reduces the influence of standing waves on pattern sidewall definition.

## Planarization

The advancement of circuits to VLSI levels has been brought about by both increased surface density and more surface layers. As the various layers are etched into patterns, the surface becomes stepped into high and low plateaus. The plateaus are delineated by steps each with dif-

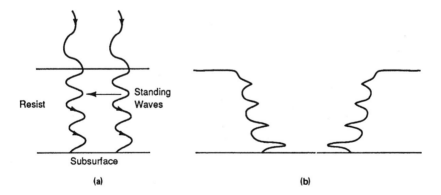

**Figure 10.21**  Standing-wave effect. (a) During exposure and (b) after develop.

ferent reflection qualities. The resolution of submicron (especially 0.5 μm and smaller) images on this type of surface is not possible in a simple one-resist layer process. One problem is depth-of-field issues. A limited-DOF lens cannot resolve images on both the high and low plateaus. Another is light reflection at the steps causing the notching of metal lines (Fig. 10.22). A number of *planarization* techniques are used to offset the effects of a varied wafer topography. Techniques include multilayer resist processing, planarization layers, reflow techniques, and physically flattening the surface by chemical-mechanical polishing. Current focus is on the global flattening of the surface (CMP) to allow patterning with only one resist layer. However, the multilayer resist processes described are still useful during the transition.

## Photoresist Process Advances

The ten-step lithography process detailed in Chapters 8 and 9 is based on a single photoresist layer, and it assumes that the layer can resolve the required images without pinholes or failing during the etch process. The nanometer era has forced the development of variations to the single resist layer process. These processes are usually used in conjunction with the advances in exposure control described previously.

### Multilayer resist/surface imaging

There are a number of multilayer resist processes. The choice of a process depends on the size of the resist opening and the severity of the surface topography. While multilayer resist processes add the penalty of more process steps in some situations, they are the only reliable means of creating the desired images. A multilayer resist process features a thicker bottom layer that serves to fill in the valleys and pla-

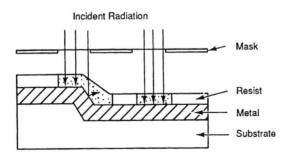

**Figure 10.22**  Light reflection at steps.

narize the surface. The image is first formed in a layer of photoresist on the top of the planarizing layer(s). This *surface imaging* allows small dimension imaging because the surface is flat, i.e., away from reflecting steps and in the same plane to avoid DOF problems.

A dual multilayer resist process uses two layers of photoresist, each with a different polarity. The process is suited to resolve small geometries on wafers with a varied topography. First, a relatively thick layer of resist is applied and baked to the thermal flow point (Fig. 10.23). A typical thickness is three to four times the highest step height on the wafer. The goal is a planar top resist surface. A typical multilayer process will use a positive-acting polymethylmethacrylate (PMMA) resist sensitive to deep ultraviolet radiation.

Next, a thin layer of positive resist, sensitive only to ultraviolet radiation, is spun on top of the first layer and processed through the development step. The thin top layer allows the resolution of the pattern without the adverse effects encountered with thick resist layers or reflections from steps in the surface. Since the top layer conforms to the shape of the bottom layer, it is referred to as a *conformal layer* or *por-*

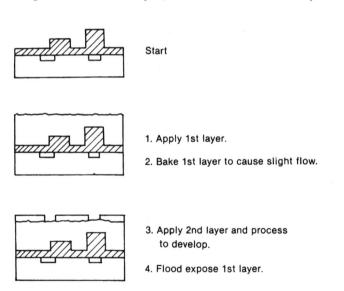

Start

1. Apply 1st layer.

2. Bake 1st layer to cause slight flow.

3. Apply 2nd layer and process to develop.

4. Flood expose 1st layer.

5. Develop 1st layer.

6. Etch and strip.

**Figure 10.23**  Dual-layer photoresist processing.

*table conformal layer.* This top layer of resist acts as a radiation block, leaving the bottom layer unpatterned. Next, the wafer is given a blanket or flood (no mask) deep ultraviolet exposure, which exposes the underlying positive resist through the holes in the top layer, thus extending the pattern down to the wafer surface. A development step completes the hole resolution and the wafer is ready for etch.

Considerations in the choice of photoresists are compatibility of the two resists through the process, reflection problems from the subsurface, standing waves, and sensitivity problems with PMMA resists.[13] In addition, the two resists must have compatible bake processes and independent developing chemistries.

Variations on the basic dual-level resist process include dyes in the PMMA and the use of antireflection layers under the first resist layer. Many variations on the basic dual-level process are practiced. One use of a dual-layer resist process is as a lift-off technique. By adjusting the development of the bottom layer, an overhang can be created that assists in the clean definition of the metal line on the surface (Fig. 10.24).

A trilevel resist process (Fig. 10.25) incorporates a "hard" layer between the two resist layers. The hard layer may be a deposited layer of silicon dioxide or other developer-resistant material. As in the dual-layer process, the image is formed in the top photoresist layer. Then, the image is transferred into the hard layer by a conventional etching step. The finishing step is the formation of the pattern in the bottom layer, using the hard layer as an etch mask. The use of a hard intermediate layer makes possible the use of nonphotoresist bottom layers such as a polyimide layer.

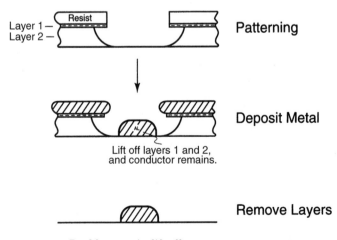

**Figure 10.24**  Dual-layer resist lift-off process.

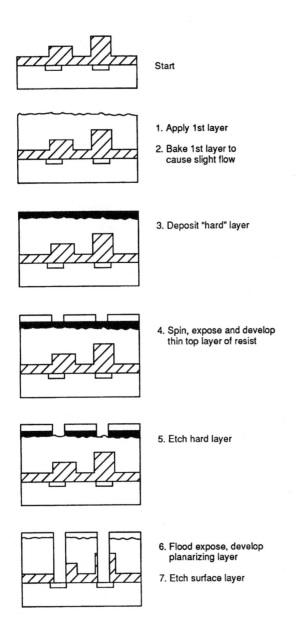

Start

1. Apply 1st layer

2. Bake 1st layer to
   cause slight flow

3. Deposit "hard" layer

4. Spin, expose and develop
   thin top layer of resist

5. Etch hard layer

6. Flood expose, develop
   planarizing layer

7. Etch surface layer

**Figure 10.25**  Trilevel resist process.

## Silylation/DESIRE process

A novel approach to surface imaging is the diffusion enhanced silylat-
ing resist (DESIRE) process[20] (Fig. 10.26). Like other multilayer resist
processes, the concept is to planarize the wafer and form the image in

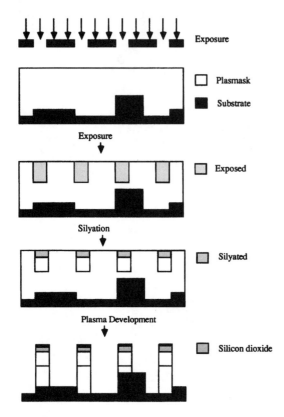

**Figure 10.26**   DESIRE process. *(Source:* Solid State Technology, *June 1987.)*

a surface layer. The DESIRE process uses on layer exposed by a standard UV exposure. In this process, the exposure is confined to the top layer of the planarizing layer. Next, the wafer is placed in a chamber (see vapor prime baking)[21] for exposure to HMDS for a *silylation process*. During this step, silicon becomes incorporated into the exposed areas. The silicon-rich areas become a hard mask, allowing the dry development and removal of the underlying material with an anisotropic RIE etch (Chapter 9). During the etch step, the silylated areas are converted to silicon dioxide ($SiO_2$), forming a more resistant etch mask. Techniques relying on defining the pattern in the top most layer are called *top surface imaged* (TSI).

### Polyimide planarization layers

Polyimides have been used for years in printed circuit board manufacturing. For semiconductor use, the polyimides offer the dielectric strength of deposited silicon dioxide films and the process advantage of application to the wafer with the same spinning equipment used for photoresist.[22]

Once applied to the wafer, the polyimides flow over the surface, making it more planar. After application and flow, the polyimide can be covered with a hard layer and patterned with chemicals much like a photoresist. A popular use of polyimide layers is as an interdielectric layer between two layers of conducting metal. The planarizing effect of the polyimide makes the definition of the second metal layer easier.

### Etchback planarization

Etchback is used for local planarization (Fig. 10.27). After metal lines are defined, a thick oxide layer is deposited and a photoresist layer spun on top of the oxide. Etch takes place in plasma etcher. First, the thinner resist etches away and starts etching the oxide. Later, the thicker resist etches away and some oxide is removed. The net result is a local flattening of the surface.

### Dual-damascene process

Increased component density has forced the use of multimetal layers. Their use required the need for connecting conductors called *studs* or *plugs*. Tungsten is the preferred metal, but there are complications in

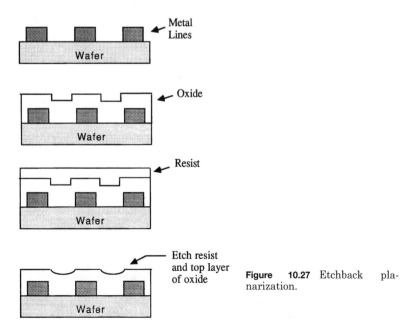

**Figure 10.27** Etchback planarization.

etching tungsten. Also, copper has become the preferred metallization system, replacing aluminum. However, copper technology introduces a whole host of process issues. One is the replacement of the traditional image and etch processes with a process called *dual-damascene*. It is an inlaid process similar in principle to the inlaid metal processes of antiquity. In that process, grooves are made in the surface of a bowl or other object. A metal is applied to the entire surface, also filling the groove. After the excess metal on the surface is removed, some remains in the grooves, leaving a decorative pattern. In the semiconductor application, the grooves are created with lithography techniques and the copper deposited by electroplating. The metal deposition process overspills the surface. A *chemical mechanical polishing* (CMP) step is used to remove the overspill, leaving the metal isolated in its trench (Fig. 10.28). Chapter 13 has a more detailed discussion on this new and important patterning technique.

### Chemical mechanical polishing (CMP)

Each of the previously described planarization techniques tends to smooth out the surface but does not produce a totally flat surface (*global planarization*). Small-dimension images with most planarization techniques are still difficult to define because of light scattering. They also can cause metal step coverage problems at the steps. The chemical mechanical polishing process used to flatten wafers in the wafer preparation stage (Chapter 3) is also used for the global planarization of in-process wafer surfaces.

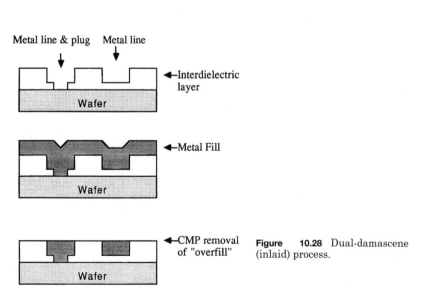

Metal line & plug    Metal line

◄─Interdielectric layer

Wafer

◄─Metal Fill

Wafer

◄─CMP removal of "overfill"

Wafer

**Figure    10.28** Dual-damascene (inlaid) process.

CMP is favored because of its ability to:[29]

- Achieve *global planarization* across the entire wafer
- Polish and remove all materials
- Work on multimaterial surfaces
- Make possible a high-definition damascene process and copper metallization
- Avoid the use of hazardous gases
- Be a low-cost process

The same basic process is used for both wafer polishing and planarization (Fig 10.29). However, the challenges are very different. In wafer polishing, several microns of silicon is removed. In metallization processes, CMP is required to remove amounts of material on the order of a micron or less. Also, these metal processes present a number of materials for removal. They have different removal rates and high expectations of uniform planarization. These issues are addressed in Chapter 13.

**Basic CMP processing steps.**    A wafer is mounted upside down on a carrier, which, in turn, is mounted, wafer face down, onto a rotating platen. The platen surface is covered with an polishing pad. A *slurry* carrying small abrasive particles is flowed onto the platen. The particles attack and remove small pieces of the wafer surface, which are carried away be the movement of the slurry across the surface. The combined actions of the two (orbital) rotations and the abrasive slurry polish the wafer surface. High plateaus on the wafer are polished first and faster than the lower areas, thus achieving planarization. These are the mechanical polishing actions. However, mechanical polishing alone is unacceptable for semiconductor processing due to excessive mechanical damage to the surface. Damage is reduced and/or managed by selecting a slurry chemistry that dissolves or etches the surface materials. Chemical removal generally requires the corrosion of the surface, usually through an oxidation mechanism. An analogy would be the rusting of iron, which is chemical corrosion. A layer of rust forms when iron is exposed to oxygen. However, the layer of rust slows down the rusting process by blocking the iron surface from the oxygen in the water or air. This is where the chemical and the mechanical work together. The abrasive particles wipe away the corrosion layer, exposing a fresh surface for corrosion, and the process repeats itself continuously. Following CMP, there is a *post-CMP cleaning* required to maintain wafer cleanliness.

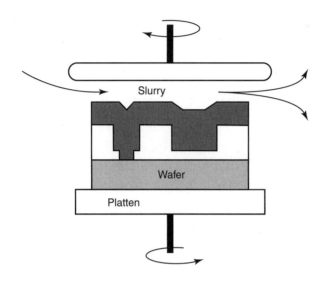

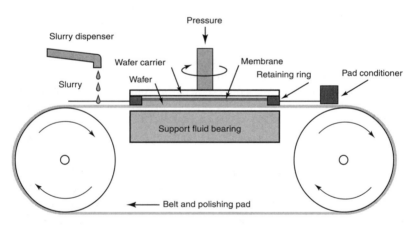

**Figure 10.29**   Chemical mechanical polishing and planarization (CMP).

Linear CMP systems are also used. In this arrangement, the wafer carriers are rotating over a moving belt instead of a rotating platen. A chief advantage of linear systems is the increased speed of the slurry movement under the wafer.

Primary measures of performance after a CMP process are:

- The surface flatness
- Surface mechanical condition
- Surface chemistry

- Surface cleanliness
- Productivity
- Cost of ownership (see Chapter 15)

## Polishing pads

Polishing pads are made of cast polyurethane foam with fillers, polyurethane impregnated felts, or other materials with special properties. Important pad properties are porosity, compressibility, and hardness.[24] Porosity, usually measured as the specific gravity of the material, governs the pad's ability to deliver slurry in its pores and remove material with the pore walls. Compressibility and hardness relate to the pad's ability to conform to the initial surface irregularities. Generally, the harder the pad, the more global the planarization. Softer pads tend to contact both the high and low spots, causing nonplanar polishing. Another approach is flexible polish heads that allow more conformity to the initial wafer surface.[31]

## Slurry

Slurry chemistry is complex and critical as a result of its dual role. On the mechanical side, the slurry is carrying abrasives. Small pieces of silica (silicon dioxide) are used for oxide polishing. Alumina ($Al_2O_3$) is a standard for metals. Multimetal surfaces found in multimetal layer schemes (Chapter 13) are challenging the industry to identify more "universal" abrasives. Abrasive diameters are kept in the 10 to 300 nm size[32] to achieve polishing, as opposed to grinding, which uses larger-diameter abrasives but causes more surface damage.

On the chemical side, the etchant would be KOH or $NH_4OH$ (basic solutions with low ph levels) for silicon or silicon dioxide. For metals such as copper, reactions usually start with an oxidation of the metal from the water in the slurry. A typical reaction is:

$$2CU + H_2O - Cu_2 + 2H^+ + 2^-$$

Following the oxidation, the basic materials chemically reduce the film, which is removed by the mechanical actions.

Various additives are found in production slurries. They perform various roles. Reducing post-CMP surface residues is addressed by balancing the ph of the slurry to control electrical charges on the $0^{28}$ abrasive particles. Silica-based slurries have high ph levels, while silica slurries have a ph below 7. Other additives are surfactants to establish desired flow characteristics and chelating agents. These latter agents interact with metal particles to reduce their redeposition on the wafer surface.

Other factors critical to a planar polish are the ph (degree of acidity or alkalinity), the flow dynamics at the wafer/pad interface, and the etch selectivity of the slurry on different surface materials and underlying layers.

### Polishing rates

A primary production measure is the polishing rate. Many factors influence the rate. Pad material parameters already described, the slurry types and size, and the chemicals and their properties used for corroding the surface are important factors. Other process factors include the pad pressure, the rotation rates, the flow rate of the slurry, the flow property (viscosity) of the slurry, and the temperature and humidity in the polishing chamber. Wafer diameter, pattern sizes, and surface materials are also factors. All of these factor must be balanced to achieve a productive polish rate (go faster) without creating an out-of-control process (go slower).

### Planarity

Global planarity is the goal of CMP, but the advent of multimetal schemes challenges that goal. Copper is a particular issue and illustrates several of the basic problems. Copper deposition into the trenches of a damascene patterning system (Fig. 10.30) results in lower density in the center. During the CPM process, the center polishes faster leaving a dish shape. Also, copper deposition density differences in dense patterns results in differential polishing across the pattern.

Tungsten plugs are also a CMP challenge (Fig. 10.31). During the initial CMP, the tungsten surface ends up recessed below the surrounding oxide. An oxide buff is required to planarize the surface.[32]

In some copper metallization schemes, tantalum is used as a diffusion barrier in trenches to keep the copper from diffusing into the silicon. However, the tantalum polishes much more slowly than copper (in a copper oriented slurry), leaving the copper surface exposed to more polishing time and more dishing.[28]

Pattern geometry variations result in differential removal rates. Larger areas tend to be removed faster, leaving dishing of low spots on the surface.

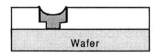

**Figure 10.30** "Dishing" of copper in trench.

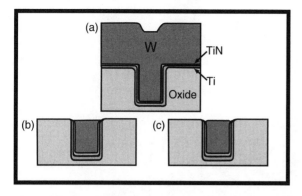

**Figure 10.31** Tungsten plug formation. (*a*) Deposit W, (*b*) CMP removal, and (*c*) buff oxide layer. *(Reprinted from the April 1998 edition of* Solid State Technology, *copyright 1998 by PennWell Publishing Company.*)

Also challenging is the presence of metals of different hardness, which polish at different rates, and interdielectric layers (IDL) of polymer materials that are soft. With all the challenges, by the year 2010, wafer surfaces will have to be flat down to the 150-nm range.

### Post-CMP clean

The critical role of clean wafer surfaces has been stressed throughout this book. It is just as important after CMP, and there are some particular challenges. CMP is the only process that intentionally puts particulates on the surface; namely, the abrasive particles. These are usually removed with mechanical brush cleaners or high-pressure water jets. Chemical cleaning generally employs the same techniques used for other FEOL cleaning.

Carefully choosing slurry surfactants and adjusting the ph can create an electrical repulsion between the slurry particles and the wafer surface. This technique can reduce contamination, particularly of the types that become electrostatically attached to the surface.

Copper contamination is a particular concern. If it gets into the silicon, it changes and diminishes electrical operations of the circuit components. Copper residues should be reduced to the $4 \times 10^{13}$ atoms/cm$^2$ range.[30]

### CMP tools

Operating a successful CMP process requires a sophisticated integrated system more than the simple polishing unit (Fig. 10.32). Production-level tools include automatic wafer handling robots, on-board

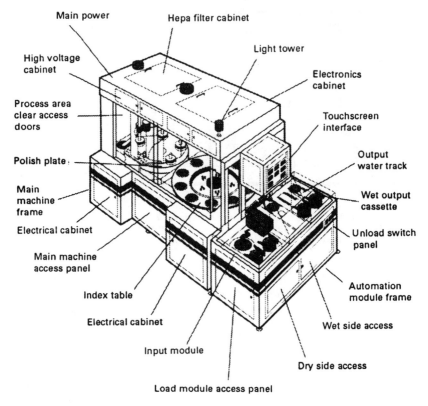

Main power          Hepa filter cabinet

Light tower

High voltage
cabinet                                                    Electronics
                                                           cabinet

Process area
clear access                                          Touchscreen
doors                                                 interface

                                                           Output
Polish plate                                               water track

Main
machine                                               Wet output
frame                                                 cassette

Electrical cabinet                                    Unload switch
                                                      panel

Main machine
access panel
                                                      Automation
Index table                                           module frame

Electrical cabinet                                    Wet side access

Input module

                                                      Dry side access

Load module access panel

**Figure 10.32** CMP System. *(Source: SpeedFam CMP-V system,* Semiconductor International, *May 1993.)*

metrology, and cleanliness detectors. Various end-point detection systems are used to signal when a particular layer material is gone or a specific removal depth is reached. Post-CMP cleaning units may be included in the main cabinet or mated to the primary CMP unit by robots in a cluster design. The goal is a "dry-in, dry-out" process.

## CMP Summary

CMP is a critical planarization process that requires a high level of integration and balancing of many process parameters. The primary parameters are: polishing pad composition, polishing pad pressure, pad rotation speed, platen rotation speed, slurry flow rates, slurry chemistry, and slurry material selectivity. In addition to planarization for lithography improvements, CMP is the enabling process for dual-damascene patterning and copper metallization. This application is further explored in Chapter 13.[25]

## Reflow

Some device schemes use a hard planarizing layer or layers. A popular layer is a deposited silicon dioxide doped with about 4 to 5 percent boron, called boron silicate glass (BSG). The presence of the boron causes the glass to flow at a relatively low temperature (less than 500°C), creating a planarized surface.

Another hard planarizing layer used is a spin-on-glass (SOG) layer. The glass is a mixture of silicon dioxide in a solvent that evaporates quickly. After spin application, the glass film is baked, leaving a planarized silicon dioxide film. The glass as spun is brittle,[26] and some formulas contain between 1 and 10 percent carbon to increase resistance to cracking.

### Image reversal

The preference for positive resists over negative resists for small geometry patterning has been discussed. One of the advantages with positive resists is the use of dark-field masks for the imaging of holes. Dark-field masks offer a lower-defect process, because the majority of the surface is covered by hard chrome that does not damage like glass. However, some mask levels require the printing of islands rather than holes. Metal mask levels are island patterns. Unfortunately, the printing of an island with a positive resist requires the use of a clear-field mask with its glass damage potential.

A process that allows the printing of islands with positive resists and dark-field masks is image reversal, which involves the formation of the image in the resist with a dark-field mask by conventional masking steps (Fig. 10.33). At the conclusion of the exposure step,

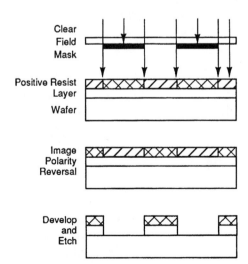

**Figure 10.33**  Image reversal.

there is an image in the resist that is reversed from the desired image. That is, if the resist was developed, a hole rather than an island would be formed.

The image reversal process involves exposing the resist-covered wafer to amine vapors in a vacuum oven. The vapors penetrate the resist, reversing its polarity. On removal of the wafers from the oven, they are given a flood exposure which completes the reversal process. The effect of the amine bake and flood exposure is to change the relative dissolution rates of the exposed and unexposed regions, thus reversing the original image when the resist is finally developed. This process is capable of the same resolution capabilities as a nonreversed positive process.[33]

### Contrast enhancement layers

Optical projection system resolution is approaching the limits imposed by the constraints of the lens and the wavelength of the exposing radiation. The two set up a condition in which the resist *contrast threshold* becomes the limiting factor. This is because, at short exposure times and ultraviolet and deep ultraviolet energies, the energy of the exposing wave varies in intensity. Thus, the image formed in the resist is fuzzy.

A method used to decrease this threshold is a contrast enhancement layer (CEL), which is a layer spun on top of the resist that is initially opaque to the exposing radiation (Fig. 10.34). During the exposure cycle, the CEL becomes bleached (transparent) and allows the radiation to pass into the underlying photoresist. The CEL responds first to the higher intensities before turning transparent, in effect storing the lower intensities before turning transparent. The result is that the resist receives a uniform exposure of high-intensity radiation, which improves its resolution capability. Another way to imagine the role of the CEL is as the top layer of a dual-layer resist system with the image being formed in the thin top layer.

Before the resist is developed, the CEL is removed by a chemical spray and development of the resist proceeds by normal processing. Positive resist processes normally capable of 1.0-μm resolution can achieve a 0.5-μm image with a CEL.

### Dyed resists

Various dyes may be added into the resist during manufacture. A dye may have one or several effects during the exposure step. One possible effect is the absorption of radiation, thereby attenuating the reflected radiation and minimizing standing wave effects. Another is a change of the dissolution rate of the resist polymer during development. This

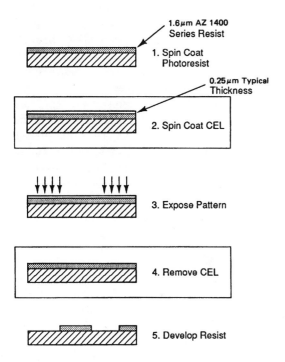

**Figure 10.34**  Contrast enhancement layer process flow.

effect creates a cleaner developed line (increased contrast).[34] An important use of dyes is the elimination of the notching that occurs in thin lines of deposited material crossing over surface steps. Addition of a dye to a resist can cause an increase in exposure time of 5 to 50 percent.[35]

## Improving Etch Definition

Forming the correct image in the photoresist is a critical step, but not the only step that defines the image in the wafer surface. Etching must also be controlled and precise. Several techniques are available that provide improved etch definition.

### Lift-off process

The final dimensions of the images in the surface layer are the result of variations in both the exposure step and the etch step. In processes where etch undercutting (resist adhesion) is a problem, such as aluminum etching, the etch component of the dimensional variation can be the dominant one.

A patterning variation that eliminates the etch variation component is lift-off (Fig. 10.35). In this variation, the wafer is processed through the development step, leaving a hole in the resist layer where a deposited layer is to be located. The exposure and development steps are adjusted to create a negative slope in the sidewall of the hole.

Next, the wafer receives the deposited layer, which covers the entire wafer and fills in the hole. Definition of the pattern comes when the wafer is processed through a photoresist removal step that lifts off the resist and unwanted metal layer. Usually, the removal step is assisted by ultrasonic agitation. This helps form a clean break of the deposited film at the resist edge. After resist and film removal, the desired pattern is left on the wafer surface.

### Self-Aligned Structures

Overetching has the effect of placing two structures closer together than intended. There is always some amount of overetching, and the alignment of some structures is absolutely critical. One solution is the *self-aligned structure,* such as a MOS gate (Chapter 16). The definition of the gate structure also defines the neighboring source/drain regions (Fig. 10.36). Opening the source/drain regions is a simple procedure of dip etching the oxide off the source/drain regions. The thinner oxide on the source/region allows a short etch time that does not allow enough time to etch the sides of the gate structure. The subsequent source/drain doping places the dopants next to the gate. This basic technique

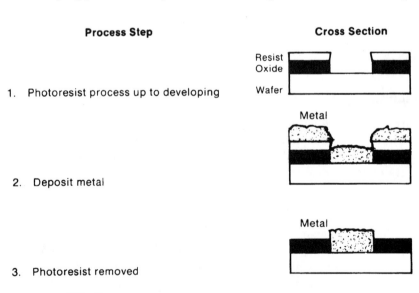

| Process Step | Cross Section |
|---|---|
| 1.  Photoresist process up to developing | Resist / Oxide / Wafer |
| 2.  Deposit metal | Metal |
| 3.  Photoresist removed | Metal |

**Figure 10.35**  Lift-off.

Gate

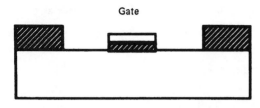

Source and Drain Self Aligned to Gate

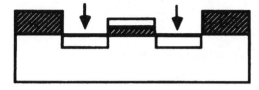

Figure 10.36   Self-aligned silicon gate (SAG) structure.

of a differential oxide thickness and dip etch is used to define/etch other structures. The design uses the gate structure as a doping block.

## Etch Profile Control

In Chapter 9, the concept of anisotropic etching was explained as a way of producing vertical (or near vertical) etched side walls. The problem becomes more complicated when the etched layer is actually a stack of different materials (Fig. 10.37). Use of an etchant that has poor selectivity can produce a stack with varying widths (Fig. 10.37b). Profile control for multilayer etches becomes a trade-off of the etchant chemistry, power levels, and the system pressure and design.

## Review Questions

1. What is a contrast effect, and how does it affect an image?
2. Name two problems resulting from subsurface reflection.

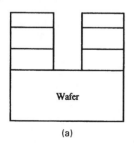

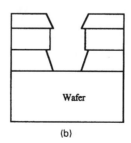

(a)                                    (b)

Figure 10.37   (a) Anisotropic and (b) isotropic etch of surface layer "stack."

3. What is the purpose of the top layer of a multilayer resist process?
4. What is the purpose of the bottom layer of a multilayer resist process?
5. What is flood exposure, and at what step is it used in a multiresist process?
6. What basic resist process step is eliminated in a lift-off process?
7. What advantages come from a planarization surface?
8. How does a contrast enhancement layer improve resolution?
9. How does a pellicle improve the wafer sort yield?
10. Explain the advantage of a self-aligned structure and give an example.

## References

1. C. Mack, "Lithography, Forecast 1993, Fitting the Pieces Together," *Semiconductor International,* Cahners Publishing, January 1993, p. 31.
2. M. McCallum, J Canning, G Shelden, Lithography Trends: Future Fab International 9, p. 145)
3. Staff, "Speeding the Transition to 0.018 μm," *Semiconductor International,* January, 1998, p. 66.
4. M. D. Levenson, "Extending Optical Lithography to the Gigabit Era," *Microlithography World,* Autumn 1994, p. 5.
5. L. Peters, "Reading Resists for the 90-nm node," *Semiconductor International,* February 2002, p. 63.
6. M. D. Levenson, "Extending Optical Lithography to the Gigabit Era," *Microlithography World,* Autumn 1994, p. 370.
7. Executive Summary, SCALPEL Process, web site.
8. S. K. Ghandhi, *VLSI Fabrication Principles,* John Wiley & Sons, Inc., New York, 1994, p. 687.
9. Y. Zhang, P. Ware, et al., Potential of KrF Scanning Lithography, Future Fab International 9, p. 14
10. S. K. Ghandhi, *VLSI Fabrication Principles,* John Wiley & Sons, Inc., New York, 1994, p. 693.
11. D. J. Elliott, *Integrated Circuit Fabrication Technology,* McGraw-Hill, New York, 1976, p. 82.
12. H. Jeong, D, Markle, G. Owen, F. Pease, A, Grenville and R. von Bunau, "The Future of Optical Lithography, *Solid State Technology,* PennWell Publishing Company, April 1994, p. 39.
13. A. Hand, Intrinsic Birefringence Won't Halt 157 nm Lithography, *Semiconductor International,* p. 42)
14. J. Reynolds, "Maskmaking Tour Video Course," Semiconductor Services, Redwood City, August 1991, Segment 10.
15. C. Spence, *Optical Proximity Photomask Manufacturing Issues,* OCG Microlithography Seminar Proceedings, 1984, p. 255.
16. R. Ixcoff, "Pellicles 1985; An Update," *Semiconductor International,* April 1985, p. 111.
17. *Micropel Product Data Sheet,* Micropel Division, EKC Technology, Hayward, CA, 1988.
18. C. H. Ling and K. L. Liauw, "Improved DUV Multilayer Resist Process," *Semiconductor International,* November 1984, p. 102.

19. F. Coopmans and B. Roland, "DESIRE: A New Route to Submicron Optical Lithography," *Solid State Technology,* PennWell Publishing Company, June 1987, p. 93.
20. B. Moffatt, "Private Conversation," *Yield Engineering Systems,* San Jose, CA.
21. K. Skidmore, "Techniques for Planarizing Device Topography," *Semiconductor International,* April 1988, p. 116.
22. J. Steigerwald, S. Muraka, and R. Gutmann, *Chemical Mechanical Planarization of Microelectronic Materials,* John Wiley & Sons, Inc., p. 4.
23. *Ibid,* p. 66.
24. K. Wijekoon, R. Lin, B. Fishkin, S. Yang, F. Redeker, G. Amico and S. Nanjangud, Tungsten CMP Process Developed. *Solid State Technology,* April 1998, p. 54.
25. J. Steigerwald, S. Muraka and R. Gutmann, *Chemical Mechanical Planarization of Microelectronic Materials,* John Wiley & Sons, Inc., p. 63.
26. M. Peterson, R. Small, G. Shaw, Z. Chen and T. Truong, Investigation CMP and Post-CMP Cleaning Issues for Dual-Damascene Copper Technology, *Micro,* January 1999, p. 31.
27. K. Wijekoon, R. Lin, B. Fishkin, S. Yang, F. Redeker, G. Amico and S. Nanjangud, "Tungsten CMP Process Developed." *Solid State Technology,* April 1998, p. 53.
28. R. Jackson, E. Broadbent, T. Cacouris, A. Harrus, M. Biberger, E. Patton, T. Walsh, "Processing and Integration of Copper Interconnects, *Solid State Technology,* March 1998, p. 49.
29. M. Peterson, R. Small, G. Shaw, Z. Chen and T. Truong, "Investigation CMP and Post-CMP Cleaning Issues for Dual-Damascene Copper Technology, *Micro,* January 1999, p. 30.
30. R. Iscoff, "CMP Takes a Global View," *Semiconductor International,* Cahners Publishing, May 1994, p. 74.
31. K. Skidmore, "Techniques for Planarizing Device Topography," *Semiconductor International,* April 1988, p. 117.
32. E. Aling and G. Stauffer, "Image Reversal of Positive Photoresist," *SPIE Journal,* vol. 539, March 11–12, 1985, p. 194.
33. J. F. Boland, H. F. Stanford, and S. A. Fine, "Dye Effects on Exposure and Development of Positive Resists," *Micro Manufacturing and Test,* August 1985, p. 18.
34. J. Housley, R. Williams, and I. Horiuchi, "Dyes in Photoresists: Today's View," *Semiconductor International,* April 1988, p. 142.
35. P. Burggraf, "Optical Lithography to 2000 and Beyond," *Solid State Technology,* February 1999, p. 31.
36. D. J. Elliott, *Integrated Circuit Fabrication Technology,* McGraw-Hill, New York, 1976, p. 168.

# 11

# Doping

## Overview

One of the unique properties of semiconductor materials is that their conductivity and the type of conductivity (N or P) can be created and controlled. In this chapter, the formation of specific "pockets" of conductive regions and N-P in and on the wafer surface is described. The principles and practice of the two doping techniques, diffusion and ion-implantation, are explained.

## Objectives

Upon the completion of this chapter, you should be able to:

1. Define an N-P junction.
2. Draw a flow diagram of a complete diffusion process.
3. List the three most common dopants used in silicon technology.
4. List the three types of deposition sources.
5. Draw a typical concentration-versus-distance curve for a deposition and drive-in.
6. List the major parts of an ion implanter.
7. Describe the principle of an ion implanter.
8. Compare the advantages and disadvantages of diffusion and ion implant processes.

## Introduction

The structure that makes transistors and diodes work is an N-P or N-P junction. A *junction* is the separation between a region that is rich in negative electrons (N-type region) and a region that is rich in holes (P-type region). The exact location of a junction is where the concentration of electrons equals the concentration of holes. This concept is explained later in the section, "Formation of a Doped Region and Junction by Diffusion."

The usual way to form junctions in the surface of semiconductor wafer is by thermal *diffusion* or by *ion implantation*. With thermal diffusion, dopant materials are introduced into the exposed top surface of the wafer, typically through a hole in the top silicon dioxide layer. With heating, they spread down into the bulk of the wafer. The amount and depth of the spread is governed by a set of rules, as explained below. These rules arise from a set of chemical rules that govern any movement of dopants in the wafer whenever the wafer is heated to a threshold temperature. In ion implantation, the dopant materials are literally shot into the wafer surface, with most of dopant atoms coming to rest below the surface. Additional movements of the implanted atoms are also governed by the rules of diffusion (Fig. 11.1). Thus, this chapter starts with a discussion of semiconductor junctions, goes on to the rules and techniques of diffusion, and ends with a description of the ion implant processes.

## Formation of a Doped Region by Diffusion

A major advance in semiconductor production was the development of diffusion doping. Diffusion, the movement of one material through another, is a natural chemical process with many examples in everyday life. Two conditions are necessary for a diffusion to take place. First, one of the materials must be at a higher concentration than the other. Second, there must be sufficient energy in the system for the higher-concentration material to move into or through the other. An example of a gas-state diffusion is the action of a common pressurized spray can (Fig. 11.2) such as a room deodorant. When the nozzle is depressed, the pressurized material leaves the can and moves into the surrounding air. Thereafter, movement of the gas into the room proceeds by the process of diffusion. The movement takes place while the nozzle is depressed and continues after it is closed. The diffusion will continue as long as the advancing spray is at a concentration higher than that in the air. As the material moves away from the can, the concentration of the material becomes progressively less. This is a characteristic of a diffusion process. Diffusion will continue until the concentration is even throughout the room.

Another example of a liquid-state diffusion is represented when a drop of ink is dropped into a glass of water. The ink is more concentrated than the surrounding water and immediately starts diffusing

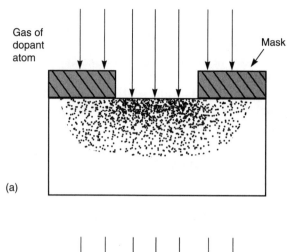

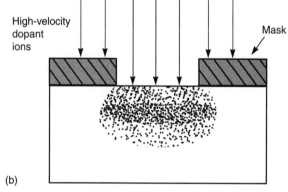

**Figure 11.1**  (*a*) Dopant concentration from diffusion and (*b*) dopant concentration from ion implantation.

**Figure 11.2**  Examples of diffusion.

into the glass of water. The diffusion will continue until the whole glass of water is the same color. This example can be used to illustrate the influence of energy on the diffusion process. If the water in the glass is heated (giving the water more energy), the ink will spread into the water faster.

The same diffusion phenomenon takes place when a doped wafer is exposed to a concentration of atoms higher than the concentration in the wafer. This is called *solid state* diffusion.

## Formation of a Doped Region and Junction by Diffusion

The formation of a doped region and junction is illustrated by examining the doping of a wafer in a diffusion process. The starting condition is depicted in Fig. 11.3. The wafer illustrated is from a P-type crystal. The + symbols in the diagram represent the P-type dopants that were incorporated into the crystal during the crystal-growing process. They are uniformly distributed throughout the wafer.

The wafer receives a thermal oxidation and a patterning process that leaves a hole in the oxide layer. In a diffusion tube, the wafer is exposed to a concentration of N-type dopants at a high temperature (the + symbols in Fig. 11.4). The N-type dopants diffuse through the hole in the oxide layer.

The effect in the wafer is illustrated by examining what happens at different levels in the wafer. The conditions in the diffusion tube are set such that the number of N-type dopant atoms that diffuse into the wafer surface are greater than the number of P-type atoms in layer no. 1. In the illustration, there are seven more N-type atoms than P-type atoms, and that level is electrically an N-type layer. This process has converted the top layer from P-type to N-type.

The diffusion process proceeds with N-type atoms diffusing from the first level down to the second level (Fig. 11.5). At the second level, there are again more N-type dopants than P-type dopants, converting level 2 to N-type. In the table (Fig. 11.6) is an accounting of the number of N-type and P-type atoms at each level. This process goes deeper into the wafer.

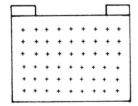

**Figure 11.3** P-type wafer ready for diffusion.

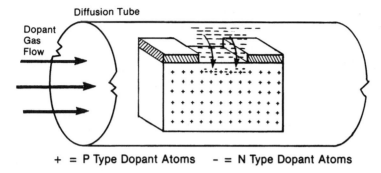

**Figure 11.4**  Start of a diffusion process.

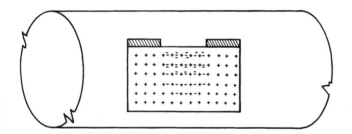

**Figure 11.5**  Cross section of wafer at conclusion of diffusion.

| Layer | # N's (-) | # P's (+) | Net (N - P) | Layer |
|---|---|---|---|---|
| 1 | 12 | 5 | 7 | N |
| 2 | 10 | 5 | 5 | N |
| 3 | 8 | 5 | 3 | N |
| 4 | 5 | 5 | 0 | Jct |
| 5 | 3 | 5 | -1 | P |
| 6 | 0 | 5 | -5 | P |

**Figure  11.6**  Dopant   amounts and level conductivity type.

### The N-P junction

At level 4, there are exactly the same number of N-type and P-type atoms. This level is the location of an N-P junction. The definition of an N-P junction is the location in the wafer where the number of N-type and P-type dopant atoms are equal. Note that, below the junction at level 5, there are three N-type atoms, which are not enough to convert that layer to N-type.

The term *N-P junction* indicates that there is a higher concentration of N-type dopants on one side of the junction. A P-N junction would indicate a higher P-type on one side of the junction. The behavior of electrical currents across semiconductor junctions gives rise to the particular performance of individual semiconductor devices and is the subject of Chapter 14. In this chapter, the emphasis is on the formation and character of doped regions in a wafer.

### Goals of solid-state diffusion

The goals of a diffusion process, thermal diffusion or ion-implantation, are threefold:

1. The creation of a specific number (concentration) of dopant atoms in the wafer surface
2. To create an N-P (or P-N) junction at a specific distance below the wafer surface
3. To create a specific distribution and concentration of dopant atoms in the wafer surface

### Lateral diffusion

The diffusion doping process depicted in Fig. 11.5 shows the incoming dopant atoms traveling straight down into the wafer. In reality, the dopant atoms move in all directions. An accurate cross section (Fig. 11.7) would show that some of the atoms have moved in a lateral direction, forming a junction under the oxide barrier. This movement is also called *lateral* or *side diffusion*. The amount of the lateral or side diffusion is approximately 85 percent of the vertical junction depth. Lateral diffusion takes place regardless of whether the introduction was through diffusion or ion implantation. The effects of side diffusion on circuit density are discussed in the introduction to ion implantation.

### Same-type doping

Some devices call for a doping with a dopant type the same as the host region. In other words, an N-type dopant will be put into an N-type

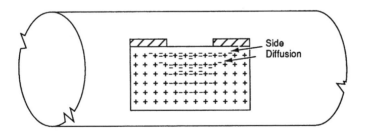

**Figure 11.7**   Side-diffused N-type dopants.

wafer, or a P-type dopant will be put into a P-type wafer (Fig. 11.8). When this situation happens, the added dopant atoms simply increase the concentration of the dopant atoms in the localized region. No junction is formed.

### Graphical representation of junctions

In a cross section (Fig. 11.9) of a semiconductor device, the N-P junctions are indicated simply as regions in the device. There is no graphical convention to represent N- or P-type areas. The drawings just show the relative location of the doped region and the junction. This type of graphical representation gives little information about the concentration of the dopant atoms and only approximates the actual dimensions of the regions. A drawing of a 2-μm-deep junction in a 20-mm-thick wafer, scaled to an 8-ft wafer, would have a junction depth only 0.4 in thick!

### Concentration versus depth graphs

Another two-dimensional graphical representation of a doped region is the concentration-versus-depth graph. The concentration is represented on the vertical axis and the depth into the wafer on the horizontal axis. An example of such a graph is illustrated in Fig. 11.10. The illustration uses the data from the doping example shown in Fig. 11.6. First, the P-type dopant concentration is plotted. In the example, there are exactly five P-type dopant atoms at every level, resulting in a straight horizontal line on the graph (Fig. 11.10b). Next, the number of N-type dopant atoms is plotted. Since the number of atoms decreases deeper into the wafer, the plotted line slopes down and to the right. At level 4, the number of N- and P-type dopants is

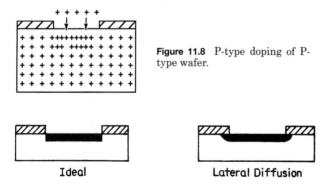

**Figure 11.8**  P-type doping of P-type wafer.

Ideal                    Lateral Diffusion

**Figure 11.9**  Cross-sectional representation of diffuse junctions.

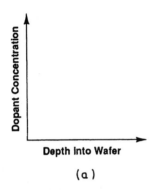

(a)

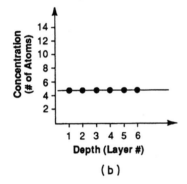

(b)

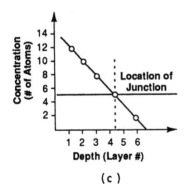

(c)

**Figure 11.10** Construction of concentration-versus-depth curve. (*a*) Axes, (*b*) P-type dopant, and (*c*) N-type and P- type dopant.

equal, and the lines cross. This is a graphical representation of the location of the junction.

The graph of an incoming dopant concentration versus depth profile for an actual process is not a straight line. They are curved lines. The shape of the curve is determined by the physics of the dopant technique. The actual shapes are discussed in the deposition and drive-in sections.

## Diffusion Process Steps

The use of *solid-state thermal diffusion* to create junctions in semiconductor wafers requires two steps. Step one is called *deposition,* and step number two is called *drive-in oxidation.* Both steps take place in a horizontal or vertical tube furnace. The equipment is the same described in Chapter 7 for oxidation.

## Deposition

The first step of a diffusion process is called *deposition;* it is also called *predeposition, dep,* or *predep.* It takes place in a tube furnace, with the wafers placed in the flat zone of the tube. A source of dopant atoms is located in the source cabinet and their vapors are transferred into the tube at a required concentration (Fig. 11.11).

In the tube, the dopant atoms diffuse into the exposed wafer. Within the wafer, the dopant atoms move by two different mechanisms: vacancy and interstitial movement. In the vacancy model (Fig. 11.12*a*) the dopant atoms move by filling empty crystal positions, called *vacancies.* The second model (Fig. 11.12*b*) relies on interstitial movement of the dopant.[1] In this model, the dopant atom moves through the spaces between the crystal sites; that is, intersite.

A deposition process is controlled or limited by several factors. One is the *diffusivity* of the particular dopant. Diffusivity is the rate (speed) of movement of the dopant through the particular wafer material. The higher the diffusivity, the faster the dopant moves through the wafer. Diffusivity increases with temperature.

Another factor is the *maximum solid solubility* of the dopant in the wafer material. It is the maximum concentration of a specific dopant that can be put into the wafer. A familiar analogy is the maximum liquid solubility of sugar in coffee. The coffee can dissolve only a certain amount of sugar before collecting in the bottom of the cup as a solid.

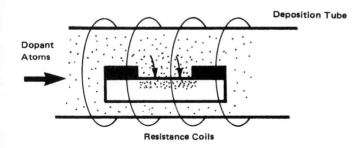

**Figure 11.11**  Deposition.

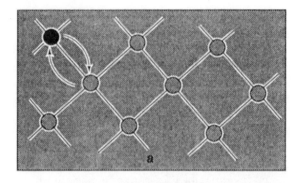

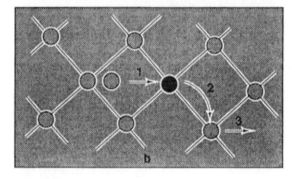

**Figure 11.12** Diffusion models. (*a*) Vacancy model and (*b*) interstitial model.

The maximum solid solubility limit increases with increasing temperature.

In a semiconductor deposition step, the concentration of the dopant is purposely set higher than the maximum solid solubility of the dopant in the wafer material. This situation ensures that the maximum amount of the dopant will be accepted by the wafer.

The amount of dopant entering the wafer surface is a function of the temperature only, and the deposition is said to take place at solid solubility. Maximum solubility levels for various dopants in silicon are shown in Fig. 11.13.

The concentration of dopant atoms at each level in the wafer is an important factor in the performance of junction diodes and transistors. A dopant concentration versus depth curve for a deposition is shown in Fig. 11.14. The shape of the curve follows a mathematical formula specifically known as an *error function*. An important factor in device performance is the concentration of the dopant at the wafer surface. This is called the *surface concentration* and is the quantity indicated where the error function curve intersects the vertical axis. Another

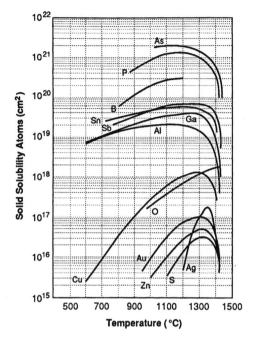

**Figure 11.13** The solid solubility of impurities in silicon.

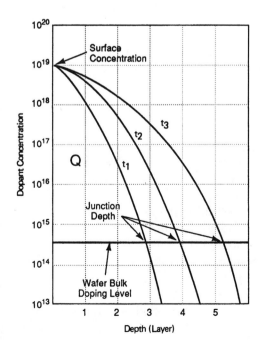

**Figure 11.14** Typical deposition (error function) dopant profile for three different deposition times.

deposition parameter is the total quantity of atoms diffused into the wafer. This amount increases with the time of the deposition. Mathematically, the quantity of atoms ($Q$) is represented by the area under the error function curve.

### Deposition steps

A deposition process requires four steps. They are:

1. Preclean and etch
2. Tube deposition
3. Deglaze
4. Evaluation

**Preclean and etch.**   Wafers coming to the deposition station first receive a preclean to remove particulates and stains. The chemicals and processes are the same as in the preoxidation process cleaning. After the preclean, the wafers will be chemically etched in an HF or HF and water solution to remove any oxide that may have grown on the exposed silicon surface. An oxide can form on the silicon from exposure to the air and from the chemical preclean process. Removal of the oxide is essential to allow the dopants to diffuse unimpeded into the wafer surface. The etch time and concentration must be balanced to prevent the surface-blocking oxide from being removed or thinned too much.

**Deposition.**   The deposition process, like oxidation, requires a minimum of three cycles. The first cycle is the loading cycle, which takes place in a nitrogen atmosphere. The second cycle is the actual doping cycle. The third cycle is the exit cycle, which also takes place in a nitrogen atmosphere.

Wafers are positioned on the boats at right angles (Fig. 11.15) or parallel to the tube axis. Right-angle placement is the highest packing density, but it can cause uniformity problems, because the wafers act as baffles to the gas flow. For uniform doping, the gas must mix uniformly between all of the wafers. Parallel placement offers the advantage of more uniform doping, since the doping gas proceeds unimpeded through the wafer boat. The disadvantage is a lower loading density. In both placement systems, dummy wafers are placed on the outsides and/or the front and back positions of the boat to create uniform doping of the interior device wafers.

**Deglaze.**   During the deposition cycle, a thin oxide can form on the exposed wafer surface. This oxide is doped and can act as an unwanted

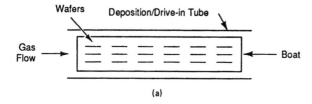

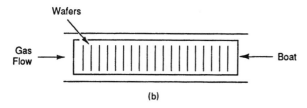

**Figure 11.15**  Boat loading patterns. (*a*) Parallel and (*b*) perpendicular.

source of dopant during the drive-in-oxidation step. Also, a deposition-created oxide can be difficult to etch, causing incomplete etch in a subsequent masking process. The oxide is removed from the surface by immersion in a diluted HF solution, followed by a water rinse and a drying step.

**Evaluation.**    Test wafers are added to the boat of device. A deposition test wafer has no patterns and has a conductivity type opposite that of the dopant. They are placed on the deposition boat in locations to sample the deposition distribution throughout the wafer batch. After deglazing, the test wafers are evaluated. The primary in-line test is of the sheet resistance with a four-point probe or contactless apparatus. The concept of sheet resistance and the measurement techniques are explained in Chapter 14. The junction depth after deposition is very thin and is generally not measured at this point.

Certification of process and tube cleanliness is determined by oxidizing test wafers and performing a capacitance-voltage measurement for mobile ionic contaminants (Chapter 14). Device wafers are 100 percent or sample-inspected for surface contamination and/or stains with high-intensity ultraviolet lamps or microscopes.

### Dopant sources

A deposition depends on the presence of a concentration of dopant atom vapors in the tube. The vapors are created from a dopant source located in the source cabinet of the tube furnace and passed into the tube with a carrier gas. Dopant sources are either in liquid, gaseous, or solid states. Several dopant elements are available in more than one state (Fig. 11.16).[2]

**Liquid sources.** Liquid sources of dopants are chlorinated or brominated compounds of the desired element. Thus, a boron liquid source is boron tribromide ($BBr_3$) and a phosphorus liquid source is phosphorus oxychloride ($POCl_3$). The liquid sources are held in temperature-controlled quartz flasks (Fig. 11.17). An inert gas, such as nitrogen, is bubbled through the heated liquid, and the gas becomes saturated with the dopant vapors. The flask is connected to a gas manifold whose valves are controlled by a microprocessor. The nitrogen carries the dopant vapors into the tube and maintains a gas volume sufficient to create a laminar flow condition in the tube. Laminar flow is required to prevent the gas from spiraling in the tube, creating nonuniform doping. Another device used to create laminar flow is a baffle

| Type | Element | Compound Name | Formula | State | Diffusion Reactions* |
|------|---------|---------------|---------|-------|----------------------|
| N | Antimony | Antimony Trioxide | $Sb_2O_3$ | Solid | |
| | Arsenic | Arsenic Trioxide | $As_2O_3$ | Solid | $2AsH_3 + 3O_2 \longrightarrow As_2O_3 + 3H_2O$ |
| | | Arsine | $AsH_3$ | Gas | |
| | Phosphorus | Phosphorus Oxychloride | $POCl_3$ | Liquid | $4POCl_3 + 3O_2 \longrightarrow 2P_2O_5 + 6Cl_2$ |
| | | Phosphorus Pentoxide | $P_2O_5$ | Solid | $2PH_3 + 4O_2 \longrightarrow P_2O_5 + 3H_2O$ |
| | | Phosphine | $PH_3$ | Gas | |
| P | Boron | Boron Tribromide | $BBr_3$ | Liquid | $4BBr_3 + 3O_2 \longrightarrow 2B_2O_3 + 6Br_2$ |
| | | Boron Trioxide | $B_2O_3$ | Solid | $B_2H_6 + 3O_2 \longrightarrow B_2O_3 + 3H_2O$ |
| | | Diborane | $B_2H_6$ | Gas | |
| | | Boron Trichloride | $BCl_3$ | Gas | $BCl_3 + 3H_2 \longrightarrow 2B + 6HCl$ |
| | | Boron Nitride | $BN$ | Solid | |
| | Gold | Gold | $Au$ | Solid (Evap.) | |
| | Iron | | $Fe$ | | |
| | Copper | | $Cu$ | | |
| | Lithium | | $Li$ | | Undesirable impurities |
| | Zinc | | $Zn$ | | from contamination |
| | Manganese | | $Mn$ | | |
| | Nickel | | $Ni$ | | |
| | Sodium | | $Na$ | | |

*Note: Only selected diffusion reactions are listed

**Figure 11.16** Deposition source table.

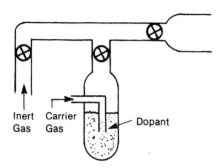

Figure 11.17  Liquid dopant source.

(Fig. 11.18) at the tube inlet. Baffles break up the incoming gas stream into a laminar form.

Also connected to the manifold is a reaction gas that is required to create the elemental dopant form in the tube. For BBr$_3$, the reaction gas is oxygen, which creates the boron oxide (B$_2$O$_3$), as shown in Fig. 11.19. At the wafer surface, a boron trioxide layer deposits on the silicon, and the boron diffuses from the oxide into the surface. Liquid sources offer the advantages of low to moderate cost and consistent doping. The disadvantages are uniformity problems (especially for larger-diameter wafers), safety considerations, and the potential of contamination associated with the opening of the flasks for recharging. Several vendors supply liquid dopants in pluggable sealed ampules that minimize contamination and safety problems.

**Gas sources.**   Many wafer manufacturers prefer gas dopant sources. These are hydrated forms of the dopant atom such as arsine (AsH$_3$ and diborane (B$_2$H$_6$). The gases are mixed in different dilutions in pressurized containers and connected directly to the gas manifold

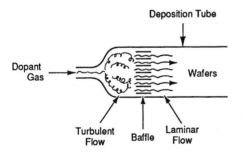

Figure 11.18  Use of baffle to create laminar flow.

$$4BBr_3 + 3O_2 \rightarrow 2B_2O_3 + 6Br_3$$

Figure 11.19  Reaction of liquid source BBR3 and oxygen in a deposition tube.

(Fig. 11.20). Gas sources offer the advantage of precise control through pressure regulators and are favored for deposition on larger-diameter wafers. The processes are in general cleaner, since the pressurized sources last longer than liquid sources. On the downside, unwanted chemical reactions in the manifold can create silica dust that can contaminate the tube and wafers.[2]

**Solid sources.**    The original deposition sources were solid. An oxide powder of the desired dopant was placed on a quartz holder, called a *spoon*, and placed in a source tube furnace attached to the main deposition tube (Fig. 11.21*a*). In the source furnace, the oxide gives off source vapors that are carried into the deposition tube where the diffusion takes place. This setup is called a *remote solid source*. Remote solid sources are economical but nonuniform. They are used primarily in the doping of discrete devices where less precision is required.

A more popular solid source is the planar source wafer (Fig. 11.21*b*). These are wafer-size "slugs" that contain the desired dopant. Boron slugs are a compound of boron and a nitride (BN). Slugs are also available for arsenic and phosphorus diffusions.

The slugs are stacked on the deposition boat, with one slug between every two device wafers. This arrangement is called a *solid neighbor source*. In the tube, the dopant diffuses out of the slug, crosses the short distance to the wafer, and diffuses into the surface. This system provides good uniformity for larger-diameter wafers, since the slug is the same size as the wafer. Slugs are safer to use, having no toxic vapors at room temperature. Drawbacks to the slug process include breakage of the slugs, lower productivity (the slugs take up space in

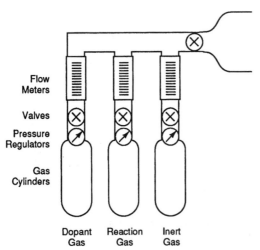

Flow Meters

Valves

Pressure Regulators

Gas Cylinders

Dopant Gas    Reaction Gas    Inert Gas

**Figure 11.20** Gas source manifold.

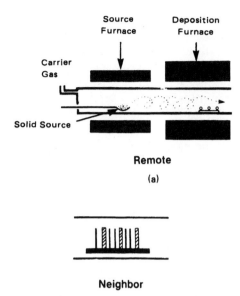

Figure 11.21  Solid source. (a) Remote furnace and (b) neighbor source.

the tube), and the necessity of cleaning the slugs. Some slug compounds require bake steps to maintain the dopant activity.

The third solid dopant source is a conformal layer spun directly on the wafer surface. The sources are powdered oxides (same as remote sources) mixed in solvents. They are spun onto the wafer surface using photoresist-type spinners and baked to evaporate the solvents. Left on the surface is a layer of doped oxide that conforms to the wafer surface. The wafers are placed on a boat and positioned in the deposition tube, where the heat drives the dopant out of the oxide and into the wafer.

Spin-on dopants have the potential of high uniformity and offer high productivity, and, like the slugs, are safe to handle. Problems with the system are the distribution of the dopant in the oxide layer; thickness variations, especially over surface steps; and the costs of the additional spin and bake steps.

## Drive-in Oxidation

The second major part of the diffusion process is the drive-in-oxidation step. It is also variously known as *drive-in, diffusion, reoxidation,* and *reox.* The purpose of this step is twofold: redistribution of the dopant in the wafer and growth of a new oxide on the exposed silicon surface.

1. The first step is redistribution of the dopant deeper into the wafer. During the deposition, a high-concentration layer and a shallow

layer of dopant are diffused into the surface. In the drive-in, there is no dopant source. The heat alone drives the dopant atoms deeper and wider into the wafer just as material from a spray can will continue to spread into the room after the nozzle is released. During this step, the total amount of atoms ($Q$) from the deposition step remains constant. The surface concentration is reduced, and the distribution of atoms takes a new shape. The distribution after the drive-in is described by mathematicians as a gaussian distribution (Fig. 11.22). The junction depth increases. Generally, the drive-in-oxidation step takes place at a higher temperature than the deposition step.

2. The second purpose of drive-in oxidation is the oxidation of the exposed silicon surfaces. The atmosphere in the tube is oxygen or water vapor, which performs the oxidation simultaneously as the dopants are being driven deeper into the wafer.

The setup, process steps, and equipment for the drive-in-oxidation step are the same as an oxidation process. After the completion of the drive-in, the wafers are again evaluated. The test wafers (from the deposition step) are again measured for surface concentration

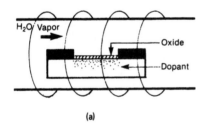

(a)

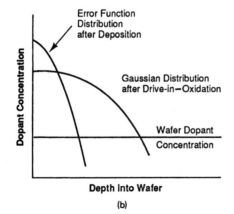

(b)

**Figure 11.22** Drive-in oxidation. (a) Cross section of wafer and (b) dopant concentration in wafer.

with a four-point probe, and the device wafers are checked for clean-liness. The test wafers are measured for junction depth and perhaps for mobile ionic contamination. After some diffusion steps, a test structure in the engineering die may be electrically probed for junc-tion parameters.

### Oxidation effects

The oxidation of the silicon surface affects the final distribution of the dopants.[3] The effects are related to the relocation of the top-level dopants after the oxidation. Recall that the silicon in the silicon diox-ide film is consumed from the wafer surface. The question to ask is, "What happened to the dopants that were in the top level?" The an-swer to that question depends on the conductivity type of the dopant.

If the dopant is an N-type, an effect called *pile-up* (Fig. 11.23a) oc-curs. As the oxide-silicon interface advances into the surface, the N-type dopant atoms segregate into the silicon rather than the oxide. The effect is to increase the number of these dopants in the new top layer of the silicon. In other words, the N-type dopants pile up in the wafer surface, and the surface concentration of the dopant is in-creased. Pile-up changes the electrical performance of the devices.

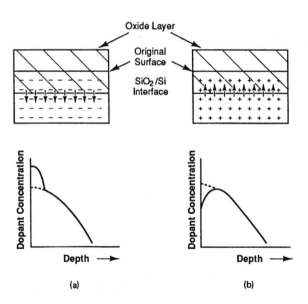

**Figure 11.23**   Pile-up and depletion of dopants during oxida-tion. (a) Pile-up of N-type dopants and (b) depletion of P-type dopants.

If the dopant is the P-type boron, an opposite effect occurs. The boron atoms are more soluble in the oxide and are drawn up into it (Fig. 11.23b). The effect on the wafer surface is a lowering of the concentration of boron atoms, which lowers the surface concentration and also affects the electrical performance of the devices. A summary of the deposition and drive-in-oxidation steps is provided in Fig. 11.24 The effect of the deposition and drive-in steps on the diffusion electrical parameters is shown in Fig. 11.25.

|  | Deposition | Drive-In |
|---|---|---|
| Goals | Introduction of Dopant | 1. Redistribution of Dopant<br>2. Reoxidation |
| Variables |  | 1. Surface Composition<br>2. Junction Depth<br>3. Time<br>4. Diffusivity<br>5. Temperature<br>6. Quantity of Atoms |
| Source Conditions | Continuous Source | No Source |
| Temperature Range | 900 – 1100°C | 1050 – 1200°C |
| Oxidation | No | Yes |

Figure 11.24   Summary of deposition and drive-in steps.

| Change | Parameter | Deposition | Drive-in/ Oxidation |
|---|---|---|---|
| Increase Temperature (same time) | Dopant concentration (Dc) | Increases | Same |
|  | Surface concentration (Co) | Increases | N type-Lower<br>P type- Higher |
|  | Junction depth (Xj) | Deeper | Deeper |
|  | Sheet Resistance | (Rs1) Lower | (Rs2) N type-Lower<br>P type- Higher |
| Increase Time (same temp.) | Dopant concentration (Dc) | Increases | Same |
|  | Surface concentration (Co) | Same | N type-Lower<br>P type- Higher |
|  | Junction depth (Xj) | Deeper | Deeper |
|  | Sheet Resistance | (Rs1) Lower | (Rs2) N type-Lower<br>P type- Higher |

Figure 11.25   Doping process parameter changes.

## Introduction to Ion Implantation

The advent of high-density circuits requires smaller feature sizes and closer spacing of the circuit components. Thermal diffusion places a limit on the production of advanced circuits. Five challenges are lateral diffusion, ultra thin junctions, poor doping control, surface contamination interference, and dislocation generation. Lateral diffusion occurs during deposition and drive-in but also continues every time the wafer is heated into a range where diffusion movement can take place (Fig. 11.26). The circuit designer must leave enough room between adjacent regions to prevent the laterally diffused regions from touching and shorting. The accumulative effect for a dense circuit can be a largely increased die area. Another problem with high temperatures is crystal damage. Every time a wafer is heated and cooled, crystal damage from dislocations occurs. A high concentration of these dislocations can cause device failure from leakage currents. One goal of an advanced process sequence is a reduced *thermal budget* to reduce these two problems.

The advent of MOS transistors created two new doping requirements: low dopant concentration control and ultra-thin junctions. Gate regions with dopant concentrations below $10^{15}$ atoms/cm$^2$ are required for efficient MOS transistors. However, this level is difficult to achieve consistently with a diffusion process. Scaling MOS transistors smaller to achieve higher packing densities also requires thin junction depths in the source/drain areas.[4] At the 0.18-μm design rule level, junctions will be in the 40 nm range; at the sub-0.10 μm range, they will be in the 20-nm range.[5]

A fourth problem is imposed by the physics/mathematics of a diffused region. As illustrated in Figs. 11.14 and 11.22, the majority of the dopant atoms are located near the wafer surface. This puts the majority of the current traveling near the surface where the dopant atoms are located. Unfortunately, this is the same location as contaminants (in and on the wafer surface) that interfere or degrade the current flow. A requirement of advanced devices is special wells in the surface (Fig. 11.27) with specific dopant gradients that cannot be achieved with diffusion technology. These wells allow high-performance transistors (see Chapter 16).

Required spacing

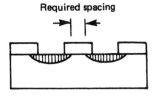

Figure 11.26   Side diffusion.

**Figure 11.27** Ion implantation analogy.

Ion implantation overcomes these limits of diffusion and also adds additional benefits. During the ion implant process, there is no side diffusion, the process takes place at close to room temperature, the dopant atoms are placed below the wafer surface, and a wide range of doping concentrations are possible. With ion implantation, there is greater control of the location and number of dopants put in the wafer. Also, photoresist and thin metal layers can be used as doping barriers along with the usual silicon dioxide layers. Given the benefits, it is not surprising that the majority of doping steps for advanced circuits are done by ion implantation. Diffusion still finds use in the doping of less-critical layers and will continue to be used for lower-integration-level circuits.

## Concept of Ion Implantation

Diffusion is a chemical process. Ion implantation is a physical process; that is, the act of implanting does not rely on chemical interaction between the dopant and the wafer material. An analogy that demonstrates the concept of ion implantation is a cannon firing balls into a wall (Fig. 11.27). Given enough momentum from the powder in the cannon, the balls will penetrate the wall and come to rest below the surface of the wall. The same events take place in an ion implantation machine. Instead of cannonballs, dopant atoms are ionized and isolated, accelerated (gain momentum), formed into a beam, and swept across the wafer. The dopant atoms physically bombard the wafer, enter the surface, and come to rest below the surface (Fig. 11.28b).

## Ion Implantation System

An ion implanter is a collection of very sophisticated subsystems (Fig. 11.28a), each performing a specific action on the ions. Ion implanters come in a variety of designs used for advanced research and/ or high-volume production. All of the machines have the same major subsystems as described below.

### Implant sources

The same dopant elements used in diffusion processes are used in ion implant processes. In diffusion processes, the dopants originate in liq-

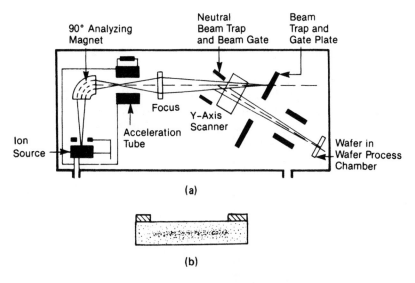

**Figure 11.28** Ion implantation. (*a*) Block diagram of ion implanter and (*b*) distribution of implanted atoms in wafers.

uid, gas, or solid sources. For ion implantation, only gas and solid sources are used.

Gases are favored for ion implantation because of their ease of use and higher control. The gases most used are $AsH_3$, $PH_3$, and $BF_3$. The gas cylinders are connected to the ion source subsystem through mass flow meters, which offer more control of the gas flow than normal flow meters. A newer system, called the *safe delivery system* (SDS) has the dopant, such as arsine, at low pressure and absorbed onto a solid surface.[6]

### Ionization chamber

The name "ion implant" implies that ions are a part of the process. Recall that ions are atoms or molecules with a negative or positive charge. The ions implanted are ionized atoms of the dopants. The ionization occurs in a chamber that is fed by the source vapors. The chamber is maintained at a low pressure (vacuum) of about $10^{-3}$ torr. Inside the chamber is a filament that is heated to the point where electrons are created from the filament surface. The negatively charged electrons are attracted to an oppositely charged anode in the chamber. During the travel from the filament to the anode, the electrons collide with the dopant source molecules and create a host of positively charged ions from the elements in the molecule. The results of the ionization of the source $BF_3$ are shown in Fig. 11.29.

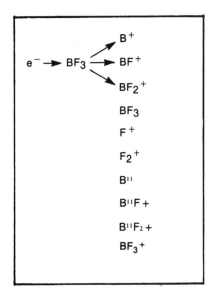

Figure 11.29  Ion species of $BF_3$.

Another ionization method uses a cold-cathode technique to generate the electrons, A high-voltage electric field is created between a cathode and anode, which creates the electrons in a self-sustaining process.

**Mass analyzing/ion selection**

At the top of the list in Fig. 11.29 is a lone boron ion. This is the atom that is desired in the wafer surface. The other species resulting from the ionization of the boron trifluoride are not wanted in the wafer. The boron ion must be *selected* from the group of positive ions. This process is called *analyzing, mass analyzing, selection,* or *ion separation.*

Selection is accomplished in a mass analyzer. This subsystem was first developed during the Manhattan Project for the atomic bomb. The analyzer (Fig. 11.30) creates a magnetic field. The species leave the ionization subsystem with voltages of 15 to 40 keV (thousand electron volts). In other words, they are traveling at a relatively high speed.

In the field, each of the positively charged species is bent into an arc with a specific radius. The radius of the arc is dictated by the mass of the individual species, its speed, and the strength of the magnetic field. At the exit end of the analyzer is a slit that will allow only one species to exit. The magnetic field is adjusted to match the path of the boron ion to the exit slit position. Thus, only the boron ion leaves the analyzing subsystem.

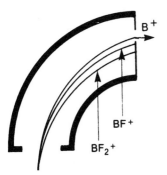

Figure 11.30   Analyzing magnet.

In some systems, the analysis also takes place after acceleration (Fig. 11.30). After acceleration, analysis is necessary if the implant species is a molecule that might separate in the acceleration process and/or to ensure an uncontaminated beam.

### Acceleration tube

On leaving the analyzing section, the boron ion moves into an acceleration tube. The purpose is to accelerate the ion to a high enough velocity, thus gaining sufficient momentum to penetrate the wafer surface. *Momentum* is defined as the product of the mass of the atom multiplied by its velocity. This section is kept at a high vacuum (low pressure) to minimize contaminants from entering into the beam. Turbo vacuum pumps (Chapter 13) are typically used for this purpose.

The required velocity is achieved by taking advantage of the fact that negative and positive charges attract each other. The tube is a linear design with annular anodes along its axis. Each of the anodes has a negative charge. The charge amount increases down the tube.

As the positively charged ion enters the tube, it immediately starts to accelerate down the tube. The voltage value is selected based on the mass of the ion and the momentum required at the wafer end of the implanter. The higher the voltage, the higher the momentum, and the faster and deeper the dopant ion can be implanted. Voltages range from 5 to 10 keV for low-energy implanters to 0.2 to 2.5 MeV (million electron volts)[6] for high-energy implanters.

Ion implanters are classified into the categories of medium- and high-current machines, high-energy devices, and oxygen ion implanters. The stream of positive ions exiting the tube is actually an electric current. The beam current level translates into the number of ions implanted per minute. The higher the current, the more atoms are implanted. The amount of atoms implanted is called the *dose*. Medium-current machines produce currents in the 0.5 to 1.7 mA (milliampere) range at energies from 30 to 200 keV (thousand electron volts). High-current machines generate beam currents of about 10 mA at energies up to 200 keV.[7] High-energy implanters are finding use in several CMOS doping applications, including retrograde wells, channel stops, and deep buried layers (see Chapter 16).

### Wafer charging

High-current implants create an unacceptable degree of electrical charging (*wafer charging*) of the wafer surface. The high-current beam carries excess positive charges that charge the wafer surface. The positive charge draws neutralizing electrons from the surface, the bulk, and from the beam. High charge levels can degrade or destroy surface dielectric layers. Wafer charging is a particular problem on thinner MOS gate dielectrics.[8] Methods used to neutralize or reduce the charge are *flood guns* specifically designed to provide electrons, a plasma bridge method of providing low-energy electrons,[9] and control of the electron path with magnetic fields.[10]

The relationships of current and energy for production-level implanters is shown in Fig. 11.31. High-energy machines accelerate ions in the 10 keV to 3.0 MeV range, with beam currents up to 1.0 mA. Oxygen implant machines are high-current implanters used to implant oxygen in SOI applications (see Chapter 16).

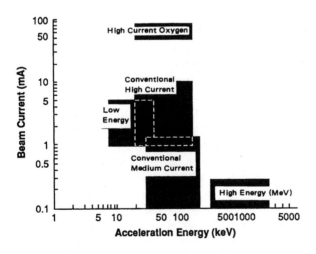

**Figure 11.31**  Ion implanters. The conventional descriptions of ion implanters were based on applications. However, some of today's more advanced implanters cannot be classified easily; several systems are capable of a broader process window than the conventional descriptions indicate. *(From Ref. 4.)*

A successful ion implant relies on the implantation of only the desired dopant atom. Single-dopant implantation requires that the system be maintained at a low pressure, greater than $10^6$ torr. The danger is that any residue molecules in the system (such as air) can become accelerated and end up in the wafer surface. Either oil-diffusion or cryogenic high-vacuum pumps are employed to reduce the pressure. The operation of these systems is described in Chapter 12.

## Beam focus

On exiting the acceleration tube, the beam separates due to repulsion of like charges. The separation (or defocus) causes uneven ion density and nonuniform layers in the wafer. For successful implantation, the beam must be focused. Electrostatic or magnetic lenses are used to focus the ions into a small diameter beam or a band of parallel beams.[11]

## Neutral beam trap

Despite the vacuum removal of the majority of the air in the system, there are still some residual gas molecules in the vicinity of the ion beam. Collisions between the ions and the residual gas atoms result in a neutralization of the dopant ion.

$$P^+ + N_2 \rightarrow P^0 \text{ (neutral)} + N_2^+$$

In the wafer, these "neutrals" cause nonuniform doping and, because they cannot be "counted" by evaluation equipment, they result in incorrect counting of the amount of dopants in the wafer. Suppression of the neutral beam is accomplished by bending the ion beam (Fig. 11.32) with electrostatic plates, leaving the neutral beam to travel straight ahead away from the wafers.

## Beam scanning

Ion beams have a smaller diameter than the wafer ($\sim$1 cm). Covering the entire wafer with a uniform doping requires scanning the beam

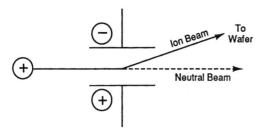

**Figure 11.32** Deflection of ion beam from neutral beam.

across the wafer. Three methods are used: beam scanning, mechanical scanning, and shuttering, alone or in combination.

A beam-scanning system (Fig. 11.33) has the beam pass between a number of electrostatic plates. Negative and positive charges can be controllably changed on the plates to attract and repel the ionized beam. By manipulating the charges in two dimensions, the beam can be swept across the entire wafer surface in a raster scan pattern.

Beam sweeping is used primarily in medium-current machines for single-wafer implanting. The procedure is fast and uniform. A drawback is the requirement that the beam be moved completely off the wafer to make the turns. For a large-diameter wafer, this procedure can lengthen the implant time by 30 percent or more. Another problem occurs with high-current machines where the high density of ions causes discharges (called *space charge forces*) that destroy the electrostatic plates. Wide beams are swept across the wafer. In some systems, the wafer is rotated 90° after each sweep of the beam to ensure uniformity.[12]

Mechanical scanning approaches the scanning problem by holding the beam in one position and moving the wafer(s) in front of it. Mechanical scanning is used primarily on high-current machines. One advantage is that there is no wasted time to turn the beam, and the beam speed is constant. If the wafer is at an angle to the beam, nonuniform implant depths can result. But, in some cases, the wafer will be oriented at an angle to the beam. Beam shuttering employs either an electronic field or a mechanical shutter to turn the beam on when it

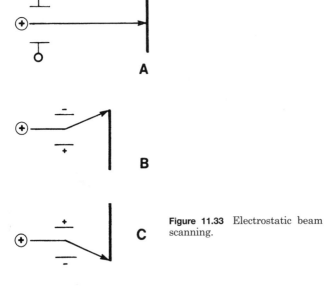

**Figure 11.33** Electrostatic beam scanning.

is on the wafer and off when it is not. Most systems use a combination of beam sweeping and mechanical movement.

### End station and target chamber

The actual implantation takes place in the target chamber of the end station. It includes the scanning system and the wafer load and unload mechanisms. There are several tough requirements for end stations. Wafers must be loaded into the chamber; a vacuum pulled; wafers individually placed on the holder; the implant completed; and the wafers dismounted, loaded into a cassette, and removed from the chamber.

Both batch and single-wafer designs are used to present the wafer surface to the beam (Fig. 11.34). A batch process is more efficient, but there are higher maintenance and alignment chores. For batch processing, the wafers are placed on a disk that rotates them in front of the beam as it is being scanned. The multiple motions increase the dose uniformity. Single-wafer designs require more time to process a group of wafers due to the additional time to load, pull the vacuum, implant, and unload.

Cryogenic pumps are favored for evacuating the end station. Contamination produced during the process consist of nitrogen outgassing from the wafers and hydrogen from the photoresist masking layer. Cryogenic pumps (see Chapter 13) are a capture type and hold the potentially dangerous hydrogen frozen in the pump.

The mechanical motions can take longer than the implant itself. Improvements include load locks to allow loading without breaking the

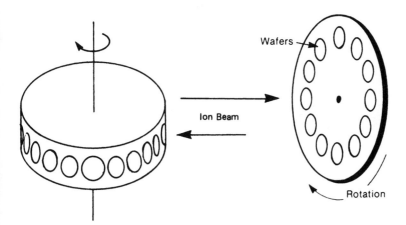

**Figure 11.34** Mechanical scanning.

chamber vacuum. A big challenge is to keep particulate generation low during all of the mechanical movements in the chamber.[13] Installation of antistatic devices in the chamber is critical. Electrostatic handlers (no mechanical clamps) are an option.[14]

Wafer breakage can cause contamination from wafer chips and dust, which in turn requires time-consuming cleaning. Contamination on the wafers causes shadowing that blocks the ion beam. Production speed must be maintained with a system that quickly evacuates the chamber for implanting and returns it to room pressure for exit and reloading. The target chamber may house a detector (called a *Faraday cup*) to "count" the number of ions impacting the surface. These detectors can automate the process by allowing beam contact with the wafer until the correct dose is achieved.

High-current implantation can cause the wafer to heat up, and these machines often have cooling mechanisms on the wafer holders. These machines may also have an *electron flood gun* (Fig. 11.35) designed to minimize a buildup of charge on the wafer surface that can electrostatically attract contamination.

**Ion implant masks**

A major advantage of ion implantation is the variety of masks that are effective blocks to the ion beam. In diffusion doping, the only effective mask is silicon dioxide. Most films employed in the semiconductor process can be used to block the beam, including photoresist, silicon dioxide, silicon nitride, aluminum, and other thin metal films.

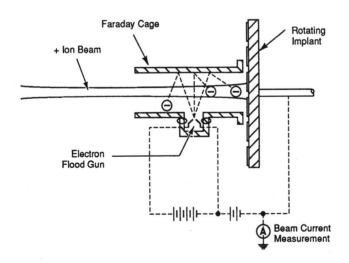

**Figure 11.35**  Electron flood gun.

Figure 11.36 compares the thicknesses required to block a 200-keV implant for various dopants.

The use of resist films as a beam block rather than an etched opening in an oxide layer offers the same dimension control advantage as the lift-off process; the etch step and its variability are eliminated. Use of resist layers also is more productive. Options to the use of silicon dioxide increase overall yield by minimizing the number of heating steps the wafers undergo.

## Dopant Concentration in Implanted Regions

The distribution of the ions in the wafer surface is different from the distribution after a diffusion process. The number and location of the dopant atoms in a diffusion process is determined by the diffusion laws and time and temperature. In an ion implant process, the number of atoms (dose) implanted is determined by the beam current density (ions per square centimeter) and the implant time.

The location of the ions in the wafer is a function of the incoming energy of the ions, the orientation of the wafer, and the stopping mechanism of the ion. The first two factors are physical ones. The heavier the incoming ion and/or the higher its energy, the deeper into the wafer it will move. The wafer orientation influences the stopping position because different crystal planes have different atom densities and the incoming ions are stopped by the wafer atoms.

Within the wafer, the ions are slowed and stopped by two mechanisms. The positive ions are slowed by electronic interactions with the negatively charged electrons in the crystal. The other interaction is the physical collision with the nucleus of the wafer atoms. All of the stopping factors are variable; the ions have a distribution of energies, the crystal is not perfect, and the electronic interactions and collisions vary. The net result is that the ions come to rest over an area in the

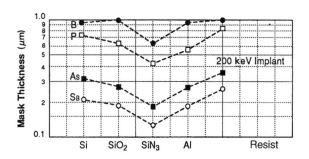

**Figure 11.36**  Barrier thickness required to block ion beam.

wafer (Fig. 11.37). They are centered about a depth called the *projected range* and fall off in density on each side of it. Additional implants create similar distribution patterns. Projected ranges for different dopants are shown in Fig. 11.38. The mathematical shape of

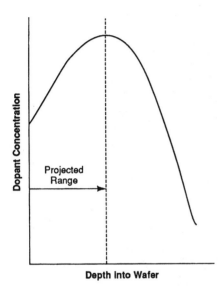

**Figure 11.37** Dopant concentration profile after an ion implant.

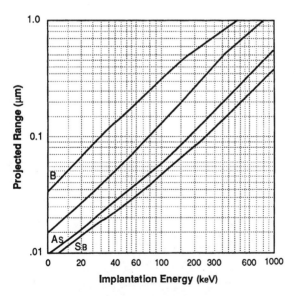

**Figure 11.38** Projected range of various dopants in silicon. *(After Blanchard, Trapp, and Shepard.)*

the ion distribution is a gaussian curve. A junction between the implanted ions and the bulk doping in the wafer takes place where the ion concentration equals the bulk doping concentration.

### Crystal damage

During the process of implantation, the wafer crystal structure is damaged by the colliding ions. There are three types of damage: lattice damage, damage cluster, and vacancy-interstitial.[15]

Lattice damage occurs when the ions collide with host atoms and displace them from their lattice site. A damage cluster occurs when displaced atoms in turn displace other substrate atoms, creating a cluster of displaced atoms. The most common implant-produced defect is a vacancy-insterstitial. This defect comes about when an incoming ion knocks a substrate atom from a lattice site and the displaced atom comes to rest in a nonlattice position (Fig. 11.39).

Light atoms, such as boron, produce a small number of displaced atoms. The heavier atoms, phosphorus and arsenic, generate a large number of displaced atoms. With prolonged bombardment, the regions of dense disorder may change to an amorphous (noncrystal) structure. In addition to the structural damage to the wafer from ion implantation, there is an electrical effect. The damaged regions do not have the required electrical characteristics, because the implanted atoms do not occupy lattice sites.

### Annealing and dopant activation

Restoration of the crystal damage and electrical activation of the dopants can be achieved by a thermal heating step. The temperature of the anneal is below the diffusion temperature of the dopant to prevent lateral diffusion. A typical anneal in a tube furnace will take place between 600 and 1000°C in a hydrogen atmosphere.

RTP techniques are also used for post-implant annealing. RTP offers fast surface heating that restores the damage without the substrate temperature rising to the diffusion level. Additionally, the anneal can take place in seconds, whereas a tube process takes 15 to 30 min. If the wafer has a heavy subsurface amorphous layer, the

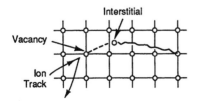

**Figure    11.39**  Vacancy-interstitial damage mechanism.

damage may be restored with a second implant of a light atom, such as oxygen or neon.

### Channeling

The crystalline structure of the wafer presents a problem during the ion implantation process. The problem comes about when the major axis of the crystal wafer is presented to the ion beam. Ions can travel down the channels, reaching a depth as much as 10 times the calculated depth. An ion concentration profile of a channeled cross section (Fig. 11.40) shows an significant amount of additional dopants. Channeling is minimized by several techniques: a blocking amorphous surface layer, misorientation of the wafer, and creating a damage layer in the wafer surface.

The usual blocking amorphous layer is simply a thin layer of grown silicon dioxide (Fig. 11.41). The layer randomizes the direction of the ion beam so that the ions enter the wafer at different angles and not directly down the crystal channels. Misorientation of the wafer 3 to 7° off the major plane also has the effect of preventing the ions from entering the channels (Fig. 11.42). Predamaging the wafer surface with a heavy silicon or germanium implant creates a randomizing layer in

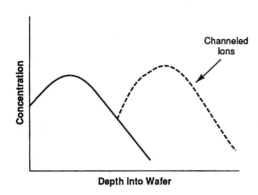

**Figure 11.40** Effect of channeled ions on total dose.

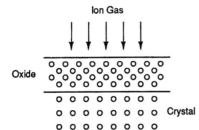

**Figure 11.41** Implant through an amorphous oxide layer.

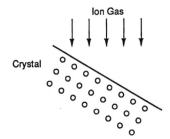

**Figure 11.42** Misorient the beam direction to all crystal axes.

the wafer surface (Fig. 11.43). The method increases the use of the expensive ion-implant machine. Channeling is more of a problem with low-energy implants and heavy ions.[16]

### Evaluation of Implanted Layers

The evaluation of implanted wafers is essentially the same as for diffused layers. Four-point probes are used to determine the sheet resistance of the layer. Spreading resistance techniques and capacitance-voltage techniques determine the concentration profile, dose, and junction depth. Junction depths can also be determined by bevel and decoration methods. These procedures are explained in Chapter 14.

For implanted layers, a special structure, called the *van der Pauw structure*, is sometimes used in place of a four-point probe (Fig. 11.44). The structure allows a determination of sheet resistance without the contact resistance problems of a four-point probe. Variation in the implanted wafers can come from many sources: the beam uniformity,

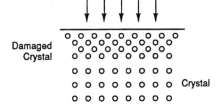

**Figure 11.43** Predamage on the crystal surface.

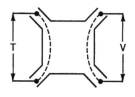

**Figure 11.44** Van der Pauw text pattern.

variations in voltage, scanning variations, and problems in the mechanical systems. These potential problems have the possibility of causing a wider sheet resistance surface variation than a diffusion process. To detect and control the sheet resistance across the wafer surface, mapping techniques are popular and, for critical implants, required. Wafer surface maps (Fig. 11.45) are drawn from four-point probe measurements that are computer corrected for proximity and edge effects.

A measurement technique unique to ion implantation is optical dosimetry. The technique requires spinning a glass disk with a layer of photoresist. Before being placed in the ion implanter, the resist film is scanned with a dosimeter that measures the absorption of the film. The information is stored in a computer. The wafer and film receive the same implant as the device wafers. The resist absorbs the ion dose and darkens. After the implant, the film is again scanned. The computer subtracts the before-implant value for each location and prints out a contour map of the surface. The spacing of the contour lines is an indication of the uniformity of the dopants in the surface (Fig. 11.46).

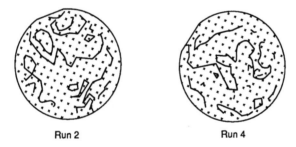

Run 2                              Run 4

**Figure 11.45** Four-point probe surface measurement pattern. *(Courtesy of Prometrics.)*

**Figure 11.46** Contour map of ion dose.

## Uses of Ion Implantation

An ion implantation can be substituted for any diffusion deposition process. The greater control and lack of side diffusion make it the preferred doping technique for dense and small-feature-size circuits. A predeposition application in CMOS devices is the creation of the deep P-type wells (see Chapter 16), called *retrograde wells,* using high-energy implants.

A particular challenge is ultra-shallow junctions. These are junctions in the sub-125-nm range. As devices are continually scaled to smaller dimensions, junction areas also become reduced. This in turn leads to lower energy ion implant processes to minimize surface damage and channeling. This leads to implanting pure boron instead of $BF_3$, which contains corrosive fluorine. All of these requirements are being met in a new generation of implanters that can deliver high-dose beams at acceptably low energies.

One of the most important uses of ion implantation is for MOS gate threshold adjustment (Fig. 11.47). A MOS transistor consists of three parts: a source, a drain, and a gate. During operation, a voltage is applied between the source and drain regions. However, no current can flow between the regions until the gate becomes conductive. The gate becomes conductive when a voltage applied to it causes a conductive channel to form in the surface and connects the source and drain. The amount of voltage required to first form the connecting channel is called the *threshold voltage* of the device. The threshold voltage is very sensitive to the dopant concentration in the wafer surface under the gate. Ion implantation is used to create the required dopant concentration in the gate region. Also, in MOS technology, ion implantation is used to alter the field dopant concentration. However, in this use, the purpose is to set a concentration level that prevents current flow between adjacent devices. In this application, the implanted layer is part of a device *isolation* scheme.

In bipolar technology, ion implantation is used to create all of the various transistor parts. The "custom" dopant profiles available from

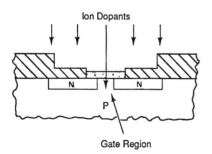

Ion Dopants

**Figure 11.47**  Ion doping of MOS gate region.

Gate Region

implantation can improve device performance. One particular application is arsenic buried layers. When the buried layer is diffused, the high concentration of arsenic atoms affects the quality of a subsequent epitaxial layer deposited on the surface. By using an arsenic ion implant, a high concentration of arsenic is possible, and an anneal process restores the damage, allowing a higher-quality epitaxial layer to be deposited.

Resistors for both MOS and bipolar circuits are good candidates for ion implantation. Diffused resistors vary in uniformity from 5 to 10 percent, whereas ion-implanted resistors have variations of less than 1 percent or better. Figure 11.48 is a table of typical applications of ion-implantation doping.

## The Future of Doping

Ion implantation also has its drawbacks. The equipment is expensive and complex. Training and maintenance take longer and have more stringent requirements. The machines present new dangers in the form of high voltages and more toxic gas use. From the process perspective, the biggest worry is over the ability of the annealing processes to completely eliminate the implant-induced damage. However, despite the drawbacks, ion implantation is the preferred doping process for advanced circuits. In addition, many new structures are possible only with the unique advantages offered by ion implantation.

Ion implantation may run out of steam beyond the shallow junction range of 40 nm.[17] A new technique under development is *plasma ion immersion* (PII). In this technique the analyzing magnet is removed from the system. Dopants leave the source section and a plasma field increases their energy.[19] This technique places the wafer in a plasma field (similar to ion milling or sputtering) in the presence of dopant atoms. With proper charges on the dopant atoms and the wafer, the dopant atoms accelerate to the wafer surface and penetrate much like an ion implantation. The difference is the lower energy of the plasma

| Typical Uses of Ion Implantation |
|---|
| • Gate threshold voltage adjustment |
| • Ultra-shallow junctions |
| • Buried layers (retrograde wells) |
| • Predeposition layer |
| • Insulation layer of SOI |

**Figure 11.48**   Ion implantation uses.

field yielding less wafer charging and giving more control of the shallow junctions.[18]

Ion implantation, in one form or another, is the doping technology that will carry the industry into the nanometer era. The benefits include:

- Precise dose control from $10^{10}$ to $10^{16}$ atoms/cm$^2$
- Uniform topping of large areas
- Dopant profile control through energy selection
- Relative ease of implanting all dopant elements
- Minimal side diffusion
- Implanting of nondopant atoms
- Can dope through surface layers
- Choice of dopant barriers for selective doping
- Tailored doping profiles in deep (retrograde) wells

## Review Questions

1. Explain the difference between a deposition and a drive-inoxidation.
2. Name the three types of sources used in a doping processes.
3. List two evaluation measurements made after a doping process.
4. Is the junction depth deeper after deposition or the drive-in oxidation?
5. What type of source is $BCl_3$?
6. Name the subsystems in an ion implantation.
7. In what subsystem is the dopant ion selected?
8. What is the difference between a diffusion and an ion-implanted dopant profile? Sketch an example of each.
9. Name two materials used as an ion implant doping barrier.
10. Why is an anneal required after ion implantation?

## References

1. P. B. Griffin and J. D. Plummer, "Advanced Diffusion Models for VLSI," *Solid State Technology*, May 1988, p. 171.
2. K. T. Robinson, "A Guide to Impurity Doping," *Micromanufacturing and Test*, April 1986, p. 52.
3. P. Guise and R. Blanchard, *Modern Semiconductor Fabrication,* Reston Books, Reston, VA, 1986, p. 46.

4. S. Felch, et al., "A Comparison of Three Techniques for Profiling Ultrashallow $p^+$-n Junctions," *Solid State Technology*, PennWell Publishing Company, January 1993, p. 45.
5. N. Cheung, "Ion Implantation," *Semiconductor International*, Cahners Publishing, January 1994, p. 411.
6. P. Lessard, Vacuum Issues in the Semiconductor Industry, Ion Implantation, Vacuum Technology and Coating, June 2003, p. 35.
7. P. Burggraaf, "Ion Implanters: Major Trends," *Semiconductor International*, April 1986, p. 78.
8. R. Iscoff, "Are Ion Implanters the Newest Clean Machines?," *Semiconductor International*, Cahners Publishing, October 1994, p. 65.
9. N. Cheung, "Ion Implantation," *Semiconductor International*, Cahners Publishing, January 1993, p. 35.
10. J. England, et al., Charge Neutralization During High-Current Ion Implantation," *Solid State Technology*, PennWell Publishing, July 1994, p. 115.
11. Japan Report, *Semiconductor International*, Cahners Publishing, November 1994, p. 32.
12. Eaton Corp., Product Video, The NV8200P, 1993.
13. *Ibid.*
14. R. Iscoff, "Are Ion Implanters the Newest Clean Machines?" *Semiconductor International*, Cahners Publishing, October 1994, p. 65.
15. "Wafer Handler for Ion Implanters, Varian Semiconductor Equipment," *Solid State Technology*, PennWell Publishing, July 1994, p. 131.
16. J. Hayes and P. Van Zant, *Doping Today Seminar Manual, Semiconductor Services*, San Jose, CA, 1985.
17. D. Zrudsky, "Channeling control in ion implantation," *Solid State Technology*, July 1988, p. 73.
18. N. Cheung, "Ion implantation," *Semiconductor International*, Cahners Publishing, January 1993, p. 35.
19. A. Braun, Ion Implantation Goes Beyond Traditional Parameters, *Semiconductor International,* March 2002, p. 48.
20. P. Singer, "Plasma Doping: An Implant Alternative?," *Semiconductor International*, Cahners Publishing, May 1994, p. 34.

# Layer Deposition

## Overview

Doped regions and N-P junctions are the electronic hearts of the active components in a circuit. However it takes various other layers of semiconductors, dielectrics, and conductors to complete the components and facilitate the integration of the components into the circuit. These layers are added to the wafer surface by a number of techniques. The principle ones are chemical vapor deposition (CVD), physical vapor deposition (PCD), electroplating, spin-on, and evaporation. This chapter describes the most commonly used CVD techniques and dielectric and semiconductor materials deposited on the wafer surface. PVD, electroplating, spin-on, and evaporation processes are described in Chapter 13.

## Objectives

Upon completion of this chapter, you should be able to:

1. Name the parts of a CVD reactor.

2. Describe the principle of chemical vapor deposition.

3. List the conductor, semiconductor, and insulator materials deposited by CVD techniques.

4. Know the difference between atmospheric CVD, LPCVD, hot-wall, and cold-wall systems.

5. Explain the difference between epitaxial and polysilicon layers.

## Introduction

Advances in photomasking technology have allowed the fabrication of ever-smaller dimensioned ULSI circuits. But as the circuits have shrunk, they also have grown in the vertical direction through increased numbers of deposited layers. In the 1960s, bipolar devices had two layers deposited by chemical vapor deposition (CVD), an epitaxial layer, and a top-side passivation layer of silicon dioxide (Fig. 12.1), while early MOS devices had just a passivation layer (Fig. 12.2). By the 1990s, advanced devices featured four levels of metal interconnects, requiring numerous deposited layers (see Fig. 13.2, Chapter 13). The added layers take a variety of roles in the device/circuit structures. They include the following:

- Deposited doped silicon layers called *epitaxial layers* (see related section in this chapter)

- Intermetal dielectrics (IMDs)

- Vertical (trench) capacitors

- Intermetal conducting plugs

- Metal conducting layers

- Final passivation layers

There are two primary techniques for layer deposition: chemical vapor deposition (CVD) and physical vapor deposition (PVD). The metallization deposition techniques of evaporation and sputtering are explained in Chapter 13. The uses of the particular films, while indicated in this chapter, are detailed in Chapters 16 and 17. Chemical vapor deposition (CVD), the subject of this chapter, is practiced in a number of atmospheric and low-pressure techniques.

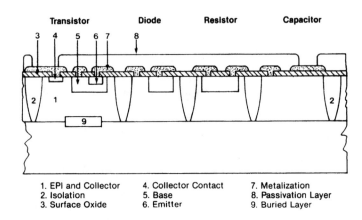

1. EPI and Collector    4. Collector Contact    7. Metalization
2. Isolation            5. Base                 8. Passivation Layer
3. Surface Oxide        6. Emitter              9. Buried Layer

**Figure 12.1** Cross section of bipolar circuit showing epitaxial layer and isolation.

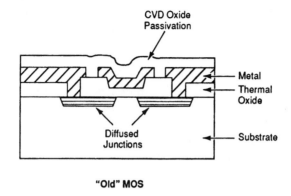

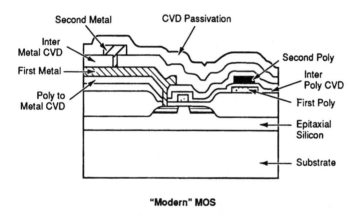

Figure 12.2 Evolution of MOS layers.

## Film parameters

Device layers must meet general and specific parameters. The specific parameters are noted in the sections on individual layer materials. General criteria that all films must meet for semiconductor use include:

- Thickness/uniformity
- Surface flatness/roughness
- Composition/grain size
- Stress free
- Purity
- Integrity

Uniform thickness is required of films to meet both electrical and mechanical specifications. Deposited films must be continuous and free of pinholes to prevent the passage of contamination and to prevent shorting of sandwiched layers. This is of great importance for thin films. Epitaxial films have shrunk from 5-micron levels to submicron thicknesses. Recall that the thickness of a layer is one of the factors contributing to its resistance. Also, thinner layers tend to have more pinholes and less mechanical strength. Of particular concern is the maintenance of thickness over steps (Fig. 12.3). Excessive thinning at a step can cause electrical shorts and/or unwanted induced charges in the device. The problem becomes very acute in deep and narrow holes and trenches, called *high-aspect-ratio patterns*. The ratio is calculated by dividing the depth by the width (Fig. 12.3). One problem is a thinning of the deposited film at the lip of the trench. Another is thinning in the bottom of the trench. Filling high-aspect trenches is a major issue in the execution of multimetal structures.

Surface flatness is as important as the thickness. In Chapter 10, the effect of steps and surface roughness on image formation was detailed. Deposited films must be flat and as smooth as the material and deposition method will allow to minimize steps, cracking, and subsurface reflections.

Deposited films must be of the desired uniform composition. Some of the reactions are complex, and it is possible that films will be deposited with something other than the intended composition. *Stoichiometry* is the methodology by which the quantities of reactants and products in chemical reactions are determined. In addition to chemical composition, grain size is important. During deposition, the film materials tend to collect or grow into grains. Varying grain size within films of the same composition and thickness will yield different electrical and mechanical properties. This is because electrical current flow is affected as it passes through the grain interfaces. Mechanical properties also change with the size of grain interface area.

Stress-free films are another requirement. A film deposited with excess stress will relieve itself by forming cracks. Cracked films cause

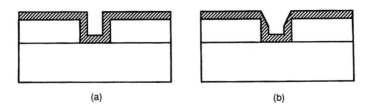

(a)                              (b)

**Figure 12.3**  Thinning of deposited layer (b) at step.

surface roughness and can allow contamination to pass through to the wafer. In the extreme, they cause electrical shorts.

Purity (i.e., a lack of unwanted chemical elements or molecules in the film) is required for the film to carry out its intended function. For example, oxygen contamination of an epitaxial film will change its electrical properties. Purity also includes the exclusion of mobile ionic contaminants and particulates.

An electrical parameter of importance to deposited films is capacitance (see Chapter 2). Semiconductor metal conduction systems need high conductivity and, therefore, low-resistance and low-capacitance materials. These are referred to as *low-k dielectrics*. Dielectric layers used as insulators between conducting layers need high capacitances or *high-k dielectrics*.

## Chemical Vapor Deposition Basics

Not surprisingly, the growth in the number and kinds of deposited films has resulted in a number of deposition techniques. Where the process engineer of the 1960s had a choice of only atmospheric chemical vapor deposition (CVD), today's engineer has many more options (Fig. 12.4). These techniques are described in the following sections.

Thus far, the terms *deposition* and *CVD* have been used without explanation. In semiconductor processing, deposition refers to any process in which a material is physically deposited on the wafer surface. Grown films are those, such as silicon dioxide, that formed from the material in the wafer surface. The majority of films are deposited by a CVD technique. In concept, the process is simple (Fig. 12.5). Chemicals (C) containing the atoms or molecules required in the final film are mixed and reacted in a deposition chamber to form a vapor (V). The atoms or molecules deposit (D) on the wafer surface and build up to form a film. Figure 12.5 illustrates the reaction of silicon tetrachloride ($SiCl_4$) with hydrogen to form a deposited layer of silicon on the wafer. Generally, CVD reactions require energy to take place.

| Atmospheric Pressure (AP) | Low Pressure (LP) and Ultra High Vacuum (UHV) |
|---|---|
| Cold wall | Hot wall |
| • Horizontal | Plasma enhanced |
| • Vertical | Vertical isothermal |
| • Barrel | Molecular beam epitaxy (MBE) |
| • Vapor phase epitaxy | |
| Metalorganic CVD | |

**Figure 12.4** Overview of deposition systems.

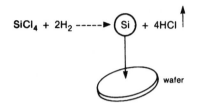

$$SiCl_4 + 2H_2 \ ----\blacktriangleright \ \boxed{Si} + 4HCl \uparrow$$

wafer

**Figure 12.5**  Chemical vapor deposition of silicon from silicon tetrachloride.

The chemical reactions that take place fall into the four categories of pyrolysis, reduction, oxidation, and nitridation (Fig. 12.6). Pyrolysis is the process of chemical reaction driven by heat alone. Reduction causes a chemical reaction by reacting a molecule with hydrogen. Oxidation is the chemical reaction of an atom or molecule with oxygen. Nitridation is the chemical process of forming silicon nitride by exposing a silicon wafer to nitrogen at a high temperature.

Deposited films grow in several distinct stages (Fig. 12.7). The first stage, *nucleation*, is very important and critically dependent on substrate quality. Nucleation occurs as the first few atoms or molecules deposit on the surface. These first atoms or molecules form islands that grow into larger islands. In the third stage, the islands spread, finally coalescing into a continuous film. This is the transition stage of the film growth with a typical thickness of several hundred angstroms. The transition region film has chemical and physical properties much different from those of the final, thicker "bulk" film.[1]

| | |
|---|---|
| Pyrolysis | $SiH_4 \rightarrow Si + 2H_2$ |
| Reduction | $SiCl_4 + 2H_2 \rightarrow SI + 4HCL$ |
| Oxidation | $SiH_4 + O_2 \rightarrow SiO_2 + 2H_2$ |
| Nitridation | $3SiH_2Cl_2 + 4NH_3 \rightarrow Si_3N_4 + pH + 6H_2$ |

**Figure 12.6**  Examples of CVD reactions.

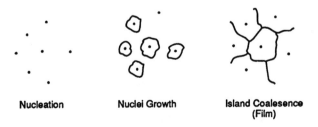

**Nucleation**          **Nuclei Growth**          **Island Coalesence (Film)**

**Figure 12.7**  CVD film growth steps.

After the transition film is formed, the bulk growth begins. Processes are design to produce three different structures: amorphous, polycrystalline, and single crystal (Fig. 12.8). These terms have been defined previously. A poorly defined or controlled process can result in a film with the wrong structure. For example, attempting to grow a single-crystal epitaxial film on a wafer with islands of unremoved oxide will result in regions of polysilicon in the bulk film.

## Basic CVD system components

CVD systems come in a wide variety of designs and options. Understanding the many variations is helped by an examination of the basic subsystems common to most CVD systems (Fig. 12.9). In most respects, a CVD system has the same basic parts as a tube furnace (described in Chapter 7): source cabinet, reaction chamber, energy source,

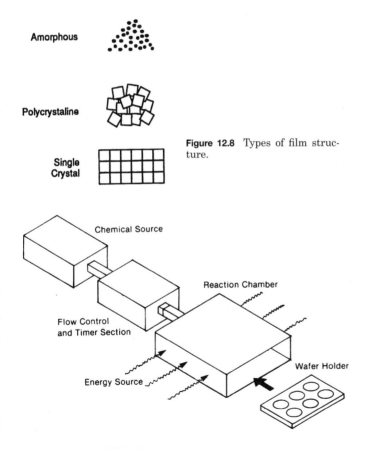

**Figure 12.8** Types of film structure.

**Figure 12.9** Basic CVD subsystems.

wafer holder (boat), and loading and unloading mechanisms. In some cases, the CVD system *is* a tube furnace identical to those used for oxidation and diffusion. The source chemicals are housed in a source section. Vapors are generated from pressurized gas cylinders or liquid source bubblers. Gas flow control is maintained by pressure regulators, mass flow meters, and timers.[2]

The actual deposition takes place on the wafers in a reaction chamber. Energy sources can be heat conduction, convection, induction RF, radiant, plasma, or ultraviolet. Energy sources are explained in the sections on particular systems. Temperatures range from room temperature to 1250°, depending on the reaction, film thickness required, and the growth parameters.

The fourth basic part of the system is the wafer holder. Different chamber configurations and heat sources dictate the style and material of the holders. Most production-level systems for ULSI circuits are automated from wafer load to unload. A full production system will have an associated cleaning section or station and a loading area.

## CVD Process Steps

A CVD process follows the same steps as an oxidation or diffusion process. For review, the steps are preclean (and etch, if required), deposition, and evaluation. Cleaning processes are those already described to remove particulates and mobile ionic contaminants. Chemical vapor deposition, like an oxidation, takes place in cycles. First, the wafers are loaded into the chamber, usually with an inert atmosphere. Next, the wafers are brought to temperature. Chemical vapors are introduced for as long as required to deposit the film. Finally, the chemical source vapors are flushed out and the wafers removed. Evaluation of the films are for thickness, step coverage, purity, cleanliness, and composition. Evaluation techniques are explained in Chapter 14.

## CVD System Types

CVD systems (Fig. 12.10) are divided into two primary types: atmospheric-pressure (AP) and low-pressure (LP). There are a number of atmospheric pressure CVD systems (APCVD). However, most films for advanced circuits are deposited in systems where the pressure has been lowered. These are called low-pressure CVD or LPCVD.

Another differentiation is cold wall versus hot wall. Cold-wall systems directly heat the wafer holder or wafers, with induction or radiant heating. The walls of the chamber remain cold (or cooler). Hot-wall systems heat the wafers, the wafer holder, and the chamber walls. The advantage of cold-wall CVD is that the reaction occurs only

| Atmospheric systems | Low-pressure systems |
|---|---|
| Hot wall | Hot Wall |
| Cold wall | Plasma enhanced |
| • Epitaxial | • Hot-wall tube |
| • Barrel | • Cold-wall planar—batch |
| • Pancake | • Cold-wall planar—single |
| • Single-wafer | |
| | High-density plasma |

**Figure 12.10**   CVD reactor designs.

at the heated wafer holder. In a hot-wall system, the reaction occurs throughout the chamber, leaving reaction products on the inside chamber walls. The reaction products build up, necessitating rigorous and frequent cleaning to avoid contaminating the wafers.

CVD systems are operated with two principal energy sources: thermal and plasma. Thermal sources are tube furnaces, hot plates, and RF induction. Plasma enhanced chemical vapor deposition (PECVD) in combination with lower pressure offers the unique advantage of lowered temperatures and good film composition and coverage.

A specialty CVD used to deposit compound films, such as GaAs, is vapor phase epitaxy (VPE). A newer technique used to deposit metals is a metalorganic (MOCVD) source in a VPE system. The last deposition method described is the non-CVD molecular beam epitaxy (MBE) used for low-temperature deposition of thin films in a very controlled process.

## Atmospheric-Pressure CVD Systems

As the name implies, atmospheric CVD system reactions and deposition take place at atmospheric pressure. There are a number of system designs that fall under this category (Fig. 12.10).

### Horizontal-tube induction-heated APCVD

The first widespread use of CVD was for the deposition of silicon epitaxial films for bipolar devices. The basic system design is still used

(Fig. 12.11). It is essentially a horizontal tube furnace, but with some significant differences. First, the tube has a square cross section. The major difference, however, is in the heating method and wafer holder.

Wafers are arranged on a flat graphite slab and positioned in the tube. Surrounding the tube are copper coils that are connected to an RF generator. The RF waves traveling in the coils pass through the quartz tube and the flowing gas in the tube without heating them. This is the cold-wall aspect of the system. When the radiant waves reach the graphite wafer holder, they "couple" with the molecules of the holder, causing the graphite to heat up. This heating method is called *induction*.

The heat of the holder is passed to the wafers by conduction. The film deposition takes place at the wafer surface (and at the holder surface). One problem with this type of system is downstream depletion of the reactants in the laminar gas flow. Laminar flow is required to minimize turbulence. But if the wafers are laid flat in the chamber, the layer of gas closest to the wafers becomes depleted. This results in successively thinner films along the wafer holder. A wafer holder tilted in the tube corrects the problem (Fig. 12.12).

### Barrel radiant-induction-heated APCVD

Larger-diameter wafers laid horizontally on a holder in a horizontal system have a low packing density. And larger wafer holders strain the uniformity capabilities of the system.

The development of the barrel radiant-heated system (Fig. 12.13) solved these problems. The reaction chamber of the system is a cylindrical stainless steel barrel with high-intensity quartz heaters placed about the inside surface. The wafers are placed on a graphite holder that rotates in the center of the barrel. The rotation of the wa-

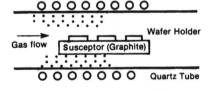

Figure 12.11  Cold-wall induction APCVD with horizontal susceptor.

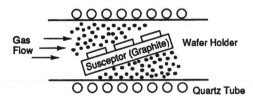

Figure 12.12  Cold-wall induction APCVD with tilted susceptor.

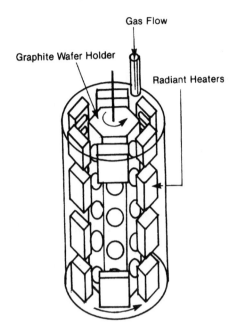

Gas Flow

Graphite Wafer Holder

Radiant Heaters

**Figure 12.13** Cylindrical or barrel system.

fers produces a more uniform film thickness as compared to horizontal systems.

Radiant heat from the lamps heats the wafer surface, where the deposition takes place. While some heating of the chamber walls occurs, the system is close to a cold-wall deposition. Direct radiant heating produces a very controlled and even film growth. In an induction-heated system, the wafers are heated from the bottom, and, as the film grows, there is some small but measurable drop in temperature at the film surface. In the barrel system, the wafer surface is always facing the lamps and receives a more uniform temperature and film growth rate.

In 1987, Applied Material introduced a jumbo barrel system for large-diameter wafers featuring an induction heating system.[3] A principal advantage of the barrel reactor is an increased productivity based on the increased number of wafers per cycle. This system configuration is popular for deposition of epitaxial silicon in the 900 to 1250°C range.

### Pancake induction-heated APCVD

The pancake or vertical-flow APCVD system has been a favorite for small fabrication lines and R&D labs (Fig. 12.14). The wafers are held on a rotating holder of graphite and heated by induction-conduction

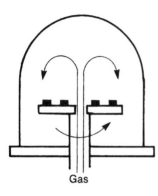

**Figure 12.14** Rotating pancake APCVD.

Gas

from an RF coil below the holder. The reaction gases are fed through a tube exiting above the wafers. Vertical gas flow offers the advantage of a continuous supply of fresh reactants to the wafers, thus minimizing downstream depletion. The combination of the rotation and vertical flow of the gases produces good film uniformity. Productivity in smaller systems is restricted, as in the horizontal tube system, by the number of wafers that the pancake can accommodate.

A production-level variation of the pancake design is produced by Gemini Research. The reactor features radiant-resistance heating and a large-capacity holder with robot autoloading.[4]

### Continuous conduction-heated APCVD

Two horizontal conduction-heated APCVD systems feature mixing the gases outside the chamber and showering them onto the wafers. In one design, a heated hot-plate wafer holder moves back and forth (Fig. 12.15) under a series of nozzles that dispense a vapor of the desired material. Another version of the system (Fig. 12.16) has wafers moving on a belt under a plenum that dispenses the reacted gases.

### Horizontal conduction-heated APCVD

One of the original CVD designs is the horizontal conduction-heated APCVD system (Fig. 12.17) used to deposit the silicon dioxide passiva-

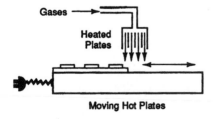

**Figure 12.15** Moving hot-plate APCVD.

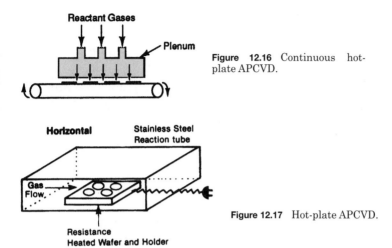

Figure 12.16 Continuous hot-plate APCVD.

Figure 12.17 Hot-plate APCVD.

tion layers. Wafers are mounted on a removable hot plate and placed in a stainless steel chamber. The hot plate heats the wafers and chamber walls (hot wall system). Reaction gases are fed into the chamber.

## Low-Pressure Chemical Vapor Deposition (LPCVD)

Uniformity and process control within atmospheric-pressure CVD systems rely on temperature control and the flow dynamics in the system. A factor influencing film uniformity and step coverage is the mean free path of the molecules in the reaction chamber. The mean free path is the average distance a molecule will travel before colliding with an object in the chamber, be it another molecule, the wafers, or wafer holder. Collisions change the direction of the particles. The longer the mean free path, the higher the uniformity of the film deposition. A major determiner of the length of the mean free path is the pressure in the system. Lowering the pressure in the chamber increases the mean free path and the film uniformity. Decreasing the pressure also allows a lowering of the deposition temperature.

These benefits became available to the industry in 1974 when Unicorp, under license to Motorola Inc., introduced an LPCVD system that operated at a few hundred millitorr.[5] The list of LPCVD system advantages includes:

- Lower chemical reaction temperature
- Good step coverage and uniformity
- Vertical loading of wafers for increased productivity and lower exposure to particles

- Less dependence on gas flow dynamics
- Less time for gas phase reaction particles to form
- Can be performed in standard tube furnaces

A vacuum pump must be added to the system to reduce the pressure in the chamber. A discussion of the types of pumps used with LPCVD systems appears in Chapter 13.

**Horizontal conduction-convection-heated LPCVD**

One production-level LPCVD system uses a horizontal tube furnace (Fig. 12.18), with three major exceptions. The tube is connected to a vacuum pump that pulls the system down to a pressure range of 0.25 to 2.0 torr.[6] A second change is a ramping of the temperature in the center zone to offset reaction depletion down the tube. The third change may be special injectors at the gas inlet end to improve gas mixing and deposition uniformity. In some systems, the injectors are positioned directly over the wafers. Disadvantages of this system design are particles formed on the inside wall surface (hot-wall reactions), uniformity along the tube axis, the use of cages around the wafers to minimize particle contamination, and the higher downtime required for frequent cleaning.

These systems are most often used for polysilicon, silicon dioxide, and silicon nitride films with typical thickness uniformities of ±5 percent. The primary deposition variables are temperature, pressure, gas flow, gas partial pressure, and wafer spacing. These variables are carefully balanced for each deposition process. The deposition rates are somewhat lower (100 to 500 Å/min) than AP systems, but productivity is enhanced by the vertical wafer-loading densities that can approach 200 wafers per deposition.

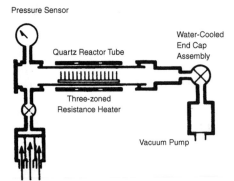

Figure 12.18  Horizontal hot-wall LPCVD system.

## Ultrahigh vacuum CVD (UHV/CVD)

Low-temperature deposition is desirable to minimize crystal damage and lower the thermal budget that in turn minimizes the lateral diffusion of doped regions. One approach is the CVD of silicon and silicon/germanium (SiGe) in ultra-low vacuum conditions. Lowering the pressure allows keeping the deposition temperature low. UHV/CVD takes place in a tube furnace where the pressure is initially reduced to 1 to 5 $\times 10^{-9}$ millibar (mbar). Deposition pressure is in the $10^{-3}$ mbar range.[7]

## Plasma-enhanced CVD (PECVD)

Replacement of silicon dioxide passivation layers with silicon nitride led to the development of plasma enhanced (PECV) techniques. A thermal silicon dioxide deposition temperature of approximately 660°C causes unacceptable alloying of aluminum interconnects into the silicon surface (see Chapter 13). A solution to this problem was a plasma enhancement of the deposition energy. The increased energy allows a temperature under the 450°C maximum level for deposition over aluminum layers. Plasma-enhanced systems are physically similar to plasma-etch systems. They feature a parallel plate chamber operated at a low pressure. A radio-frequency-induced glow discharge or other plasma source (see Chapter 9) is used to induce a plasma field in the deposition gas. The combination of low pressure and lower temperatures provides good film uniformity and throughput.

PECVD reactors have the capability of also using the plasma for etching and cleaning the wafer prior to the deposition step. This step is the same as the dry-etch processes described in Chapter 9. This *in situ* cleaning prepares the deposition surface, eliminating the problem of added contamination picked up during the loading step.

**Horizontal vertical-flow PECVD.**   This system follows the design of a bottom-heated pancake vertical-flow CVD (Fig. 12.19). The plasma region is created in the top of the chamber by a radio frequency (RF) feed to an electrode or other plasma source.Wafer heating comes from radiant heaters mounted below the wafer holder, creating a cold-wall deposition system. With PECVD systems, there are several additional critical parameters to control in addition to those in a standard LPCVD reactor. They are the RF power density, the RF frequency, and the duty cycle. Film deposition speed is generally increased, but it must be controlled to prevent film stress and/or cracking.

Another design (developed by Novellus) has the wafers sitting on a series of resistance heated holders. The wafers are indexed around the chamber with the film built up in increments.

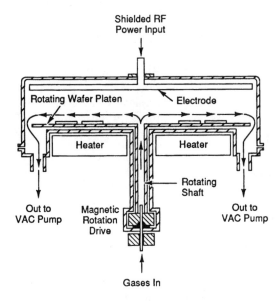

**Figure 12.19**  Vertical-flow pancake PECVD.

A single-wafer chamber PECVD system (Fig. 12.20), where the chamber is small and each successive wafer is exposed to identical conditions, addresses most of the control needs. Single-wafer systems are generally slower than batch processes. The productivity trade-off with the larger-chamber batch machines lies in fast mechanisms to feed wafers in and out of the chamber and how quickly the vacuum can be established and released. Load-lock systems enhance the productivity by moving the wafers into an antechamber, pumping it down to the required pressure, and moving the wafers into the deposition chamber as it becomes available. System designs include fixed and rotating wafer holders.

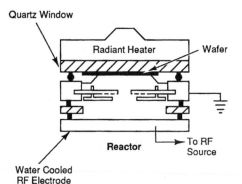

**Figure 12.20**  Single-chamber planar PECVD.

**Barrel radiant-heated PECVD.**    This system is a standard barrel radiant heated system with low-pressure and plasma capabilities. It is favored for the deposition of tungsten silicide.

## High-density plasma CVD (HDPCVD)

Intermetal dielectric (IDL) layers are essential for multimetal structures. They also present a challenge of filling high-aspect-ratio (greater than 3:1) holes. One approach is a deposition and an *in situ* etch sequence. The first deposition exhibits the usual thinning in the bottom. Etching away the shoulder and redeposition creates an uniform layer thickness and a more planar surface.

A system to accomplish this process is *high-density plasma* CVD.[8] A plasma field is created inside a CVD chamber that contains oxygen and silane for the deposition of silicon dioxide. Also included is argon that becomes energized by the plasma and is directed to the wafer surface. This is a sputtering (see "Dry Etching," Chapter 13) action that removes material from the surface and trench. HDPCVD has the potential of depositing a variety of materials for uses as IMD layers, etch stops, and final passivation layers.

CVD methods are summarized in Fig. 12.21.

## Atomic Layer Deposition (ALD)

At some point on the roadmap, CVD and its many variations will run out of steam. A candidate for the next generation is *atomic layer deposition* (ALD). Based on a basic CVD approach, ALD uses a pulsing technique. The film is grown in stages with each reactant type separated by a purge gas. Unlike CVD, each of the reactant types chemically reacts with the surface. This makes each individual "deposition" self-limiting as the film grows, much like silicon dioxide growth. Since each film stage is growing at a monolayer rate, the control is very precise. Also, the slow rate facilitates high levels of conformity to the wafer surface and dense film composition.

| Level | Temperature Range | Methods |
|---|---|---|
| High Temp.: | 600–1250°C | R.F. Induction (Cold Wall) Radiant Heat (Cold Wall) Resistance Coils (Hot Wall) |
| Mid Temp.: | 200–600°C | Hot Plates Plasma Enhanced LPCVD |
| Low Range. | 22–200°C | Hot Plates P.E. CVD Photochemical |

**Figure 12.21**  Summary table of CVD methods.

Candidate uses include very thin silicon dioxide gates, filling deep trenches with materials like aluminum oxide, and created barrier metal layers for copper metallization processes.[9]

## Vapor Phase Epitaxy (VPE)

VPE differs from the CVD systems described in its ability to deposit compound materials, such as gallium arsenide. A VPE system[10] is a combination of a standard liquid source tube furnace and a two-zone diffusion furnace (Fig. 12.22). An example is the particular arrangement in Fig. 12.22 used to deposit epitaxial gallium arsenide. The creation of the GaAs layer on the wafer in the main chamber proceeds in two stages. $AsCl_3$ is bubbled into the first section of the tube where it reacts with a solid source of gallium that is sitting in a boat. The arsenic trichloride reacts with the hydrogen in the first section to form arsenic by the reaction

$$4AsCl_3 + 6H_2 \rightarrow 12HCl + As = 4$$

The arsenic deposits on the gallium, forming a crust. The hydrogen passing over the crust reacts in the first section to form three gases that pass into the wafer section.

$$\frac{GaAs}{(solid)} + \frac{HCl}{(gases)} \leftrightarrow \frac{GaCl}{(gas)} + \frac{1/2H_2}{(gas)} + \frac{1/4As_4}{(gas)}$$

This section is at a somewhat lower temperature, and the reaction proceeds in reverse, depositing GaAs on the wafers. The technique offers the advantages of clean films, since the gallium and arsenic trichloride are available in very pure forms and have higher production rates than the MBE technique. On the downside, the film structures produced are not the quality of MBE films.

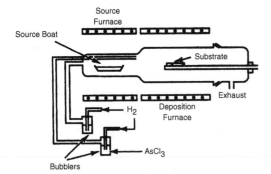

**Figure 12.22**  Diagram of gallium arsenide VPE deposition system.

## Molecular Beam Epitaxy (MBE)

Deposition rate control, low deposition temperature, and controlled film stoichiometry are always goals in film-deposition systems. Molecular beam epitaxy (MBE) has emerged from the laboratory to claim production status as these issues have become more important. MBE is an evaporation rather than a CVD process. The system consists of a deposition chamber (Fig. 12.23) that is maintained at a low pressure to $10^{-10}$ torr. Within the chamber can be one or more cells (called *effusion cells*) that contain a very pure sample of the target material desired on the wafer. Shutters on the cells allow exposure of the wafer to the source material(s). An electron beam[10] is directed into the center of the target material, which it heats to the liquid state. In this state, atoms evaporate out of the material, exit the cell through an opening, and deposit on the wafers. If the material source is a gas, the technique is called *gas source MBE* or *GSMBE*. For most applications, the wafer in the chamber is heated to give additional energy to the arriving atoms. The additional energy fosters epitaxial growth and good film quality.

If the wafer surface is exposed, the depositing atoms will assume the orientation of the wafer and grow an epitaxial layer. MBE offers the intriguing option of *in situ* doping by the inclusion of dopant sources in the chamber. The usual silicon dopant sources are not usable in MBE systems. Solid gallium is used for P-type doping and antimony for N-type doping. Phosphorus deposition is virtually not possible with MBE.[12]

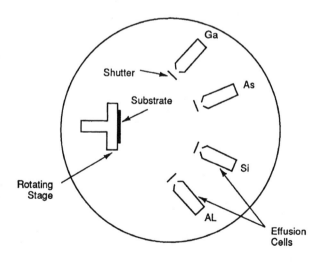

**Figure 12.23**  Diagram of MBE deposition system.

The primary advantage of MBE for silicon technology is the low temperature (400 to 800°C), which minimizes autodoping and outdiffusion. Perhaps the biggest advantage of MBE is the ability to form multiple layers on the wafer surface during one process step (one pump down). This option requires the mounting of several effusion cells in the chamber and shutter arrangements to direct the evaporant beams to the wafer in the right order and for the correct time.

An advantage and disadvantage of MBE is the low film growth rate of 60 to 600 Å/min.[12] On the plus side, the films produced are very controllable. Films can be grown (and mixed) in one monolayer increments. However, most semiconductor layers do not need this level of control and quality, making the low productivity and expense of the system an expensive luxury.

A bonus possibility with MBE is the incorporation of film growth and quality-analyzing instruments in the chamber. With these instruments, the process can become very controlled and produce uniform films from wafer to wafer. MBE has found production use in the fabrication of special microwave devices and for compound semiconductors such as gallium arsenide.[14]

## Metalorganic CVD (MOCVD)

MOCVD is one of the latest options for CVD of compound materials. Whereas VPE refers to a compound material deposition system, MOCVD refers to the sources used in VPE systems (Fig. 12.24). Two chemistries are used, halides and metalorganic. The reactions described for the VPE deposition of gallium arsenide above constitute a

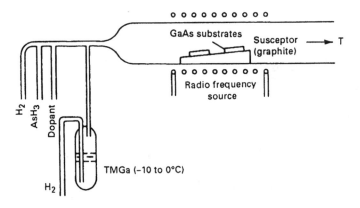

**Figure 12.24**   MOCVD system. *(Source:* VLSI Fabrication Principles, *S. K. Ghandhi, Wiley-Interscience, 1994.)*

halide process. A group III halide (gallium) is formed in the hot zone, and the III–IV compound is deposited in the cold zone. In the metalorganic process[15] for gallium arsenide, trimethylgallium is metered into the reaction chamber along with arsine to form gallium arsenide by the reaction

$$(CH_3)_3Ga + AsH_3 \rightarrow GaAs + 3CH_4$$

Where MBE processes are slow, MOCVD processes can meet volume production requirements and accommodate larger substrates.[16] MOCVD also has the capability of producing multiple layers with very abrupt changes in composition. This characteristic is critical to compound semiconductor structures. Also MOCVD, unlike MBE, can deposit phosphorus as in InGaAsP devices. Common devices made by MOCVD processes are photocathodes, high-poser LEDs, long-wavelength lasers, visible laser, and orange LEDs (see Chapter 16).

MOCVD is a broad term referring to the metalorganic chemical vapor deposit of semiconductor films. When metalorganic sources are used in a vapor phase epitaxial system to grow epitaxial layers, the method is called MOVPE.[17]

## Deposited Films

The types of films deposited by CVD techniques are divided into their electrical classifications of semiconductors, dielectrics, and conductors. In the following sections, the deposition of the various films is examined. For each film, the principal use(s) in semiconductor devices is presented along with the particular film properties for the described use. The uses are treated in general. For a more detailed explanation of film roles in particular devices, see Chapter 16. Deposition methods for conductive metal films are discussed in Chapter 13.

## Deposited Semiconductors

So far in this text, we have discussed the formation of wafers as the base of semiconductor devices and circuits. However, there are several drawbacks to using bulk wafers for high-quality devices and circuits. Crystal quality, doping ranges, and doping control all limit bulk wafer use. These factors placed a limit on the fabrication of high-performance bipolar transistors. A solution was found by the development of a deposited silicon layer, called an *epitaxial layer*. It is one of the major advances in the industry. Epitaxial layers were a part of the technology as early as 1950.[17] Since then, deposited silicon layers have found additional uses in advanced bipolar device design, a high-quality base for CMOS circuits, and silicon epitaxial layers deposited on sapphire and other substrates (Chapter 14). Gallium arsenide and other III–IV and II–VI films are also deposited in epitaxial films. Epitaxial films that are the same material as the substrate (silicon on silicon) are *homoepitaxial*. When the deposited material is different from that of the substrate (GaAs on silicon), the film is called *heteroepitaxial*.

### Epitaxial Silicon

The term *epitaxial* comes from the Greek word meaning "arranged upon." In semiconductor technology, it refers to the single crystalline structure of the film. The structure comes about when silicon atoms are deposited on a bare silicon wafer in a CVD reactor (Fig. 12.25). When the chemical reactants are controlled and the system parameters set correctly, the depositing atoms arrive at the wafer surface with sufficient energy to move around on the surface and orient themselves to the crystal arrangement of the wafer atoms. Thus, an epitaxial film deposited on a ⟨111⟩-oriented wafer will take on a ⟨111⟩ orientation.

If, on the other hand, the wafer surface has a thin layer of silicon dioxide, an amorphous surface layer, or contamination, the depositing atoms have no structure to which they can align. The resulting film structure is polysilicon. This condition is useful for some applications, such as MOS gates, and is unwanted if the goal is to grow a single-crystal film structure.

**Silicon tetrachloride source chemistry.**    A number of different sources are used for the deposition of epitaxial silicon (Fig. 12.26). Deposition

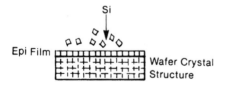

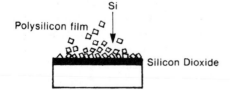

**Figure 12.25**  Epitaxial and polysilicon film growth.

| | |
|---|---|
| Silicon tetrachloride | $SiCl_4 + 2H_2 \leftrightarrow Si + 4HCl$ |
| Silane | $SiH_4 + heat \rightarrow Si + 2H_2$ |
| Dichlorosilane | $SiH_2Cl_2 \leftrightarrow Si + 2HCl$ |

**Figure 12.26**  Epitaxial silicon chemical sources.

temperature, film quality, growth rate, and compatibility with a particular system are factors in choosing a silicon source. An important process parameter is the deposition temperature. The higher the temperature, the faster the growth rate. Faster growth rates create more crystal defects and film cracking and stress. Higher temperatures also cause higher levels of autodoping and out-diffusion. (These effects are described in the following text.)

Silicon tetrachloride ($SiCl_4$) is the favored source of silicon for deposition of silicon. It allows a high formation temperature (growth rate) and has a reversible chemical reaction. In Fig. 12.26, there is a double-headed arrow, which indicates that the reaction creates silicon atoms in one direction and removes (etches) silicon in the other direction. Within the reactor, these two reactions compete with each other.

Initially, the silicon surface is etched preparing it for the deposition reaction. In the second stage, the deposition of silicon is faster than the etch, with the net result of a deposited film.

The graph in Fig. 12.27 shows the effect of the two reactions. With an increasing percentage of $SiCl_4$ molecules in the gas stream, the deposition rate first increases. At the 0.1 ratio, the etching reaction starts to dominate and slows down the growth rate. This latter reaction is actually one of the first events in the reactor. Hydrogen chloride (HCl) gas is metered into the chamber, where it etches away a thin layer of the silicon surface, preparing it for the silicon deposition.

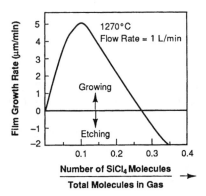

**Figure 12.27**  Growth-etch characteristics of $SiCl_4$ epitaxial deposition.

**Silane source chemistry.**    The second-most-used silicon source chemistry is silane ($SiH_4$). Silane offers the advantage of not requiring a second reaction gas. It forms silicon atoms by decomposing when heated. The reaction takes place several hundred degrees lower than a silicon tetrachloride deposition, which is attractive from an autodoping and wafer warping perspective. Also, silane does not produce pattern shift (see "Epitaxial film quality," below). Unfortunately, the reaction occurs at all locations in the system, creating a powdery film inside the reactor which, in turn, contaminates the wafers. Silane finds more use as a source for polysilicon and silicon dioxide depositions.

**Dichlorosilane source chemistry.**    Dichlorosilane ($SiH_2Cl_2$) is also a lower-temperature silicon source that is used for thin epitaxial films. The lower temperature reduces autodoping and solid-state diffusion from previously diffused buried layers and provides a more uniform crystal structure.

**Epitaxial film doping.**    One of the advantages of an epitaxial film is the precise doping and doping range available by the process. Silicon wafers are manufactured in a concentration range of approximately $10^{13}$ to $10^{19}$ atoms/cm$^3$. Epitaxial films can be grown from $10^{12}$ to $10^{20}$ atoms/cm$^3$. The upper limit is close to the solid solubility of phosphorus in silicon.

Doping in the film is achieved by the addition of a dopant gas stream to the deposition reactants. The sources of the dopant gases are exactly the same chemistries and delivery systems used in deposition doping furnaces. In effect, the CVD deposition chamber is turned into a doping system. In the chamber, the dopants become incorporated into the growing film, where they establish the required resistivity. Both N- and P-type films can be grown on either N- or P-type wafers. The classic epitaxial film in bipolar technology is an N-type epitaxial film on a P-type wafer.

**Epitaxial film quality.**    Epitaxial film quality is a prime concern of the process. In addition to the usual considerations over contamination, there are a number of faults specifically associated with epitaxial growth. Contaminated systems can cause a problem called *haze*.[19] Haze is a surface problem that varies from a microscopic disruption to severe cases that are observable as a dull matte finish. Haze comes about from residual oxygen in the reactant gases or from leaks in the system.

Contaminants on the surface at the start of the deposition result in an accelerated growth known as *spikes* (Fig. 12.28). The spikes can be

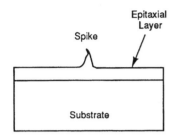

Figure 12.28 Epitaxial growth spike.

as high as the film thickness. They cause holes and disruption in photoresist layers and other deposited films.

During the growth, a number of crystal problems can occur. One is *stacking faults*. A stacking fault is due to the inclusion of an extra atomic plane with a corresponding "dislocation" of the atoms around the plane. A stacking fault begins at the surface and "grows" to the surface of the film. The shape of the stacking fault depends on the orientation of the film and wafer. Faults in ⟨111⟩-oriented films have a pyramidal shape (Fig. 12.29), whereas ⟨100⟩-oriented wafers form rectangular-shaped stacking faults. The faults are detected by either X-ray or etching techniques.

A growth problem associated with ⟨111⟩ wafers is *pattern shift*. This problem occurs when the deposition rate is too high and the film planes grow at an angle to the surface. Pattern shift is a problem when alignment to a subsurface pattern relies on locating it from a film surface step (Fig. 12.30). Another major growth problem is *slip*. This condition comes about from poor control of the deposition parameters and results in a "slippage" of the crystal along plane interfaces (Fig. 12.31).

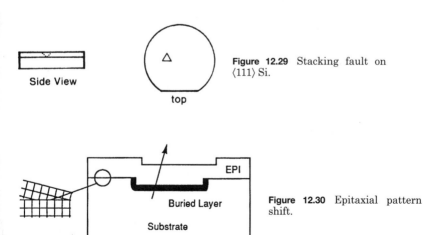

Side View

top

Figure 12.29 Stacking fault on ⟨111⟩ Si.

EPI

Buried Layer

Substrate

Figure 12.30 Epitaxial pattern shift.

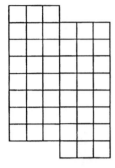

**Figure 12.31** Crystal slip.

There are two issues associated with the temperature of the deposition: autodoping and out-diffusion. Autodoping of the growing film occurs when dopant atoms from the back of the wafer diffuse out from the wafer (Fig. 12.32), mix in the gas stream, and become incorporated into the growing film. In the film, they change the resistivity and the conductivity level. *Autodoping* in a P-type film, grown over an N-type wafer, will be less P-type than intended and have a lower P-type concentration as the autodoped atoms neutralize a number of the P-type atoms in the film.

*Out-diffusion* causes the same effect but at the epitaxial layer-wafer interface. The source of the out-diffused atoms is doped regions diffused into the wafer before the epitaxial deposition. In bipolar devices,

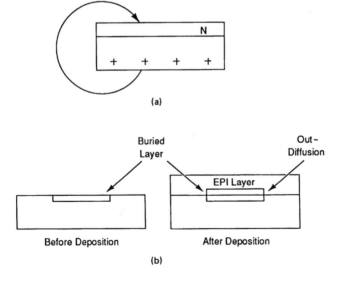

**Figure 12.32** (*a*) Epitaxial autodoping and (*b*) out-diffusion.

the regions are called *buried layers* or *subcollectors*. In the usual format, the buried layer is an N-type region in a P-type wafer, over which is grown an N-type epitaxial layer. During the deposition, the N-type atoms diffuse out and become incorporated into the bottom of the epitaxial film, changing the concentration. In the extreme, the buried layer can out-diffuse up into the bipolar device structure causing electrical malfunctions.

**CMOS epitaxy.** Until the late 1970s, the dominant use of epitaxial films was as the collector region of bipolar transistors. The technique provided a quality substrate for device operation and a clever means of isolating adjacent devices (see Chapter 16). A newer and perhaps more dominant use of silicon epitaxial films is for CMOS circuit wafers. The need for an epitaxial layer was driven by a CMOS circuit problem called *latch-up* (see Chapter 16).

**Epitaxial process.** A typical epitaxial process starts with a complete and rigorous cleaning of the wafer surface prior to loading the reactor. Within the deposition chamber, a number of steps take place to correctly deposit the film. A typical $SiCl_4$ epitaxial process is shown in Fig. 12.33. The first several steps are a gas-phase cleaning of the wafer surface. Deposition follows the cleaning with a cool-down cycle at the end. During all the steps, control of the temperature and gas flows is critical.

**Selective epitaxial silicon.** Advancements in epitaxial deposition systems have introduced the selective growth of epitaxial films. Whereas the epitaxial films for bipolar and CMOS substrates are deposited on the entire wafer, in selective growth they are grown through holes in either silicon dioxide or silicon nitride films. The wafer is positioned in

| Cycle | Temperature | Gas | Purpose |
|---|---|---|---|
| 1 | Room | $N_2$ | Purge air from system |
| 2 | Room | $H_2$ | Reduce any organic contaminants of wafers in system |
| 3 | (Heating) | $N_2$ | Bring system to deposition temperature |
| 4 | Deposition Temperature | HCl | Etch wafer to prepare surface for epi deposition |
| 5 | Deposition Temperature | Source + Dopant + Carrier | Grow epitaxial film |
| 6 | (Heat Off) | $N_2$ | Purge system of reactant gases |

**Figure 12.33** Typical $SiCl_4$ epitaxial deposition process.

the reactor chamber, and the epitaxial film grows directly on the silicon exposed at the bottom of the hole (Fig. 12.34). As the film grows, it takes on the crystal orientation of the underlying wafer. An advantage of such a structure is that devices formed in the surfaces of the epitaxial regions are isolated from each other by the oxide or nitride regions.

If the deposition is allowed to continue onto the isolating surface, the structure of the film switches to a polysilicon structure. Another outcome of extended deposition is that the overlaying deposited layer becomes entirely epitaxial in nature. All of these outcomes add attractive structure options for advanced device design.

## Polysilicon and Amorphous Silicon Deposition

Until the advent of silicon-gate MOS devices (Fig. 12.35) in the mid-1970s, polysilicon layers had little or no use in device structures. Silicon-gate device technology drove the need for reliable processes to deposit thin layers of polysilicon. By the mid-1980s, polysilicon seemed to be the workhorse material of advanced devices. In addition to MOS gates, polysilicon finds use as load resistors in SRAM devices, trench fills, multilayer poly in EEPROMs, contact barrier layers, emitters in bipolar devices, and as part of silicide metallization schemes (see Chapters 13 and 16).

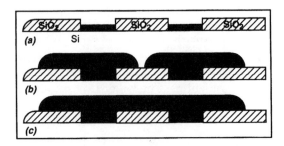

**Figure 12.34**   Steps in selective epitaxial growth.

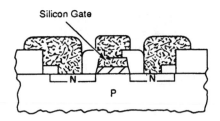

**Figure 12.35**   Cross section of silicon gate MOS transistor.

Early processes involved simply placing the oxide-covered wafers in a horizontal APCVD system and letting the polysilicon deposit on the oxide. The major difference of early polysilicon depositions from epitaxial depositions was the use of silane sources. While silane is not favored for epitaxial film deposition, it is more than adequate for polysilicon depositions.

Typical polysilicon deposition processes take place in the 600 to 650°C range. The deposition may be from either 100 percent silane or from gas streams containing $N_2$ or $H_2$. The structure of polysilicon was previously described as a total nonarrangement of the silicon atoms. In the case of deposited polysilicon, the structure is somewhat different. During the early stages of deposition, at temperatures below 575°C, the structure is amorphous (no structure). The polysilicon structure formed by deposition techniques consists of small pockets (crystallites or grains) of single-crystalline silicon separated by grain boundaries. This structure is called *columnar poly*.

The importance of grain size and grain boundary consistency shows up in the electrical current flow characteristics of the films. Current resistance comes as the current crosses the grain boundaries. The larger the grain boundaries, the higher the resistance. The achievement of consistent current flow from device to device and within a device is dependent on a well controlled polysilicon structure. One of the advantages claimed for the use of $H_2$ in the gas stream is the reduction of surface impurities and moisture, which in turn results in a reduced grain size. Moisture or oxygen impurities in the system cause the growth of silicon dioxide within the structure. The oxide increases the resistance of the film and its etchability in subsequent masking steps.

All of the system's usual operating parameters (temperature, silane concentration, pump speed, nitrogen flow, and other gas flows[19]) affect the deposition rate and the grain size. Often, the wafers will receive a postdeposition anneal in the 600°C range to further crystallize the film. The process of recrystallization goes on whenever the wafers go through a high-temperature process. The grain size and electrical parameters of the polysilicon film on the finished device or circuit are never the same as the deposited film.

Also influencing the grain size is the presence of dopants in the gas stream. In many devices or circuits, a strip of polysilicon functions as a conductor which requires doping to decrease its resistivity. Doping can be done by diffusion before or implantation after the deposition.

*In situ* doping takes place by adding gas dopant sources in the source cabinet and metering them into the chamber. When diborane (boron source) is added, there is a large increase in the deposition rate. An opposite effect takes place when phosphine (phosphorus source) or arsine (arsenic source) is the dopant gas. Undesirable effects of *in situ* doping are a loss of film uniformity, doping uniformity, and control of the deposition rate.

Doped polysilicon film resistivities are less than those of equally doped epitaxial or bulk silicon. The lower resistivities are due to dopants being trapped in the grain boundaries.

Most polysilicon layers are deposited with LPCVD systems that provided good productivity and lower deposition temperatures. LPCVD provides good step coverage (Fig. 12.36), a requirement because polysilicon layers are usually deposited later in the process, and the surface has become varied in its topography. Single-chamber polysilicon LPCVD systems offer the advantage of higher deposition rates without raising the temperature.[20]

## SOS and SOI

These two acronyms stand for *silicon on sapphire* and *silicon on insulator*. Both refer to the deposition of silicon on a nonsemiconductor surface. The need for such structures came about from the limits placed on some MOS devices by the presence of a semiconducting substrate under the active device. These problems are resolved by forming a silicon layer on an insulating substrate. The first substrate used for this purpose was sapphire (SOS). As different substrates were investigated, the term was expanded to the more general silicon on insulator (SOI).

One technique is a direct deposition on the substrate followed by a recrystallization process (laser heating, strip heaters, oxygen implan-

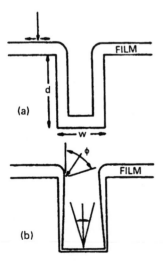

**Figure 12.36** Step coverage. (*a*) Good step coverage and (*b*) nonconformal coverage.

tation) to create a usable film.[22] Another approach is a selective deposition through holes in a surface oxide with an overgrowth to form the continuous film.

Another SOI method is *SIMOX*. In this process, the top layer of a wafer is converted to oxide with a heavy oxygen implant. An epitaxial layer is grown on top of the oxide. There is some exploration of *bonded wafers*. This approach has two wafers bonded together followed by the thinning (grinding and polishing) of one to device layer thickness.[22]

## Gallium Arsenide on Silicon

Gallium arsenide is a great III-V semiconductor material. However it is fragile and limited to wafer sizes of 4 inch (102 mm) diameter. Attempts to grow GaAs films on silicon wafers have been thwarted by the mismatch of the lattice sizes of each. During the growing process, the mismatch cause dislocations that degrade device performance. A new approach is to first grow a thin layer of strontium titanate on the silicon wafer. It reacts with the silicon to form a amorphous silicon dioxide layer. When a film of GaAs is deposited, the silicon dioxide layer acts as a cushion that absorbs the mismatch, thus allowing a single crystalline layer.[24]

## Insulators and Dielectrics

CVD is the favored method of depositing films that will function in the device or circuit as insulators or dielectrics. The two films in widespread use are silicon dioxide and silicon nitride. In general, the two films find a multiplicity of uses in device and circuit designs. While they have processing and quality differences, they must meet the same general requirements as other deposited films.

**Silicon dioxide.**   Deposited silicon dioxide films are best known from their long-term use as a final passivation layer covering the completed wafer. In this role, they provide physical and chemical protection to the underlying circuit devices and components. Deposited silicon dioxide films used as a protective top layer are known by the proprietary terms *Vapox*, *Pyrox*, and *Silox*®. Vapox (vapor-deposited oxide) is a term coined by Fairchild engineers. Pyrox stands for pyrolitic oxide. Silox is a registered trademark of Applied Materials, Inc. Sometimes, the layer is simply called a *glass*. This protective role has expanded, and deposited silicon oxide layers are used as interdielectric layers in multimetallization schemes, as insulation between polysilicon and metallization layers, as doping barriers, as diffusion sources, as isolation regions. Silicon dioxide has become a major part of silicon gate structures.

There are gate stacks, consisting of thermal oxide/silicon dioxide or oxynitride/silicon dioxide (TEOS deposited), and various silicon dioxide fillers for plugs in multimetal designs.[25]

CVD-deposited silicon dioxide films vary in structure and stoichiometry from thermally grown oxides. Depending on the deposition temperature, deposited oxides will have a lower density and different mechanical properties, such as index of refraction, resistance to cracking, dielectric strength, and etch rate. These factors are highly affected by the addition of dopants to the film. In many processes, the deposited film will receive a high-temperature anneal, a process called *densification*. After the densification, the deposited silicon dioxide film is close to the structure and properties of a thermal oxide.

The need for a low-temperature-deposited $SiO_2$ was dictated by the unacceptable alloying of aluminum and silicon at temperatures above 450°C. The early deposition process used was a horizontal conduction heated APCVD system from silane and oxygen by the reaction

$$SiH_4 + O_2 \rightarrow SiO_2 + 2H_2$$

This process produced films that were of unacceptable quality for use in advanced device designs and on larger wafers due to the poorer film quality produced by the 450°C deposition temperature.

The development of LPCVD systems made possible higher-quality films, especially for the factors of step coverage and lower stress. LPCVD processes are the preferred deposition techniques from both quality and productivity considerations. High-temperature (900°C) LPCVD of silicon dioxide is performed with a dichlorosilane reaction with nitrous oxide.

$$SiCl_2 + 2NO_2 \rightarrow SiO_2 + 2N_2 + HCl$$

**Tetraethyl orthosilicate (TEOS).**  By far, the majority of silicon dioxide are deposited from $Si(OC_2H_5)$ sources. The source is known as tetraethyl orthosilicate, or TEOS. TEOS history goes back to the 1960s.

Early systems relied on the simple pyrolysis of the TEOS in the 750°C range. Current depositions are based on the hot-wall LPCVD systems established in the 1970s, with temperatures the 400°C+ range. TEOS sources used with plasma-assistance (PECVD or PE-TEOS) allowed deposition temperatures in the sub-400°C range.[24] This process faces limits on conformal coverage of high aspect ratio patterns in 0.5-μm devices. Step coverage is improved by the addition of ozone ($O_3$) to the gas stream.[26]

Another option is the reaction of silane with nitrous oxide in an argon plasma.

$$SiH_4 + 4N_2O \rightarrow SiO_2 + 4N_2 + 2H_2O$$

**Doped silicon dioxide.**    Silicon dioxide layers are doped to improve their protective characteristics and flow properties, or for use as dopant sources. The earliest dopant used with deposited oxides was phosphorus. The phosphorus source is phosphine ($PH_3$) gas added to the deposition gas stream. The resultant glass is called *phosphorus silicate glass* or PSG. Within the glass, the phosphorus is in the form of phosphorus pentoxide ($P_2O_5$), making the glass a dual compound or, more correctly, a binary glass.

The role of the phosphorus is threefold. The added dopant increases the moisture-barrier property of the glass. Mobile ionic contaminants become attached to the phosphorus and are prevented from traveling into the wafer surface. This action is called *gettering*. The third result is an increase of the flow characteristics (Fig. 12.37), which aid the planarization of the glass surface after a heating step in the 1000°C range. The phosphorus content is limited to about 8 weight by percent. Above this level, the glass becomes hydroscopic and attracts moisture. The moisture can react with the phosphorus, form phosphoric acid, and attack underlying metal lines.

Boron is often added to the glass from a diborane ($B_2H_6$) source. The purpose of the boron is to also aid the flow characteristics (Fig. 12.37).

The resultant glass is called a borosilicate glass (BSG). The boron and phosphorus are often used together in the glass. The result is referred to as BPSG (borophosphorus silicate glass).

**Silicon nitride.**    Silicon nitride is a replacement for silicon dioxide uses, especially for top layer protection. Silicon nitride is harder, which provides better scratch protection, is a better moisture and sodium barrier

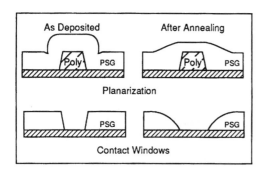

**Figure  12.37** Planarization of surface by flowing glass.

(without doping), has a higher dielectric strength, and resists oxidation. The latter property has led to its use in the local oxidation of silicon (LOCOS) for isolation purposes. Figure 12.38 illustrates the process, where patterned islands of silicon nitride prevent oxidation under the islands. After thermal oxidation and removal of the nitride, there are wafer surface regions ready for device formation, separated by isolating regions of oxide. A disadvantage of silicon nitride is that it does not flow as easily as silicon oxide and is more difficult to etch. The etch restriction has been overcome with the development of plasma etch processes.

An early limit on the use of silicon nitride protective films was the lack of a low-temperature deposition process. In APCVD systems, a temperature of 700 to 900°C is required for the deposition of silicon nitride from silane or dichlorosilane (Fig. 12.39). The result is a film with the composition $Si_3N_4$. The reactions also take place in LPCVD reactors but at a temperature low enough for deposition over an aluminum metallization layer. The advent of PECVD has opened up the use of different source chemistries. One use is silane reacted with ammonia ($NH_3$) or nitrogen in the presence of an argon plasma.

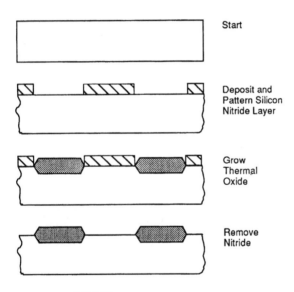

Start

Deposit and
Pattern Silicon
Nitride Layer

Grow
Thermal
Oxide

Remove
Nitride

**Figure 12.38** LOCOS process.

| Silane | $3SiH_4 + 4NH_3 \rightarrow Si_3N_4 + 12H_2$ |
|---|---|
| Dichlorosilane | $3SiCl_2H_2 + 4NH_3 \rightarrow Si_3N_4 + 6H_2 + 6HCl$ |

**Figure 12.39** Silicon nitride deposition reactions.

## High-k and low-k dielectrics

In addition to the dielectric films mentioned above, there are numerous other types deposited for special applications. They fall into the broad categories of high-k and low-k. In Chapter 2, the basics of a capacitor were described. For review, the k value of a material is called its *dielectric constant*. It relates to the level of capacitance a capacitor will have. High-k materials produce capacitors with high capacitance that are desirable for charge storage. Uses of high-k dielectrics are discussed in Chapter 16.

Low-k materials are used in metallization systems as barriers between the wafer surface and the primary metal system. In this case, the capacitor function should be low to facilitate conduction. These low-k materials are discussed in Chapter 13.

## Conductors

The traditional metal conductors of aluminum and aluminum alloys are deposited by evaporation or sputtering techniques. The advent of silicon gate MOS transistors added doped polysilicon as a device conductor. Plus, the advent of multimetal structures and new conducting materials has thrust CVD and PVD techniques into the conducting metal business. The techniques and use of these deposited metals are explained in the next chapter.

## Review Questions

1. Sketch and name the major subsystems of a basic CVD system.
2. Describe the differences between APCVD, LPCVD, and PECVD.
3. Define an epitaxial film.
4. Why is the deposition temperature of a silicon dioxide passivation layer limited to 450°C?
5. List an advantage of a horizontal vertical-flow CVD reactor.
6. Write the reaction equation for the deposition of silicon from silicon tetrachloride.
7. Describe the difference between MBE, VPE, and MOCVD systems.
8. What wafer surface condition is necessary for the deposition of a polysilicon layer?
9. Describe the advantages of plasma-assisted depositions compared to APCVD systems.
10. Why, and with what, are deposited silicon dioxide layers doped?

# References

1. J. Hayes and P. Van Zant, *CVD Today Seminar Manual, Semiconductor Services,* San Jose, CA, 1985, p. 9.
2. S. Wolf and R. Tauber, *Silicon Processing for the VLSI Era,* Lattice Press, Sunset Beach, CA, 1986, p. 165.
3. M. L. Hammond, "Epitaxial Silicon Reactor Technology: A Review," *Solid State Technology,* May 1988, p. 160.
4. M. L. Hammond, "Epitaxial Silicon Reactor Technology: A Review," *Solid State Technology,* May 1988, p. 160.
5. J. Hayes and P. Van Zant, *CVD Today Seminar Manual,* Semiconductor Services, San Jose, CA, 1985, p. 13.
6. S. Wolf and R. Tauber, *Silicon Processing for the VLSI Era,* Lattice Press, Sunset Beach, CA, 1986, p. 165.
7. B. Meyerson, H. Kaiser, and S. Schultz, "Extending Silicon's Horizon through UHV/CVD," *Semiconductor International,* Cahners Publishing, March 1994, p. 73.
8. P. Singer, "The Future of Dielectric CVD: High Density Plasma?" *Semiconductor International,* July 1997, p. 127.
9. A. Braum, "ALD Breaks Materials, Conformity Barriers," *Semiconductor International,* October, 2001, p. 52.
10. R. Williams, *Gallium Arsenide Processing Techniques,* Artech House, Dedham, MA, 1984, p. 44.
11. S. Wolf and R. Tauber, *Silicon Processing for the VLSI Era,* Lattice Press, Sunset Beach, CA, 1986, p. 157.
12. P. Burggraaf, "The Status of MOCVD Technology," *Semiconductor International,* Cahners Publishing, July 1993, p. 81.
13. P. Singer, "Molecular beam epitaxy," *Semiconductor International,* Oct. 1986, p. 42.
14. P. Burggraaf, "The Growing Importance of MOCVD," *Semiconductor International,* November 1986, p. 47.
15. P. Burggraaf, "The Growing Importance of MOCVD," *Semiconductor International,* November 1986, p. 48.
16. P. Burggraaf, "The Status of MOCVD Technology," *Semiconductor International,* Cahners Publishing, July 1993, p. 81.
17. A. Thompson, R. Stall, and B. Droll, "Advances in Epitaxial Deposition Technology," *Semiconductor International,* Cahners Publishing, July 1994, p. 173.
18. M. L. Hammond, "Epitaxial Silicon Reactor Technology: A Review," *Solid State Technology,* May 1988, p. 159.
19. R. P. Roberge, et al., "Gaseous Impurity Effects in Silicon Epitaxy," *Semiconductor International,* January 1988, p. 81.
20. S. M. Sze, *VLSI Technology,* McGraw-Hill, New York, 1983, p. 119.
21. M. Venkatesan and I. Beinglass, "Single-wafer Deposition of Polycrystalline Silicon," *Solid State Technology,* PennWell Publishing, March 1993, p. 49.
22. L. Jastrzebski, "Silicon CVD for SOI: Principles and Possible Applications," *Solid State Technology,* September 1984, p. 239.
23. K. Yallup, "SOI Provides Total Dielectric Isolation," *Semiconductor International,* Cahners Publishing, July 1993, p. 134.
24. P. Singer, "GA-As-on-Silicon, Finally!" *Semiconductor International,* October, 2001, p. 36
25. P. Singer, "Directions in Dielectrics in CMOS and DRAMs," *Semiconductor International,* Cahners Publishing, April 1994, p. 57.
26. B. L. Chin and E. P. van de Ven, "Plasma TEOS Process for Interlayer Dielectric Applications," *Solid State Technology,* April 1988, p. 119.
27. K. Maeda and S. Fisher, "CVD TEOS/O3: Development History and Applications," *Solid State Technology,* PennWell Publishing, June 1993, p. 83.

# 13

# Metallization

## Overview

Fabrication of circuits is divided into two major segments. First the active and passive parts are fabricated in and on the wafer surface. This is called the *front end of the line* (FEOL). In the *back end of the line* (BEOL), the metal systems necessary to connect the devices and different layers are added to the chip. In this chapter, the materials, specifications, and methods used to complete the metallization segment is presented along with other uses of metals in chip manufacturing. Vacuum pumps (used in CVD, evaporation, ion implant, and sputtering systems) are explained at the end of the chapter.

## Objectives

Upon completion of this chapter, you should be able to:

1. List the requirements of a material for use as a chip surface conductor.

2. Draw cross sections of single and a two layer metal schemes.

3. Describe the purpose and operation a low-k dielectric layer.

4. Make a list of three materials used in the metallization of semiconductor devices and identify their specific use(s).

5. Describe the principle of sputtering.

6. Draw and identify the parts of a sputtering system.

7. Describe the principle and operation of turbo and cryogenic high-vacuum pumps.

## Introduction

The most common use of metal films in semiconductor technology is for surface wiring. The materials, methods, and processes of "wiring" the component parts together is generally referred to as the *metallization process*. Depending on device complexity and performance requirements, the circuit may require a single-metal system, or a multiple-level system. It may use an aluminum alloy or gold as the conducting metal.

## Conductors—Single-Level Metal

In the MSI era metallization was relatively straightforward (Fig. 13.1), requiring only a single-level metal process. Small holes, called *contact holes* or *contacts*, are etched through the surface layers, exposing areas on the device/circuit component parts. Following contact masking, a thin layer (10,000 to 15,000 Å) of the conducting metal is deposited by vacuum evaporation, sputtering, or CVD techniques over the entire wafer. The unwanted portions of this layer are removed by a conventional photomasking and etch procedure or by lift-off. This step leaves the surface covered with thin lines of the metal that are called *leads*, *metal lines*, or *interconnects*. Generally, a

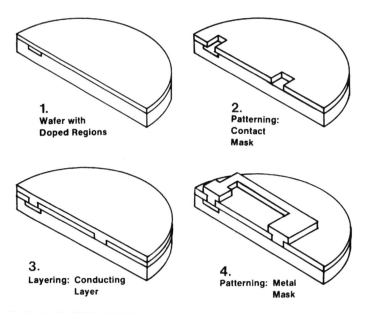

**1.**
Wafer with
Doped Regions

**2.**
Patterning:
Contact
Mask

**3.**
Layering: Conducting
Layer

**4.**
Patterning: Metal
Mask

**Figure 13.1**  Metallization sequence.

heat-treatment step, called *alloying*, is performed after metal patterning to ensure good electrical contact between the metal and the wafer surface.

Regardless of the structure, a metal system must meet the following criteria:

- Good electrical current-carrying capability (current density)

- Good adhesion to the top surface of the wafer (usually $SiO_2$)

- Ease of patterning

- Good electrical contact with the wafer material

- High purity

- Corrosion resistance

- Long-term stability

- Capable of deposition in uniform void- and hillock-free films

- Uniform grain structure

## Conductors—Multilevel Metal Schemes

Increasing chip density has placed more components on the wafer surface, which in turn has *decreased* the area available for surface wiring. The answer to this dilemma has been multilevel metallization schemes with two to four individual metal layers (Fig. 13.2). By 2012, it is expected that chips will carry up to nine wiring levels.[1] A typical two-metal stack is shown in Fig. 13.3. The stack starts with a *barrier layer* formed by silicidation of the silicon surface to produce a lowered electrical resistance between the surface and the next layer. Barrier layers also prevent alloying of aluminum and silicon if aluminum is the conducting material. Next comes a layer of an dielectric material, called an *intermetallic dielectric layer* (IDL or IMD) that provides the electrical isolation between metal layers. This dielectric may be a deposited oxide, silicon nitride, or a polyimide film. This layer receives a masking step that etches new contact holes, called *vias* or *plugs*, down to the first-level metal. Conducting plugs are created by depositing conducting material into the hole. Next, the first-level metal layer is deposited and patterned. The IMD/plug/metal deposition/patterning sequence is repeated for the subsequent layers. A multilevel metal system is more costly, of lower yield, and requires greater attention to planarization of the wafer surface and intermediate layers to create good current-carrying leads.

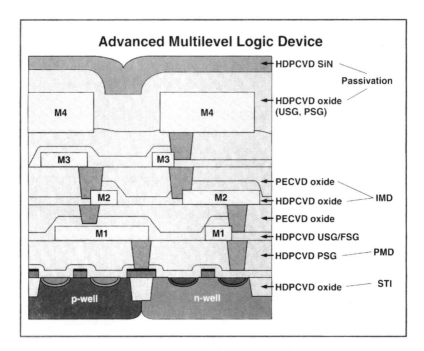

**Figure 13.2**  Multi-metal level structure. *(Courtesy of* Semiconductor International, *July 1997.)*

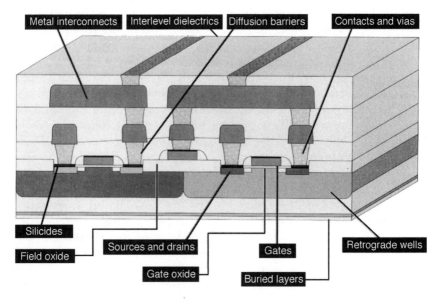

**Figure 13.3**  Two-metal structure. *(Courtesy of* Semiconductor International, *January 1998.)*

# Conductors

## Aluminum

This section addresses the three primary materials used for the metal interconnection layers. Prior to the development of VLSI-level circuits, the primary metallization material was pure aluminum. The choice of aluminum and its limitations are instructive to the understanding of metallization systems in general. From an electrical conduction standpoint, aluminum is less conductive than copper and gold. Copper, if used as a direct replacement for aluminum, has a high contact resistance with silicon and raises havoc with device performance if it gets into the device areas. Aluminum emerged as the preferred metal because it avoids the problems just mentioned. It has a low enough resistivity (2.7 $\mu\Omega$-cm),[2] and good current-carrying density. It has superior adhesion to silicon dioxide, is available in high purity, has a naturally low contact resistance with silicon, and is relatively easy to pattern with conventional photolithography processes. Aluminum sources are purified to 5 to 6 "nines" of purity (99.999 to 99.9999 percent).

## Aluminum-silicon alloys

Shallow junctions in the wafer surface presented one of the first problems with the use of pure aluminum leads. The problem came with the need to bake aluminum-silicon interfaces to stabilize the electrical contact. This type of contact is called *ohmic* because the voltage-current characteristics behave according to Ohm's law. Unfortunately, aluminum and silicon dissolve into each other and, at 577°, reach a eutectic formation point. A eutectic formation occurs when two materials heated in contact with each other melt at temperatures much lower than their individual melting temperatures. Eutectic formations occur over a temperature range, and the aluminum-silicon eutectic starts to form at about 450°, also the temperature necessary for good electrical contact. The problem is acute with shallow junctions. If the alloy region is deep, it can extend completely through the junction, shorting it out (Fig. 13.4).

Two solutions to this problem are employed. One is a barrier metal layer (see section on barrier metals) that separates the aluminum and silicon and prevents the eutectic alloy from forming. The second is an alloy of aluminum with 1 to 2 percent silicon. During the contact heating step, the aluminum alloys more with the silicon in the alloy and less with the silicon from the wafer. This process is not 100 percent effective, and some alloying between the aluminum and wafer always occurs.

Excessive Alloy

**Aluminum / Silicon**
**melted into Wafer**

**Aluminum with Si**

**Barrier Metals**

**Figure 13.4** Eutectic alloying of aluminum and silicon contacts.

### Aluminum-copper alloy

Aluminum suffers a problem called *electromigration*. The problem occurs when long, skinny leads of aluminum are carrying high currents over long distances, as is the situation in VLSI/ULSI circuits. The current sets up an electric field in the lead that decreases from the input side to the output. Also, heat generated by the flowing current sets up a thermal gradient along the lead. The aluminum in the lead becomes mobile and diffuses within itself along the direction of the two gradients. The first effect is a thinning of the lead. In the extreme, the lead can become completely separated. Unfortunately, this event usually takes place after the circuit is in operation in the field, causing a failure of the chip. Prevention or moderation of electromigration is achieved by depositing an alloy of aluminum and 0.5 to 4 percent[3] copper or an alloy of aluminum and 0.1 to 0.5 percent titanium. Aluminum alloys containing both copper and silicon are often deposited on the wafer to resolve both alloying and electromigration problems.

Drawbacks of aluminum alloys are an increased complexity for the deposition equipment and process, and different etch rates, as well as an increase in film resistivity compared to the pure aluminum. The amount of the increase varies with the alloy composition and heat treatments but can be as much as 25 to 30 percent.[4]

### Barrier metals

A method of preventing the eutectic alloying of silicon and aluminum metallization is by using a barrier layer. Both titanium-tungsten

(TiW) and titanium nitride (TiN) layers are used. TiW is sputter-deposited onto the wafer into the open contacts before the aluminum or aluminum alloy deposition takes place. The TiW deposited on the field oxide is removed from the surface during the aluminum etch step. Sometimes, a first layer of platinum silicide is formed on the exposed silicon before the TiW is deposited.

Titanium nitride layers can be placed on the wafer by all the deposition techniques: evaporation, sputtering, and CVD. It can also be formed by the thermal nitridation of a titanium layer at 600°C in an $N_2$ or $NH_3$ atmosphere.[5] CVD titanium nitride layers have good step coverage and can fill submicron contacts. A layer of titanium is required under TiN films to provide a high-conductivity intermediate with silicon substrates.

With copper metallization, the barrier is also critical. Copper inside the silicon ruins device performance. Barrier metals used are TiN, tantalum (Ta), and tantalum nitride (TaN).[6]

### Refractory metals and refractory metal silicides

Although the limitations of electromigration and eutectic alloying have been made manageable by aluminum alloys and barrier metals, the issue of contact resistance may prove to be the final limit on aluminum metallization. The overall effectiveness of a metal system is governed by the resistivity, length, thickness, and total contact resistance of *all* the metal-wafer interconnects. In a simple aluminum system, there are two contacts: the silicon/aluminum interconnect and the aluminum interconnect/bonding wire. In a ULSI circuit with multilevel metal layers, barrier layers, plug fills, polysilicon gates and conductors, and other intermediate conductive layers, the number of connections becomes very large. The addition of all the individual contact resistances can dominate the conductivity of the metal system.

Contact resistance is influenced by the materials, the substrate doping, and the contact dimensions. The smaller the contact size, the higher the resistance. Unfortunately, ULSI chips have smaller contact openings, and large gate array chip surfaces can be as much as 80 percent contact area.[7] These two factors make the contact resistance the dominant factor in VLSI metal system performance. Aluminum-silicon contact resistance, along with the alloying problem, have led to the investigation of other metals for VLSI metallization. Polysilicon has a lower contact resistance than aluminum and is in use in MOS circuits (Fig. 13.5).

Refractory metals and their silicides offer lower contact resistance. The refractory metals of interest are titanium (Ti), tungsten (W), tantalum (Ta), and molybdenum (Mo). Their silicides form when they are

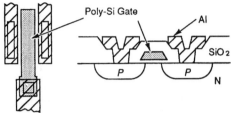

Poly-Si Gate

Al

SiO$_2$

P     P

N

**Figure 13.5** Silicon gate electrode extended for metallization lead.

Metallization
and Metal Mask

alloyed on a silicon surface (WSi$_2$, TaSi$_2$, MoSi$_2$ and TiSi$_2$). The refractory metals were first proposed for metallization in the 1950s, but they stayed in the background due to a lack of a reliable deposition method. That situation has changed with the development of LPCVD and sputtering processes.

All modern circuit designs, especially MOS circuits, use refractory metals or their silicides as intermediate (plugs), barrier, or conducting layers. The lower resistivities and lower contact resistances (Fig. 13.6) make them attractive for conducting films, but impurities and deposition uniformity problems make them less attractive for MOS gate electrodes. The solution to the problem has been the polycide and silicide gate structures, which are combinations of a silicon gate topped by a silicide. The details of this structure are explained in Chapter 16.

A popular use of refractory metals is the filling of via holes in multilevel metal structures. The process is called *plug filling,* and the filled via is called a *plug* (Fig. 13.2). The vias are filled by either selective tungsten deposition through surface holes onto the first layer metal or by CVD techniques.[8] Of the available refractory metals, tungsten finds a lot of use as aluminum-silicon barriers, MOS gate interconnects, and for via plugs.

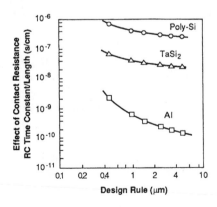

Effect of Contact Resistance
RC Time Constant/Length (s/cm)

10$^{-6}$

10$^{-7}$

10$^{-8}$

10$^{-9}$

10$^{-10}$

10$^{-11}$

Poly-Si

TaSi$_2$

Al

0.1    0.2    0.4    1    2    4    10

Design Rule (μm)

**Figure 13.6** Effect of contact resistance on RC time constant.

One of the attractions of copper metallization is that copper can be the plug material, creating a monometal system that minimizes intermetal resistances.

## Copper

Aluminum metallization ran into a performance barrier as circuits reached hundred-megahertz speeds. Signals must move fast enough through the metal system to prevent processing delays. Also, longer leads and smaller cross sections required for larger chips increase the resistance of the metal wiring system. As the number of contact holes increases, the small contract resistance between aluminum and silicon surfaces adds up to become significant. While aluminum provides a workable resistance, it is difficult to deposit in via holes with aspect ratios up to 10:1. To date, barrier metal schemes, stacks, and refractory metals have been employed to reduce aluminum metal system resistance. Additional resistance reductions needed for 0.25-$\mu\Omega$-cm (and smaller) devices have renewed interest in copper as a conducting metal. Copper is a better conductor than aluminum, with a resistance of 1.7 $\mu\Omega$-cm, compared to a 3.1 $\mu\Omega$-cm value for aluminum. Copper is resistant to electromigration and can be deposited at low temperatures.[9] It also can be used as a plug material. Deposition can be by CVD, sputtering, electroless plating, and electrolytic plating. Drawbacks, besides lack of a learning curve, include etching problems, vulnerability to scratching, corrosion, and the requirement of barrier metals to keep the copper out of the silicon. Nevertheless, the overall benefits of copper led IBM, followed quickly by Motorola, to announce the availability of production copper-based devices in 1998.[10] Novellus Systems estimates that copper-based circuits will operate four times faster than aluminum-based ones and realize a 20 to 30 percent decrease in cost.

All current circuits are being developed with copper metallization and low-k dielectrics. The primary benefits are increased performance and a reduction in the number of metal layers required.

## Low-k dielectric materials

In the dual-damascene illustration, the dielectric separating the two metals is silicon dioxide. However, this material presents a problem for high-performance circuits. The slowing of circuit signals results from the combination of the metal resistance (R) and capacitance (C). It is called the *RC constant* of the system. A major contribution to the capacitance factor is the dielectric constant of the material used to separate the metal layers, the intermediate metal dielectric (IMD).

Silicon dioxide has a dielectric constant (k) in the 3.9 range. Success-ful circuits will require k values dropping to the 1.5 to 2.0 range in the 2006 time frame (Fig. 13.7), according the SIA *International Technol-ogy Roadmap for Semiconductors*. In addition to the dielectric prop-erty the IMD must have a number of chemical and mechanical properties. They include thermal stability (subsequent metal pro-cesses can take an initial film through a number of heat steps up to the 450°C), good etch selectivity, being pin-hole free, enough flexibility to withstand on chip stresses, and compatibility with the other pro-cesses.

A number of low-k dielectric materials have been developed to meet ULSI circuit needs. They are listed in Figure 13.8 along with their di-electric constants. The major categories are oxide based materials, or-ganic based and variations of each. The organics, based on poly(alylene)thers (PAE) or hydrido-organic siloxane polymers (HOSP), offer the advantage of spin-on applications. Spin-on processes can provide great uniformity and planarity, and they are less expen-sive than CVD processes.

| Year | 1997 | 1999 | 2001 | 2003 | 2006 |
|---|---|---|---|---|---|
| Metal levels | 5–6 | 6–7 | 7 | 7 | 7–8 |
| Interconnect length (m/chip) | 820 | 1480 | 2160 | 2840 | 5140 |
| Dielectric constant (k) | 3.0–4.1 | 2.5–3.0 | 2.0–2.5 | 1.5–2.0 | 1.5–2.0 |

**Figure 13.7**  Trends in metal and low-k dielectric requirements.

| Metal system | Low-k material |
|---|---|
| Aluminum | parylene |
|  | HSQ |
|  | methyl silesequioxane |
|  | F-doped oxide |
|  | F-doped amorphous carbon |
|  | Parylene-F (AF$_4$) |
|  | Xerogel |
| Gold | polyimide |
|  | BCB |
|  | Xerogel |

**Figure 13.8**  Low-k materials. *(Source:* Future Fab International, *no. 8, pp. 180.)*

## The dual-damascene copper process

Transitioning from aluminum to copper metallization is not a simple matter of switching materials. Copper has its own set of problems and challenges. It is not easy etched by wet or dry techniques. Copper has a high electrical resistance connection with silicon. It diffuses easily through silicon dioxide and can enter the silicon structure. There, it can degrade device performance and create junction leakage problems. Copper does not adhere well to silicon dioxide surfaces, causing structural problems. These challenges led to the development of a unique process specifically designed to overcome the copper problems and produce a high-production process. It features a lithography process, the development of a low-k barrier/liner process, copper electrochemical plating, and a chemical mechanical polishing process.

In Chapter 10, a basic damascene process was introduced. The damascene concept is simple. A trench is formed in a surface dielectric layer using a photolithography process, and the required metal is deposited into it. Usually, the trench is overfilled, requiring a CMP step to replanarize the surface (Fig. 13.9). This process offers superior dimensional control, because it eliminates the variation introduced in a typical metal etch process.

In practice the process is a bit more complicated. Figure 13.10 illustrates a typical dual-damascene process that connects two metal levels. It starts with the first metal already in place. A layer of a low-k dielectric is deposited and planarized with a CMP process. A patterning step creates a via hole in the dielectric layer. A second patterning step results in the lowering of the dielectric and a "step back" on the

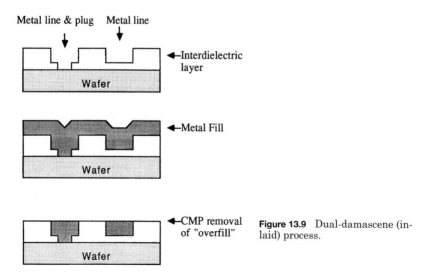

**Figure 13.9** Dual-damascene (inlaid) process.

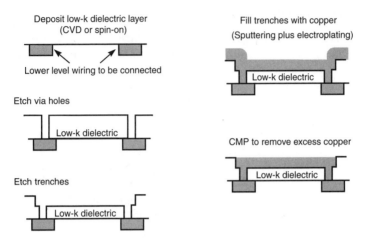

**Figure 13.10**   Typical dual-damascene process.

surface to allow a wider trench width. This pattern leaves a wider opening in the top layer to allow enough width for the copper stripe to carry the required current levels. This sequence offers the advantage of a one-step process to fill the via and form the copper metal lead. There are a number of variations on this basic dual-damascene process, each ending up with a narrow via and a wider trench opening ready for metal fill.

### Barrier/liner deposition

As previously mentioned Copper diffuses easily through silicon dioxide and can cause electrical performance problems if it gets in the circuit components. This problem is addressed by depositing a "liner" layer in the via hole bottom and sides (Fig. 13.11). Typically, the materials used are tin (Ti) or tantalum (Ta), such as TiN, Ta, TaN, and TaSiN. Depending on the material, either sputtering or CVD deposition is used to create the barrier/liner. These vias are very high aspect and challenge the process to produce an uniform film over the entire via/trench inside surface.

### Seed deposition

While copper can be deposited by sputtering, or CVD deposition, electrochemical plating (ECP) has emerged as the preferred deposition method. Producing a uniform, void-free copper film with ECP requires a starting "seed" layer in the via/trench hole. PVD techniques are used to deposit the copper seed in the via hole. The challenge, as in

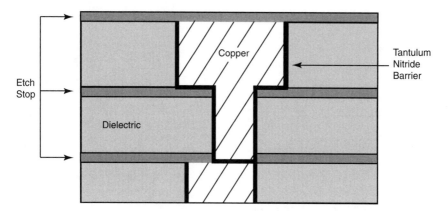

**Figure 13.11**  Single-level dual-damascene with tantalum nitride barrier. *(From Wolf, and Tauber,* Microchip Manufacturing, *Lattice Press.)*

the barrier/liner deposition, is producing a uniform layer in very high-aspect vias.

## Electrochemical Plating (ECP)

Electroplating has emerged as a production copper deposition method due to its low temperature and low cost.[11] Low temperature is necessary due if low-k dielectric layers are used. The seed layer must uniformly coat the bottom and sides of the via/trench to ensure uniform physical and electrical properties of the copper metal lead. Electroplating of copper has been a mainstay of printed circuit board processing for decades (Fig. 13.12). The wafer is suspended in a bath

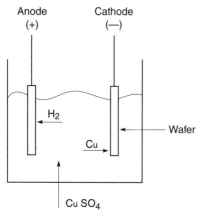

**Figure 13.12**  Schematic of electroplating of copper.

$$2\ Cu^{++} + 2H_2O \longrightarrow 2Cu(s) + O_2\ (G)$$

containing copper sulfate ($CuSO_4$) and is connected to an cathode (negative pole). With the application of a current, the bath components disassociate. The copper plates out on the wafer and hydrogen gas is liberated at the anode. One concern is the uniformity of the film across the wafer. The varied materials and structures on the wafer surface mitigate against uniform current distribution. Nonuniform film growth and density can be the result. Another concern is buildup of a bevel at the lip of the opening. This is addressed by a separate cleaning step after the plating. Most processes also include a post-ECP to stabilize the film prior to CMP. Nonuniform areas across the wafer surface will have different removal rates in the CMP process. Production level ECP systems will include wafer preclean, the plating sections, bevel removal, and annealing.

## Chemical Mechanical Processing

Chemical mechanical processing (CMP) is used in several steps in the semiconductor process. In Chapter 3, its use for planarizing raw silicon wafers was described. In Chapter 10, we describe its use for planarizing in-process wafers to achieve a flat surface for lithography accuracy. The post-copper CMP is a similar process but with a different surface to be flattened. During copper plating, the via/trench hole is overfilled to ensure complete filling of the trench. Before proceeding to the next step, it is necessary to reflatten the surface by removing the copper overfill. The process and details are discussed in Chapter 10.

### Doped polysilicon

The advent of silicon-gate MOS technology turned deposited polysilicon lines on the chip into conductors. For use as a conductor, the polysilicon has to be doped to increase its conductivity. Generally, the preferred dopant is phosphorus, due to its high solid solubility in silicon. Doping is by either diffusion, ion implantation, or *in situ* doping during an LPCVD process. Each of the methods produces a different doping result. The differences relate to the doping temperature's effect on the grain structure. The lower the temperature, the greater the amount of dopant trapped in the polygrain structure, where it is unavailable for conduction. This is the situation with ion implantation. Diffusion doping results in the lowest film sheet resistivity. *In situ* CVD doping has the lowest dopant carrier mobility due to grain boundary trapping.

Doped polysilicon has the advantage of a good ohmic contact with the wafer silicon and can be oxidized to form an insulating layer. Polysilicon oxides are of a lower quality than thermal oxides grown on single-crystal silicon because of the nonuniformity of the oxide grown on the rougher polysilicon surface.

Although polysilicon has a low contact resistance with silicon, it still exhibits too high a resistance the metal material(s). Creating a multimetal stack of the polysilicon and a silicide (such as titanium silicide). These are called polycide structures (see Chapter 16).

## Metal Film Uses

### MOS gate and capacitor electrodes

Most electrical devices depend on the passage of an electrical current to operate. Capacitors are an exception. These devices (see Chapter 16) require two conductive layers, called electrodes, separated by a dielectric. In most designs, the top electrode is a section of the conductor metal system. A discussion of the relationship of capacitor parameters is in Chapter 2.

MOS transistors are a capacitor structure, and the top electrode, called a *gate*, is a critical structure in MOS circuits.

### Fuses

The development of thin-film fuse technology allowed creation of the programmable read-only memory (PROM) circuit. The fuse allows field programming of data in the memory section of the chip. In this role, the fuse is not a protective device, as in most electrical circuits, but is included specifically to be "blown" or disabled.

In the memory section of the chip, called the *array*, are a number of memory cells, each with a fuse between the cell and the main metallization system. The array is essentially a blank blackboard (Fig. 13.13). Information can be coded into the array in digital form (on/off) by having some of the cells operating and others not operating. Nonoperating cells can be created by *blowing* the fuse, thus removing it from the circuit. The same system is used to program logic arrays. (Fig. 13.14). Once the fuse is blown, the associated memory cell is permanently removed (electrically) from the circuit.

There are two primary fuse configurations. One consists of thin films of nichrome, titanium-tungsten, or polysilicon lying under two metal leads, which are patterned with thin "necks" that can be blown by a current pulse directed through the metal lines. Another fuse scheme employs a thin film of polysilicon or oxide in the contact hole. The fuse is blown when a high current is passed through the layer and destroys it by heating.

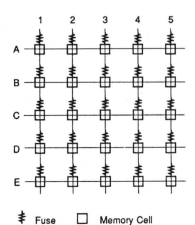

Figure 13.13  Schematic of fused memory array.

≹ Fuse    ☐ Memory Cell

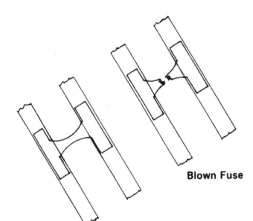

**Blown Fuse**

Figure 13.14  Thin film fuses.

### Backside plating

Gold is sometimes evaporated onto the entire back of the wafer just prior to wafer sort. The gold functions as a solder in certain packaging processes (see Chapter 18).

### Deposition Methods

Metallization techniques, like other fabrication processes, have undergone improvements and evolution in response to new circuit requirements and new materials. The mainstay of metal deposition up to the mid-1970s was vacuum evaporation. Aluminum, gold, and the fuse metals were all deposited by this technique. The needs of depositing multi-metal systems and alloys, along with the need for better step coverage, led to the introduction of sputtering as the standard deposition technique for VLSI circuit fabrication. Refractory metal use has added the third technique, CVD, to the arsenal of the metallization engineer. Some of the basic issues of the deposition of metals, by all techniques, are discussed in the following section on vacuum evaporation.

### Vacuum evaporation

Vacuum evaporation is used for the deposition of metals on discrete devices and circuits of lower integration levels. It is also used for the deposition of gold to the back of a wafer for die adhesion into a package.

Vacuum evaporation takes place inside an evacuated chamber (Fig. 13.15). The chamber can be a quartz bell jar or a stainless steel enclosure. Inside the chamber is a mechanism to evaporate the metal source, wafer holders, a shutter, thickness and rate monitors, and heaters. The chamber is connected to a vacuum pump(s) (see "Vacuum Pumps," p. 426).

Since aluminum is the most critical of the materials evaporated, we shall focus on its deposition. The vacuum is required for a number of reasons. First is a chemical consideration. If any air (oxygen) molecules were in the chamber when the high-energy aluminum atoms were coating the wafer, they would form aluminum trioxide ($Al_2O_3$), a dielectric that, if incorporated into the deposited film, would compromise aluminum's role as a conductor. A second requirement for vac-

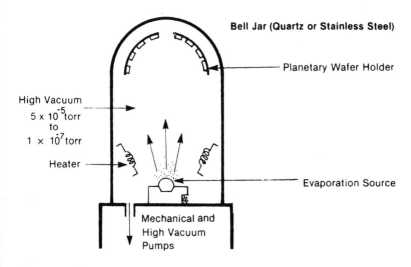

**Figure 13.15**  Vacuum evaporator.

uum deposition is uniform coating. The vacuum required for successful evaporation of aluminum is from $5 \times 10^{-5}$ to $1 \times 10^{-9}$ torr of pressure. (See Chapter 2 for a discussion on vacuum and pressure.)

There are systems that operate in the ultra-high vacuum (below $10^{-9}$ torr) range.[13] They offer the advantage of lowered background contamination in the system hardware, which reduces contamination of the wafers.

**Evaporation sources.**    Before describing the various methods of causing a metal to evaporate, a review of basic evaporation theory is in order. Most of us are familiar with the evaporation of a liquid from a beaker. This happens because there is sufficient energy (heat) in the liquid to cause the molecules to escape into the atmosphere. Over time, some of them stay in the atmosphere; we call this *evaporation*. The same process of evaporation can be made to occur in solid metals. The requirement is to heat the metal to a liquid state so that the atoms or molecules evaporate into the surrounding atmosphere. Filaments and electron beam sources are the usual methods of evaporation. Filament evaporation is the simplest and is used for noncritical evaporations such as backside gold layers. The material, in wire form, is either wrapped around a coiled tungsten (or other metal that is able to withstand high temperatures) wire. A high current is passed through the tungsten wire, where it heats the deposition metal to a liquid and evaporates into the chamber, coating the wafers. Another version uses a flat filament with a dimple to hold pieces of the deposition material.

The need for evaporation control and low contamination led to the development of the electron beam evaporation source (Fig. 13.16) for aluminum. The system is called an *e-beam gun* or just *e-gun*. This evaporation source consists of a water-cooled copper crucible with a center cavity to hold the aluminum. At the side of the crucible is a

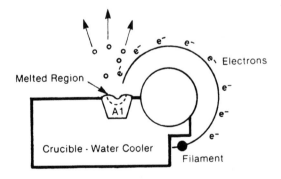

**Figure 13.16**    Electron gun evaporation source.

high-temperature filament. A high current is passed through the filament, which, in turn, "boils" off electrons. The negative electrons are bent 180° by a magnet so that the electron beam strikes the center of the charge in the cavity. The high-energy electrons create a pool of liquid aluminum in the center of the charge. Aluminum evaporates from the pool into the chamber and deposits on the wafers in holders at the top of the chamber. The water cooling maintains the outer edges of the charge in a solid state, thus preventing contaminants from the copper crucible from evaporating. Film thickness is controlled by shutters and by rate and thickness monitors. In-chamber monitors, located near or above the wafer holders, feed back information to the e-gun power supply, which controls the evaporation rate.

A major goal of any metal deposition system is good step coverage (Fig. 13.17). This is a challenge for vacuum evaporators, because the source is essentially a point source. The problem comes when material from a point source is shadowed by steps on the wafer surface. The result can be that one side of the openings in the surface oxide can be too thin or have a void. Planetary wafer "domes" rotating in the chamber are used to create uniform thicknesses (Fig. 13.18). Quartz heaters in the chamber aid step coverage by maintaining evaporant mobility as the atoms arrive at the surface. They "fill in" the steps by a capillary action.

### Sputter deposition (PVD)

Sputter deposition (sputtering) is another old process adapted to semiconductor needs. It is a process first formulated in 1852 by Sir William

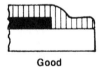

Good

Thin at Step

Step Shadowed

**Figure 13.17**   Step coverage.

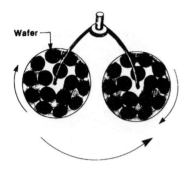

Wafer

**Figure 13.18** Planetary wafer holder.

Robert Grove.[12] Sputtering (in general) can deposit any material on any substrate. It is widely used to coat costume jewelry and put optical coatings on lenses and glasses. Discussion of the benefits of sputtering to the semiconductor industry is best left until the principles and methods of sputtering have been covered. Sputtering, like evaporation, takes place in a vacuum. However, it is a physical, not a chemical, process (evaporation is a chemical process) and is referred to as physical vapor deposition (PVD).

Inside the vacuum chamber is a solid slab, called a *target*, of the desired film material (Fig. 13.19). The target is electrically grounded. Argon gas is introduced into the chamber and is ionized to a positive charge. The positively charged argon atoms are attracted to the grounded target and accelerate toward it. During the acceleration, they gain momentum, which is force, and strike the target. At the target, a phenomenon called *momentum transfer* takes place. Just as a cue ball transfers its energy to the other balls on a pool table, causing them to scatter, the argon ions strike the slab of film material, causing its atoms to scatter (Fig. 13.20). The argon atoms "knock off" atoms and molecules from the target into the chamber. This is the sputtering activity. The sputtered atoms or molecules scatter in the chamber with some coming to rest on the wafer. A principal feature of a sputtering

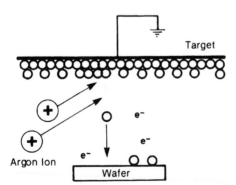

Target

Argon Ion

$e^-$

Wafer

**Figure 13.19** Principle of sputtering.

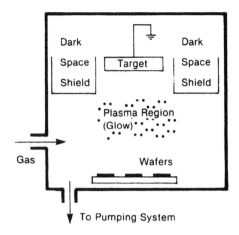

**Figure 13.20** Typical sputtering equipment.

process is that the target material is deposited on the wafer without chemical or compositional change.

There are several advantages of sputtering over vacuum evaporation. One is the aforementioned conservation of target material composition. A direct benefit of this feature is the deposition of alloys and dielectrics. The problem of evaporating alloys was described in the preceding section. In sputtering, an aluminum and 2 percent copper target material yields an aluminum and 2 percent copper film on the wafers.

Step coverage is also improved with sputtering. Whereas evaporation proceeds from a point source, sputtering is a planar source. Material is being sputtered from every point on the target, with material arriving at the wafer holder with a wide range of angles to coat the wafer surface. Step coverage is further improved by rotating the wafer holder and by heating the wafer.

Adhesion of the sputtered film to the wafer surface is also improved over evaporation processes. The higher energy of the arriving atoms makes for a better adhesion, and the plasma environment inside the chamber has a "scrubbing" action of the wafer surface that enhances adhesion. Adhesion and surface cleanliness can be increased by grounding the wafer holder and sputtering the wafer surface for a brief time prior to the deposition. In this mode, the sputter system is functioning as an ion-etch (sputter-etch, reverse-sputter) machine, as described in Chapter 10.

Another technique to improve step coverage and uniform film formation in deep holes is a collimated beam (Fig. 13.21). Atoms come off of the target at many angles and tend to fill the sides of holes before filling the bottom. A collimator is a physical barrier plate similar to a honeycomb with round or hexagonal holes. It is grounded for electrical

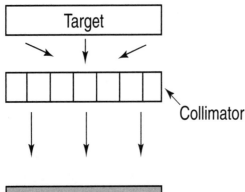

**Figure 13.21** Sputtering with a collimator.

neutrality. Atoms arriving at the collimator at steep angles are caught on the sides, while straighter-angle atoms continue onto the wafer surface. The thickness of the collimator is a factor in the degree of collimation of the atom beams.

Perhaps the greatest contribution of sputtering is the control of film characteristics available by the balancing of the sputtering parameters of pressure, deposition rate, and target material. Sandwiches of material can be sputtered in one process with multiple target arrangements.

Clean and dry argon (or neon) is required to maintain film composition characteristics, and low moisture is required to prevent unwanted oxidation of the deposited film. The chamber is loaded with the wafers, and the pressure is reduced by pumps (pumped down) to the $1 \times 10^{-9}$-torr range. The argon is introduced and ionized. Control of the argon amount entering the chamber is critical due to its effect of raising the pressure in the chamber. With the argon and sputtered material in the chamber, the pressure rises to a level of about $10^{-3}$ torr. Chamber pressure is a critical parameter in the deposition rate of the system. After liberating material from the target, the argon ions, the sputtered material, gas atoms, and electrons generated by the sputtering process form a plasma region in front of the target. The plasma region is evident by its purple glow. The plasma region is separated from the target by a darkened region, known as the *dark space.*

Four sputtering methods are used in semiconductor applications. They are

- Diode [direct current (dc)]
- Diode [radio frequency (RDI)]

- Triode

- Magnetron

The first two methods, known as *diode sputtering*, are simple in concept. The target is connected to a negative potential with a positively charged anode present in the chamber. The negatively charged target ejects electrons, which accelerate toward the anode. Along the way, they collide with the argon gas atoms, ionizing them. The positively ionized argon atoms then accelerate to the target, initiating the sputtering process. The ionized argon (+) and the target (–) form a diode.

A secondary effect of the ionization process is the impact of the electrons on the gas atoms, resulting in the plasma region that is visible as the glowing purplish region just in front of the target. Dark spaces exist just in front and to the sides of the target. Sputtering efficiency is enhanced when the plasma is confined to the region between the target and wafers. This condition is enhanced by placing "dark space" shields to the sides of the target. The shields prevent target material from being sputtered from the sides—material that will never deposit on the wafers.

Another problem arises from the outgassing of contamination from the chamber walls while the chamber is under vacuum. This condition is called a *virtual leak*, as opposed to an actual leak of atmosphere into the system. Besides compromising the pressure level in the chamber, the contamination can be incorporated into the deposited film.

This latter problem is addressed by placing a small negative bias (charge) on the wafer holder. The bias creates ions on the wafer surface and has the effect of dislodging stray outgassed atoms from the growing film. Direct-current diode sputtering is used primarily to deposit metals.

Improved sputtering is gained by connecting the target to the negative side of a radio-frequency (RF) generator. The ionization of the gas takes place near the target surface without requiring a conductive target. Radio-frequency sputtering is necessary to sputter nonconducting materials (dielectrics) and is used also for conductors. Biasing is also used with the radio-frequency sputtering to achieve a cleaning effect at the wafer surface. Radio-frequency biasing offers the advantage of etching and cleaning of the exposed wafer surface. Etching and cleaning are achieved by putting the wafer holder at a different field potential from that of the argon, causing the argon atoms to impinge directly on the wafer. This procedure is called *sputter etch*, *reverse sputter*, or *ion milling*. The process removes contamination and a small layer from the wafer. Removal of contamination improves electrical contact between the exposed wafer regions and the film and improves adhesion of the film to the rest of the wafer surface.

In diode sputtering, a number of processes occur at or near the wafer surface. On impact of the argon atom, a number of electrons are created. These electrons cause heating of the substrate (as high as 350°C), which in turn can cause uneven film deposition. The electrons also create a radiation environment that can damage sensitive devices.

The heating produced with diode sputtering causes a serious problem with the deposition of aluminum. The heating causes residual oxygen in the target and in the chamber to combine with the aluminum to form aluminum oxide. The aluminum oxide is a dielectric and can compromise the conductive property of the deposited aluminum. More serious, a layer of aluminum oxide can form on the target surface, and the impinging argon atoms do not (in diode sputtering) have enough energy to break through the layer. In effect, the target becomes sealed, and the sputtering stops. Triode sputtering avoids some of the problems of diode sputtering.

The electrons necessary to ionize the argon are created by a separate high-current filament. In designs where the filament is located outside the deposition chamber, the wafers are protected from radiation damage. Films deposited by a triode method are more dense.

Another problem with diode sputtering is the electrons that escape into the chamber and do not contribute to the establishment of the plasma necessary for deposition. The situation is resolved in *magnetron* sputtering systems, which use magnets behind and around the target (Fig. 13.22). The magnets capture and/or confine the electrons to the front of the target. Magnetron systems are more efficient for increased deposition rates. The resulting ion current (density of ionized argon atoms hitting the target) is increased by an order of magnitude over conventional diode sputtering systems. Another effect is a lower pressure required in the chamber, which contributes to a cleaner deposited film. Magnetron sputtering leaves a lower target temperature,

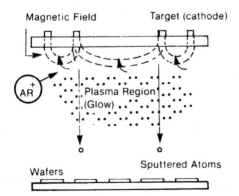

**Figure 13.22**  Magnetron sputtering.

which makes it a favorite for the sputtering of aluminum and aluminum alloys.

Production-level sputtering systems come in a variety of designs. Chambers are either batch systems or single-wafer in-line designs. Most production machines have load-lock capabilities. A load lock is an antechamber where a partial vacuum is created so that the deposition chamber can be maintained at vacuum. The advantage of a load lock is a higher production rate. Production machines are usually dedicated to one or two target materials, while development machines have a wider range of capability.

### CVD metallization

Advances in LPCVD offer the deposition department a third choice for metal depositions. LPCVD offers the advantages of not requiring expensive and maintenance-intensive high-vacuum pumps, conformal step coverage, and high production rates. Perhaps the most often deposited CVD refractory metal film is tungsten (W).

Tungsten is used in a variety of structures, including contact barriers, MOS gate interconnects, and via plugs. The filling of via holes is a key to effective multi-metal systems (see Fig. 13.2). The dielectric layer is relatively thick and the via holes have to be relatively thin (high aspect ratio). These two factors make for a difficult continuous metal deposition to fill the vias without thinning of the metal in the via. Selective CVD-deposited tungsten plugs fill the entire via and present a planar surface for a subsequent conducting metal layer. For use as a barrier metal, tungsten can be deposited selectively by the silicon reduction of the gas tungsten hexafluoride ($WF_6$) by the reaction

$$2WF_6 + 3Si \rightarrow 2W + 3SiF_4$$

Tungsten can also be deposited selectively over aluminum and other materials from $WF_6$. The processes are called *substrate reduction*. Tungsten is also deposited from $WF_6$ by hydrogen reduction; the reaction is

$$WF_6 + 3H_2 \rightarrow W + 6HF$$

All of the depositions are performed in LPCVD systems at temperatures of about 300°C, which makes the processes compatible with aluminum metallization.

The depositions of tungsten silicide and titanium silicide proceed by the following reactions:

$$WF_6 + 2SiH_4 \rightarrow WSi_2 + 6HF + H_2$$

$$TiCl_4 + 2SiH_4 \rightarrow TiSi_2 + 4HCl + 2H_2$$

## Vacuum Pumps

LPCVD, ion implantation, evaporation, and sputtering processes all take place in reduced-pressure (vacuum) chambers (see Chapter 2). Vacuum chambers provide process conditions free of contaminating gases. In the deposition processes, the vacuum increases the mean free path of the depositing atoms and molecules, which in turn results in more uniform and controllable deposited films. LPCVD takes place in the pressure range down to $10^{-3}$ torr (medium range) while the other processes take place at pressure ranges down to $10^{-9}$ torr high to ultra-high range). Medium range is reached with mechanical vacuum pumps. These same pumps are used to initially reduce the pressure in the high-vacuum process chambers. In this role, they are called *roughing pumps*. Additionally, mechanical vacuum pumps are used on the outlet end of high-vacuum pumping systems to assist in the removal of gas molecules from the pump to the exhaust system.

After the rough vacuum is established, a high-vacuum pump takes over to establish the final vacuum. The pumps used for this purpose are *oil diffusion, cryogenic, ion,* or *turbomolecular.* Whatever the type, all pumps are constructed of materials that will not *outgas* into the system and compromise the vacuum. Materials used are typically type 304 stainless steel, oxygen-free high-conductivity copper (OFHC), Kovar, nickel, titanium, borosilicate glasses, ceramics, tungsten, gold, and some low-vapor-pressure elastomers.[13] Pumps used to evacuate corrosive and toxic gases or reaction by-products must have corrosion-free inside surfaces. Also, care must be taken in servicing pumps with these types of applications.

Pumps are selected and used based on a number of criteria, including

- Vacuum range required
- Gases to be pumped (lighter gases such as hydrogen are more difficult to pump)
- Pumping speed
- Overall throughput
- Ability to handle impulsive loads (periodic outgassing)
- Ability to pump corrosive gases

- Service and maintenance requirements
- Downtime
- Cost

Recall from Chapter 2 that pressure in a system results from the activity of gas atoms or molecules in an enclosure striking the chamber walls with some force. Reduction of the pressure in a system requires the removal of the gas in the chamber. This is generally accomplished by the pump establishing a lower pressure, first within itself, which allows gas material in the process chamber to flow into the pump, where it is removed entirely from the system. At very low pressures, there is not much material in the chamber, and continued pressure reduction requires that the system be leak-free and not add to the pressure by its own outgassing. Some systems require traps to prevent material from the pump from *backstreaming* into the chamber. Cold traps are explained in the section on oil diffusion pumps.

### Mechanical pumps

Mechanical oil rotary vacuum pumps are a basic design that traces to the 1640s, when Galileo and Torricelli were investigating the theory that air had weight. They are referred to as *blower*-type pumps. Air is removed from the pump by an eccentric rotating vane in a cavity (Fig. 13.23). As the vane rotates, it compresses and sweeps out the gas in front of it in the cavity, simultaneously leaving a reduced-pressure region behind it. The "pushed" material exits through a valve, while another valve opens to the cavity from the chamber and allows material in the chamber to flow into the cavity. As the vane rotates, more and more material is removed from the cavity, thus reducing the pressure in the chamber.

A critical part of the pump is the efficiency of the exit valve. An ineffective valve that leaks atmosphere back into the system will limit the

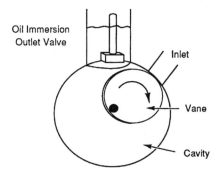

Oil Immersion
Outlet Valve

Inlet

Vane

Cavity

**Figure 13.23** Mechanical rotary oil pump.

ultimate pressure level of the pump. This style of pump uses an oil-immersed valve that prevents backstreaming of the atmosphere.

Oil-filled pumps met their match with the advent of LPCVD and etch applications. These processes produce a lot of physical residues that contaminate the oil quickly. *Oil-sealed, oilless, scroll pumps,* and *roots pumps* (called *roots blowers*) are all push-type pumps that have replaced oil-mechanical pumps.[16]

**Oil diffusion pumps**

The oil diffusion pump (Fig. 13.24) was the mainstay of most semiconductor vacuum processes. The pump requires the services of an oil mechanical pump to first reduce the pressure in the chamber to the $10^{-3}$-torr level. Either the same or a second mechanical pump is required at the outlet end. High vacuum, in the $10^{-8}$-torr range (under production conditions), is achieved by a clever momentum transfer system. A low-vapor-pressure, hydrocarbon-based oil is heated in the base of the pump, where it rises up a structure called the *stack*. At the top of the stack is a series of downward-facing baffles. The hot oil molecules, which have gained speed and energy from the boiling, exit the stack in a downward direction. Outside the stack, they collide with gas from the chamber, causing them to be propelled toward the bottom of

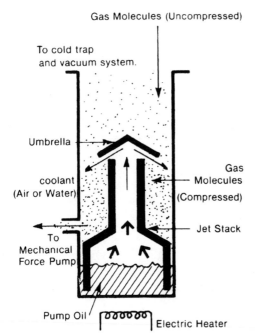

**Figure 13.24** Oil diffusion pump.

the pump, where they are removed by a mechanical pump. The oil molecules return to the heated reservoir.

Two problems with oil diffusion pumps are the migration of oil molecules back up into the chamber and the inability of the pump to handle water vapor from the chamber. Both problems are resolved by the use of a *cold trap* between the pump and the chamber. A cold trap is similar in design to a liquid source bubbler. The fluid in the trap is liquid nitrogen, which reduces the temperature to −96°C. At this temperature, oil, contaminant and water vapor molecules are frozen to the inside walls of the trap and do not add to the pressure in the system.

### Cryogenic pumps

Even with cold-trap technology, some processes cannot stand contamination from hydrocarbon oils such as are used in oil diffusion pumps. This situation has led to the use of cryogenic (cryo) pumps. A cryogenic pump (Fig. 13.25) uses the fact that gas molecules will "freeze" out on cold surfaces. The cold trap described in the preceding section and the frost that collects on the insides of refrigerators are examples of cryogenic activity.

Cryogenic pumps are designed with a central finned stack. At the low temperatures, gas from the chamber collects on the vanes, removing material from the system, which in turn reduces the pressure. The central stack, called an *expander*, is cooled as a compressor releases the liquid helium or nitrogen into it from the bottom. Cooling is by a phenomenon called *adiabatic expansion.* This is the same phenomenon that causes a pressurized can to cool when the nozzle is opened and the gas expands rapidly into the lower-pressure atmosphere. The top of the expander is at a higher temperature than the bottom, resulting in different gas molecules freezing out at different levels. Cryogenic pumps can efficiently pump water vapor, which turbo pumps cannot.

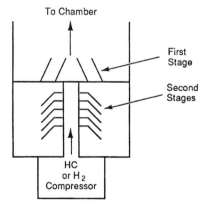

**Figure 13.25** Cryogenic pump.

Cryopumps operate without cold traps or mechanical roughing pumps. Due to their action of trapping gas material on the fins, they are of a type known as *capture pumps* rather than the displacement-type pumps that move the chamber material to the atmosphere. The total capture nature of the pump and its oilless operation drastically reduce the possibility of contamination. However, when the pump is brought to room temperature, by mistake or for maintenance, the frozen gases are released, and care must be exercised to vent any toxic or flammable gases trapped on the vanes. The buildup of gases also affects the pumping speed. The pump's speed must be monitored and the system cleaned when the speed falls. Cryopumps are simpler to operate and maintain, since no cold trap liquids or messy oils are needed. Additionally, cryopumps can handle bursts of outgassing from the process chamber and feature a fast pumping speed.

### Ion pumps

Another capture-type pump is the ion pump, also known as a *sputter ion pump* or a *getter pump* (Fig. 13.26). An ion pump operates in a manner similar to an ionization section in an ion implanter or sputter machine; only in this application, the atoms and molecules come from the chamber. A portion of those that drift into the ionization chamber are ionized to a positive charge by bombardment with electrons and attracted to a titanium cathode (negative potential). On collision with the titanium, some of the titanium is sputtered away and travels into the pump. The titanium atoms are chemically active enough to combine with other gases in the pump, which also accumulate on the pump walls. Again, material is removed from the chamber which reduces the pressure. Ion pumps are capable of pressures down to $10^{-11}$ torr, which is the ultra-high vacuum range.

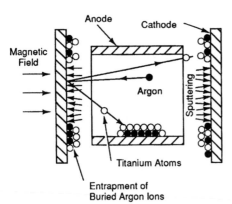

**Figure 13.26**  Principle of ion vacuum pump.

## Turbomolecular pumps

Turbomolecular pumps are similar in design to a jet turbine engine. A series of blades (Fig. 13.27) with openings are mounted and rotated at very high speeds (24,000 to 36,00 rpm[14]) on a central shaft. Gas molecules from the chamber encounter the first blade and gain momentum from the collision with the rotating blade. The momentum direction is downward to the next blade, where the same thing happens. The net result is a removal of gas from the chamber. The use of a momentum transfer makes the pumping principle the same as an oil diffusion pump. Major advantages of turbomolecular pumps are a lack of back-streaming from oils, no need to recharge, high reliability, and pressure reduction into the high vacuum range. Drawbacks are a slower pumping speed compared to oil diffusion and cryogenic pumps and vibration and wear due to the high rotational speeds.

An addition to turbo pumps is a *drag type* pump. Molecules are bounced off a rotating drum or disk rather than vanes or stators.[15] These *combination* pumps can exhaust at high pressures. Use of turbo pumps with corrosive gas processes requires coating the rotors and staters and/or heating the pump to keep the gases from forming solid particles that deposit on the pump parts.

## Summary

Figure 13.28 is an overview of the metals most used, their uses, and deposition methods.

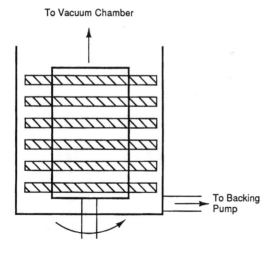

**Figure 13.27** Turbomolecular pump.

| Metal | Use in Device | | | | | Deposition Method | | | |
|---|---|---|---|---|---|---|---|---|---|
| | Conductor | Fuse | Barrier | Plug | Other | Evaporation | Sputtering | CVD | Plating |
| Aluminum | X | | | | | X | X | | |
| Aluminum/Silicon | X | | | | | X | X | | |
| Aluminum/Copper | X | | | | | X | X | | |
| Aluminum/Silicon/Copper | X | | | | | X | X | | |
| Copper | X | | | X | | | | X | X |
| Titanium | | | | | | | | | |
| Tungston | | | | | | | | | |
| Titanium/Tungston | | X | | | | | | | |
| Nichrome | | X | | | | | | | |
| Doped Polysilicon | X | | | | | | | X | |
| Molybdenum | | | X | | | | X | X | |
| Platinum Silicide | | | X | X | | | X | X | |
| Titanium Silicide | | | X | X | | | X | X | |
| Tungsten Silicide | | | X | X | | | X | X | |
| Molybdenum Silicide | | | X | X | | | X | X | |
| Gold | | | | | Die Attach Life Time-Control | | | | |

**Figure 13.28**  Summary of metallization materials and processes.

## Review Questions

1. Discuss the properties required for a microchip metal system and how they apply to aluminum and copper metallization.

2. Explain electromigration and how it is controlled.

3. Explain aluminum and silicon eutectic alloying and two processes used to prevent it.

4. Why are refractory metal and their silicides favored for VLSI circuit metallization?

5. What is a silicide and how is it formed?

6. Draw a diagram of a two-metal system, identifying all parts.

7. Why are low-k dielectrics used with copper metallization systems?

8. Draw a diagram of and label the parts of a sputter machine.

9. What are the advantages of a sputter process compared to an evaporator process?

10. List the four types of high-vacuum pumps used in semiconductor processes and their principles of operation.

11. Discuss the RC constant effect on circuits and techniques to reduce it.

12. List three advantages of copper metallization over aluminum.

13. Sketch the steps in a dual-damascene process.

## References

1. Staff, "Speeding the Transition to 0.18 µm," *Semiconductor International*, January 1998, p. 66.
2. S. Wolf and R. Tauber, *Silicon Processing for the VLSI Era*, Lattice Press, Sunset Beach, CA, 1986, p. 332.
3. P. Riley, S. Peng, and L. Fang, "Plasma Etching of Aluminum for ULSI Circuits," *Solid State Technology*, PennWell Publishing, February 1993, p. 47.
4. S. M. Sze, *VLSI Technology*, McGraw-Hill, New York, 1983, p. 347.
5. P. Singer, "New Interconnect Materials: Chasing the Promise of Faster Chips," *Semiconductor International*, November 1994, p. 53.
6. P. Singer, "Copper Goes Mainstream: Low-k to Follow," *Semiconductor International*, November 1997, p. 67.
7. P. Singer, "New Interconnect Materials: Chasing the Promise of Faster Chips," *Semiconductor International*, November 1994, p. 54.
8. D. M. Brown, "CMOS Contacts and Interconnects," *Semiconductor International*, 1988, p. 110.
9. D. Pramanikm and V. Jain, "Barrier Metals for ULSI," *Solid State Technology*, PennWell Publishing, January 1993, p. 73.
10. P. Singer, "Copper Goes Mainstream: Low-k to Follow," *Semiconductor International*, November 1997, p. 68.
11. A. Braun, ECP Technology, *Semiconductor Technology*, May 2000, p. 60

12. Y. Pauleau, "Interconnect Materials for VLSI Circuits," *Solid State Technology*, February 1987, p. 61.
13. G. Tisdale, et al., "Next-Generation Aluminum Vacuum Systems," *Solid State Technology*, May 1998, p. 79.
14. A. J. Aronson, "Fundamentals of Sputtering," *Microelectronics Manufacturing and Testing*, January 1987, p. 22.
15. J. Ballingall, "State-of-the-art Vacuum Technology," *Microelectronics Manufacturing and Testing*, October 1987, p. 1.
16. S. Wolf and R. Tauber, *Silicon Processing for the VLSI Era*, Lattice Press, Sunset Beach, CA, 1986, p. 95.
17. P. Singer, "Vacuum Pump Technology Leaps Ahead," *Semiconductor International*, Cahners Publishing, September 1993, p. 53.

# 14

# Process and Device Evaluation

## Overview

The wafer-fabrication process requires a high degree of precision in process control, equipment operation, and material manufacture. One process mistake can render the wafer completely useless. One "killer" defect can ruin a die. Throughout the process, a variety of tests and measurements are made to determine both wafer quality and process performance. The tests take place on in-process wafers, test die and production die, and the finished circuit. Individual tests are described in this chapter. Statistical process control programs are addressed in Chapter 15.

## Objectives

Upon completion of this chapter, you should be able to:

1. Explain the difference between resistance, resistivity, and sheet resistance.

2. Draw a sketch of the parts and current flow in a four-point probe.

3. Compare the principles and uses of color interference, fringe counting, spectrophotometers, ellipsometers, and stylus for film thickness measurements.

4. Compare the principles and uses of groove and stain, SEM, and spreading resistance for junction depth measurements.

5. List the methods and advantages of microscope and SEM inspection of wafer surfaces.

6. Draw sketches of diodes in forward and reverse bias and their companion current-voltage curves.

7. Explain the effect of surface current leakage on a junction performance characteristic.

8. Draw sketches of a bipolar and MOS transistor in operation and their companion current-voltage characteristics.

9. List the process steps for a capacitance-voltage measurement and the principle of contamination detection.

10. Describe the principle and use of atomic force microscopes.

## Introduction

Characterization of processes and circuit parameters is required for production-line control and product stability. *Metrology* is the general term applied to the measurement of physical surface feature. Interests include pattern widths, film depths, defect identification and location, and pattern registration errors. Good characterization can warn of a process that is about to go out of control, and device characterization is essential to analyze circuit performance and conformance to customer specifications. Consequently, every process step has a rigid set of equipment and process parameters that are controlled (temperature, time, and so forth). After every significant process step, there is an evaluation of the result on the wafer or a test wafer. *Test wafers* are blank wafers or wafer pieces that are included in the process step for post-process measurements. Many of the tests are destructive and cannot be performed on the device wafers or cannot be performed on the actual components in the chip. In the process chapters, the important parameters for each process were identified (e.g., film thickness, resistivity, cleanliness). Here, the basic theory, applicability, and range of sensitivity of the test methods are examined.

Some are direct measurements, and some are indirect. One group includes electrical measurements of test wafers and on the actual devices. They measure the direct effect of some of the processes, such as ion implantation. Device performance measurements are usually inclusive of several processes, and the results are used to infer individual process parameter control. Another group directly measures physical parameters such as layer thicknesses and widths, composition, and others. This group includes defect detection. A third group measures contamination in and on the wafer and in materials.

Not surprisingly, test and measurement methods have changed along with the levels of integration and smaller image sizes. ULSI technology is ushering in nanometer- and angstrom-level inquiry, called the *nanoanalysis era.*[1] And the price of in-line testing is going up. Larger wafers and more dense circuits require more tests to properly characterize processes. High-volume processing requires real-time testing and analysis to guard against scrapping volumes of production wafers. Data management systems for ULSI circuits usually include on-board statistical analysis and data base management capabilities.

## Wafer Electrical Measurements

### Resistance and resistivity

The object of the fabrication process is to form, in and on the wafer surface, solid-state electrical components (transistors, diodes, capacitors, and resistors) that are wired together to form the circuit. Each of the components must meet individual electrical performance specifications if the entire circuit is to function. Throughout the process, electrical measurements are performed to judge the process and predict electrical device performance.

### Resistivity measurements

The addition of dopants to the wafer, both during crystal growth and during the doping processes, alters the electrical characteristics of the wafer. The altered parameter is its resistivity, which is a measure of a material's specific "resistance" to the flow of electrons (Fig. 14.1). Whereas the resistivity of a given material is a constant, the resistance of a specific volume of the same material is a function of its dimensions and resistivity. This relationship parallels that of density and weight. For example, the density of steel is a constant, whereas the weight of a particular piece depends on its volume.

The units of resistance ($R$) are ohms ($\Omega$) and the units of resistivity ($\rho$) are ohm-centimeters ($\Omega$-cm). Because adding dopants to a wafer will alter its resistivity, measurement of resistivity is actually an indirect measure of the amount of dopants added.

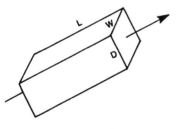

$$R = \rho \frac{L}{A} = \rho \frac{L}{WXD}$$

**Figure 14.1** Relationship of resistance to resistivity and dimensions.

## Four-point probe

The parameters of resistance, voltage, and current are governed by Ohm's law. The three parameters are related mathematically in the following way:

$$R = V/I = (\rho)L/A = (\rho)L/(W \times D)$$

where  $R$ = resistance
$V$ = voltage
$I$ = current
$\rho$ = resistivity of sample
$L$ = length of sample
$A$ = cross-sectional area of sample
$W$ = width of sample
$D$ = depth of sample

Theoretically, the resistivity of a wafer can be measured with a multimeter (Fig. 14.2) by measuring the voltage at a constant current through a sample of known dimensions and calculating the resistivity. However, the resistance between the probes and the wafer material is too great to accurately measure the resistivity of semiconductors with their relatively low quantity of dopants.

The four-point probe is the instrument used to measure resistivity on wafers and crystals. It employs four thin, in-line probes connected to a power supply and voltmeter. The four-point probe consists of four thin metal probes arrayed in a line. The two outside probes are connected to a power supply, and the inside probes are connected to a voltage meter. During operation, the current passes between the outer probes, and the voltage drop (change) is measured between the inner

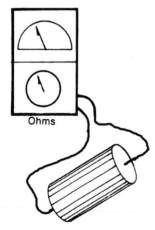

Ohms

**Figure 14.2**  Multimeter.

probes (Fig. 14.3). The relationship of the current and voltage values is dependent on the resistance of the space between the probes and the resistivity of the material. The four-point probe cancels out the effects of probe-wafer contact resistance on the measurement.

## Process and Device Evaluation

Using a four-point probe, the voltage and current are related to the resistivity by the following relationship:

$$\rho = 2\pi s V/I$$

where $s$ is the distance between probes when $s$ is less than the wafer diameter and less than the film thickness.

### Sheet resistance

The four-point probe measurement just described is used to measure the resistivity of wafers and crystals. It is also used to measure the resistivity of thin layers of dopants added into the wafer surface by the dopant processes. When a four-point probe measurement is made on a thin layer of added dopants, the current is confined in the layer (Fig. 14.3). A thin layer is defined as a layer thinner than the probe spacing (distance between probes).

The electrical quantity measured on a thin layer is called *sheet resistance, $R_s$*. This quantity has the units of ohms per square ($\Omega/\square$). The concept of ohms per square can be understood by considering the resistance of two squares of the same thin material of equal thickness (Fig. 14.4). Since the resistivity of $\rho$ is the same for each piece, and $T_1 = T_2$, the sheet resistance is the same for each piece. Or, the resistance of the thin sheet is a constant for any square of the same material.

The formula relating sheet resistance to the voltage and current is

$$R_s = 4.53 V/I$$

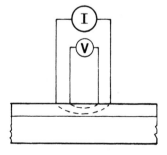

**Figure   14.3** Four-point   probe measurement of a thin layer.

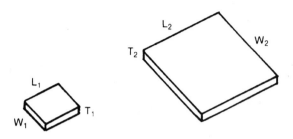

**Figure 14.4** Resistance of a "square."

where 4.53 is a constant that arises from the probe spacing. Some companies elect to drop the constant 4.53 from the formula and just measure the $V/I$ of a wafer as in Fig. 14.5.

### Four-point probe thickness measurement

The thickness of uniform conducting layers on an insulating layer can be determined using a four-point probe. For thin films, the formula is

$$T = \rho_s/R$$

where $T$ = layer thickness
$\rho_s$ = resistivity
$R_s$ = sheet resistance

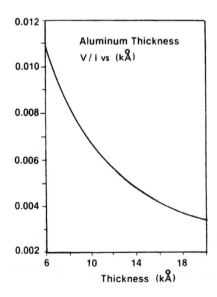

**Figure 14.5** Voltage/current (V/I) versus thickness of aluminum.

Since the resistivity is a constant for pure materials such as aluminum (Fig. 14.5), the sheet resistance measurement is actually a measurement of the film thickness. This formula does not calculate the thickness of a doped layer, since the dopants are not evenly distributed throughout the layer.

### Concentration/depth profile

The distribution of dopant atoms in the wafer (see Chapter 11) is a major influence on the electrical operation of a device. The distribution (or dopant concentration profile) is determined by several techniques. One is spreading resistance. After doping, a test wafer sample is prepared by the bevel technique. After the junction is exposed by the beveling, a series of electrical two-point probe measurements are made sequentially down the bevel (Fig. 14.6). At each point, the vertical drop of the probes is recorded and a resistance measurement made. The resistance value at each point changes with the change in dopants at each level.

A computer is used to perform calculations that relate the depth and resistance values to the dopant concentration at each level. The computer uses the data to construct a dopant concentration profile for the sample. This measurement is usually made periodically off-line or when electrical device performance indicates that the dopant distribution may have changed.

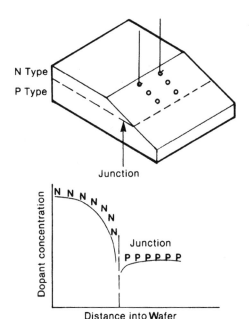

**Figure 14.6** Spreading resistance.

Another method is to incrementally remove thin layers from the wafer surface with the anodic oxidation (see Chapter 7) technique. After the oxide is grown, it is removed by etching, and a four-point probe measurement is made on the new surface. The distance down into the wafer is a function of the oxide thickness and is related to the dopant concentration four-point probe sheet resistance values by a computer program.

### Secondary ion mass spectrometry (SIMS)

SIMS is essentially a combination ion milling and secondary ion detection method. Ions are directed at the sample surface, removing a thin layer. Secondary ions are generated from the removed material, which contains the wafers material and the dopant atoms. The ions are collected and analyzed providing a calculation of the amount of dopant at each level, which in turn allows the construction of a dopant profile.[2]

### Differential Hall effect (DHE)

Like the other two profiling methods, DHE requires sequential removal of the doped layer, down to the junction. As each layer is removed, the resistivity and the Hall coefficient (which relates to carrier mobility) are measured. Dopant concentration for each layer is calculated from the two measured parameters. A drawback to this method is the additional time to prepare a specially shaped test structure (a van der Pauw structure).[3]

### Oxide rupture ($BV_{ox}$ or rupture voltage)

An electrical measurement, $BV_{ox}$, is used as a measure of oxide quality. The test structure used is the same as for capacitance-voltage ($C/V$) analysis. But, in this case, the voltage is continually increased until the oxide is physically destroyed and current flows freely from the aluminum dot to the silicon. The maximum voltage that the oxide can withstand before breakdown is a function of its thickness, structural quality, and purity. Perhaps the most critical oxide layer is in the gate of a MOS transistor. This test is often called *gate oxide integrity* (GOI). With device geometries getting smaller and oxides thinner, high quality is essential for device operation.

### Physical Measurement Methods

Product reliability and maintenance of yields requires on-line detection of defects, mistakes, and so forth to allow removal from the line of suspect material. Process control requires measuring the process results at each of the process steps and knowing the quantity, density, location, and nature of the various problems. This data comes from a series of measurements and evaluations that vary with the sophistication of the circuit in terms of image size, sensitivity to contamination, and density.

Tests are performed on test wafers or directly on product wafers. Product wafer tests may require revealing the subsurface structure by beveling, microsectioning, or using a focused ion beam (FIB) to remove portions of the circuit.

## Layer Thickness Measurements

### Color

Both silicon dioxide and silicon nitride layers exhibit different colors on the wafer. We know that, while silicon dioxide is transparent (glass is silicon dioxide), an oxidized wafer has a color. The color seen is actually the result of an interference phenomenon—the same phenomenon that creates the colors of rainbows.

The silicon dioxide layer on a silicon wafer is actually a thin transparent film on a reflecting substrate. Some of the light rays impinging on the wafer surface reflect off the oxide surface, while others pass through the transparent oxide and reflect off the mirrored wafer surface (Fig. 14.7). When the light rays exit the film, they combine with the surface-reflected ray, giving the surface an appearance of having a color. This phenomenon is the reason oxidized wafers change color as the angle of viewing is changed.

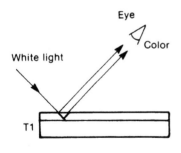

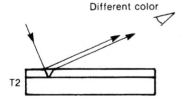

**Figure 14.7**  White light interference.

The exact color is a function of three factors. One, which is a property of the transparent film material, is the *index of refraction*. A second factor is the *viewing angle*. The third factor is the *thickness of the film*. The color of a thin transparent film becomes an indication of the thickness when the nature of the viewing light is specified (i.e., daylight, fluorescent), along with the viewing angle. The classic color versus thickness chart (Fig. 14.8) is a regular feature at oxidation and diffusion stations. Color alone is not an exact indication of thickness because of the consequences of the interference phenomenon.

As the film gets thicker, the colors change in a specific sequence and then repeat themselves. Each repetition of the color is called an *order*. To determine the exact film thickness, a knowledge of the color order is necessary. A principal use of color charts is for process control.

| Film thickness, μm | Color* and Comments |
|---|---|
| **Order I** | |
| 0.050 | Tan |
| 0.075 | Brown |
| 0.100 | Dark Violet to Red Violet |
| 0.125 | Royal Blue |
| 0.150 | Light Blue to Metallic Blue |
| 0.175 | Metallic to Very Light Yellow |
| 0.200 | Light Gold or Yellow-Slightly Metallic |
| 0.225 | Gold with Slight Yellow Orange |
| 0.250 | Orange to Melon |
| 0.275 | Red Violet |
| 0.300 | Blue to Violet Blue |
| 0.310 | Blue |
| 0.325 | Blue to Blue Green |
| 0.345 | Light Green |
| 0.50 | Green to Yellow Green |
| **Order II** | |
| 0.365 | Yellow Green |
| 0.375 | Green Yellow |
| 0.390 | Yellow |
| 0.412 | Light Orange |
| 0.426 | Carnation Pink |
| 0.443 | Violet Red |
| 0.465 | Red Violet |
| 0.476 | Violet |
| 0.480 | Blue Violet |
| 0.493 | Blue |
| **Order III** | |
| 0.502 | Blue Green |
| 0.520 | Green (Broad) |
| **\*When viewed perpendicularly in daylight fluorescent light** | |

**Figure 14.8**  Silicon dioxide thickness color chart.

Each oxidation or silicon nitride process is set up to produce a specified thickness. Naturally, the thickness will vary from run to run. Operators quickly become sensitive to the wafer color. When a variation occurs, a quick check of the chart will indicate if the film thickness is out of specification. Rarely is a process so far off that the film thickness is a whole order (same color, different thickness) out of specification. The accuracy of color chart thickness determination is limited to the accurate perception of the colors (what exactly is red-orange?). A typical chart is accurate to ±300 Å.

### Fringes

When the order is not known, a fringe-counting technique can be used. When a test wafer edge is dipped in hydrofluoric acid for a few seconds, the acid quickly eats through the oxide at an angle, leaving the oxide exposed to view (Fig. 14.9). When the wafer is viewed in white light, colored fringes are formed between the wafer surface and the top of the film. Thickness determination is made by first determining the order of the film thickness. It is easy to see the repeated sequence of colors. If three blue-red fringes exist, the thickness corresponds to the surface color in the third order on the color-thickness chart.

A more accurate fringe-counting method uses monochromatic light as the viewing light. Monochromatic light consists of one color (wavelength), whereas white light is polychromatic (many wavelengths). The sample is prepared in the same way as it is for color fringe counting. However, in a microscope eyepiece, using monochromatic light, the fringes appear as alternating, evenly spaced black and white stripes (Fig. 14.10) with each fringe separation representing a specific vertical distance. Film thickness is determined by counting the number of fringes and multiplying by a correction factor. The correction factor is determined by the wavelength of the monochromatic light used. For sodium light, the wavelength is 5890 Å.

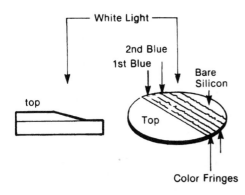

**Figure 14.9**  Color fringes.

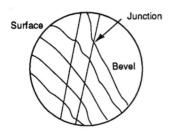

**Figure 14.10** Junction depth determination by monochromatic fringes.

## Spectrophotometers/reflectometry

Film thickness interference/reflectance measurement techniques can be automated. To understand the method, let's review the interference effects. Light is actually a form of energy. The interference phenomenon can also be described in terms of energy. White light is really a bundle of rays (different colors), each with different energies. When the rays interfere through the transparent film, the result is a ray of one color, one wavelength, and one energy level. It is our eyes that interpret the energy as a color.

In a spectrophotometer, which is an automatic interference instrument, a photocell takes the place of the human eye. Monochromatic light in the ultraviolet range is reflected off the sample and analyzed by the photocell. To ensure accuracy, readings are made under different conditions. The conditions are changed by either using another monochromatic light (to change wavelength) or changing the angle of the wafer to the beam. Spectrophotometers specifically designed for use in semiconductor technology have on-board computers to alter the measurement conditions and calculate the film thickness. With visible and ultraviolet (UV) light sources, these machines can measure films down to the 100-Å level.[4]

Index of refraction measurements are also made with these instruments. Spectrophotometers are also used to measure silicon film thickness. Because silicon is opaque to ultraviolet light, an infrared source is used.

## Ellipsometers

Ellipsometers are film-thickness instruments that use a laser light source and operate on a different principle from that of a spectrophotometer. The laser light source is polarized. Polarization is the creation of a wave with all the rays traveling in only one plane. Polarization can be imagined by considering looking into the beam of a flashlight. In an ordinary beam, light rays come to your eyes in many planes, like an arrow with many feathers. A polarized beam has all of the light in only one plane, or an arrow with only one feather (Fig. 14.11).

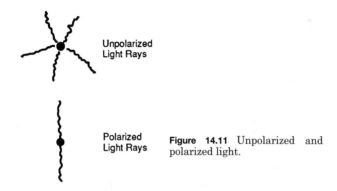

Figure 14.11 Unpolarized and polarized light.

In the ellipsometer, the polarized beam is directed to the oxide-covered wafer at an angle. The beam enters the transparent film and reflects off the reflective wafer surface. During its passage through the film, the angle of the beam plane is rotated. The amount of rotation of the beam is a function of the thickness and index of refraction of the film. A detector in the instrument measures the amount of rotation, and an on-board computer calculates the thickness and index of refraction.

Ellipsometers are used to measure thin oxides (50 to 1200 Å) and the index of refraction of the film. Their accuracy in this range is unequaled by any of the other techniques. The technique is also used for multilayer thin-film stacks. Ellipsometer accuracy and range is enhanced with the addition of multiple-viewing-angle capabilities, multiple-wavelength sources, and reduction in beam spot size.[5]

Oxides thinner than 50 Å can be measured by the corona-oxide-semiconductor (COS) technique described in the electrical measurement section.

### Stylus (surface profileometers)

Some thin films, such as aluminum, cannot be measured by optical techniques. And in the case of aluminum and other very thin conductive films, the four-point probe thickness measurement is not sufficiently accurate. In these situations, a mechanical moving-stylus apparatus is used (Fig. 14.12). The method requires that a portion of the film be removed, creating a step on a test-wafer surface. This is normally done by a masking and etching step. The prepared sample is mounted and leveled on the pivoting stylus instrument stage.

After leveling, the measuring stylus is lowered gently onto one of the surfaces. The measurement is made as the stage is slowly moved under the stylus. As the stylus goes over the step, its physical position

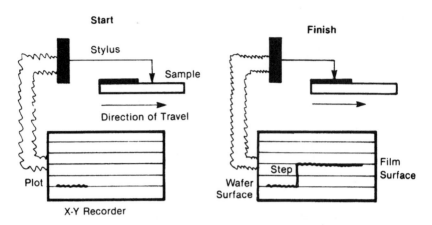

**Figure 14.12**  Step height measurement.

is changed. The stylus is linked to an inductor that generates an electrical signal in response to the stylus vertical position. This signal is amplified and fed into an x-y recorder.

While the leveled wafer is moving under the stylus, it does not move in the vertical direction, and no change in signal is produced. The trace on the x-y plotter is a straight line. When the stylus reaches the surface step, it changes position, causing a change in the signal output. This change is evidenced by a change of pen position on the x-y chart trace. The change in position is relative to the step height, which is read directly from the calibrated x-y chart.

### Photoacoustic

A nondestructive thickness test relies on photoacoustic principles. In 1877, Alexander Graham Bell discovered that, under certain circumstances, the interruption of a light wave will cause a sound. In the semiconductor thickness application, a laser beam is converted to tiny sounds, which are in turn reflected off two surfaces on the wafer surface. By measuring the reflection delay between the two pulses, the thickness can be calculated.

### Junction Depth

A critical device parameter is the junction depth of the various doped regions. This parameter is measured after each of the doping steps. The measurement methods described are all performed off line; that is, the test wafers or device wafers have to be taken to a measurement station or laboratory for the measurement.

## Groove and stain

The traditional method of junction depth measurement is by the groove (or bevel) and stain technique. Grooving or beveling is a mechanical method of exposing the junction for viewing and measurement from the horizontal plane (Fig. 14.13). The extremely shallow depth of the junction requires either grooving or beveling of the wafer to expose the junction.

The junction itself is not visible to the naked eye. Two techniques, called *junction delineation,* are available to make it visible. Both techniques utilize the electrical differences between N-type and P-type regions. The first technique, the etch technique, starts with the placement of a drop of hydrofluoric acid and water mixture over the junction (Fig. 14.14). A heat lamp is directed onto the exposed junction. The heat and light cause holes or electrons to flow in each region. As a result of the flowing current, the etch rate of the $HF-H_2O$ mixture is higher on the N-type region, making it appear darker.

The second delineation technique is electrolytic staining. A mixture containing copper is dropped on the exposed junction. Again, the heat lamp is directed onto the junction. In effect, a battery is formed, with the poles of the junction being the poles of the battery and the copper solution being the electrolytic connection. The current flowing in the liquid drop causes the copper to plate out on the N-type region side of the junction.

The final step, after exposure and delineation of the junction, is depth measurement. A number of methods may be used, including optical interference and SEMs.

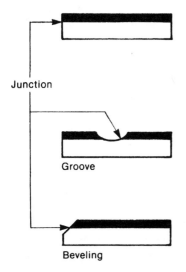

Junction

Groove

Beveling

**Figure 14.13** Exposure of junction by groove or beveling.

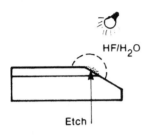

Etch

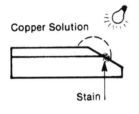

Copper Solution

**Figure 14.14** Etching or staining of junction.

Stain

### Scanning electron microscope (SEM) thickness measurement

The SEM technique (described below) can also be used to measure junction depths and film thickness. The wafer is cleaved with the break position over the junction. The exposed wafer junction is delineated by one of the methods described.

The exposed cross section is positioned to the SEM beam at right angles to the wafer surface and a photograph is taken. The depth is determined from the photograph and the scale factor of the SEM (Fig. 14.15). SEM and the groove and stain methods provide a visual look at the junction area and lateral diffusion that other methods do not.

### Spreading resistance probe (SRP)

Spreading resistance is a technique also used for measuring the junction depth. As the probes pass through the junction, they sense the change in conductivity type (N or P). This information, when plotted on the profile curve, also gives the junction depth (Fig. 14.16).

### Secondary ion mass spectrometry (SIMs)

A SIMs instrument bombards the wafer surface with a source that produces secondary ions from species on the surface. It can detect the dopant(s) in the junction region. Coupled with careful removal of the surface through the junction and monitoring, when the dopant atoms disappear from the stream can be translated into a junction depth.

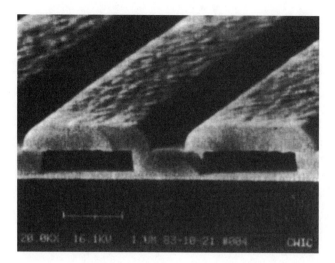

**Figure 14.15**  SEM declination of device cross section.

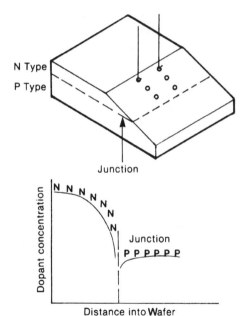

**Figure 14.16**  Spreading resistance.

## Scanning capacitance microscopy (SCM)

One of the options with atomic force microscopy (AFM, see description below) is a probe that measures capacitance. Since the capacitance of a doped layer changes as the junction is approached, it can be used for junction depth detection. Sample preparation is the same as for spreading resistance measurements.

SCM offers the advantage of nanometer precision[6] and may be the concentration profile and junction depth measurement technique for submicron junctions.

### Critical Dimensions (CD) and Line Width Measurements

The exact dimensions required of each component in the circuit are controlled and influenced by *all* processes. Vertical dimensions are set by the doping and layering processes. The horizontal surface dimensions are produced in the lithography process. As part of that process, the critical dimensions are measured at both develop inspection and final inspection with microscopes.

### Filar and image shearing dimension measurement

Two manual microscope methods are used to measure surface pattern dimensions: filar eyepiece and image shearing. Both have limited accuracy for small dimension lines but are found in laboratories and training labs. The filar measuring eyepiece is a dimension-measuring instrument that is fitted to a microscope. The eyepiece features a movable hairline that requires calibration to an outside standard. A calculation is made to "correct" the filar measurements to the actual dimensions on the standard.

Filar systems are easily automated. The hairline movement mechanism can be motorized and the correction factor programmed into an onboard computer, resulting in a direct digital readout of the actual width. Operator fatigue is minimized by a video monitor rather than requiring the operator to view the wafer through a high-power microscope.

An image-shearing attachment on a microscope is another method of critical-dimension measurement. A control on the unit allows the operator to optically separate (shear) the pattern into two images. To start the measurement, the two images are butted against each other (Fig. 14.17), then moved until the sheared images exactly overlap. The

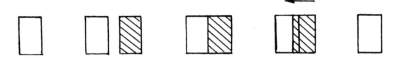

**Figure 14.17**   Single-image shearing.

difference between the starting and ending values is the width of the pattern. Like the filar unit, an image shearing unit must be calibrated to a standard.

### Reflectance

Both the filar and image-shearing techniques require some operator decision, which can be a source of error. A third type of dimension measurement instrument is based on reflectance. Like the other two, the operator locates the pattern to be measured on a monitor. One edge of the pattern is positioned under a marker on the screen. The measurement is made automatically as a laser beam is swept along the direction of measurement and the reflection energy of the beam recorded (Fig. 14.18). When the beam comes to the edge of the pattern, it steps up (or down) to a new surface. The new surface causes a different reflectance of the beam, which is recorded by the detector. The width of the pattern is the difference between the starting point and the point at which there is a change in reflectance. This value is read out automatically. Operator decision is limited to determining the starting point, resulting in a more constant accuracy.

The limitation of this method is with patterns that have a number of steps in them, such as metal steps. The instrument is designed to read to the first change in reflectance that it senses. Another limitation is in measuring pattern edges that are sloped. The reflecting beams change gradually rather than abruptly, making a width determination difficult.

Some dimension-measuring microscopes use transmitted light for the measurement of mask and reticle critical dimensions. A light beam is passed from the bottom through the mask. The pattern to be measured is moved in front of the beam, and a detector above the mask senses when the light is blocked by the opaque pattern. The horizontal distance is measured until the light again is detected by the sensor.

### Shape metrology and optical critical dimension (OCD)[7]

The SEM inspection tool described below is also used for very accurate measurement of line widths. In the nano-era, especially when using

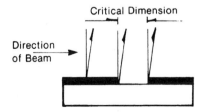

**Figure 14.18** Reflectance CD measurement.

copper metallization, there is also interest in knowing and controlling the cross-sectional shape (3D shape metrology) of the hole or surface island. This is accomplished with sophisticated SEMs that direct beams off the top surface, the sides, and the bottom surface to reconstruct the exact shape. Measurements are made directly on the wafer. A draw back is OCD's inability to measure isolated lines. The great goal is to have OCD systems integrated directly into process tool, providing real-time measurement and process control (Fig. 14.19).

A summary of the various surface-inspection techniques and their use in a fabrication process is tabulated in Fig. 14.20.

## Contamination and Defect Detection

Detection of contamination and visual defects is essential to high yields and process control. Particulate contamination is detected primarily by visual techniques including high-intensity lights, micro-

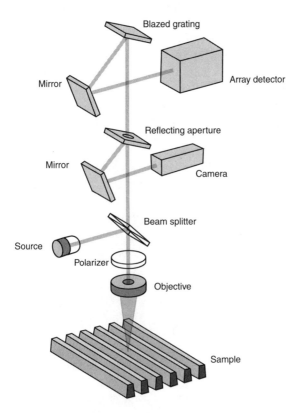

**Figure 14.19**  Schematic of OCD optics. (*Courtesy of* Semiconductor International.)

| Method | Visual contamination | Surface Defects | Alignment | Contamination Elements | Contamination Compounds | C.D.'s |
|---|---|---|---|---|---|---|
| 1x Incident White Light | X | X | | | | |
| 1x Incident U.V. Light | X | X | | | | |
| Microscope– Light field | X | X | X | | | |
| Dark field | X | X | X | | | |
| S.E.M. | X | X | X | | | X |
| Auger | | | | X | | |
| E.S.C.A. | | | | | X | |
| Filar | | | | | | X |
| Image Shearing | | | | | | X |
| Reflectance | | | | | | X |

Figure 14.20  Overview of surface-inspection techniques.

scopes, SEMs, and automatic machines. Chemical contamination is detected and identified by both Auger and ESCA techniques.

Mobile ionic contamination in the wafer is detected by capacitance-voltage plotting and by interpretation of the transistor and diode electrical tests. Many other sophisticated techniques can be used to make these inspections. The ones described here are those employed in a typical wafer-fabrication line.

### 1 × visual surface inspection techniques

The first line of defense is to look at the wafers, which is a magnifying power of 1, or, in microscope terminology, a 1x power. Operators quickly become sensitive to the way "normal" wafers look. Even minor changes in the surface appearance are picked up by the experienced eye.

### 1 × collimated light

The resolving power of the naked eye (1x) can be assisted by using a high-intensity white light, such as the beam of light from a slide projector (Fig. 14.21). Particulate contamination is highlighted in the light beam when the wafer surface is viewed at an angle. The effect is similar to the highlighting of dust in the air by light streaming through a window.

### 1 × ultraviolet

In actuality, the eye cannot see ultraviolet light, but ultraviolet light from a mercury-vapor lamp emits blue, green, and even some red

Eye

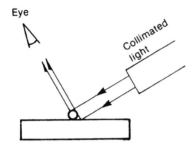

**Figure 14.21**  Collimated light inspection.

light. Because ultraviolet is harmful to the retina, a filter is frequently placed over the light source to block out the ultraviolet. The primary benefit of the ultraviolet lights used in fabrication areas is that they are very bright, which means that the intensity of the scattered light is greater, therefore increasing the detection of surface contamination.

**Microscope techniques**

**Light-field microscope.**    The metallurgical microscope is the workhorse tool of surface inspection. The term *metallurgical* differentiates it from the standard microscopes found in biology labs. A biological microscope illuminates transparent samples by shining the light up through the sample. In a metallurgical microscope, the light is passed down to the nontransparent sample through the microscope objective (Fig. 14.22). The light reflects off the sample surface and is transmitted back up through the optics to the eyepieces. With white light illumination, the picture in the field of view exhibits the surface colors, which helps identify particular components on the wafer. The use of filters will change the surface colors.

A typical fabrication microscope is fitted with 10× or 15× eyepieces and a range of objectives from 10 to 100×. Increasing the total viewing power (eyepiece power times the objective power) reduces the field of

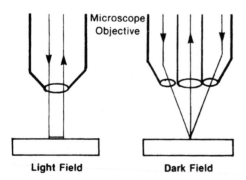

Light Field       Dark Field

**Figure 14.22**  Light- and dark-field inspection.

view. This reduction requires more inspection time for the operator to look at the required sample inspection area on the wafer. The consequence is a slower inspection process. A trade-off power level, when inspecting LSI and VLSI devices, is 200 to 300× magnification.

The industry typically uses a microscope inspection procedure requiring the operator to look at three to seven specific locations on the wafer. This procedure is easily automated with motorized stages. Most of the automated microscope inspection stations feature automatic wafer placement on the stage, automatic focusing, and automatic binning of the completed wafers. At the touch of a button, the operator can direct each wafer into a boat for passed wafers or into one of several boats for different reject categories. Obviously, a microscope inspection procedure is used to judge surface and layer quality and (in masking) pattern alignment.

Just as image resolution in a patterning process is limited by the wavelength of light, so is bright light inspection. With a broadband white light source, the theoretical resolution limit is 0.30 μm.[8] Use of UV light sources and image processing can bring the resolution limit down to 80 nm.

**Dark-field inspection.**    Dark-field illumination is achieved by fitting a metallurgical microscope with a special objective (Fig. 14.22). In this objective, the light is directed to the wafer surface through the outside of the objective body. It impinges on the surface at an angle and passes up through the center of the objective. The effect on the "picture" in the eyepieces is to render all flat surfaces black. Any surface irregularities, such as a step or pieces of contamination, appear as bright lines. Dark-field illumination is more sensitive than light-field to any surface irregularity. It has the drawback of limiting the ability to discern the nature of the surface irregularity. A passable surface dimple may look the same as a rejectable piece of contamination.

Dark-field resolution of defects is enhanced with the use of laser light[8] sources and multiple sources.

**Confocal microscopes.**    Distinguishing fine detail with a normal microscope is hampered by stray light rays interfering as they bounce off the various surfaces and from the different plane depths on a wafer surface. Confocal light sources minimize the scatter and provide greater detail by limiting the returning light rays to a narrower plane. This is done by passing an intense white light or laser beam through holes in a disk (positioned between the light source and the wafer) spinning at high rpm. Only light rays from the inspection plane of interest bounce back through the holes. Confocal systems are capable of imaging submicron dimensions.[9] Laser beams used in a confocal system cause the sample to fluoresce (glow). The resultant image, after $x$-$y$ scanning and computer processing, has resolution in the nanometer range and can display layers of registered translucent samples.

**Other microscope techniques.**    Optical technology is capable of providing many evaluation techniques beyond simple light- and dark-field viewing, such as phase contrast and fluorescence microscopes. Each allows the viewer to determine additional visual information about the surface. Phase contrast brings out surface irregularities in the vertical plane, and fluorescence-illuminated microscopes use ultraviolet illumination sources. In the ultraviolet light, organic residues (photoresist, cleaning chemicals) not easily visible in white light are brought into view. Their use and interpretation generally require technicians trained beyond the level of production operators.

**Scanning electron microscopes (SEMs).**    Conventional optical microscopes are limited in their ability to provide accurate information about the wafer surface. First, their resolving power is limited by their optical light source. The ability of a viewing system to distinguish detail is related to the wavelength of the light (radiation used). The shorter the wavelength, the smaller the detail that can be seen.

Depth of field is another viewing factor. It relates to the ability of the system to keep two planes in focus simultaneously. A conventional photograph, with the subject in focus and the background out of focus, has a background beyond the depth of field limit of the camera. In a microscope, the depth of field decreases as the power (magnification) of the system is increased. If the power is increased to see the surface "closer," the operator may not be able to see the top and bottom surfaces in focus. Constant refocusing results in loss of information and a longer inspection time.

Magnification is the third limiting factor of optical microscopes. An optical system with white light illumination is limited to about 1000× magnification with conventional objectives. The oil-immersion technique pushes the limit up, but it is unacceptable, because it is too slow, too messy, and a possible source of contamination to the wafer.

All three limitations are overcome by using a scanning electron microscope. The microscope varies from an optical one in many aspects. The "illumination" source is an electron beam scanned over the wafer or device surface. The impinging electrons cause electrons on the surface to be ejected. These secondary electrons are collected and translated into a picture of the surface (Fig. 14.23) on either a screen or a photograph.

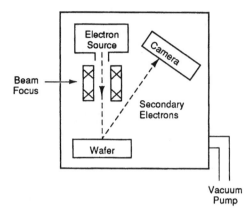

Figure 14.23   SEM analysis.

SEM analysis requires that the wafer and beam be in a vacuum. The electron beam has a much smaller wavelength than white light and allows the resolution of surface detail down to submicrometer levels. Depth of field problems do not exist; every plane on the surface is in focus.

Magnification is similarly very high, with a practical upper limit of 50,000×. A tilting wafer holder in an SEM allows the viewing of the surface at angles, which enhances the three-dimension perspective (Fig. 14.24). Surface details and features can be viewed at advantageous angles.

Some materials, such as photoresist, do not give off secondary electrons in response to e-beam bombardment. To inspect a photoresist

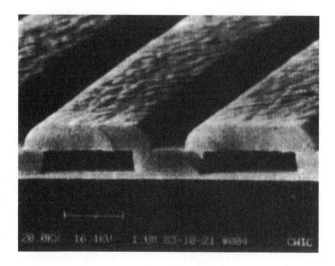

Figure 14.24   SEM declination of device cross section.

layer in an SEM, the resist layer is covered with a thin layer of evaporated gold. The gold layer conforms to the topography of the photoresist layer. Under e-beam bombardment, the gold gives off secondary electrons, thus resulting in an SEM picture that is an accurate reproduction of the underlying photoresist layer.

**Transmission electron microscope (TEM).**    A standard SEM instrument has a resolution range of 20 to 30 Å,[10] which presents a barrier for inspection of ULSI devices. However, passing (transmission) the electron beam through a thin sample increases the resolution to 2 Å. As attractive as the increased resolution may be, sample preparation is time consuming and requires precision. TEM is best used as an offline laboratory tool.

This principle is used to measure film thicknesses. In an SEM, the energetic e-beam causes X-rays to leave the surface. Fortunately, the X-ray energies are reflective of the materials (elements) presence on the surface and can be analyzed for additional information. This is accomplished when an X-ray spectrometer is added to the SEM to detect dispersed X-rays. This additional information is about the chemistry of the materials.

## Automated in-line defect inspection systems

**Automatic defect detection.**    The detection of ever smaller-sized particles on the wafer surface has led to the use of laser beam and e-beams as the detection illumination. Two advantages accrue from the use these beams. First is the obvious ability to detect smaller particle sizes (Fig. 14.25) due to the high intensity of the reflected light, which makes very small surface particles detectable (helium and neon lasers are the usual sources). E-beams have even smaller spot size and are the system of choice for nano-sized patterns, especially for copper/low-k dielectric metallization systems.

The second advantage is automation. Laser inspection equipment is easy to automate such that the inspection is automated from cassette to cassette, and the quantity and size of the contamination on the surface can be determined. The information is used to produce a map of the wafer surface showing the size, location, and density of surface contamination. This last factor is a big help in process control. Laser inspection of masks has become a standard production technique.

Some systems mix bright- and dark-field viewing and process the information into maps or data bases (Fig. 14.26). Advances in image processing have allowed automatic defect and pattern distortion detection instruments. The machines mate image processing and computer tech-

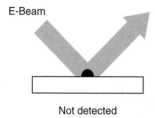

E-Beam

Not detected

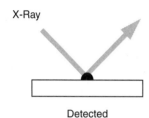

X-Ray

Detected

**Figure 14.25** X-ray detection of small surface particles.

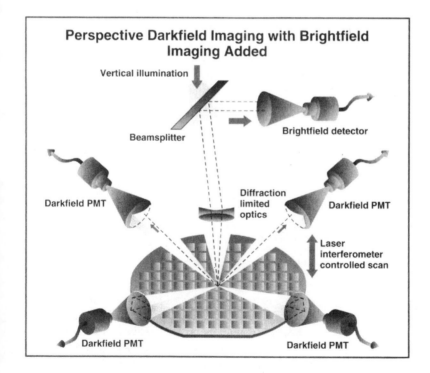

**Figure 14.26** Mixed dark-field and light-field system. *(Courtesy of* Semiconductor International, *May 1997.)*

nology. They have scanners with laser or optical light sources that move over the wafer surface. In one version, the computer is preprogrammed with the design pattern for the circuit. In the *die-to-database* system, each die is scanned and the results compared to a stored image of the mask/reticle for the particular layer. The scanner looks for added or missing parts from the expected pattern. Anything missing from or on the die that is not in the data base is flagged for further inspection. The presumption is that, if an image is not in the data base, it is probably a defect of some kind. Defect locations are recorded and surface maps printed out. Engineers can go back to the wafer or mask and determine the exact nature of the problem.

Another system, called die-to-die inspection, compares adjacent die on the wafer or mask. A die is scanned and the pattern recorded in the computer memory. A second die is also scanned, and any deviations between the two are recorded. This system will not detect any repeated pattern defect that occurs on every die, but it will pick up random defects that have a very low probability of occurring in exactly the same spot on two adjacent die. In both machine types, the information from the surface is captured by charge coupled device (CCD) cameras or photomultiplier tubes.[12] For in-line inspection, a grave concern is calibration and standardization.

In addition to on-board verification of the electronics, most systems use a standard wafer to verify machine operation. Several inspection processes are used. While an automatic machine can find defects, the determination of which ones are "killer" defects is more difficult. With overall defect counting, it is possible that the machines may report a rise in defect count, yet the increase may contain only minor, non-killer types. Verification by humans is still a critical part of any defect detection and management system.

The development of copper/low-k dielectric combinations has introduced a whole new host of defects to the process. Figure 14.27 lists a sampling of defects associated with these metal systems.

## General Surface Characterization

### Atomic force microscopy (AFM)

Several of the measurement techniques are multiuse, such as SEMs. A newer entry in the general category is *atomic force microscopy* (AFM). It is a surface profiler that scans a delicately counterbalanced probe over the surface (Fig. 14.28). Probe and surface separation is so small (about 2 Å) that atomic forces between the surface and probe materials actually affect the probe. Sophisticated electronics measure the relative position of the probe as it moves across the surface, thus compiling a three-dimensional map of the surface[13] (Fig. 14.29).

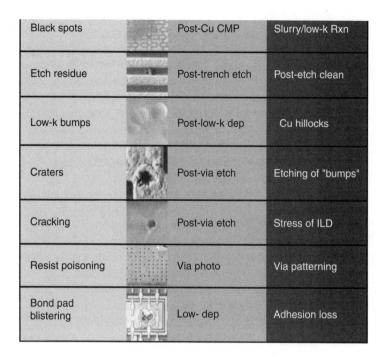

| Black spots | | Post-Cu CMP | Slurry/low-k Rxn |
|---|---|---|---|
| Etch residue | | Post-trench etch | Post-etch clean |
| Low-k bumps | | Post-low-k dep | Cu hillocks |
| Craters | | Post-via etch | Etching of "bumps" |
| Cracking | | Post-via etch | Stress of ILD |
| Resist poisoning | | Via photo | Via patterning |
| Bond pad blistering | | Low- dep | Adhesion loss |

**Figure 14.27**   Sources of low-k defects. *(Courtesy of* Semiconductor International.*)*

AFM sensitivity is in the 1-Å range and can be operated in contact or noncontact modes. The first AFM production instrument was introduced in 1990. Projections are for it to become an integral part of the inspection arsenal. By its nature, AFM can characterize grain sizes, detect particles, measure surface roughness, and provide critical dimension measurements—all in three dimensions. In one novel use, an AFM probe has been combined with an optical microscope objective.[14]

### Scatterometry

The search for fast, accurate, and nondestructive surface inspection tools led to the *scatterometry* metrology. Accuracy in optical systems is limited by the wavelength of the light. Roughly, particles or surface features smaller than the wavelength used cannot be detected. However, a scattered beam can give information about surface features smaller than the wavelength. A scatterometry system has the wafer at the center of curvature of a screen (Fig. 14.30). An incident laser beam is scanned over the surface and is reflected and "scattered" from the surface onto the screen. A camera with a microprocessor captures the screen image to reconstruct the surface that produced the particular

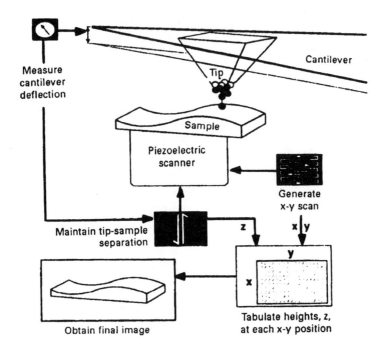

**Figure 14.28**  Atomic force microscope (AFM). *(Source:* Semiconductor International, *August 1993, p. 66.)*

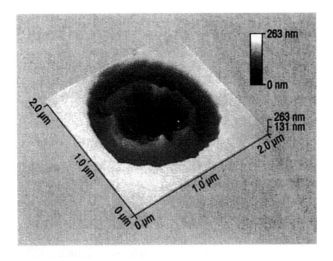

**Figure 14.29**  Atomic force microscope (AFM) surface map.

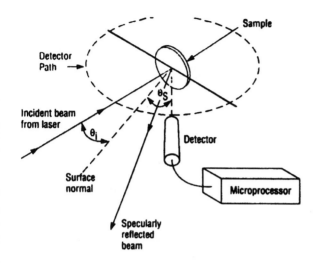

**Figure 14.30**   Arrangement of a scatterometer. *(Source: Solid State Technology, March 1993, p. 30.)*

pattern on the screen. Like AFM, this technique has the potential of measuring grain sizes, contours, and CDs. It can also measure latent images in undeveloped photoresist and characterize phase shift masks.

## Contamination Identification

Pushing into the 0.18-μm era is requiring more detailed information about the nature of the various contaminants that end up on the wafer surface or in the deposited layers. Contamination types, forms, amounts, and other data are needed to maintain clean processes and product. In this section, the instruments used to collect this data are examined. In general, all of the techniques are based on a common phenomenon. Whenever a surface is excited with energy, there will be energy given off. The energy given off will have characteristics reflecting the interaction of the incoming energy wave with the material(s) on the wafer surface.

### Auger electron spectroscopy (AES)

In an SEM, a range (spectrum) of secondary electrons is released by the impinging electron beam. One portion of this spectrum is the electrons that are released from the top several nanometers of the surface. These electrons, known as *Auger electrons,* have energies characteristic of the element that emits them. Thus, sodium and chlorine each give off different Auger electrons.

The collection and interpretation of Auger electrons allows the identification of the surface materials, including contamination. In operation, the e-beam is scanned across the wafer. The ejected Auger electrons are analyzed for their energies (wavelengths) and printed out on an $x$-$y$ plotter (Fig. 14.31). Energy peaks at specific wavelengths indicate the presence of specific elements on the surface.

Scanning Auger microanalysis (SAM) is limited to the identification of elements. This technique cannot identify the chemical state of the element, what it may be combined with, or its quantity on the surface. Salt (NaCl) contamination is identified only as the presence of sodium and chlorine on the surface.

### Electron spectroscope for chemical analysis (ESCA)

Solving surface contamination problems often requires a knowledge of the chemical state of the contaminant. An Auger detection of chlorine on the surface does not reveal whether the chlorine is present as hydrochloric acid or a trichlorobenzene. Knowledge of the form of the chlorine expedites locating and eliminating the process source of contamination.

The electron spectroscope for chemical analysis (ESCA) is an instrument used to determine surface chemistry. The instrument works on principles similar to the Auger technique. However, X-rays instead of electrons are used for the bombarding radiation. Under bombardment, the surface gives off photoelectrons. Analysis of the photoelectron information leads to the determination of the chemical formula of the contamination. Unfortunately, the ESCA X-ray beam is wider than most integrated circuit features. The beam diameter limits the tech-

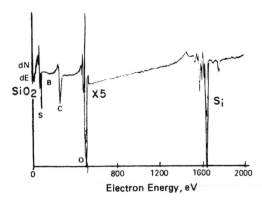

**Figure 14.31**  Typical Auger trace.

nique to a macro-surface analysis. By contrast, an Auger electron beam can zero in on specific bits of contamination.

### Time of flight secondary ion mass spectrometry (TOF-SIMS)

TOF-SIMS surface analysis can use a number of incident radiation sources, such as an Nd YAG laser, a Cs ion beam, or a Ga ion beam.[15] For this method, the impinging energy pulses and ionizes surface material and produces secondary ions that are accelerated to a mass spectrometer. In the spectrometer, each ion's time of flight from the surface is measured. The time of flight is indicative of the species on the surface. An SIMS analysis can also determine the junction depth.

TOF-SIMS samples material depths of only several tenths of a nanometer.[16] Along with the ability to characterize both organic and inorganic contaminants, it is a mainstay of an analytical lab.

### Vapor phase decomposition/atomic absorption spectroscopy (VPD-AAS)

VPD-AAS is used to detect trace inorganic contaminants. The sample is exposed to vapor phase hydrogen fluoride gas, which chemically reduces (decomposes) the contaminant to its fluoride salt. Next, the reduced material is dissolved in water or an acid and analyzed for elements with standard atomic absorption spectroscopy.

### Evaluation of stack thickness and composition

The top of the wafer is becoming more and more complicated with stacks of various materials. Each material can be measured for thickness and composition from monitor wafers included in the deposition process. However, their effect in and on the circuit comes from the combination of thicknesses and compositions. A combination machine to determine these parameters, *in situ*, uses an electron beam and X-ray spectrometers (Fig. 14.32). An e-beam is directed at the stack, which in turn kicks off X-rays. A series of X-ray spectrometers arrayed around the sample analyze the rays. Because each thickness and material composition creates a unique X-ray signature, the machine's onboard computer can run an analysis and determine the thickness and composition of each component in the stack.[17]

### Device Electrical Measurements

During the process, it is necessary to make direct measurements of the actual device parameters. These measurements are usually made

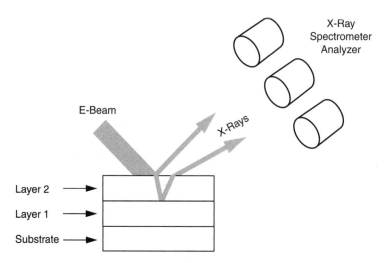

**Figure 14.32**   E-beam/X-ray detection of material stacks.

on special devices in the test die or on special structures in the scribe line areas.

A great deal of information about the process is available from these electrical tests. In this section, we explain the basic tests and several of the more obvious and frequent device failures. Often, these tests are referred to as *parametric testing*. The reference is to the measurement of the device electrical parameters as opposed to testing the overall function of the device/circuit, which occurs at wafer sort at the end of the fab process. A complete device and process troubleshooting guide is beyond the scope of this text.

### Equipment

The basic equipment required to perform device electrical tests are (1) a probe machine (Fig. 14.33) with the capability of positioning needle-like probes on the devices, (2) a switch box to apply the correct voltage, current, and polarities to the device, and (3) a method of displaying the results. If many devices are to be measured, the probes may be fixed on a printed circuit board-like structure called a *probe card*. They are the same cards used in the wafer sort process. Display may be by a curve tracer or by a special digital voltmeter. A curve tracer is a specially designed oscilloscope set up to display voltage on the horizontal scale and current on the vertical scale. They show the characteristic voltage/current traces for each type of measurement. Example displays are shown in the following sections for each test. In high-production situations, the tests are performed digitally and the results

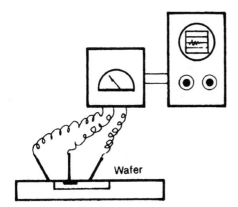

**Figure 14.33** Device measurement equipment.

Wafer

displayed on a readout. These measurement systems have the capability of storing and correlating data.

In advanced systems, the probe station may be automated to sequentially test several die. The switching can also be automated to allow the equipment to perform a series of tests in a predetermined sequence. Automated systems also include hard-copy printouts of the test results as well as analysis of the data. The individual tests will be explained, as performed on manual equipment. The shape and relationship of the traces displayed on the oscilloscope are very helpful for understanding device operation.

All the device measurements are made in basically the same way. A voltage is applied to the component contact probe, and the resultant current flowing between the contacts is measured. The results are displayed on the oscilloscope screen or display screen. The shape of the trace is governed by the dimensions of the device, the doping, and the presence of junctions. The current is modified by the resistance of the component or the presence of junctions. Properly working components will display known traces. The fundamental relationship is that of Ohm's law:

$$R = V/I$$

### Resistors

Resistor measurements are made by contacting each end of the resistor and applying a voltage. Current passage between the probes will be governed by the nature of the material or, in electrical terms, its resistivity. During the test, the applied voltage is varied from zero to some higher value. On the screen, the voltage value is displayed on the $x$ axis, with current measured on the $y$ axis (Fig. 14.34).

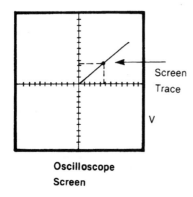

**Oscilloscope**
**Screen**

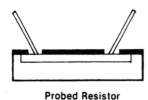

**Probed Resistor**

**Figure 14.34**  Resistor.

The value of the resistance is calculated by dividing one of the volt-age values by the corresponding current value. One might ask, "Why not determine the resistance by simply measuring the voltage and current values with a meter?" In other words, why display the values on the screen? The answer lies in the quality of information gained from the trace. A resistor's $V/I$ relationship should be a linear one (straight line). Any deviation from linearity could indicate a process problem, such as high contact resistance or a leaking junction.

### Diodes

Diodes function as switches in a circuit. This means that a diode can pass current in one direction (forward bias) and not in the other (re-verse bias). Checking diode operation in either direction requires mea-suring the diode with the proper polarities, as shown.

As the voltage is increased in the forward direction, current immedi-ately starts flowing across the junction and out of the diode (Fig. 14.35). The initial resistance to that flow comes from contact re-sistance and a small resistance at the junction. After the resistances are overcome, there occurs a "full" flow of current through the diode. A diode is designed to have this condition occur at some minimum volt-age value. If the diode forward voltage value exceeds the design value, it is out of specification. This voltage is called the *forward voltage*.

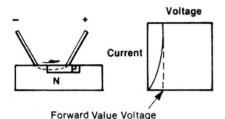

**Figure 14.35** Diode forward bias.

In the reverse direction, the diode is designed to block current flow as long as the voltage stays below a specified value. In the reverse condition, a small current, called a *leakage current,* always flows across the junction (Fig. 14.36). Increasing the voltage increases the leakage current to a level at which the junction breaks down, allowing "full" current flow. This value of the voltage where full flow begins is called the *breakdown voltage* (Fig. 14.37). Circuits are designed to operate at a voltage level below the designed breakdown voltage of the diode to take advantage of the blocking nature of the junction. Lowered breakdown voltages are generally the result of processing mistakes or excessive contamination.

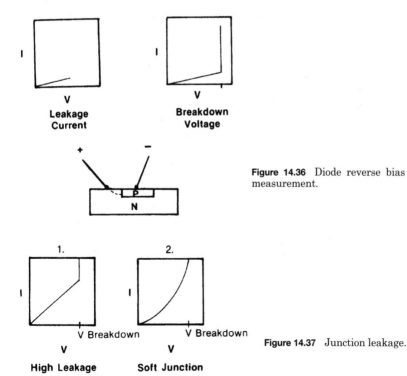

**Figure 14.36** Diode reverse bias measurement.

**Figure 14.37** Junction leakage.

Exceeding the breakdown voltage does not (normally) permanently damage the junction. However, if the applied voltage is extremely high, the diode (junction) can sustain permanent physical damage from a high current flow.

A second parameter determined during this test is the current at the breakdown voltage. A small amount normally occurs as illustrated above. Contamination and/or improper processing can result in additional leakage current.

Trace 1 in Fig. 14.37 shows a diode with a small amount of leakage. The amount of current increases as the voltage is increased. Eventually, the breakdown voltage is reached, and the diode becomes fully conducting. In trace 2, gross leakage is demonstrated. The junction leaks current with every increase in voltage, and the problem is so severe that a breakdown level is never reached, and the diode never operates as a current block.

**Bipolar transistors**

Bipolar transistors, as explained in Chapter 16, are three-region, two-junction devices. Electrically, they can be thought of as two diodes back to back. Many tests are performed to characterize bipolar transistors. The individual junctions are characterized separately and the whole transistor operation measured. The junctions are probed for forward and reverse characteristics. The breakdown voltage (BV) tests are followed by the probing of the entire transistor.

Individual junction probe measurements are designated by the letters BV followed by lowercase letters that indicate the particular junction being probed. $BV_{cbo}$ indicates the breakdown voltage measured between the collector and base regions. The "o" indicates that the emitter region is "open"—it has no voltage applied to it. $BV_{cbo}$ is the breakdown measured between the collector and emitter.

The forward voltages of the two bipolar structure junctions are also measured. $V_{be}$ is the forward voltage of the emitter-base junction, and $V_{bc}$ is the forward voltage of the base-collector junction. $BV_{iso}$ probes the collector-isolation junction for leakage currents.

A principal electrical measurement of a bipolar transistor is the beta (Fig. 14.38) measurement. This is a measurement of the amplification characteristic of the transistor. In a bipolar transistor, the current flows from the emitter to the collector, through the base (see Chapter 16). The base current is varied to change the resistance in the base region. The amount of current flowing out of the collector (from the emitter to base) is regulated by the base resistance.

The amplification of the transistor is defined as the collector current divided by the base current. This number is known as *beta*. Thus, a

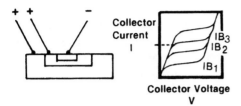

Figure 14.38  NPN transistor beta measurement.

beta of 10 means that a 1-mA (milliampere) base current will give rise to a 10-mA collector current. The beta of a transistor is determined by junction depths, junction separations (base width), doping levels, concentration profiles, and a host of other process and design factors. Measurement of beta is done by a variation of the $BV_{ceo}$ measurement. A $BV_{ceo}$ measurement at a specific base current is performed. In this mode, the emitter base junction is forward biased.

The collector characteristic of the transistor is displayed on the oscilloscope screen. The almost horizontal lines represent increasing base current values ($IB_1$, $IB_2$, and so on). With every increase in base current, a corresponding increase in collector current occurs. Calculation of the beta value takes place from the data displayed on the screen.

Collector current is determined from the vertical axis (dotted line). The base current is calculated by multiplying the number of horizontal lines (steps) by the scale value for each step (from the oscilloscope).

### MOS transistors

MOS circuits are also made up of resistors, diodes, capacitors, and transistors. The first three are measured by the same methods used to measure bipolar circuit components. Like the bipolar transistor, the MOS transistor is composed of three regions, in this case called the source, gate, and drain (see Fig. 14.39 and Chapter 16). Measurement of this type of transistor consists of determining the reverse and forward values of the source and drain junctions. The functioning of the gate is determined by the threshold voltage test.

A MOS transistor has the source region forward biased. Because of the high resistivity of the gate region, the forward current does not

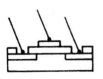

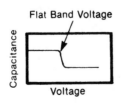

Figure 14.39  Threshold voltage measurement.

reach the drain. A voltage applied to the gate at a specified level (threshold) will cause enough charges to appear in the gate region to form a conducting channel that allows the source current to reach the drain region. Every MOS transistor is designed to operate at a specific threshold voltage. This value is measured using the capacitance-voltage technique. The gate voltage is continuously increased, while the capacitance of the gate structure is monitored.

A capacitor is a storage device. Initially, during the voltage-increase portion of the measurement, the capacitance does not change. At the threshold voltage level, the inversion layer forms and acts like a capacitor. Because two in-series capacitors have a combined lower capacitance than the sum of the two, the result is a drop to a combined lower capacitance. MOS transistors also exhibit amplification characteristics. The gain is defined as the source-drain current divided by the gate current. The source-drain characteristic for various gate currents is shown (Fig. 14.40).

### Capacitance-voltage plotting

A variation of the threshold voltage test is used to test for the presence of mobile ionic contamination in the oxide. The test is performed on specially prepared test wafers. A thin oxide is grown on a "clean" silicon wafer. After oxide growth, aluminum dots are formed on the wafer by evaporation through a mask (Fig. 14.41). Dot evaporation is usually followed by an alloy step to ensure good electrical contact between the aluminum and oxide.

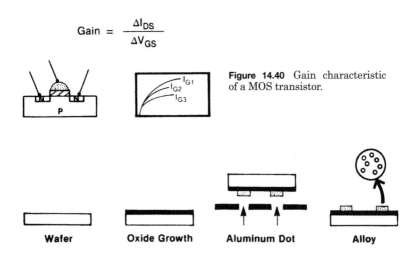

$$\text{Gain} = \frac{\Delta I_{DS}}{\Delta V_{GS}}$$

**Figure 14.40** Gain characteristic of a MOS transistor.

Wafer        Oxide Growth        Aluminum Dot        Alloy

**Figure 14.41**   Preparation of C/V test wafer.

The "dotted" wafer is placed on a chuck, and a probe is placed on the aluminum dot. This structure is actually a MOS capacitor. A voltage is applied to the dot and gradually increased as the capacitance of the structure is simultaneously measured. The results are printed out on an $x$-$y$ plotter with capacitance on the $y$ axis and voltage on the $x$ axis (Fig. 14.42).

At a voltage level known as the *threshold voltage* (or *inversion voltage*) charge starts to build up at the silicon surface. The charges "invert" the conductivity type from N-type to P-type. The inverted layer has a capacitance of its own. Electrically, the structure now has two capacitors in series. The total capacitance value of the two is less than the sum of the two by the relationship,

$$1/C_{total} = 1/C_{oxide\ layer} + 1/C_{inversion\ layer}$$

The trace on the $x$-$y$ plotter drops vertically to the new capacitance level.

The second step in the process is to force the mobile positive ions in the oxide to the $SiO_2$-silicon interface. This is done by simultaneously heating the wafer to the 200 to 300°C level and placing a positive 50-V bias on the structure (Fig. 14.43). The elevated temperature increases the mobility of the ions, and the positive bias "repels" them to the oxide-silicon interface.

The last step in the process is a repetition of the initial $C/V$ plot. However, as the voltage increases, inversion does not start at the

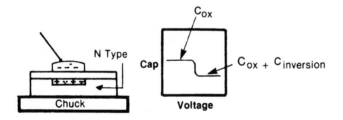

**Figure 14.42**  $C/V$ plotting—first test.

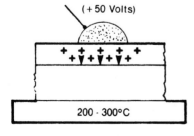

**Figure 14.43**  $C/V$ plotting—ionic charge collection.

same level as in the initial test (Fig. 14.44). The positive charges at the interface require additional negative voltage to "neutralize" them before inversion can happen. The result is a $C/V$ plot identical to the original but displaced to the right. The additional voltage required to complete the plot is known as the *drift* or *shift*.

The amount of the shift is proportional to the amount of mobile ionic contamination in the oxide, the oxide thickness, and the wafer doping. $C/V$ analysis cannot distinguish the element (Na, K, Fe, and so forth) that was in the oxide—only the amount. Neither can this test determine where the contamination came from. It may have come from the wafer surface, any cleaning step, the oxidation tube, the evaporation process, the alloy tube, or any other process(es) the wafer has been through.

$C/V$ analysis is usually a part of the evaluation of any process changes that may contaminate a wafer, such as a new cleaning process. To make the evaluation, $C/V$ wafers are divided into two groups. One group receives normal processing as detailed above. The second group goes through the proposed cleaning process, usually between the oxidation and aluminum steps of the test. The drift on the standard processed wafers is compared with the drift of the experimental group. An increased drift on the experimental wafers would indicate that the proposed cleaning process actually contaminated the wafers with mobile ionic contamination.

Acceptable $C/V$ drifts vary between 0.1 and 0.5 V, depending on the sensitivity of the device being made. $C/V$ analysis has become a standard test in a fabrication area. The test is made after any process change, equipment maintenance, or cleaning that could have the potential of contaminating the wafers. The $C/V$ plot provides a wealth of other information, such as flat-band voltage and surface states. The thickness of the gate oxide can also be calculated from the $C/V$ test data.

**Contactless $C/V$ measurement.**    $C/V$ monitoring as described above requires rigorous test wafer preparation, which is time consuming and expensive. Another method to determine unwanted charges (drift) and the other MOS gate parameters is with a contactless method called COS (Fig. 14.45). The MOS transistor method requires two electrodes separated by the gate oxide. A voltage on the top electrode (the metal)

1. Original Plot
2. Second Plot
3. Voltage Shift
   or Drift

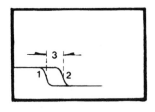

**Figure 14.44**  $C/V$ replot.

produces a charge buildup at the metal/oxide interface. A similar re-sult can be obtained by using a corona source (the *C* of COS) to build a charge directly on the oxide surface. Increasing the charge puts the MOS transistor through the same paces as a metal electrode device and the same information, charge (drift), flat-band voltage, surface states and oxide thickness, can be calculated. The method correlates with the standard measurements.[16]

### Device failure analysis—emission microscopy

When a semiconductor device is operating, it gives off certain emis-sions of visible light. When there are certain problems, spots of visible light are given off at the trouble spot. For example, contamination-in-duced junction leakage will reveal itself as a light spot on the junction (at the surface). Microscopes fitted with sensitive detectors and charge coupled imaging devices (CCD) can locate and image these trouble spots. This method is particularly useful when electrical measure-ments indicate a failure in a sector of a circuit, but it cannot pinpoint the exact device(s) causing the failure.[17]

### Review Questions

1. Changing the size of a resistor will change which of the following?

   a. Resistance

   b. Resistivity

   c. Both

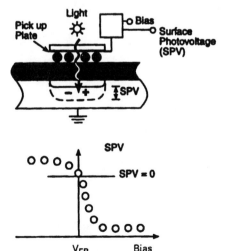

**Figure 14.45** Noncontact surface charge measurement. *(Source: Keithley Instruments Quan-tox—TM.)*

2. In a four-point probe, the current is flowing in the inside probes (true or false).

3. The color of an oxide film on silicon is an indication of its exact thickness (true or false).

4. Which is the most accurate determination of a thin oxide layer (less than 1,000 Å)?

   a. Color

   b. Fringes

   c. Ellipsometer

   d. Spectrophotometer

5. Explain why an oxidized silicon wafer changes color when rotated.

6. Name a method to delineate an N-P junction for the groove and stain technique.

7. A collimated light is used to detect sodium contamination (true or false).

8. An SEM uses which magnifying power as compared with a light-field microscope?

   a. Higher

   b. Lower

   c. The same

9. Name two methods of measuring CDs.

10. If an oxide layer is contaminated with sodium, what is the effect on the $C/V$ drift?

   a. $C/V$ drift will be higher.

   b. $C/V$ drift will be lower.

   c. $C/V$ drift will remain the same.

## References

1. B. McDonald, "Analytical Needs," *Semiconductor International,* Cahners Publishing, January 1993, p. 36.
2. S. Felch, "A Comparison of Three Techniques for Profiling Ultrathin p$^+$-n Junctions," *Solid State Technology,* January 1993, p. 45.
3. S. Felch, "A Comparison of Three Techniques for Profiling Ultrathin p$^+$-n Junctions," *Solid State Technology,* January 1993, p. 45.
4. P. Burggraaf, "Thin Film Metrology: Headed for a New Plateau," *Semiconductor International,* Cahners Publishing, March 1994, p. 57.
5. Thin Film Measurements, Rudolph Engineering, product brochure, 1985.

6. R. McDonald, "How will we examine IC's in the year 2000?" *Semiconductor International,* Cahners Publishing, January, 1994, p. 46.
7. A. Braun, "Metrology adapts to meet CD measurement needs," *Semiconductor International,* February 2002, p. 73.
8. A. Braum, "Optical Microscope Continues to Meet High Resolution, Defect Detection Challenges," *Semiconductor International,* December 1997, p. 59.
9. J. Baliga, "Defect Detection on Patterned Wafers," *Semiconductor International,* May 1997, p. 64.
10. *Product Description Bulletin,* Technical Instrument Company, San Francisco, CA, 1995.
11. S. Wolf and R. Tauber, *Silicon Processing for the VLSI Era,* Lattice Press, Sunset Beach, CA, 1986, p. 95.
12. P. Burggraaf, "Patterned Wafer Inspection Now Required!" *Semiconductor International,* Cahners Publishing, December 1994, p. 57.
13. L. Peters, "AFMs: What Will Their Role Be?" *Semiconductor International,* Cahners Publishing, August 1993, p. 62.
14. "Lithography News," *Semiconductor International,* December 1994, p. 24.
15. "Time-of-Flight Secondary Ion Mass Spectrometer," brochure, Charles Evans & Associates, Redwood City, CA, 1994.
16. A. Braum, E-Beam Techniques Measures Product Wafer Composition, Thickness, *Semiconductor International,* November, 2001.
17. P. Chu and R. Hockett, "New Ways to Characterize Thin Films," *Semiconductor International,* Cahners Publishing, June 1994, p. 143.
18. M. Peters, *COS Testing Combines Expanded Charge Monitoring Capabilities with Reduced Costs,* Keithley Instruments, Inc., product description paper.
19. T. Adams, "IC Failure Analysis: Using Real-time Emission Microscopy," *Semiconductor International,* Cahners Publishing, July 1993, p. 148.

Chapter

# 15

# The Business of Wafer Fabrication

## Overview

From its humble beginnings in laboratory-like production facilities to the automated wafer fabs of today, the semiconductor industry continues to evolve as the demands of new circuits, business factors, and global competition grow. While Moore's law still sets the product goalposts, fiscal considerations are setting the goals for productions facilities, equipment, and processes. Overall productivity and cost-of-ownership are the factors determining the bottom line. Strategies to increase the bottom line include maximizing equipment cost of ownership, automation, cost control, computer-automated manufacturing, computer-integrated manufacturing, and statistical process control.

## Objectives

Upon completion of this chapter, you should be able to:

1. List the major cost factors that influence fabrication costs.
2. Describe the intent and factors of cost of ownership (COO) models.
3. List the advantages of statistical process control.
4. Identify the parts and use of a control chart.
5. List and discuss the different levels of automation.
6. List the factors that enter into an evaluation of a particular piece of equipment.
7. Sketch a "product manager"-oriented organizational chart.
8. Define the terms "CIM" and "CAM" and their use in a manufacturing setting.

## Moore's law and the new wafer fabrication business

The semiconductor industry started supplying commercial products in the late 1940s. The manufacturing lines were little more than laboratories, and the workers were primarily trained technologists. By the 1970s, the manufacturing scene had changed to cleanrooms with highly specialized equipment attended by skilled production workers. A fabrication area of 2000 to 3000 ft$^2$ could be built for 2 to 3 million dollars.[1]

The VLSI era ushered in class 10 and 1 cleanrooms and newer, even more specialized and automated equipment and process control systems. The cost shot up to some 100 million dollars. By the mid 1990s, the industry was shipping large memory and microprocessor devices and facing 1 to 2 gigadollar (i.e., billion dollar) fab facility costs. The price tag is not so surprising, considering the factors driving the processes.

Advancements in integrated circuits progress along the prediction of Moore's law. Simply stated, it predicts a doubling of the number of transistors in a chip every 24 months. This leads to chips with higher capabilities and operating at faster speeds, as demanded by the marketplace. In reaction, the industry has developed strategies to keep pace with the higher challenges to the entire semiconductor production stream, from incoming materials to packaging.

### The nano- and 300-mm era

Capability and speed advances are accomplished with driving chip component dimensions to the nano-level (one-billionth of a meter). However, smaller components crowd the chip surface, forcing additional layers of metal connectors above the chip surface, and more process steps. Smaller component sizes on the chip surface require shallower doping layers and thinner dielectric layers. These requirements also increase the number of process steps. Along with higher-density chips, capacity improvements lead to larger chip size. However, larger chips on the same diameter wafer result in lower productivity. The natural solution to this problem is larger-diameter wafers. The companion wafer diameter to the nano-era has become the 300 mm wafer (approximately 12 in).

The 300-mm diameter wafers have brought new production challenges and forced higher levels of automation. A number of factors drive the need for automation. At a practical level, the weight and value of a 25-wafer batch of 300-mm wafers in a carrier is very high. Asking operators to manually process this size wafer puts many dollars at risk. Maintaining productivity and yields with larger wafers creates higher costs. The equipment to process larger wafers with tighter tolerances becomes more expensive. Improvements are needed to maintain uniformity across a larger wafer (or more tools processing single wafers). Of course, all of the materials and processing environment must become cleaner, since the smaller, more closely packed devices are more sensitive. And don't forget that, at these levels, the number of tests and characterizations increases to maintain product quality and process control in a wafer fabrication line that is rapidly moving large volumes of wafers. Of special note is the increase in process steps at the ULSI/nano level. Squeezing the devices closer together and making them smaller has introduced problems that were solved by the addition of new process steps, such as planarizing techniques to overcome topography generated imaging problems. The additional steps drive up the process and inventory expense. Some of these factors are noted in Fig 15.1, which shows the Semiconductor Industry Association's *International Technology Roadmap for Semiconductors*.

All of these cost increases have come as individual chip sale prices continue to erode through improved productivity and competition. To stay profitable, chip manufacturers have to continually improve their efficiency, yields and control costs at all levels. Reaching cost-effective operations generally means a throughput capability of 20,000 to 30,000 wafers per month. When the move to 300-mm wafers takes place, an advantage will be the option of smaller fabs (or *mini-fabs*) running 10,000 wafers per month and producing the same number of chips as a 200-mm fab.[2] This option comes from the larger area of the 300-mm wafers.

Despite all the changes in processes, costs, and markets, the overall financial measurement of a fabrication area has remained the same: it is *the cost per functioning die shipped out of fabrication*. When extended to a complete merchant facility with assembly capabilities, the

| Year of Production | 2001 | 2006 | 2012 |
|---|---|---|---|
| Line width (nm) | 150 | 100 | 50 |
| Memory size | 1 Gb | 16 Gb | 64 Gb |
| Logic Bits/cm$^2$ | 380M | 2.2B | 17B |
| Chip Size-DRAM (mm$^2$) | 445 | 790 | 1580 |
| Max wiring levels | 7 | 7-8 | 9 |
| Mask layers | 23 | 24/26 | 28 |
| Defect density-DRAM (D/m) | 875 | 490 | 250 |
| Chip conections-I/Os | 1195 | 1970 | 3585 |
| Wafer diameter (mm) | 300 | 300 | 450 |

**Figure 15.1**   Projection of wafer and chip parameters.

measure becomes *the cost per die shipped*. In the world of the mega-chips, the cost per transistor is becoming an indicator parameter.[3] These measurements apply to in-house wafer fabrication operations, foundry operations, and fabless chip companies.

The remainder of this chapter examines the various factors and strategies used to increase process/factory performance while controlling costs.

## Wafer Fabrication Costs

A number of factors contribute to the cost of producing a functioning die (Fig. 15.2). They are generally divided into fixed and variable categories. Fixed costs are those that exist regardless of whether any die are being made or shipped. Variable costs are those that go up or down with the volume of product being produced.

Fixed costs include:

- Overhead—administration, facilities, and research
- Equipment
- Variable costs
- Materials—direct and indirect
- Labor
- Yield

### Overhead

Overhead costs are all those incurred by the administrative and executive staff plus the cost of providing and maintaining a facility. A curious fact of company growth is that, beyond a certain level, the number of administrative personnel grows faster than the manufacturing

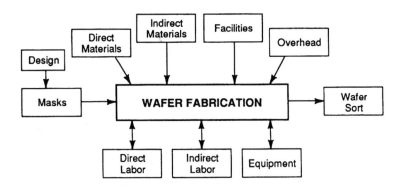

**Figure 15.2**   Fabrication cost factors.

workers. As companies grow, more information is generated internally, and more information must be handled from customers and suppliers.

To be effective, the information must be available to an ever growing staff. The two needs result in more and more staff processing "information" rather than product. Also, decision-making becomes more formalized (and costly) as more departments become stakeholders in the outcomes. Currently, some 50 percent of the workforce of the industrialized economies is involved in information processing.[4] A primary overhead expense is the design activity. With expensive CAD systems and a large professional design team, the cost of circuit design is considerable. Design cost goes up for ASIC fabrications, where a large number of individual circuits are designed and produced with a fast-turnaround requirement.

The facility cost and maintenance of the facility are major contributing costs. A fabrication area occupies only about 20 percent of the total facility area, yet it creates the majority of the expense. Air conditioning, chemical storage and delivery, and the cost of the cleanroom are all major expenses. Fabrication cleanroom costs for a ULSI facility are thousands of dollars per square foot. Actual floor costs are a factor of the cleanliness strategy chosen (see Chapter 4). A total cleanroom layout is more expensive than a hybrid/mini-environment approach. But in the latter, there is more expense for the equipment.

Co-development projects are emerging as the cost of product and process development increases. These partnerships are more important for processes whose effectiveness is heavily influenced by production factors. Often, a vendor cannot duplicate high-volume production parameters and needs access to wafer fab partners. Another trend is for semiconductor companies to co-develop new processes. As the industry has matured, more research is coming from universities and consortiums such as Semetech.

### Materials

Manufacturing materials are divided into the categories of direct and indirect. Direct materials are those that go directly in or on the chip. This includes the wafer materials, the materials and chemicals needed to form the deposited layers and doped layers, and the packaging material costs. Indirect materials are the masks and reticles, chemicals, stationery supplies, and other materials that support the process but do not enter into the product.

### Equipment

This cost is the equipment used directly in the fabrication of the devices and wafers. It shows up in the cost calculations as fixed overhead or as depreciation. Depreciation is the loss of value of the machines as they wear out or become obsolete.

The transition from 200-mm wafers to 300-mm wafers has bumped up equipment costs. At the transfer stage, 300-mm processing is essentially the same as 200-mm processing. This means that the process tools are the same, except for size capability. However, the larger wafers generally require more handling time and more process time, resulting in productivity losses. These losses equate to more tools to maintain production quotas and more expense. Figure 15.3 compares the productivity of the primary process tools for the two diameters. New to the 300-mm process is greater need for in-line metrology measurement and monitoring. The downside of larger wafer diameters is bigger losses if wafers are misprocessed or have low yields. Now, 300-mm processing is proceeding with copper metallization, which brings surface factors and new low-k materials into the picture. Monitoring and controlling each of these steps at the process tool is critical if the whole system is to work at the end of the process. Critical dimension measurements and e-beam defect inspection systems have become in-process requirements and add to the equipment cost.

Interestingly, the move to larger wafers and more sophisticated processes has increased the typical life cycle of new processes. Figure 15.4 shows the history and projections of wafer size life cycles.

**Labor**

Labor also has a direct and indirect component. Direct labor takes in those workers actually handling the wafers and equipment. Indirect

| Tool | 200-mm baseline (1.0x) | 300-mm productivity in 2001 | Expected productivity in 2002, 2003 | Major upgrade required |
|---|---|---|---|---|
| Photo | 1.0x @ wafer passes/day | 0.6x | 0.9x | Yes |
| Etch | 1.0x @ wph | 1.0x | 1.2x | No |
| Wet bench | 1.0x @ wph | 0.8x | 0.9x | No |
| CVD | 1.0x @ preventive maintenance | 0.3x | 1.0x | No |
| PVD | 1.0x @ available time | 0.9x | 1.0x | No |
| CMP | 1.0x @ available time | 0.8x | 1.0x | No |
| Furnace | 1.0x @ wafers/day | 0.7x | 0.9x | No |
| Implant | 1.0x @ wafers/day | 1.0x | 1.0x | No |
| Defect inspection | 1.0x @ wph | 0.6x | 1.0x | Yes |
| Metrology | 1.0x @ wph | 0.7x | 1.0x | Yes |

**Figure 15.3** 300-mm tool productivity factors. (*Courtesy of* Semiconductor International, *August 2001.*)

Diameter

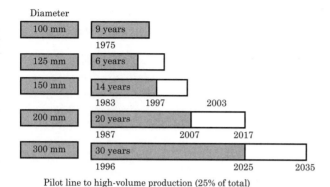

Pilot line to high-volume production (25% of total)

**Figure 15.4**  Wafer size life cycles. *(Courtesy of Future Fab International.)*

labor refers to support personnel such as supervisors, engineers, facility technicians, and office workers. Ironically, the new demands of process accuracy and productivity, using very sophisticated process tools, have returned the industry to the requirement that operators now have more technical training. Some companies, such as Intel, require all of their fabrication workers to have technician-level education. Coupled with higher levels of technicians and engineers to set up and maintain the tools, labor costs have increased.

## Production cost factors

Typical contributions of each of the factors to the overall unyielded die projected for 300-mm diameter wafer costs are shown in the following table. The percentages will vary with the wafer and feature size, degree of automation, and number of process steps. Depreciation in the

| Factor | Percent contribution | Costs ($) |
|---|---|---|
| Depreciation | 35 | 1189 |
| Labor | 7 | 232 |
| Maintenance | 7 | 232 |
| Consumables: | | |
| direct materials | 12 | 405 |
| test wafers | 6 | 203 |
| indirect materials | 26 | 890 |
| Other | 7 | 226 |
| Total | 100 | $3378 |

chip industry is high as a result of the short useful lifetime of the expensive equipment. Direct materials include wafers. Others include the overhead costs.

In any discussion about wafer cost factors, one is naturally curious about the actual costs. This number changes from line to line, with the complexity of the devices and with the position of the product in the maturity cycle (Fig. 15.5). The more mature the product and process, the higher the yields and the lower the equipment depreciation factor. Newer products suffer from operator learning curves, equipment shakedown times, and development of new processes. Wafer volume and type are major influences on cost. High-volume products, such as DRAMS, generally have the lowest per-transistor or wafer cost because of high manufacturing efficiencies. ASIC wafers, on the other hand, have higher costs as a result of smaller production runs and the higher design and processing costs required for a varied product mix. The range of costs is from about $100 for discrete and simple ICs to several thousand dollars for advanced products.

### Yield

The overall fabrication yield (see Chapter 6) determines how the various costs affect the final die cost. If the die yield is low, the cost per die goes up. Not only are the fixed costs distributed over fewer die, but the variable costs go up as more materials are required to get out the die. When the yield is calculated into the cost, the term used is *yielded die cost*. The cost of producing the wafers without considering die yield is the *unyielded die cost*.

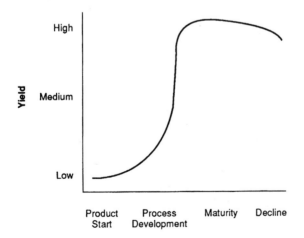

**Figure 15.5**   Wafer production loss versus product maturity.

Die costs are a function of the wafer size, die size, and wafer sort yield. A $3000 wafer-manufacturing cost for a wafer with 300 die will translate into an unyielded die cost of $10 each. If the die sort yield is 50 percent, the die cost rises to $20 each ($3000 wafer cost divided by 150 functioning die). Increasing the sort yield to 90 percent reduces the die cost to $11.11.

Market pressures require that wafer fabrication operations reach wafer sort yields in the low 90 percent range at ever faster rates. Figure 15.6 shows the yield ramps for different levels of DRAM memory devices as they progress from R&D to full production.

## Yield improvements

While it is true that increased attention is being given to traditional business factors, no less attention is paid to process and wafer yields. The effect of a yield improvement can be significant in terms of dollars. Consider the yield figures in Fig. 15.7. Line 2 shows that a wafer fabrication yield improvement of five percentage points increases the overall yield from 38 to 40.4 percent, a 1.2 percent increase. If the fabrication area starts 10,000 wafers per month and there are 350 die per wafer and the selling price is $5 per die, the increased revenue is

10,000 wafer starts × 350 die per wafer × 0.012 (1.2%)  =  $210,000/month

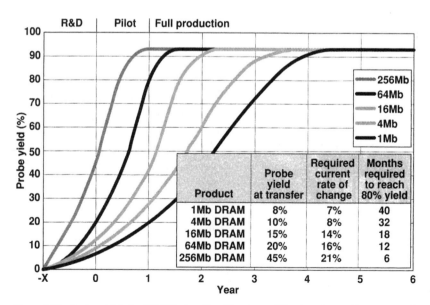

Figure 15.6   Probe yields for DRAM circuits. (*Courtesy of* Semiconductor International, *January 1998.*)

| | | Yield | | | |
|---|---|---|---|---|---|
| Process | Change | Fab | Sort | Assembly | Overall |
| Base Line | | 0.80 | 0.50 | 0.95 | 0.38 (38%) |
| | Fab to 0.81 | 0.81 | 0.50 | 0.95 | 0.384 (38.4%) |
| | Fab to 0.85 | 0.85 | 0.50 | 0.95 | 0.404 (40.4%) |

**Figure 15.7**   Yield impact of fabrication yield improvement.

Whether this amount of increased income is significant depends on the cost of effecting the improvement.

## Yield and productivity

Yield has been the traditional measure of wafer fab success. High yields translated to lower production costs and higher profits. With wafer sort yields at the 90 percent level, the next cost-reduction factor is productivity. Improved production efficiency is measured by the number of wafers per square foot of the total fab area (or number of wafers per piece of equipment), or the number of wafers produced per operator. Increasing these factors means a lower manufacturing cost. Throughput is another factor. Increasing the number of wafers out per hour means a more efficient process with lower costs. These factors increase the bottom line if the cost of achieving them is less than the advantage.

## Book-to-bill ratio

On the business side, one of the most watched industry factors is the book-to-bill ratio (b/b). *Book* refers to bookings, which is the dollar amount of sales orders received in a period. *Bill* refers to the dollar amount of billing invoices sent out. In a normal operation, chip orders are shipped and billed some time after the order is received. In good times, the delay reaches into months or longer if shortages occur. A high book-to-bill means that unfilled orders are accumulating, generally indicating a healthy economy. A low book-to-bill means that orders are drying up, and the factory is shipping out to inventory rather than building new product. This indicator is used to project inventory and production levels. A b/b ratio greater than 1 indicates a rising market, and less than 1 indicates a declining market. When the market becomes very good, a high b/b ratio is worrisome, since shipping times become stretched, and the user end of the electronic industry can become starved for chips.

## Manufacturing strategies

Minimizing the cost per wafer (or transistor) involves a host of factors that are the result of decisions about how the fab area is laid out, contamination control systems, equipment/automation levels, process control programs, and material management schemes. These are discussed in the following sections.

## Contamination control systems

There are two major contamination control systems in use (see Chapter 4). Some manufacturers elect to build and maintain a total cleanroom environment with class 1 or better air quality. This approach has the highest facility expense and requires rigorous maintenance and contamination control programs. Wafer isolation technology (WIT), or the use of mini-environments, is the other approach. It requires a large investment in clean boxes and equipment interfaces, but the fab area can be at a lower class value.

## Equipment

Over the years, semiconductor processing has changed from a laboratory activity to full manufacturing using sophisticated and dedicated equipment. The development of semiconductor-specific equipment has allowed the advancement into the ULSI era and the spread of the technology throughout the world. IBM coined the term *tool* for semiconductor process machines, and it has become part of the industry vocabulary. As the tools have become more sophisticated, the prices have risen to the point where they are a major expense in any new facility. An ASA/Dataquest calculation shows a worldwide chip business of $804 billion and an equipment/materials business of $77 billion.[6]

### Equipment factors

A number of considerations (Fig. 15.8) go into the selection of a particular process tool. They can be roughly divided into two broad categories: performance and economic. Performance factors relate to the ability of the tool to produce the required results. Economic factors relate to price, costs, and support factors. Figure 15.7 shows how these four factors have shifted as the industry has moved into the manufacturing era.

Performance has fallen as a selection criteria; nevertheless, any tool must have the basic capability to meet the process requirements. If the machine cannot routinely produce the right product, it is of no use. Other performance factors are repeatability, flexibility, upgrad-

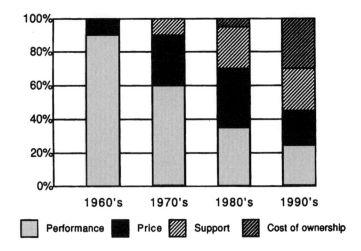

Performance ☐   Price ■   Support ▨   Cost of ownership ▨

**Figure 15.8**   Changes in equipment purchase factors. *(Source:* Semiconductor International, *May 1993, p. 58.)*

ability, ease of operation, setup considerations, and reliability/downtime factors. Repeatability is the ability to produce the same result every time and over long periods of time. Flexibility relates to the ease with which a machine can be switched to run a variety of products and processes. This is an important factor for ASIC and other small-run fab lines. Upgradability is the ability of the machine to handle future process requirements, such as increasing wafer diameters. For example, a lot of 300-mm process tools are basic 200-mm versions, retrofitted to handle the larger diameter. This factor is very important for fabrication lines doing ASIC work, where staying abreast of the latest product technology is vital. In fabrication areas dedicated to long production runs of the same (or similar) products, the ability to upgrade is less important.

Ease of operation and setup factors address the issues of minimizing operator mistakes by good design and user friendly controls. Setup issues relate to the amount of time required to bring the tool on line (tests, calibrations, and so on) and the related loss of production time.

Most companies run their equipment two to three shifts per day, sometimes six or seven days a week. Scheduled maintenance and unscheduled breakdowns stall product flow and run up expenses. The factors are: *scheduled maintenance frequency and time, mean time to failure* (MTF), and *mean time to repair* (MTR). Given the expense of most machines, having backups is a costly luxury. It falls on the vendor to provide a machine that runs for long periods of time (long MTF), can be repaired quickly (short MTR) when failure occurs, and does not need frequent and lengthy routine maintenance. Cost also includes the price of materials needed for the machine process. Machines that waste materials also waste money.

Vendor support has emerged as a critical factor. The advancement of process technology and equipment sophistication has forced close cooperation and alliances between the chip manufacturers and their suppliers. Reducing the cost of machine development requires input from the operating process engineers, and the detailed nature of the machines requires on-time backup from the vendors. Management guru Tom Peters sees modern corporations as having "fuzzy edges." By this he means that, more and more, a corporation must rely on its vendors and customers for information and co-development of needed equipment and materials. This is certainly true of the semiconductor industry.

### Cost of Ownership

Shifting equipment purchase decisions away from purely technical to business-related factors has driven the development of cost of ownership (COO) models. These models attempt to bring together all of the relevant factors driving the total cost of ownership of a tool, process, or facility over its expected lifetime. Beyond the initial equipment purchase price, tools differ in the amount of expensive floor space they occupy, the amount of power and materials they require for operation, the yield of in-spec wafers, maintenance, repair and failure rates, and so forth. The COO formula is one developed by Semetech to evaluate equipment purchases.[7]

$$C_W = \frac{\$F + \$V + \$Y}{L \times \text{TPT} \times Y_{\text{TPT}} \times U}$$

where   $C_W$ = cost of the finished wafer, based on the particular cost associated with the tool/process for the lifetime of the tool

  $\$F$ = fixed costs, which include the equipment purchase prices, facility costs, initial modifications, and so forth

  $\$V$ = all material, labor, and process costs generated when the tool/process is operating

  $\$Y$ = the cost of wafers scrapped due to tool scrap and defect induced losses

  $L$ = the lifetime of the tool in hours

  TPT = the wafer throughput rate as reduced from the maximum by maintenance requirements, setup, test wafer monitoring, and so on, expressed in wafers per hour

  $U$ = the tool utilization factors that reduce available process time from the maximum

Each of these equation terms is calculated from formulas that consider the subfactors for a particular process. The COO formula provides the method to determine trade-offs for various tool factors. For example, a tool may have a high initial cost that is offset by lowered operating costs or a long time between failures. Or, a tool may provide a high yield but require so much adjustment and calibration that additional machines would be required to meet production schedules.

## Automation

As most industries mature, the technology becomes stabilized, and the market drives up demand. These two conditions are precursors to process automation. Since 1940, automation of oil refineries has reduced the number of workers by a factor of 5.[8] Automation of semiconductor processes has been in progress since the 1970s, when process tools were designed to accept wafers in cassettes. Since then, automation has marched along toward the fabled dream of the total "peopleless lights-out" fab. Automation stages start with the process tool and extend to the factory level.

### Process automation

The first level of automation is of the process itself. Most semiconductor equipment, by definition, automates a part of a process. Photoresist spinners automatically dispense the primer and resist at the correct speeds and for the correct time. Automatic gas-flow controllers dispense gases to the tool in the right amount, at the right pressure, and for the right time. Process automation brings consistency to the process and the product by reducing reliance on operator skills, training, morale, and fatigue.

Most tools are controlled by a set of instructions programmed in an on-board computer. The program is called a *recipe*. The recipe is loaded into the machine by the operator or from a central host computer.

### Wafer-loading automation

The next level of automation is the loading and unloading of the wafers. The industry has settled on the wafer cassette as the primary wafer holder and transfer vehicle. Cassettes are placed on the machine by various mechanisms. Elevators and/or wafer extractors or robots feed the wafers into the particular process chamber, spin chuck, and so on. In some processes, such as process tubes, the entire cassette is placed in the process chamber. This level of automation is referred to as the "one-button" operation. With one button, the operator activates the loading system, and the wafers are processed and returned to the cassette. At the end of the cycle, the machine sounds an alarm or turns on a light, and the operator removes the cassette.

Some machines have buffer storage systems that maximize the machine efficiency by always having fresh wafers (or reticles for the imaging tools) available for processing. These are called *stockers*. The operator places the cassette(s) on the machine loader and pushes a start button, after which the machine takes over the processing.

At the 300-mm diameter wafer level, and with single-wafer processing, wafers are transported in individual holders.

### Clustering

Mating two or more process steps in a single unit is another level of automation. Generally, this level is called *clustering*. The industry has been "clustering" for a long time. Photoresist spinners were long ago mated to soft-bake modules and other track equipment groupings.

Recent clustering designs (Fig. 15.9) have been driven by both technical and economic forces. On the technical side, some clustered processes make better product by, e.g., keeping a silicon wafer clean after etch and before metal deposition. Another process advantage is sequential deposition of different materials in the same chamber. In these cases, the deposition process is better, because the wafer is not exposed to air between steps. Certainly, anytime a wafer loading/unloading step can be eliminated, both cleanliness and cost are favorably effected. For vacuum processes, time and cleanliness factors are affected when two or more processes can be performed with only one vacuum pump down. Clustering in which two or more sequential processes are performed is called *integrated processing*.[9] On the economic

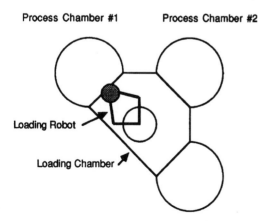

Process Chamber #1      Process Chamber #2

Loading Robot

Loading Chamber

**Figure 15.9**  Three-chamber cluster tool.

side, some types of processes that are clustered for increased through-put are called *parallel processing.*

Despite the obvious attractions of clustering, there are drawbacks. Clustering for critical process advantages is easy to justify. But a cluster of same type processes requires interlocks, electronics, and software more sophisticated than individual tools. And a shutdown for maintenance or repair idles a larger part of the production capacity. Another barrier occurs when cluster modules cannot be provided by the same vendor. Customers prefer one responsible vendor, and vendors are slow to pair up when responsibilities are clouded.

### Wafer-delivery automation

The third level of automation is when wafers are automatically brought to, loaded on, and removed from the machines. In addition to the production advantages of automation, there are ergonomic, safety, and cost benefits. These accrue from the weight of a cassette of larger diameter wafers, which can reach the 18 to 20 lb (6.7 to 7.5 kg) level. Operators injuries are a possibility, and the financial loss of dropping a cassette of expensive wafers can be staggering. Early delivery systems used traveling robotic carts that duplicated human delivery (Fig. 15.10). Called *automated guided vehicles* (AGVs), the carts travel along the aisles and dispense wafer cassettes when the machines need them. Another version is the rail guided vehicle (RGV), where carts follow a track laid out between the process tools and the stockers. These approaches have the advantage of being retrofittable to fabrication lines where the equipment is lined up in rows.

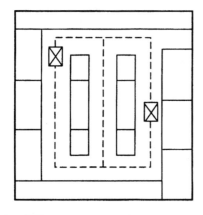

AGV

AGV Path

**Figure 15.10** Automated guided vehicle wafer delivery system.

Another approach is the use of an overhead rail[10] (also called a gantry). The wafer cassettes arrive at the process tool area (Fig. 15.11) where a secondary system (usually a robot) removes them from the overhead rail and places them in the tool buffer. This system works best with machines that are grouped in bays rather than in the traditional linear layout.

### Closed-loop control-system automation

The industry is just embarking on the last level of total automation, the closed-loop feedback system. There are two aspects. Some tools have on-board sensors that measure critical process parameters. Through feedback electronic circuits, the tool adjusts itself to maintain process operation specifications. This stage of automation is difficult to achieve for many processes. It requires measuring sensors that operate in hostile environments, such as heated-tube furnaces and sputtering chambers. One area of progress is in the area of mask-making, where recognition systems are advanced enough to compare the manufactured plates with the design criteria.

Second are tools that have, or are connected to, automatic inspection or measuring subsystems. The subsystem measures important parameters in real time (as they are happening), compares them with a standard, and feeds the information back. It may trigger a shutdown of the out-of-spec tool or make process parameter adjustments. The development of closed-loop process and machine control is necessary before the goal of a "peopleless, lights-out" wafer fab can be achieved.

### Factory-Level Automation

The advent of higher levels of process and tool automation and inventory control systems requires higher levels of centralized control and information sharing. Most companies have computer-based management information systems (MISs) handling the paperwork and details

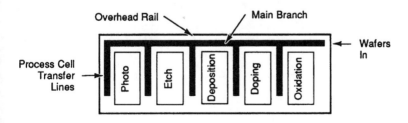

**Figure 15.11** Overhead rail wafer delivery system.

of employment and finances. These systems are being expanded to the entire manufacturing environment in a process called *computer-integrated manufacturing* (CIM).

CIM is the computerization of all plant operations and the integration of those operations into one computer design, control, and distribution system. The processes involved in CIM are all related and interdependent, as illustrated in Fig. 15.12. The major activities of CIM are business functions, product design (mask and circuit), manufacturing planning (inventory, shop floor priorities, etc.), manufacturing control, and the fabrication processes. A complete CIM system is interactive at all levels. This means that each of the five functional areas input data to the system in real time and that the information is available to all who need it.

Computer-aided design (CAD) and computer-aided manufacturing (CAM) are two subsystems within the CIM system. The role of CAD has been discussed in terms of mask and circuit design. CAM is the part of the system that does the planning and control of the manufacturing operation. A CAM system includes a computer network and the automated process tools and material delivery systems.

In concept, a CIM system kicks into operation when a customer order is received. The computer logs the order and initiates the CAD system to start the design (if it is a custom order). It also triggers the ordering of needed materials in the right quantities and schedules

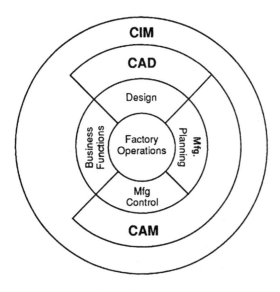

**Figure 15.12** Provinces of factory computer control systems.

their delivery times. This subsystem goes by the name *computer-aided process planning* (CAPP). Through the CAM program, the individual process recipes are downloaded to the individual equipment computers. Once processing begins, the CAM system controls the WIP and makes necessary priority decisions to meet shipping schedules. It also keeps track of equipment performance and schedules repairs and maintenance.

An important feature of the CAM system is yield monitoring and reporting. Important measuring systems are connected directly to the factory computer. If a poor yield problem appears, the system reports it to the engineering and facility staff and (if necessary) will reorder materials to make up the losses. At the end of the production run, the system calculates costs and yields and schedules shipment to the customer.

Other packages included in CIM systems are facility monitoring, process modeling, and security systems. Facility monitoring might include power consumption and environmental factors inside the plant. Some facility CAM systems include monitoring the levels of liquids and gases in storage with automatic alerting and/or reorder. Monitoring and precise control of these factors can save appreciable amounts of money. Process modeling is a system of testing a particular design against the known process variations. The computer can run many variations that simulate the changes expected in fabrication. Good modeling can identify weak points in the design or process before wafers are committed to the line. Security monitoring may include the entering and exiting of employees (and/or intruders) and the securing of expensive finished product or materials. A security system may also include fire and other hazard control.

## Equipment Standards

Given the many suppliers of materials and equipment and the very stringent technical demands of chip manufacturers, the need for standards is obvious. In many industries, different manufacturers attempt to establish their "standard" as the industry standard, as a way of maintaining a competitive edge. Fortunately, the semiconductor industry informally has settled on many standards—for example the, use standardized cassettes.

In 1973, Semiconductor Equipment and Materials International (SEMI) established a standards program. Supplier and user personnel come together and establish standards by a consensus process.

At the automation level, SEMI has published communication protocols for equipment interfaces.

## Fab floor layout

The tradition fab floor layout of lining up the equipment in rows has given way to the bay or tunnel system. This system groups sections of the process in separate rooms, which minimizes contamination from personnel traffic and cross contamination from other process areas. Automation needs is forcing a rethinking of the most efficient way to move material while maintaining cleanliness.

Robots and overhead gantries have become the material mover of choice. Also driving robot use is the increasing weight of wafer lots (larger diameters), especially in SMIF type transfer modules.[12] Robotic design and performance challenges come to the forefront in vacuum tools, especially if the process uses corrosive gases.

One equipment layout option is the process island. The various tools for a particular process segment are grouped around a single loading/unloading robot (Figure. 15.13). The tools may be single or clusters.

## Batch versus single-wafer processing

For the first two decades of process development, the drive was to ever larger batches of wafers. Simple economies of scale were behind the drive, i.e., gaining more product output per process sequence. However, the arrival of larger-diameter wafers and VLSI/ULSI-level circuits has changed the picture. Larger wafers require larger process chambers, and this, in turn, pushes the limits of uniformity. Many processes, particularly the deposition steps, are easier to control in a smaller (single-wafer) chamber. Fast cycle time fabs (ASIC and small-

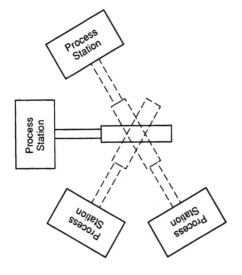

**Figure 15.13** Robot wafer delivery in process island.

volume devices) can use the flexibility of single-wafer processing to good advantage.[13]

Offsetting improved process control is a loss of productivity compared to batch processing. Larger-diameter wafers help somewhat (300-mm diameter wafers have almost four times the area of 150-mm wafers). Single-wafer process productivity is improved by tool designs that, e.g., allow one vacuum pump down for an entire batch while wafers are processed individually, and other improvements.

On the downside, single-wafer processing requires higher levels of repeatability. Assuming the same size wafers, a single-wafer processor will have to perform 25 times compared to a one-batch 25-wafer batch process. The equipment is also more expensive. Each single-wafer processor requires essentially the same components as a batch tool.

### Green fabs

Additional pressure on wafer fabrication processes is generated by environmental laws and environmental concerns. A wafer fabrication operation uses many hazardous chemicals and produces waste products. In-plant controls of chemical handling, storage, and use are a part of fab *environmental, safety, and health* (ESH) programs. Active research is aimed at development of processes that use less chemistry and chemicals that are less harmful. The SIA roadmap shows water consumption dropping from 30 gal/wafer to 2 gal/wafer in 2012.[14] The roadmap calls for lowered energy use per wafer and the elimination of polyvinyl chlorides (PVCs). Along with safety, there are cost savings related to lower-hazard chemicals in terms of the costs of on-site treatment and storage, transportation, and acceptance at special waste facilities.

### Statistical Process Control (SPC)

This text, for the most part, has presented the process technology with few mathematical formulas. The next few sections on statistical process control will break from that approach. SPC is a powerful and needed tool to maintain process control and improve yields. Unfortunately, SPC is based on statistics, which is applied with the language of mathematics. We will attempt to convey the background, use, and interpretation of commonly used statistical tools with illustrative charts and graphs. No formulas will be presented.

The first question to address is, "What is the object of process control?" The answer is simple: to produce a product that falls within its design and operational specifications and do it at a high enough rate to be profitable. Challenging this simple goal are processes that are influenced by a multitude of parameters, including the conditions already on the wafer.

Processes do not run in control without monitoring and adjustments. Process control provides the information to make the necessary changes. Statistical process control does it in a way that relies on established mathematical principles that govern the type of processes used to make chips. Process control techniques can vary from simple to very complex. The simplest (and most familiar) is the calculation of the average of a group of numbers (called a *population*). We also know that the extremes of a population (range), along with the average, give us an idea of the distribution of the data within the population.

For example, the two groups in columns A and B of Fig. 15.14 have the same average. If the numbers were the sheet resistances of wafers from furnace A and B, we could easily come to the decision that furnace A has the most in-control process because of its tighter distribution. This fact is illustrated when the data is plotted (Fig. 15.15). The plots are called *histograms* and visually display data distributions that a simple average calculation will not reveal. Average calculations and histograms are statistics in action. Histograms are usually the first step in determining whether a process is in control.

Their power comes from a mathematical distribution known as the Gaussian distribution. Named after the famed mathematician Karl Friedrich Gauss (1777–1855), its origin is interesting. Gauss set out to reconcile the different star positions reported by different astronomers. His approach was to make all the necessary corrections to the observations, taking into account the time of the year and from where on the Earth the observations were made. He expected that, when all the corrections were made, there would be agreement from all the position calculations on the position of a particular star. After

| Reading no. | Run A | Run B |
|:---:|:---:|:---:|
| 1. | 25 | 28 |
| 2. | 24 | 25 |
| 3. | 26 | 23 |
| 4. | 23 | 26 |
| 5. | 25 | 25 |
| 6. | 26 | 22 |
| 7. | 25 | 23 |
| 8. | 24 | 25 |
| 9. | 25 | 27 |
| 10. | 26 | 25 |
| Average | 24.9 | 24.9 |
| Range | 26 − 23 = 3 | 28 − 22 = 6 |

**Figure 15.14**  Sheet resistance reading ($\Omega/\square$).

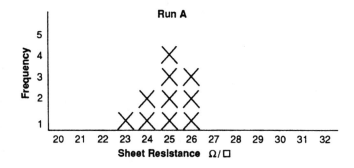

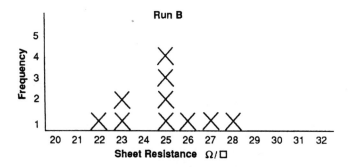

Figure 15.15   Frequency distribution of sheet resistances.

all, reason dictates that a star can occupy only one position at a time, and we should be able to determine that position.

However, the final data did not confirm his hypothesis. After all corrections were made, there were still a number of locations for each star. Fortunately, Gauss did not scrap the project but went on to establish the basis for the field of statistics and distribution probabilities. If the math-aversion readers will hang on, they will find that the concept of probability is not so difficult. The way Gauss analyzed the data was to plot the various calculated locations for a star. He calculated the center point (average) and drew a circle that encompassed the star location the farthest from the center. He reasoned that there was a 100 percent probability (a probability of 1) that the real star location was within that circle. He also reasoned that the probability of finding the star within the confines of smaller circles was less than 1. In fact, the smaller the circle, the lower the probability that the star was within those boundaries.

It also turns out that processes produce data distributed like the example given. The mathematical distribution is known as the Gaussian distribution. A good example is the height of blades of grass in a lawn.

If all the blade heights are measured and plotted on a histogram, the distribution will be the familiar bell-shaped curve, also called a *normal curve* (Fig. 15.16). From a probability consideration, it predicts that there is a higher probability that any given blade of grass will have a height closer to the average (center values) and that there is a lower probability that any given blade will be very short or very tall. The mathematical conditions that result in the distribution include human heights, IQ distributions, and (in most cases) semiconductor process parameter distributions, such as sheet resistance.

A first step in process control is to make a histogram of the particular process parameter and determine if the distribution is a normal distribution. If it is not, the chances are good that there is something wrong in the process. If the distribution is a normal one, the next step is to compare the range of the distribution with the design limits for the particular parameter (Fig. 15.16). This comparison is made to determine if the natural process distribution limits fall within the design limits. If they do not, the process must be fixed or some percentage of the parameter readings (and the wafers) will always be out of specification.

Another useful statistical tool is the Pareto chart. This is a form of the histogram, but with the $x$ axis divided into unrelated sections rather than a continuum of some parameter such as sheet resistance. Defect inspection results are good candidates for Pareto charts. On the $x$ axis is a list of the defect categories. A mark or vertical column over each defect type indicates the frequency of occurrence of each of the defect types. The operator or process engineer can see from the chart which defects are occurring most often and what processes must be improved.

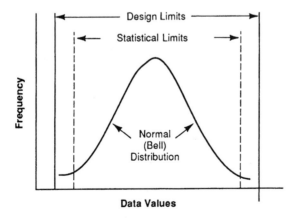

**Figure 15.16**   Frequency distribution of sheet resistances.

So far, the statistical methods explained are the after-the-fact histories of a process. A most powerful statistic of real-time process control is the *X-R* control chart (Fig. 15.17). The chart is constructed in two plots. The lower plot shows the moving range *R,* which is descriptive of process stability and the parts, with the *y* axis representing the parameter values. The top graph has a horizontal line for the historic average of the parameter. On each side of the average are control limit lines that are calculated from the historic data.[15] A control limit represents the limits that the individual data values will range between when the process is in control. Also on the chart are the process or design limits that represent the extremes the individual data points may have before being rejected. The bottom graph is constructed by calculating and plotting the amount that each data point varies from the average. When plotted, these values give further visual evidence of the amount of control in the process.

The value of the *X–R* bar control charts is their predictive powers. A process in control will produce data points that tend to vary in a regular pattern about the average (top of Fig. 15.17). The mathematics of a controlled process predicts this regular fluctuation. It also predicts *when* a process is going out of control *before the data points exceed the control limits* (bottom of Fig. 15.17). The data points in part B have shifted to the top of the control limit range. This is an unnatural pattern for an in-control process. When this situation occurs, the produc-

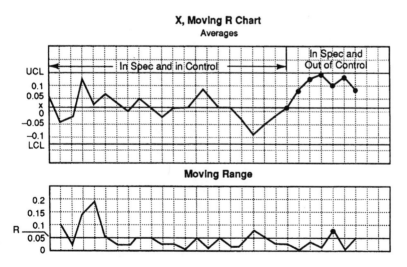

**Figure 15.17**  Moving *R* chart contains the averages of measurements, *x,* in the upper plot. The lower plot shows the moving range, *R,* which is descriptive of process stability.

tion operators, who maintain the charts as the data is produced, alert the proper personnel so the process can be brought back into control *before the data points exceed the control or design limits* and wafers have to be scrapped. A number of more sophisticated controls used in processing are beyond the scope of this text.

Another powerful statistical tool is multivariable experiment analysis. Most measured quality control parameters (sheet resistance, line width, junction depth, and so on) are influenced by a number of variables in the process. Line width, for example, varies with the resist solution, film thickness, exposure radiation time and intensity, baking temperatures, and etch factors. Any one or all can contribute to an out-of-spec condition. Multivariable evaluations allow the engineer to run tests that separate and identify the contribution(s) of each of the individual variables.

Designing an SPC system for a process requires selection of the proper statistical tool. Another decision revolves around the proper "indicator" population. Profit demands that all the die on all the wafers from all the batches, day in and day out, meet the specifications. However, picking the parameter population is not always an easy chore. Depending on the process, there are variations across the wafer, there are variations from wafer to wafer within a batch, and there are variations from tool to tool. Since every die cannot be measured, selecting the right sample point and sampling level becomes a demanding task.[16]

## Inventory Control

A critical issue in fabrication cost control and yield is inventory level and control. As the number of processing steps has increased, so has the length of the processing time and the number of wafers in process (WIP). The problem is that the company pays for the wafers when purchased and doesn't receive payment until the finished devices are shipped. This period can vary from two to eight months for a production line making similar circuits. Fabrication lines doing ASIC circuits have an even heavier burden when many different circuit types are going through the system. To get an idea of the burden, consider a CMOS-type process with 50 major steps and 4 substeps each, for a total of 200 processes. The high cost of the equipment generally requires some buffer inventory to ensure that the machines are operating at maximum efficiency. If each buffer has 4 cassettes of 25 wafers each, the total inventory of WIP is 40,000 wafers. At a cost of $100 per wafer (large diameter), the total inventory burden becomes $4 million.

Excessive WIP affects productivity by having the capability of hiding process and equipment problems.[17] With a lot of inventory at the stations, wafers can keep flowing out the back end while parts of the process are shut down. With a lower WIP, these problems are readily apparent and force the solution of problems. WIP also influences the overall fabrication yield. The collective experience of the industry is that the longer wafers are in the process the lower their wafer sort yield.

### Just-in-time (JIT) inventory control

Just-in-time inventory control is a philosophy based on the objective of, "Make only what is required, only as required."[18] The system is simple in concept. All buffer inventories are reduced to an absolute minimum, from the storeroom to the machine buffers. To work effectively, excellent vendor relationships must be established. The incoming materials have to be of the highest quality, as JIT leaves little cushion for extensive incoming inspections and returns. Second, the vendor is asked to maintain ready-to-ship materials at its facility. In effect, the vendors are asked to hold the inventory previously held by the chip manufacturer, a situation that they are not always happy about. The chip manufacturers have to be very efficient in assessing their quantity, quality, and delivery needs, and must have a system that expedites material to the proper process tool and detects quality problems quickly.

JIT also has applications in the process flow. Some companies will commit wafers to a process only if the work can flow unimpeded through all of the substeps. This system gives higher yields and, if on a properly balanced line, increases throughput time, even though parts of the line are idle. This system is called *demand-pull*. Wafers "upstream" are worked on only when there is demand from downstream process stations about to run dry of wafers.

An effective JIT procedure can reduce the number of operators in the fabrication area, since a good proportion of their time is spent sorting, staging, and delivering work to the process tools. Fabrication-area layouts can change from the traditional linear arrangement of process stations. A linear layout benefits a line making only a few different product types. Wafers are introduced into the front of the line and physically move on a first in-first out (FIFO) basis. Problems arise when particular lots of wafers must be moved quickly. Usually tagged as *hot lots*, they are given priority processing at each station. Besides the control problems associated with keeping track of the lots, hot lots have the effect of making regular product sit in a queue waiting for the hot lots to clear. This layout is particularly cumbersome for ASIC lines where shipping dates and product types have constantly changing priorities.

JIT-CAM systems offer the advantage of knowing where all the lots are in the production cycle. The computer can stage the work to minimize disruption and keep a steady flow of product. Furthermore, JIT-CAM systems teamed with automatic delivery systems can be more efficient when the process stations are grouped rather than strung out in a line. This concept, called *work cells*, is more efficient, especially when a cell can run a number of different products. When a machine is inoperable in a linear layout, the line comes to a halt, and inventory builds up in front of the machine that is down.

## Quality Control and Certification—ISO 9000

SPC and other product/process/personnel quality programs fall under the general category of quality control. In the semiconductor industry, there are two broad quality segments: quality control (QC) and quality assurance (QA). QC generally refers to all of the techniques used to monitor and controls the *process* and wafer quality while they are in the process. QA refers to the efforts to monitor and assure that the customer is receiving product that meets required specifications.

In 1987, a system of quality requirements for organizations was introduced: the ISO 9000 series. Developed by the International Organization for Standardization (ISO), the requirements are more like guidelines, sometimes called an "umbrella standard,"[19] rather than a strict cookbook set of standards.

The ISO 9000 guidelines/standards guide the development and critique of a complete quality system for product, process, and management. Individual companies decide on how best to comply and implement programs. With certification, customers and vendors know that there is a comprehensive quality program backing the product. The European Community (EC) requires that all companies doing business in the EC have ISO 9000 certification. In the U.S., the American National Standards Institute (ANSI) and American Society for Quality Control (ASQC) have developed equivalent standards. Assisting semiconductor suppliers is a program sponsored by the Semiconductor Equipment and Materials International (SEMI).[20]

## Line Organization

Most fabrication areas are organized around the product-line concept. In this concept, fabrication areas are built to accommodate products with similar processing needs. Thus, there are bipolar lines and CMOS lines, and so forth. This arrangement makes for more efficient processing, since most of the machines are in use most of the time, and the staff can gain experience in processing a few products.

The staff of these lines are also fairly self-contained. The primary responsibility falls to the fabrication or product manager (Fig. 15.18). Reporting to this individual are an engineering supervisor, a production manager (or general supervisor), a design department, and the equipment maintenance group. The production manager is responsible for producing the finished wafers to specification, to cost, and to schedule. The engineering group is responsible for the developing of high-yield processes, documentation of the processes, and the daily sustaining of the line process. Both the production and engineering staffs are divided into groups focusing on a particular part of the process. This organization has the virtue of high focus on the fabrication area's primary goal of producing chips at a profitable level.

As the processes become more automated and arranged in process cells, small group organizational teams and responsibilities are emerging. A cell is attended by the operator(s), the equipment technicians, and the process engineer(s). These small groups make floor-level decisions with the information provided by the CIM system. However, few companies have formalized this arrangement with an organization structure, and the teams tend to exist as cross-department cooperatives.

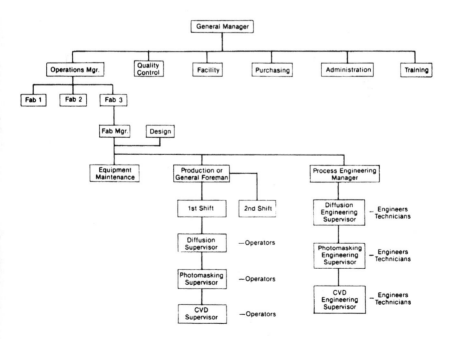

**Figure 15.18**  Typical semiconductor produce line organization.

## Review Questions

1. List the major factors determining the manufacturing cost of a wafer.
2. Rank the factors according to their level of influence.
3. Describe how WIP impacts profit.
4. Discuss how statistical process control can enhance wafer yields.
5. List the four levels of automation in a fabrication process.
6. Describe the philosophy and operation of a JIT inventory system.
7. What is the difference between CIM and CAM?
8. What is a process cell, and how does it compare with a traditional line layout?
9. List the production, process, and yield advantages of a totally automated fabrication area.
10. Draw an organizational chart of a product-line-managed fabrication area.

## References

1. J. G. Harper and L. G. Bailey, "Flexible Material Handling Automation in Wafer Fabrication," *Solid State Technology,* July 1984, p. 94.
2. D. Lam, "Minifabs Lower Barriers to 300 mm," *Solid State Technology,* January, 1999, p. 72.
3. W. Maly, "Growth of Silicon Cost and Resulting Changes in Manufacturing Paradigms," *Semiconductor International,* Cahners Publishing, June 1994, p. 158.
4. S. M. Sze, *VLSI Technology,* McGraw-Hill, New York, 1983, p. 5.
5. D. Seligson, "The Economics of 300 mm Wafers," *Semiconductor International,* January 1998, p. 52.
6. Jim Owens, "Semetech," Industry Strategy Symposium, *Semiconductor Equipment and Materials Institute (SEMI),* 1994 October 1994.
7. P. Burggraaf, "Applying Cost Modeling to Stepper Lithography," *Semiconductor International,* Cahners Publishing, February 1994, p. 40.
8. S. Shinoda, "Total Automation in Wafer Fabrication," *Semiconductor International,* September 1986, p. 87.
9. P. Singer, "The Thinking Behind Today's Cluster Tools," *Semiconductor International,* August 1993, p. 46.
10. *Ibid.*
11. *Ibid.*
12. *Ibid.*
13. M. Moslehi, "Single-Wafer Processing Tools for Agile Semiconductor Production," *Solid State Technology,* PennWell Publications, January 1994, p. 35.
14. R. Kerby and L. Novak, "ESH: A Green Fab Begins with You," *Solid State Technology,* January 1998, p. 82.
15. D. M. Campbell and Z. Ardehale, "Process Control for Semiconductor Manufacturing," *Semiconductor International,* June 1984, p. 127.
16. W. Levinson, "Statistical Process Control in Microelectronics Manufacturing," *Semiconductor International,* Cahners Publishing, November 1994, p. 95.

17. K. Levy, "Productivity and Process Feedback," *Solid State Technology,* July 1984, p. 177.
18. *Ibid.*
19. E. Hnatek, "ISO 9000 in the Semiconductor Industry," *Semiconductor International,* July 1993, p. 88.
20. P. Dunn, "The Unexpected Benefits of ISO 9000," *Solid State Technology,* March 1994, p. 55.

# 16

# Introduction to Devices and Integrated Circuit Formation

## Overview

Integrated circuits are composed of individual conductors, fuses, resistors, capacitors, diodes, and transistors. The operation and formation of the basic each is explored. The formation of the major integrated circuits from the components is also presented in this chapter.

## Objectives

Upon completion of this chapter, you should be able to:

1. Sketch and identify the structural parts of the individual components of an integrated circuit.

2. Explain the role and different isolation structures used for integrated circuits.

3. Sketch and identify the operation of a bipolar and MOS transistor.

4. List the types and advantages of the different MOS gate structures.

5. Sketch and identify the parts of a bi-MOS circuit.

## Semiconductor-Device Formation

The previous chapters have focused on the individual processes used to make semiconductor devices (also referred to as *components* or *circuit components*) and integrated circuits. It is assumed that the reader has already read about (or is familiar with) the processes and has a good understanding of the basic structure and electrical performance of the individual components as explained in Chapter 14. There are literally thousands of different semiconductor device structures. They have been developed to achieve specific performances, either as discrete components or in integrated circuits. However, there are basic structures required for each of the major device and circuit types. In this chapter, these basic structures are examined. Mastering them is essential to understanding the many variations and innovative structures that abound in the semiconductor world. The circuit components are:

- Resistors
- Capacitors
- Diodes
- Transistors
- Fuses
- Conductors

## Resistors

Resistors have the effect of limiting current flow. This is accomplished by the use of dielectric materials or high-resistivity portions of a semiconductor wafer surface. In semiconductor technology, resistors are formed from isolated sections of the wafer surface, doped regions, and deposited thin films.

The value of a resistor (in ohms) is a function of the resistivity of the resistor and its dimensions (Fig. 16.1). The relationship is

$$R = \rho LA$$

where  $\rho$ = resistivity
$L$ = length of resistive region
$A$ = cross-sectional area of the resistive region

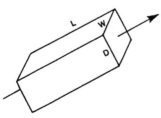

$$R = \rho \frac{L}{A} = \rho \frac{L}{W \times D}$$

Figure 16.1 Relationship of resistance to resistivity and dimensions.

The area *(A)* becomes $W \times D$, where $W$ = width of the resistor and $D$ = depth of the resistive region. For doped resistors, the length and width are the surface pattern openings and the depth is the junction depth.

It should be obvious that every doped region is also a resistor, and the basic resistor formula governs electrical flow. A conductor is simply a resistor with a low resistance. The conceptual importance of Ohm's law is that the electrical resistance of any region in the device or circuit is altered by any change in dimensions or change in the doping level.

**Doped resistors.** Most of the resistors in integrated circuits are formed by a sequence of an oxidation, masking, and doping operation (Fig. 16.2). A pattern is opened in the surface oxide. Typical resistor shapes are dumbbells (Fig. 16.3) with the square ends serving as contact regions and the long skinny region in between serving the resistor function. The resistance of this region is calculated from the sheet resistance of the region and the number of squares (□s) contained in the region. The number of squares is calculated by dividing the length by the width.

After doping and a subsequent reoxidation, contact holes are etched in the square ends to contact the resistor into the circuit. A resistor is a two-contact, no-junction device. The term *no-junction* means that

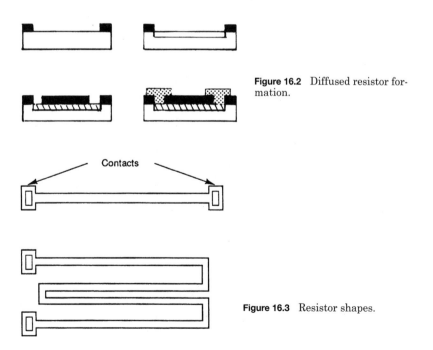

**Figure 16.2** Diffused resistor formation.

**Figure 16.3** Resistor shapes.

the current flows between the contacts without crossing an N-P or P-N junction. But the junction serves to confine the current flow in the resistive region.

Resistors doped by ion implantation have more controlled values than those in diffused regions. Doped resistors can be formed during any of the doping steps performed during the fabrication process. A bipolar base mask will have the base pattern and a set of resistor patterns. In MOS circuits, resistors are formed along with the source/drain doping step. The resistor has the same doping parameters (sheet resistance, depth, and dopant quantity) as the transistor part.

**EPI resistors.**    A resistor can be formed by isolating a section of an epitaxial region (Fig. 16.4). After surface oxidation and contact hole masking, what is left is a three-dimensional region functioning as a resistor.

**Pinch resistors.**    Ohm's law shows that the cross-sectional area of the resistor is a factor in its value (Fig. 16.5). One way to reduce the cross-sectional area (and increase the resistance) is to dope the resistor region and then do another doping of the opposite conductivity type. This occurs in bipolar processing when a resistor region is formed during the P-type base doping with a "pinched" cross section formed from a subsequent N-type region formed along with the emitter doping.

**Thin film resistors.**    Doped resistors don't always have the resistance control needed in some circuits and are poor performers in radiation environments. Radiation, such as found in space, generates unwanted holes and electrons that allow the current to leak across the confining junction. Resistors formed from deposited thin films of metal do not have this radiation problem.

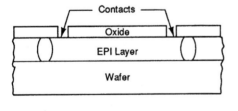

**Figure 16.4** Epitaxial layer resistor.

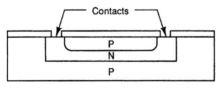

**Figure 16.5** N-type resistor "pinched" with P-type doped region.

The resistors (Fig. 16.6) are formed by either film deposit and masking sequences or by lift-off techniques. After resistor formation of the wafer surface, it is "wired" into the circuit by contact between the resistor ends and the leads of the conducting metal. Nichrome, titanium, and tungsten are typical resistor metals.

## Capacitors

**Oxide-silicon capacitors.**    Silicon planar technology is based on a silicon wafer with a grown silicon dioxide layer on top or on a silicon dioxide layer grown and an epitaxial layer. If a conducting metal lead lies on top of the oxide, a simple capacitor is formed (Figure 16.7). Recall that a capacitor is formed with a dielectric layer sandwiched between two electrodes. In fact, this structure is a metal-oxide-metal (MOS) capacitor structure. However, the oxide thickness has to be thin enough (about 1500 Å[1]) for the structure to act as a capacitor. The top electrode is also called a *cell plate*. A bottom electrode is also called the *storage node*.

A capacitor is a device that stores charge. A battery is a capacitor. When a voltage is applied to the metal, a charge builds up in the wafer layer under the oxide (Fig. 16.7). The amount of charge is a function of the oxide thickness, the dielectric constant of the oxide, and the area, as defined in the metal top plate. Capacitors of this structure are called parallel-plate, monolithic, or MOS (after the metal oxide semiconductor materials of the sandwich).

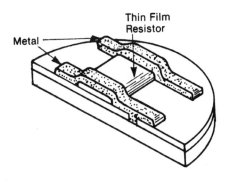

**Figure 16.6**  Formation of a thin film resistor.

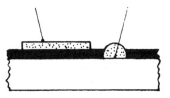

**Figure 16.7**  Monolithic capacitor.

In dense circuits, an oxide/nitride/oxide (ONO) dielectric sandwich is used. The combination film has a lower dielectric constant, allowing a capacitor area smaller than a conventional silicon dioxide capacitor. Capacitor function acts two ways in ICs. In some circuits, capacitors are formed specifically to store charge. However, capacitor structures are formed whenever metal lines lie over a dielectric on top of a layer of silicon (or other semiconductor material). In this situation, we do not want the capacitor to store charge that may interfere with the circuit operation. In this situation, the dielectric layer needs to be thick enough to prevent capacitor charge storage or the use of high dielectric materials (high-k). See Chapter 12 for reference.

**Junction capacitors.**   A capacitor is formed at every junction in a device. When a voltage is applied across any junction, carriers on each side of it move away from the junction, leaving a depleted region (Fig. 16.8). This depleted region acts as a capacitor in the device or circuit.

The value of this junction capacitance must be taken into account when the circuit is designed. Some circuits actually use junction capacitors as part of the circuit design. In some circuits, the natural junction capacitance has the effect of slowing up the circuit operation. This is due to the time required to "fill up," or charge, the depleted region before current flows. A finite time is also required for the various junction capacitors to discharge. Both of these times affect circuit switching and operational speeds.

**Trench capacitors.**   Preservation of wafer surface area is always a design criterion. One of the problems with oxide-metal capacitors is their relatively large area. Trench (or buried) capacitors solve the problem by creating a capacitor in a trench etched vertically into the wafer surface (Fig. 16.9). The trenches are etched either isotropically with wet techniques or anisotropically with dry etch techniques. The trench sidewalls are oxidized (the dielectric material) and the center of the trench filled with deposited polysilicon. The final structure is "wired" from the surface, with the silicon and polysilicon serving as the two electrodes with the silicon dioxide dielectric between them. Other dielectric materials may be used in place of the silicon dioxide to increase performance.

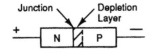

**Figure 16.8**  Depletion layer junction capacitor.

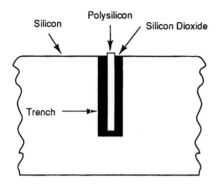

**Figure 16.9**  Trench capacitor.

**Stacked capacitors.** Another alternative for conserving surface real estate is to build *stacked capacitors* on the wafer surface. This effort has been driven by the need for small, high-dielectric capacitors for dynamic random access memory (DRAM) circuits. The storage part of a DRAM cell is a capacitor. In DRAM cells, the bottom electrode is usually polysilicon or hemispherical grain polysilicon (HSG). Capacitor shapes can be planar, cylindrical, or with "fins."[2]

A typical structure is shown in Figure 16.9. Capacitor dielectric materials under consideration include tantalum pentoxide ($Ta_2O_5$) and barium strontium titantate ($BaSrTiO_3$ or BST). The latter is a *ferroelectric* material.[3] Ferroelectric refers to iron containing materials and, in the world of electronics, represent dielectrics with improved speeds over conventional silicon-compatible materials. Another ferroelectric material is $PbZ_{1-x}T_xO_3$ or PZT.

The top electrode materials may be TiN, WN, Pt, polysilicon, or one of the other semiconductor conducting materials.

### Diodes

**Doped diodes.** A diode is a two-region (two-contact) device separated by a junction. A diode either allows current to pass easily or acts as a current block. Which function it performs is determined by the voltage polarity, called *biasing* (Fig. 16.10). When the current voltage is the same as in the diode region, the diode is in forward bias, and the current flows easily. When the polarities are reversed, the diode is reverse-biased, and the current is blocked. A reverse-biased diode can be forced into a conducting state by raising the current voltage until the junction goes into *breakdown*. This condition is temporary; when the voltage is reduced, the diode once again becomes a blocking device (see Chapter 14). Diodes are used in circuits to steer the current around the circuit. By proper choice of the circuit current polarities

**Forward Bias**

**Reverse Bias**

**Figure 16.10** Forward and re-verse biasing.

and the correct diode polarities, the current is allowed to pass into some branches of the circuit and is blocked out of others. A planar diode is formed from a doped region and two contacts on either side of the junction where it intersects the surface (Fig. 16.11). Diodes are usually formed along with transistor doping steps. Thus, in bipolar circuits, there are base-collector diodes and emitter-base diodes. In MOS circuits, most of the diodes are formed with the source-drain doping step.

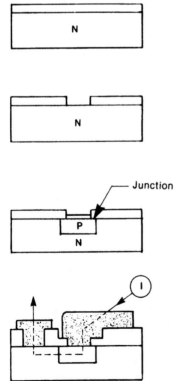

**Figure 16.11** Formation of a P/N planar diode.

**Schottky barrier diodes.**    In 1938 (ten years before the invention of the transistor), W. Schottky[4] discovered that whenever a metal is in contact with a lightly doped semiconductor, a diode is formed (Fig. 16.12). This diode has a faster forward time (it responds faster) and operates with a lower voltage than a doped silicon junction diode. Metal contacts to highly doped regions (greater than $5 \times 10^{17}$ atoms per cubic centimeter) are regular ohmic contacts. This is the situation for the majority of contacts in a silicon circuit. This Schottky diode effect is taken advantage of in some NPN bipolar transistors. The structure and effect are explained in the section on bipolar transistors.

### Transistors

**Transistor operational analogy.**    A transistor is a three-contact, three part, two-junction device that performs as a switch or an amplifier. An often used analogy to explain the role of the parts and the operation of a transistor is the water flow system in Fig. 16.13. The flowing water represents current flow. In this system, one part is the source of the water (the tank), the valve controls the flow, and the bucket collects the water. The system can be operated as a switch simply by turning the valve on and off. It can even be imagined in an amplifier role. Consider the valve as a high-mechanical advantage miniature water wheel activated by a small external stream of water. A small trickle onto the valve wheel could open the valve to allow a large flow through the system. If the whole system was enclosed so that an observer saw only the trickle going in and a large flow coming out, such an observer might conclude that the system was *amplifying* the water trickle.

**Bipolar transistors.**    The same basic parts and functions are present in solid-state transistors. A bipolar transistor is shown as both a simple block arrangement and in a double-doped planar form in Fig. 16.14. The current flows from the emitter region (tank) through the base (valve) into the collector (bucket). When there is no current to the base, the transistor is turned off. When it is on, the current flows. It only takes a small current to turn the base on enough to allow current

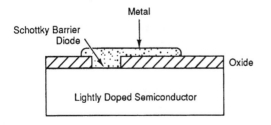

**Figure 16.12**   Schottky barrier diode.

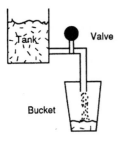

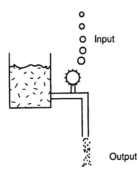

**Figure 16.13** Water analogy of transistor operation.

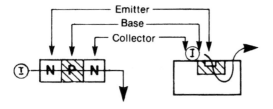

**Figure 16.14** Bipolar transistor operation.

flow through the whole transistor. The size of the base current regulates the larger amount of current flowing through the transistor (called the *collector current*). There is an amplification of the base current to the collector current. The base current in effect changes the resistance of the base region. In fact, the term *transistor* comes from an early term for a bipolar transistor: *transfer resistor*. During operation, both positive and negative currents flow in the base, hence the term *bipolar*—literally, two polarities.

The amplification property, called *gain* or *beta*, is numerically the result of dividing the collector current by the base current (see Chapter 14). For efficiency, the emitter region has a higher doping concentration than the base, and the base doping is higher than the collector. A typical doping concentration versus distance plot is shown in Fig. 16.15.

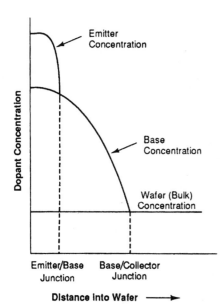

**Figure 16.15** Dopant concentration profile of bipolar transistor.

Most bipolar circuits are designed with NPN transistors. NPN represents the respective conductivity types of the emitter, base, and collector. Some applications require PNP transistors, with many of them being formed laterally (Fig. 16.16). NPN transistors are more efficient because of the ease (higher mobility) of electrons in the N-type regions.

Bipolar transistors feature fast switching speeds. The speed is governed by a number of factors, of which the most important is the base width. Applying common sense, it stands to reason that the shorter the distance an electron or hole has to travel, the less time it will take.

Bipolar transistors can switch on and off in as little as a billionth of a second. To achieve this speed, the transistor is maintained at an "on"

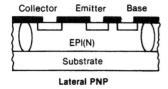

**Lateral PNP**

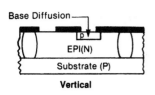

**Figure 16.16** Lateral and vertical PNP transistors.

**Vertical**

state. This means that the base always has power applied, which is a downside of bipolar-based circuits. Another penalty for this necessary condition is a buildup of heat in the transistor. This heat will eventually affect circuit operation and is the reason for the cooling fans and air-conditioned rooms of earlier bipolar-based computers.

**Schottky barrier bipolar transistors.** The Schottky barrier diode principle mentioned in a previous section is put to use in some bipolar transistors (Fig. 16.17). The construction requires that the base contact be extended into the collector region. When covered with metal, a Schottky diode is formed between the base and collector, which results in a faster-responding transistor. In a circuit, the time required to turn the transistor on and off (switching speed) is critical when millions of transistors are operating.

### Field-effect transistors (FETs)

**Metal-gate MOSFET.** As early as 1948, William Shockley noted another type of transistor operation—a field-effect-actuated transistor current flow. That effect was developed into the field-effect transistor, with the MOS structure now the most popular design. A MOS transistor, like a bipolar (Fig. 16.18), has three regions, three contacts, and two junctions, but in a different structure. There is a similar analogy to the water system described previously. Current travels from the source region (tank), through the dielectric gate material (valve), and into the drain (bucket) before exiting the device.

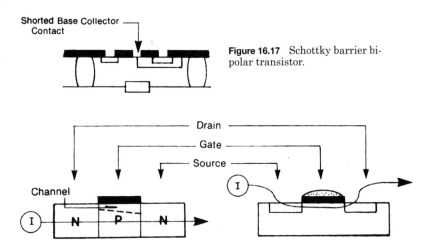

Shorted Base Collector Contact

**Figure 16.17**   Schottky barrier bipolar transistor.

Drain

Gate

Source

Channel

**Figure 16.18**   MOS transistor operation.

A MOSFET gate controls current flow by a different mechanism from that of a bipolar base. The MOS structure shown in Fig. 16.19 is a simple metal-gate type capacitor and operates the same as a capacitor. When a voltage (the gate voltage) is applied to the gate through the gate metal, a *field effect* takes place in the surface of the semiconductor. The effect is either a buildup of charge or a depletion of charges in the wafer surface under the top plate. Which event occurs depends on the doping conductivity type in the wafer under the gate and the polarity of the gate voltage.

The buildup or depletion of charge creates a channel under the gate that connects the source and drain. The surface of the semiconductor is said to be *inverted*. The source is biased with a voltage, and the drain is grounded relative to the source. In this condition, a current starts to flow as the inverted surface creates an electrically connecting channel. The source and drain are essentially shorted together. Applying more voltage to the gate increases the size of the channel, allowing more current to flow through the transistor (see Chapter 14). By controlling the gate voltage, a MOS transistor can be used as a switch (on/off) or as an amplifier. However, MOS transistors are voltage amplifiers, unlike the current amplification of bipolar transistors.

If the source and drain are N-type formed in a P-type wafer, the channel must be of N-type for conduction to occur. This type of MOS transistor is called N-channel. MOS transistors with P-type sources and drains are P-channel. Most high-performance MOS circuits are built around N-channel transistors due to the higher mobility of electrons in silicon. The mobility makes N-channel transistors faster, and they consume less power than P-channel circuits. They often are referred to as NMOS transistors. Figure 16.20 shows the major steps in the formation of an N-channel metal-gate MOS transistor.

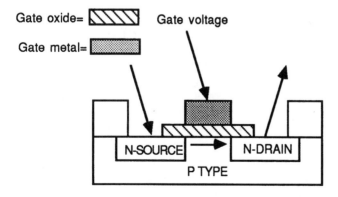

**Figure 16.19**   Metal gate MOS transistor.

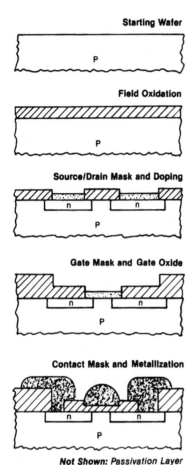

**Starting Wafer**

P

**Field Oxidation**

P

**Source/Drain Mask and Doping**

n          n

P

**Gate Mask and Gate Oxide**

n          n

P

**Contact Mask and Metallization**

n          n

P

*Not Shown: Passivation Layer*

**Figure 16.20** N-channel metal gate MOS process.

**Silicon gate MOS.**    A certain amount of voltage must be applied to the gate metal before the channel forms. This voltage is called the *threshold voltage* or *V*. The value of the threshold voltage is an important and critical circuit parameter. A lower *V* means fewer power supplies and faster circuits.

A primary parameter that determines the threshold voltage is the *work function* between the gate material and the doping level in the semiconductor. The work function can be thought of as a kind of electrical compatibility. The lower the work function, the lower the threshold voltage, the lower the power required to run the circuit, and so on.

Deposited doped polysilicon has a lower work function than aluminum as a MOS gate material and has become the standard gate electrode material for MOS transistors. The formation of the transistor is shown in Fig. 16.21. The polysilicon is heavily doped N-type to reduce its resistance. Thus doped, it serves as the gate electrode and as a circuit conduction line. A polysilicon gate can withstand subsequent high-temperature processing without degradation.

An additional benefit of the silicon-gate process is the self-aligned gate. In the metal-gate process sequence, a hole for the gate oxide must be patterned between the source and drain. To ensure that the source and drain are bridged by the gate, overlap for alignment tolerances must be allowed. This results in some overlap of the gate into the source and/or drain. The overlap becomes a region of unwanted capacitance. In the silicon-gate process, the gate is formed first and acts as a mask to locate the source and drain. Thus, whatever the gate placement, the source and drain *self-align* to it.

Other factors, besides the gate metal material, affecting the gate threshold voltage and operation are:

- Gate oxide thickness

- Gate material (dielectric constant)

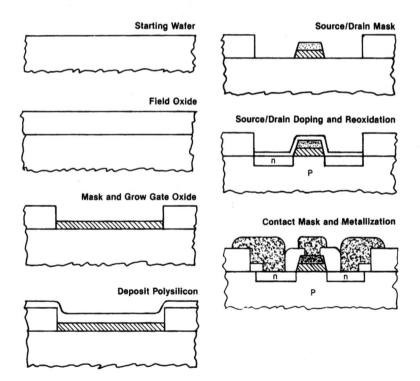

**Figure 16.21**  Silicon gate MOS process steps.

- Source-drain separation (channel length)
- Gate doping level
- Sidewall capacitance of the doped source and drain regions

In the section on capacitors, the development of high-dielectric-constant materials for higher performance in a smaller area were discussed. These same factors apply to MOS gate materials. The drive to 0.18 μm design rules will require gate thicknesses in the 90- to 60-Å range.[5]

The thinner the gate oxide, the faster the device and the lower the threshold voltage. Current gate oxide thicknesses for production devices are in the 100- to 300-Å range. The technology roadmap projects that, as the industry enters the 0.35- and 0.18-μm design rule era, gate thicknesses will fall into the 90- to 60-Å range.[5] In this range, maintaining thickness control, pinhole-free gates, and small-area gates has driven the development of new gate materials. They are the same materials discussed in the section on capacitors. The new materials are oxide/nitride sandwiches (ONO), oxide tantalum pentoxide (OTa2O3), and amorphous silicon film stacks.

Channel lengths also affect the speed. Channel lengths have continually shrunk to the submicrometer range. In self-aligned structures, the channel length is established by the gate width. The gate doping level influences the threshold voltage by modifying the work function difference between the gate metal and surface. Ion implantation, with its ability to dope through thin oxides, is often used to set the gate doping level. Sidewall capacitance of the doped source and drain also serves to slow up the operation of the device, as enough charge must be built up to overcome the junction capacitance.

**Polycide-gate MOS.** MOS development was impeded by the inability of the industry to grow noncontaminated and thin oxides in the 1960s. Contamination, especially the mobile ionic variety, interferes with the field effect, making very unreliable gates. In fact, clean gate oxides and the oxide-silicon interface are so well understood and effective that gate oxide replacement is proceeding very cautiously.

Consequently, the quest for a higher-efficiency gate has led to the polycide structure (Fig. 16.22). The gate sandwich retains the thin oxide on the wafer surface topped by a layer of polysilicon. The polysilicon provides a low work function (and lower threshold) gate, and the reliable polysilicon-oxide interface is preserved. The new layer is a refractory metal silicide on top of the polysilicon. The silicide makes a low contact resistance with the polysilicon (as compared with aluminum) and reduces the overall sheet resistance of the polyside sandwich.

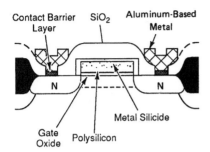

Contact Barrier   SiO₂   Aluminum-Based
Layer                          Metal

N                          N

Metal Silicide

Gate
Oxide   Polysilicon

**Figure 16.22**  Polycide gate structure.

**Salicide-gate MOS.**    The self-aligned process that uses the polycide-gate structure is called a *salicide* gate. Its formation is illustrated in Fig. 16.23. The process combines the best features of a polysilicon gate with self-alignment. The source and drain are lightly diffused around the polysilicon gate. Then a layer of silicon dioxide is deposited and anisotropically etched to form spacers on the side of the gate. These spacers act as ion implantation masks for a subsequent heavier doping of the source and drain. The more lightly doped "finger" under the gate is called a lightly doped drain extension (LDD) and is necessary for channel lengths of less than 2 μm.[6] After ion implantation, the refractory metal is deposited, and the silicide is formed by reaction with the underlying polysilicon layer by an alloy step. The final step is the removal of the unreacted refractory metal from the wafer surface.

The LLD process illustrated puts a lightly doped fingers in both the source and drain areas. There are processes designed to create asymmetrical LLD structures that place the finger only in the drain region.[7]

**Diffused MOS (DMOS).**    DMOS refers to a diffused MOS structure (Fig. 16.24). The channel length is established by two diffusions through the same opening. As the second diffusion is taking place, the

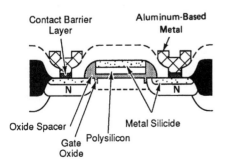

Contact Barrier              Aluminum-Based
Layer                                  Metal

N                          N

Oxide Spacer            Metal Silicide

Gate   Polysilicon
Oxide

**Figure 16.23**  Salicide-gate structure.

Diffusion #1

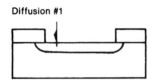

Diffusion #2    Source    Gate

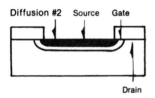

Figure 16.24    DMOS structure.

Drain

first moves laterally to the sides. The second diffusion functions as the source, and the bulk semiconducting material of the wafer functions as the drain. The difference between the two diffusion widths is the channel length of the transistor. DMOS structures feature narrow and well controlled gate widths.

**Memory MOS (MMOS).**    MMOS is a structure that provides a more or less permanent storage of the charge in the gate region. The storage is provided by a thin layer of silicon nitride between the wafer and the gate oxide (Fig. 16.25). When the gate is charged to store data, the silicon nitride layer traps and retains it. This type of transistor is used in nonvolatile circuits where protection against memory loss is important (see Chapter 17).

**Junction field-effect transistors (JFETs).**    A junction field-effect transistor (Fig. 16.26) is similar in construction to a MOSFET but has a junction formed under the gate. During operation, the current flows *under*

Metal
Oxide

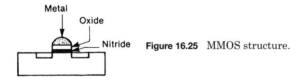
Nitride    Figure 16.25    MMOS structure.

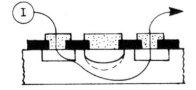

Figure 16.26    Junction field-effect transistor.

the diffused region from the source to the drain. As the gate voltage is increased, a region depleted of charge (the depletion region) spreads under the junction toward the N-type and P-type interface. The depleted region will not support current flow and has the effect of restricting the current flow as it increases in depth.

A JFET operates opposite from a MOS transistor. In the MOS version, increasing the gate voltage *increases* the current flow. In a JFET, increasing the gate voltage *decreases* the current flow. JFETs are a standard gallium arsenide device. The transistor is formed in an N-type GaAs layer that is on top of a semi-insulating GaAs wafer (Fig. 16.27).

## Alternative (Scaled) Transistor Designs

A standard approach to denser circuits is *scaling*, also called a *die shrink*. Scaling starts with a proven design and reduces the dimensions. However, scaling can introduce new problems, such as increased leakage. Many of the new materials are designed to address leakage and other scaling issues. Alternative transistor structures are also being explored. Ideas include double gates (DGs) and ultra-thin body (UTB) for MOSFET devices. Schematics are illustrated in Fig. 16.28. Other ideas include vertical transistor structures and a fin structure, called a FinFet. In this structure, the gate is build vertically, and the source and drain build up on either side of the fin.

**Metal semiconductor field-effect transistor (MESFET).**    The MESFET is the basic GaAs transistor structure (Fig. 16.29). The MESFET operates in the same manner as the JFET, but the gate metal is deposited directly onto the N-type GaAs layer.

### Fuses and conductors

The formation of fuses and surface wiring conductors were detailed in Chapter 14. In very dense circuits, precious surface area is preserved

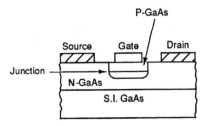

**Figure 16.27**   GaAs JFET.

**Bulk MOSFET**

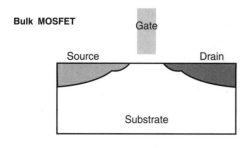

**Double-gate MOSFET**

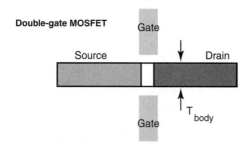

**Ultra-thin Body MOSFET**

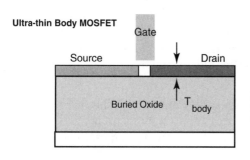

**Figure 16.28** Alternate MOSFET devices. *(Courtesy of Semiconductor International.)*

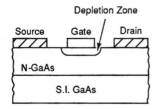

**Figure 16.29**  GaAs MESFET.

by the use of subsurface *underpass conductors*. These are created from heavily doped regions formed under surface wiring leads.

The issues and challenges are depositing and patterning multiple metal layers. Key technologies are planarization, step coverage, and plug filling. Development of low-contact-resistance metal film and film sandwiches will be required in the 0.5-micron design rule era.

## Integrated-Circuit Formation

Integrated circuits contain all the components described in the previous sections. The components are formed in specific sequences with the process flows designed around the transistor in the circuit. The process designer will attempt to form as many component parts as possible with each doping step.

Circuits are designated by the transistor type. A bipolar circuit means that the circuitry is based on bipolar transistors. MOS circuits are based on one of the MOS transistor structures. For the first 30 years of the semiconductor industry, the bipolar transistor and bipolar circuits were the structures of choice. Bipolar transistors had fast speeds (switching times), control of leakage currents, and a long history of process development. These qualities fit nicely into the logic, amplifying, and switching circuits that were the first offerings of the industry. These circuits handled the computational requirements of the growing computer industry. The internal memory functions of the early computers were handled by core memories. These memories were limited in capacity and slow. Much of the information needed was stored outside the computer on tape, disks, or punch cards. While bipolar memory circuits were available, they could not compete economically with core memories.

MOS transistor circuits held the promise of fast, economical, solid state memory, but early metal-gate MOS circuits suffered from high leakage currents and poor parameter control. Even so, the built-in advantages of MOS transistors drove the development of MOS memory circuits. The advantages are smaller dimensions, which allow denser circuits, and relatively faster switching speeds. Yields tend to be higher on smaller-dimensioned circuits, since a given defect density will affect fewer transistors and components.

Perhaps the biggest density factor advantage of MOS components is the smaller area required for isolation of adjacent components. The various isolation schemes used are discussed in the next sections. Another advantage is low-power operation. First, MOS transistors sit in the circuit in the "off" mode, not soaking up power or generating heat like bipolar transistors, which must be all the time to be in a "ready" state. Second, MOS transistors, being voltage-controlled devices, require a lower power to operate. CMOS circuits are an integrated circuit design that reduces power requirements even further.

An initial advantage of MOS circuits was fewer processing steps and smaller die sizes, which made for lower processing costs and higher yields. These advantages have disappeared as MOS circuits have evolved to VLSI/ULSI size with the additional steps required to fabricate CMOS circuits. In general, the faster switching speed of bipolar circuits has made them favored for logic circuits. MOS circuits, with their smaller component dimensions and lower power requirements, have been incorporated into memory circuits. By the 1980s, these traditional uses blurred, with CMOS technology the preferred system for most circuit designs. These topics are addressed further in Chapter 17.

### Bipolar circuit formation

**Junction isolation.** The bipolar transistor structure and basic performance have been illustrated. However, combining transistors with the other devices to form a circuit requires additional structures. They include an isolation scheme and a low-resistance collector contact. If two transistors or other devices are fabricated next to each other, they will not work, because they are electrically shorted together (Fig. 16.30). An early challenge of bipolar IC designers was a way to *isolate* the various circuit components. This need led to the epitaxial (epi) layer bipolar structure (Fig. 16.31).

The process starts with a P-type wafer into which an N-type diffusion is made (the flow diagram does not show the oxidation and masking steps required to create the diffused layer). After the diffusion step, an N-type epitaxial layer is deposited, leaving the N-type diffused region "buried" under the epitaxial layer. The N-type region is known as a *buried layer* or as the subcollector of the transistor. Its function is to provide a lower-resistance path for the collector current as it flows out of the base region on its way to the surface collector contact.

After the deposition of the epitaxial layer, it is oxidized, and a hole is opened up on each side of the buried layer. A P-type doping step is performed deep enough to reach the P-type wafer surface. The doping step divides (*isolates*) the epitaxial layer into N-type islands, each surrounded on the sides (the doped regions) and the bottom (the P-type wafer) by P-type doped regions. Components formed on the surface of each of the islands are electrically isolated from each other (Fig. 16.32). The electrical isolation occurs because the N-P junction is "wired" into the circuit to function in the reverse bias mode; that is, no

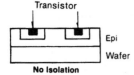

**Figure 16.30** Adjacent bipolar transistors with common collectors.

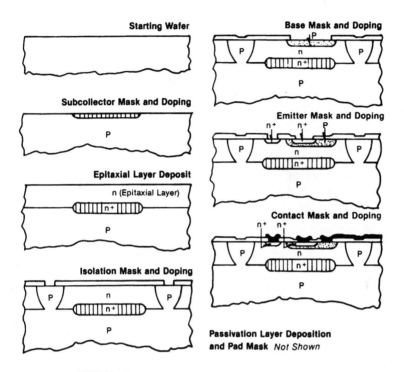

**Figure 16.31**  NPN bipolar process.

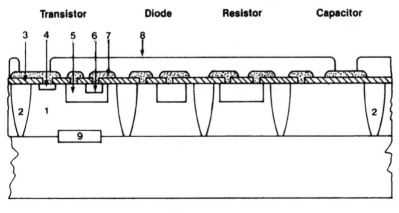

| 1. EPI and Collector | 4. Collector Contact | 7. Metalization |
| 2. Isolation | 5. Base | 8. Passivation Layer |
| 3. Surface Oxide | 6. Emitter | 9. Buried Layer |

**Figure 16.32**  Bipolar structures.

current crosses the junction. This scheme is called *junction isolation* or *doped junction isolation.*

**Dielectric isolation.**  In high-radiation environments, such as outer space or in the vicinity of atomic weapons, doped junctions produce holes and electrons that flow in the device and compromise the junction functions. Besides causing circuit component failure, the radiation swamps out the isolation protection of the doped regions. Dielectric isolation schemes provide the necessary electrical isolation and radiation protection.

The process (Fig. 16.33) starts with the etching of pockets or trenches in a wafer surface. The etching can be either an isotropic wet etch or an anisotropic dry etch. Isotropic etch profiles follow the orientation structure of the wafers. Dry etch processes allow the shaping of

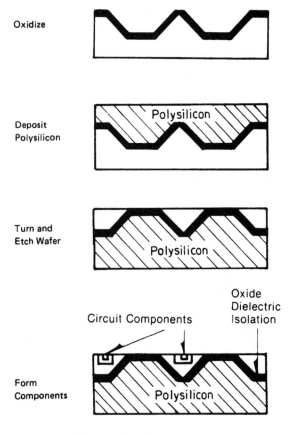

**Figure 16.33**  Dielectric isolation.

the trenches. One goal of the etch step is to minimize the area of the pocket on the wafer surface. Wide pockets limit the packing density of the circuit.

After etch, the pocket sides are oxidized and backfilled with deposited polysilicon. Next, the wafer is turned over, and the silicon of the wafer is lapped until the oxide layer is reached. These steps leave a wafer surface containing oxide-isolated pockets of the original single silicon material. The circuit components are fabricated in the silicon pockets, with each pocket being isolated on three sides by the layer of silicon dioxide. The dielectric property of the silicon dioxide prevents leakage currents in both normal and radiation environments.

**Localized oxidation of silicon (LOCOS).**    Junction isolation takes up valuable surface real estate, and dielectric isolation is area consuming and requires extra processing steps. A popular alternative is LOCOS (Fig. 16.34). The process starts with a layer of silicon nitride deposited and etched on the wafer surface. Active devices will be formed in the area defined by the silicon nitride layer. In the partially recessed version, an oxidation follows. Oxygen will not penetrate the silicon nitride to cause the oxide to grow on the exposed silicon surface. The silicon for the silicon dioxide comes from the wafer surface and, because silicon dioxide is less dense than silicon, the oxide layer forms slightly above the original silicon surface. It is *partially recessed* rela-

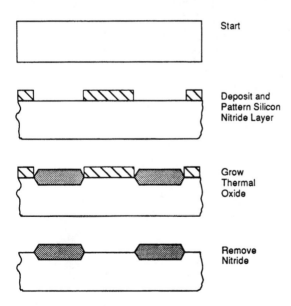

Start

Deposit and
Pattern Silicon
Nitride Layer

Grow
Thermal
Oxide

Remove
Nitride

**Figure 16.34**   LOCOS process.

tive to the wafer surface. After the oxidation, the silicon nitride is removed, leaving the area free for the formation of circuit components. A variation of the LOCOS isolation process is illustrated in Fig. 16.35*b*. In this process, the silicon surface is etched *before* the oxidation. By calculating the proper removal amount, the subsequent oxidized layer is *fully recessed* to the original surface. Bipolar schemes using LOCOS isolation are shown in Fig. 16.35.

**Collector contact.**   Note that, in the cross section (Fig. 16.32) of a circuit integrated bipolar transistor, there is a doped region under the collector contact. This doped region is put into the surface along with the emitter N-type doping. The emitter is usually designated $N^+$ to indicate that it is highly doped. The $N^+$ region under the collector contact is present to create a lower resistance between aluminum metallization and the silicon of the collector.

### MOS integrated circuit formation

**MOS LOCOS isolation.**   While MOS transistors are somewhat self-isolated (because they don't share common electrical parts), there is some electrical leakage between devices, especially as the spacing becomes very close. Isolation is necessary to block leakage currents. The structures are generically called *channel stops*.

LOCOS is the preferred isolation technique. However, there are several problems to overcome to make a LOCOS structure effective in advanced circuits.[8] One problem is the *bird's beak* spur that grows under the edge of the blocking silicon nitride layer (Fig. 16.36). The beak takes up real estate, effectively enlarging the circuit. At a performance level, it induces stress damage in the silicon during the oxidation step. The stress comes from the mismatch in thermal expansion properties between silicon nitride and silicon. A solution for the stress problem is

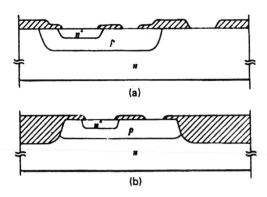

**Figure 16.35** (*a*) Conventional bipolar transistor and (*b*) LOCOS isolated bipolar transistor. (*Source:* VLSI Fabrication Principles, *Ghandhi.*)

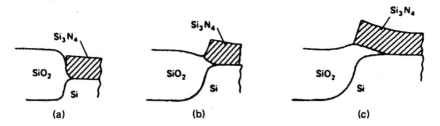

**Figure 16.36** Bird's beak growth. (*a*) No pre-etch, (*b*) 1000-Å pre-etch, and (*c*) 2000-Å pre-etch. (*Source:* VLSI Fabrication Principles, *Ghandhi.*)

to grow a thin oxide layer under the silicon nitride film. This is called a *pad oxide.*

Minimizing the bird's beak and reducing the stress in the active device region has spurred a number of variations on the LOCOS process. One, called SWAMI, was developed by Hewlett Packard (Fig. 16.37).[9] The process starts the same as a standard LOCOS. After the nitride

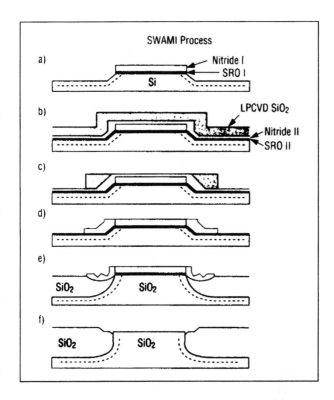

**Figure 16.37** SWAMI process. (*From* SST, *June 1993, p. 97.*)

and pad oxide, grooves (or trenches) are etched with a orientation sensitive etchant. On ⟨100⟩-oriented material, the groove side walls are at a 60° angle, which reduces stress in the silicon. Next, another stress relieve oxide (SRO) layer is grown and covered by a layer of silicon nitride, which provides conformal coverage. An LPCVD silicon dioxide completes the sandwich before etching (c). This oxide protects the silicon nitride layer from removal. Finally, the field oxide (FOX) is grown. The length of the nitride layer governs the size of the bird's beak encroachment. Removal of the original nitride/SRO and the second nitride leaves a somewhat planarized surface for device formation. LOCOS isolation schemes usually include an ion implanted layer between active regions to provide further channel stop capabilities.

**Trench isolation.**  Trench isolation is also used for MOS circuits (Fig. 16.38). The procedure is the same as forming trench capacitors. One version, called *shallow trench* isolation, addresses the bird's beak problem that comes with standard LOCOS isolation. In this structure, a shallow trench is etched between the components and filled with a deposited dielectric (Fig. 16.39). After sidewall oxidation and dielectric fill of oxide, a CMP step replanarizes the surface.

**CMOS.**  Complementary MOS (CMOS) is a MOS circuit formed with both N-channel and P-channel transistors. CMOS has become the

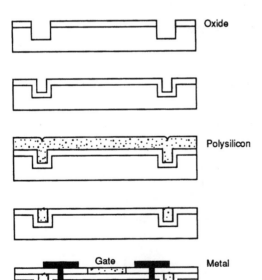

Oxide

Polysilicon

Metal

**Figure 16.38** MOS trench isolation.

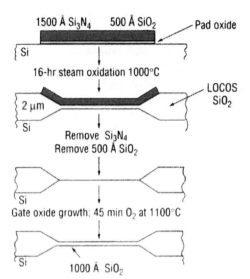

**Figure 16.39**  Shallow trench process. *(Reprinted from the September 1997 edition of* Solid State Technology, *copyright 1997 by PennWell Publishing Company.)*

standard circuit for many applications. It is the CMOS circuit that has made possible digital watches and hand-held calculators. It allows circuits on one chip that would require several chips using N-channel and P-channel only circuits. CMOS circuits also use lower amounts of power than comparable circuits.

CMOS structures (Fig. 16.40) are formed by first fabricating an N-channel MOS transistor in a deep P-type well formed in the wafer surface. After N-channel transistor formation, a P-channel transistor is fabricated. The transistor structures are silicon gate or other advanced structures. CMOS processing uses the most advanced techniques, because smaller, more densely packed, and higher-qual-

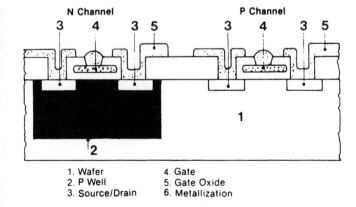

1. Wafer          4. Gate
2. P Well         5. Gate Oxide
3. Source/Drain   6. Metallization

**Figure 16.40**  CMOS structure.

ity components all increase the advantages inherent in the CMOS design.

An isolation problem particular to CMOS structures is latch-up. Figure 16.41 shows a cross section of a portion of a CMOS chip. The side-by-side MOS transistors form a lateral bipolar (NPN) transistor. During circuit operation, the lateral bipolar transistor can function as an unwanted amplifier, increasing its output to a point where it causes the memory cell to go into a state in which it cannot be switched. This is *latch-up*. In this state, the cell cannot be addressed for its information. One solution to preventing latch-up is a low-resistivity epitaxial layer that shunts (shorts) the emitter of the bipolar transistor so that it does not "turn on," preventing the latch-up.

Another solution is LOCOS and other structures. One is a hardened-well design using a buried layer that effectively breaks up the lateral bipolar transistor (Fig. 16.42).

Silicon Gate

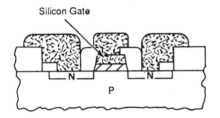

**Figure 16.41**  Cross section of silicon gate MOS transistor.

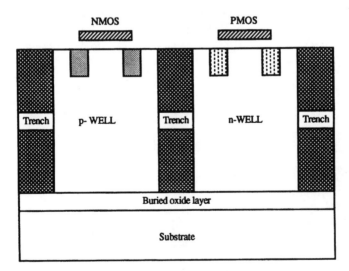

**Figure 16.42**  CMOS well structure silicon on insulator (SOI) with buried layer and trench isolation. *(Source:* Semiconductor International, *July 1993, p. 34.)*

Formation of the deep P-well by diffusion techniques requires a large thermal budget that contributes to lateral diffusion problems and stress growth. Ion-implanted wells have become common. Additionally, with ion implantation, the wells can be dopant graded in the vertical direction to maximize performance. Called *retrograde wells,* they are formed by high-energy implants (MeV) such that the bottom of the well has a higher dopant concentration. Implanted wells offer the advantage of fewer steps compared to conventional well formation[10] (Fig. 16.43). Also, because the N and P wells are formed in the same surface, planarity is preserved. In conventional well processing, the two well surfaces are offset and can cause depth-of-focus problems.

## Bi-MOS

The unique advantages of bipolar and CMOS transistors and their respective circuits come together in bi-MOS (or bi-CMOS) circuits. The circuits (Fig. 16.44) have in them bipolar, P-channel, and N-channel transistors along with memory cells (see Chapter 17). The low-power advantage of CMOS is used in logic and memory sections of the circuit, while the high-speed performance of bipolar circuitry is used for signal processing.[11] These circuits represent great challenges to the processing area with their large die size, small component size, and large number of processing steps.

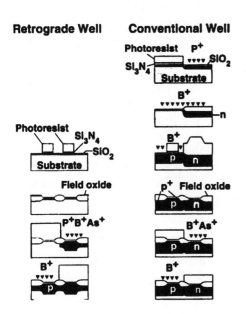

Figure 16.43  Implanted and conventional wells. *(Source:* Semiconductor International, *June 1993, p. 84.)*

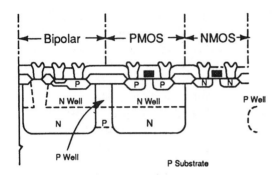

**Figure 16.44**  Bi-MOS structure.

### Silicon on insulator (SOI) isolation

There has been renewed interest in forming CMOS and bi-CMOS circuits in thin epitaxial silicon films deposited over an insulating substrate (see Chapter 12). Elimination of a conductive substrate minimizes leakage and latch-up problems.[12]

### Superconductors

Much interest has been generated by developments in superconductivity materials. Superconductivity is a phenomenon that occurs in certain materials when they are cooled close to absolute zero (−273°C). In ordinary metals, current is passed along by the flow of electrons. In an at-rest (nonconducting) state, the electrons exist in orbits around the nuclei of the atoms. To become conducting, they must gain energy to overcome the internal resistance of the particular material. The energy must be continuously supplied to maintain the current flow.

In a superconducting material, electrons exist in a "conducting state" and can support an electrical current with little or no additional energy input. The prospect of a resistanceless material has the potential of revolutionizing electronic devices.

Semiconductor researchers have investigated the superconducting effect for years. In 1962, B. D. Josephson described an effect that is now named after him (Josephson effect). When a thin oxide separates two superconducting materials, electrons will pass through the oxide with zero resistance. The structure is called a *Josephson junction*. This effect (called *tunneling*) is very complicated and requires quantum physics concepts to understand (well beyond the scope of this text!). The result is that the oxide has the functional aspects of a gate, and Josephson junctions can perform basic switching, logic, and memory functions.[13]

## Microelectromechanical systems (MEMS)

Semiconductor junctions have the property that almost anything done to them produces a change in their functioning. The anything can be a physical strain, exposure to light, radiation, or heating. These properties make possible semiconductor devices that respond to physical inputs rather than electrical signals. Miniature devices or machines can be constructed using wafer fabrication techniques. These devices are called *microelectromechanical systems* (MEMS).[14] MEMS devices made with wafer fabrication techniques include motors, motion sensors (as in air bag sensors), spectrometers, and miniature scanning tunneling microscopes.

## Strain gauges

A device that takes advantage of a junction reaction to physical pressure is the strain gauge. The gauge is made by back-etching through the wafer until only a very thin membrane is left. A junction and supporting circuitry are formed in the membrane. When the membrane is deflected by some force, as in a weight scale or pressurized gas line, the semiconductor circuit produces an output proportional to the deflection. The output of the circuit is correlated with the amount of pressure on the membrane and displayed on the appropriate meter.

## Batteries

Thin film rechargeable batteries have been made from lithium and vanadium pentoxide layers. They could be formed on a package and provide backup for CMOS devices in the event of power failures.[15]

## Light-emitting diodes

One effect when certain compound junctions are reversed-biased is the production of photons. Photons are a form of radiation that humans see as light. The devices are called *light-emitting diodes* (LED). These are the displays that are used in consumer electronics equipment and automobile displays.

The devices (Fig. 16.45) are made on gallium-arsenic-phosphide (GaAsP) wafers that are covered with thousands of diodes, wired so that they can be turned on or off individually. Groups of diodes are turned on in groups to form letters and numbers. GaAsP material produces the familiar red displays, with other colors being produced when dopants are added to the wafer.

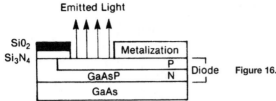

Figure 16.45    LED structure.

## Optoelectronics

On the other side of LEDs are chips that react to light. One use sought for years is integrated opto/electronics connection in local area networks (LANs). In this use, the opto devices (sensitive to laser and other lights) are connected to a fiber optic network or waveguide. Like the LEDs, the receivers are mostly III–V based ICs. An on-board conventional IC processes the data and sends it on through an output LED or laser.[16]

## Solar cells

Not only can semiconductor junctions emit light, they respond to it. This property is taken advantage of in the solar cell (Fig. 16.46). The cell is composed of diodes formed in a thin layer of semiconducting material such as amorphous silicon. When sunlight strikes the junction region, a current passes across it. The current is captured by on-board circuitry.

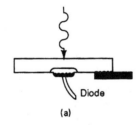

(a)

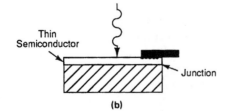

(b)

Figure 16.46   Light-sensitive semiconductor structures. (a) Photodiode and (b) solar cell.

## Temperature sensing

Semiconductor junctions are temperature sensitive. Heating a device will produce more current across the junction. This effect is taken advantage of in a variety of devices, such as solid-state medical thermometers and industrial control units.

## Acoustic wave devices

Acoustic wave devices (Fig. 16.47) are nonsilicon solid-state components used in microwave communication systems. They serve the function of converting microwaves into electrical impulses. A compound material such as $Be_{12}GeO_{20}$ has the property of reacting to the wave and setting up an electrical response in a solid-state circuit formed on the chip by normal semiconductor processes.

## Micromachines

With semiconductor technology, it is possible to create miniature machines with moving parts. It easy to understand the definition of surface patterns. Etching underneath the surface to free up a part is more difficult and the greatest challenge of this application. Nevertheless, micromachines are being produced. For example, Texas Instruments has produced an array consisting of numerous cells, each with a floating thin layer panel suspended from a "hinge" at each corner. A thin oxide layer gives the panel the ability to refract light. On-board electronic circuitry allows control of signals to the hinges. The signal causes a physical force to tilt the layer in specific directions that creates specific colors through the refraction process. The whole array functions as a high-resolution solid-state screen that can be used in projectors. Projecting the image onto a mirror provides a larger view-

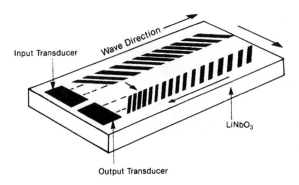

**Figure 16.47**  Acoustic wave device.

ing screen. Another recent micromachine is an actuator used to drive an optical shutter. Due to the light weight of the parts, speeds of up to 200,000 revolutions per minute have been made.[17] Shown below (Fig. 16.48) is a MEMS lens/laser system used as an optical bench. One intriguing potential is miniature space satellites that can be put into space with low-power rockets.

### Flat panel displays

Flat panel displays (FPDs) are most often seen as laptop computer screens. Their future is the replacement of cathode ray tubes (CRTs), where they provide a thin and more rugged option. However, improvements in clarity, brightness, and cost are necessary before they become regular replacements for CRTs.[18]

The majority of FPDs are the color-active matrix liquid crystal displays (AMLCDs). Both color and monochrome LCD displays are made using microchip fabrication technology. A display is a sandwich of two glass plates with a liquid crystal material in between (Fig. 16.49). On the bottom plate is an array of simple MOS devices with on-board driver circuitry. The transistors are formed by standard process steps, deposition, patterning, and so forth. The top panel is a color filter (for

**Figure 16.48**  MEMS optical guide system. *(Reprinted from the September 1997 edition of* Solid State Technology, *copyright 1997 by PennWell Publishing Company.)*

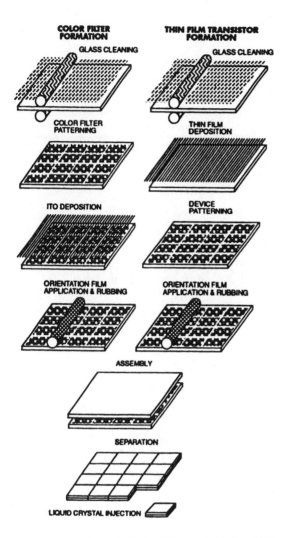

**COLOR FILTER FORMATION**

GLASS CLEANING

COLOR FILTER PATTERNING

ITO DEPOSITION

ORIENTATION FILM APPLICATION & RUBBING

**THIN FILM TRANSISTOR FORMATION**

GLASS CLEANING

THIN FILM DEPOSITION

DEVICE PATTERNING

ORIENTATION FILM APPLICATION & RUBBING

ASSEMBLY

SEPARATION

LIQUID CRYSTAL INJECTION

**Figure 16.49** Active matrix liquid crystal display (AM-LCD) process flow. *(Source: O'Mera Associates.)*

the color display) formed by multiple photoresist steps. In between is sandwiched the liquid crystal.

The image displayed on the screen is first formed as a pattern in the MOS transistors (on/off conditions).[19] This pattern is replicated by the liquid crystal, which can change states in response to an electrical signal. Lastly, for the color displays, the top filter transmits the proper color. Black-and-white displays simply transmit the on/off condition of the LCD.

## Field emission devices (FEDs)

In CRTs and FPDs, the information required to put a pattern on the screen has to come from electronic signals (such as a computer) and be transformed into a physical entity. In conventional CRTs, the entity is a stream of electrons from an electron gun. An alternative device to produce electrons is an FED (Fig. 16.50). Formed from semiconductor materials using wafer fabrication techniques, the end result is an etched microtip. When a voltage is placed between the cathode and gate, electrons are emitted. FPDs made with FEDs have the promise of brighter displays, a wider viewing angle, and improved speed.[20]

## Process flows

Increased device and circuit performance has created new structures with many new "parts." A typical structure circa 2000 is shown in Fig. 16.51. Creating the structures requires more and more process steps. Figure 16.52 shows the SIA roadmap projections for the chip and process in the year 2112.[21] Keeping track and making sense of this structure and process complexity requires a good grounding in the basic electrical operations of the individual components and basic processes.

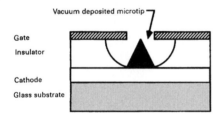

**Figure 16.50** Cross section of field emission device.

## Review Questions

1. Sketch and name the three parts and functions of a bipolar and MOS transistor.

2. Explain the function of a diode, resistor, and capacitor.

3. What is the operational advantage of a Schottky barrier diode?

4. Sketch and name the layers in a metal, polycide, and salicide MOS gate structure.

5. What is the structural difference between FETMOS, JFET, and MESFET gate structures?

6. Describe why component isolation is required in integrated circuits.

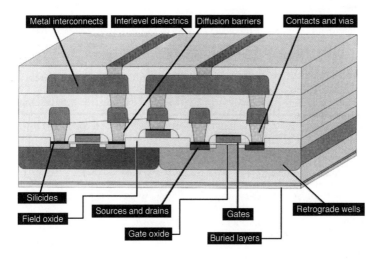

**Figure 16.51** Two-metal structure. *(Courtesy of Semiconductor Services, January 1998.)*

| Year of Production | 2001 | 2006 | 2012 |
|---|---|---|---|
| Line width (nm) | 150 | 100 | 50 |
| Memory size | 1 Gb | 16Gb | 64 Gb |
| Logic Bits/cm2 | 380M | 2.2B | 17B |
| Chip Size-DRAM (mm2) | 445 | 790 | 1580 |
| Max wiring levels | 7 | 7-8 | 9 |
| Mask layers | 23 | 24/26 | 28 |
| Defect density-DRAM (D/m) | 875 | 490 | 250 |
| Chip conections-I/O's | 1195 | 1970 | 3585 |
| Wafer diameter (mm) | 300 | 300 | 450 |

**Figure 16.52** SIA roadmap projections. *(Semiconductor International, January 1998.)*

7. What are the differences and advantages of junction and dielectric isolation schemes?

8. Sketch the process steps used to create an LOCOS isolation scheme.

9. Name three ways that a semiconductor junction in operation can be influenced.

10. How does a CMOS structure differ from an NMOS structure?

# References

1. H. R. Camenzind, *Electronic Integrated Systems Design,* Van Nostrand Reinhold, Princeton, NJ, p. 85.
2. P. Singer, "Gearing Up for Gigabits," *Semiconductor International,* November 1994, p. 34.
3. S. De Ornellas, et al., Plasma Etch of Ferroelectric Capacitors in FeRAMs and DRAMs," *Semiconductor International,* September 1997, p. 103.
4. H. R. Camenzind, *Electronic Integrated Systems Design,* Van Nostrand Reinhold, Princeton, NJ, p. 141.
5. P. Singer, "Directions in Dielectrics in CMOS and DRAMs," *Semiconductor International,* April 1994, p. 56.
6. Y. Pauleau, "Interconnect Materials for VLSI Circuits," *Solid State Technology,* April 1987, p. 157.
7. "Industry News," *Semiconductor International,* Cahners Publishing, April 1994, p. 16.
8. S. K. Ghandhi, *VLSI Fabrication Principles,* John Wiley & Sons, Inc., New York, 1994, p. 717.
9. S. Wolf, "A Review of IC Isolation Technologies—Part 8," *Solid State Technology,* PennWell Publishing, June 1993, p. 97.
10. L. Peters, "High Hopes for High Energy Ion Implantation," *Semiconductor International,* Cahners Publishing, June 1993, p. 84.
11. C. B. Yarling, "M. I. Current, Ion Implantation for the Challenges of ULSI and 200 mm Wafer Production," *Microelectronic Manufacturing and Testing,* March 1988, p. 15.
12. K. Yallup, "SOI Provide Total Dielectric Isolation," *Semiconductor International,* Cahners Publishing, July 1993, p. 134.
13. P. W. Anderson, "Electronic and Superconductors," in E. Ante'bi (Ed.), *The Electronic Epoch,* Van Nostrand Reinhold, Princeton, NJ, p. 66.
14. K. Gabriel, "Engineering Microscopic Machines," *Scientific American,* September 1995, p. 150.
15. J. Bates, "Rechargeable Thin-Film Lithium Microbatteries," *Solid State Technology,* PennWell Publishing, July 1993, p. 59.
16. P. Singer, "The Optoelectronics Industry: Has It Seen the Light?," *Semiconductor International,* Cahners Publishing, July 1993, p. 70.
17. J. Sniegowski, "Moving the World With Surface Micromachining," *Solid State Technology,* February 1996, p. 83.
18. P. Singer, "Flat Panel Display: An Interesting Test Case for the U.S.," *Semiconductor International,* Cahners Publishing, July 1994, p. 79.
19. *Ibid.*
20. K. Derbyshire, "Beyond AMLCDs: Field Emission Displays?" *Solid State Technology,* PennWell Publishing, November 1994, p. 55.
21. L. Peters, "Speeding the Transition to 0.18 μm," *Semiconductor International,* January 1998, p. 66.

# Chapter

# 17

# Introduction to Integrated Circuits

## Overview

The solid-state semiconductor components wired into integrated circuits perform many different functions in electronic instruments and machines. In this chapter, the general integrated circuit families and their different functions are explained.

## Objectives

Upon completion of this chapter, you should be able to:

1. Explain the concept of binary numbering.
2. List the three major integrated circuit functions.
3. Compare the basis of analog and digital logic circuits.
4. List the user and production advantages of logic gate arrays and PROM circuits.
5. Explain the two major memory circuit types.
6. Make a list of the four nonvolatile memory circuits.
7. Compare the performance and cost factors of DRAM and SRAM memory circuits.

## Introduction

The primary products of the semiconductor industry are integrated circuits. Countless numbers and types of circuits can be created using the processes described in this text. A circuit catalog from a major integrated circuit (IC) producer such as National Semiconductor or Motorola is as large as a New York City phone book. IBM estimates that their internal circuit catalog lists over 50,000 separate circuits!

Becoming familiar with integrated circuits is not as awesome a task as these high numbers might imply. The burden is eased by the fact that most circuits fall into three primary families: logic, memory, and microprocessors (logic and memory) (Fig. 17.1). Within each circuit family, there are a few principal designs and functions. The multiplicity of circuits comes from the many parameter variations required for specific uses.

The major functional circuit categories and their circuit designs are explained in this chapter. In the last section, we look at the future of IC circuitry from the perspective of the industry today. What the circuits will actually be like in 2010 can only be imagined, just as, in 1950, no one predicted the megabit RAM or the microprocessor.

## Circuit Basics

The question of how an integrated circuit actually works is the subject of other texts. But all circuits are based on the processing of data in binary notation. The binary system is a way of representing any number with just two digits—for example a zero and a one. It is actually an accounting system that keeps track of the place and value of the components of a number. Numbers can be expressed as the sums of numbers. For example,

$$1 = 1 + 0$$
$$3 = 2 + 1$$
$$7 = 4 + 2 + 1$$
$$10 = 8 + 2$$

Another way to express numbers is as powers of their factors. Yet another way is to express numbers as powers of their roots.

The basis of binary notation is the powers of the number 2. Figure 17.2 shows the numbers 1, 2, 4, 8 expressed as powers of 2. We could also represent those numbers just as the power of 2. Thus, 1 becomes 0, 2 becomes 1, 4 becomes 2, and 8 becomes 3. Now, for the clever part, any number can be expressed as the sum of some numbers that are powers of 2. The number 25 is the sum of $16(2^4)$ + $8(2^3)$ + $1(2^0)$. In the number 25, there is one $2^4$, one $2^3$, zero $2^2$, and one $2^0$. Or

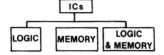

**Figure 17.1**   IC circuit functions.

$$1 = 2^0$$
$$2 = 2^1$$
$$4 = 2^2$$
$$8 = 2^3$$

**Figure 17.2** Powers of 2.

the number 25 can be represented by the code 1101, each of the digits representing the presence or absence of a particular power of 2. The chart in Fig. 17.3 lists some numbers in binary notation.

Translating numbers into binary notation is easily accomplished by establishing a grid with each column representing a power of 2. The actual number is represented by a string of zeros or ones that indicate the presence or absence of the particular powers of 2 that make up the number.

Binary notation has been known for centuries. Buckminster Fuller, in his book, *Synergistics,* has an amusing account of the use of binary coding by the ancient Phoenicians to keep track of cargo amounts. He claims that the Phoenician sailors were considered stupid because they could not count in the system of the day, when actually they were accurately keeping track of large amounts of cargo with only two "numbers"; they were counting in binary notation.

With binary notation, only two numbers are necessary—a one and a zero. In the discussion above, binary notation was represented by the numbers zero and one. In the physical world, binary numbers can be represented by any system that has two conditions. Figure 17.4 shows several different ways to code the number 7. The last row could represent binary coding by the off-on states of a transistor or memory cell.

Inside a circuit, numbers are coded, stored, and manipulated in their binary code. The numbers can be stored and manipulated because capacitors can be charged to have a charge or not have a charge, and transistors can be either on or off. The smallest piece of information in a circuit is called a "binary digit" or "bit." The binary coding system is simple. The problem of how the coded numbers could be added, subtracted, and multiplied was solved by George Boole, a nine-

| Standard | 32 | 16 | 8 | 4 | 2 | 1 |
|---|---|---|---|---|---|---|
| Power of 2 | $2^5$ | $2^4$ | $2^3$ | $2^2$ | $2^1$ | $2^0$ |

| Number Standard | = | Binary Number | | | | |
|---|---|---|---|---|---|---|
| 1 | 0 | 0 | 0 | 0 | 0 | 1 |
| 7 | 0 | 0 | 0 | 1 | 1 | 1 |
| 18 | 0 | 1 | 0 | 0 | 1 | 0 |
| 33 | 1 | 0 | 0 | 0 | 0 | 1 |

**Figure 17.3** Binary notation.

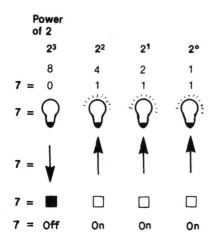

**Power of 2**

| $2^3$ | $2^2$ | $2^1$ | $2^0$ |
|-------|-------|-------|-------|
| 8     | 4     | 2     | 1     |

7 = 0    1    1    1

7 =

7 =

7 = ■    □    □    □

7 = Off    On    On    On

**Figure 17.4** Binary representations of the number 7.

teenth-century mathematician. He developed a logic system capable of handling numbers in binary notation. Until the development of computer logic, his Boolean logic (or Boolean algebra) was an academic curiosity.

Chips and computers are designed to handle a specific size of binary number or word. An eight-bit machine manipulates numbers with eight binary bits at a time. A 32-bit machine can handle a number composed of 32 binary bits. The more bits a machine can handle at one time, the faster and more powerfully it processes data. Every eight bits is known as a byte. Thus, an eight-megabyte (8 MB) storage capacity can hold eight million bits of information.

## Integrated Circuit Types

A solid-state integrated circuit is composed of a number of separate functional areas. Each chip, regardless of the circuit function, has an input and encode section where the incoming signals are "coded" into a form that the circuit can understand. The majority of the circuit area contains the circuitry required to perform the circuit function, either memory or logic. After the data is manipulated by the circuit, it goes to a decode section where it is changed back into a form that is usable by the machine's output mechanism. The circuit's output section sends the data to the outside world.

Although this is an overly simplified explanation of a circuit, it illustrates the fact that the interior of a chip is composed of definite separate functional areas. In many circuits, these areas perform the same functions as the main parts of a computer. Circuit types fall into three broad categories: logic, memory, and logic and memory (microprocessors).

*Logic circuits* perform a specified logical operation on the incoming data. For example, pushing the plus (+) key on a calculator instructs the logic portion of the chip to add the numbers presented to it. An onboard automobile computer goes through a logical operation to direct the signal from a sensor indicating an open door to light up the correct warning light on the dashboard.

*Memory circuits* are designed to store and give back data in the same form in which it is entered. Pushing the pi (π) key on a calculator activates the memory part of the circuit where the value of π (3.14) is stored. The value 3.14 is displayed on the screen. Every time that key is activated, that value is displayed.

*Microprocessors* are a third circuit type that combines both logic, arithmetic functions, and memory. In 1972, Intel Corporation introduced the first practical microprocessor. It was the microprocessor that allowed the design of powerful personal computers, digital watches, and one-chip calculators and the transfer of so many business machines to solid-state electronics, from phone systems to vending machines. Microprocessors can be programmed to perform many different circuit functions. To accomplish this, they contain logic and memory circuitry as well as the necessary encode, decode, input, and output sections. The microprocessor has been dubbed "a computer on a chip." While it contains all the functional areas of a computer, it is not truly a complete computer. Even simple computers require vast amounts of memory capacity that microprocessors do not have. Within many personal computers, microprocessors function as the central processing unit (CPU). Additional memory chips have to be included to make the computer of practical use.

Actually, every integrated circuit contains both logic capability and memory sections. For example, the logic circuitry of a calculator must have certain constants stored in a memory section to perform calculations. And memory circuits must have some logic functions to direct the flow of electrons and holes to the right parts of the circuit for storage.

## Logic circuits

**Analog-digital logic circuits.**  Logic circuits fall into two main categories: analog and digital (Fig. 17.5). Analog logic circuits were the earliest logic circuits developed. An analog circuit has an output that is proportional to the input. Digital circuits, on the other hand, feature a

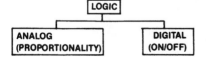

**Figure 17.5**  Logic circuit types.

predetermined output in response to a variety of inputs. A wall light dimmer is an analog device. Turning the control varies the voltage to the dimmer, which in turn varies the brightness of the light. A standard on-off light switch is a digital device. Only two brightness conditions are possible: on or off. Most audio circuits are of the analog type. Changing the level setting of the volume produces a proportional change in the sound coming out of the speaker.

**Analog logic circuit types.**    Analog circuits were the first type designed in integrated form. The home computer hobby kits of the 1950s were the analog type. These simple circuits were based on Ohm's law $(R = V/I)$. The circuit contains a resistance meter and a means for generating a current and measuring a voltage. The three quantities are related by Ohm's law. Any other three variables similarly related can be represented by the resistance, voltage, and current. Varying one changes the other two. The circuit thus becomes a computer for solving any equation of the form $A = B/C$.

The accuracy of analog circuits is dependent on the precision of the relationship between the input and the output. In the simple computer illustrated, accuracy is dependent on the precision of the components in the circuit, the clarity of the meters for setting the input and reading the output, and the immunity of the circuit to outside "noise." Unless the circuit contains a section to regulate incoming voltage levels, a change in the line voltage would alter the output, and hence the accuracy.

Both simple and complex analog circuits are vulnerable to variations in the incoming signal and to internal noise. Analog circuits are also dependent on precise control of the resistor values. Unfortunately, diffused resistors cannot be fabricated with a resistance variation from design value better than 3 to 5 percent, which is unacceptable for many applications. Greater resistance precision is gained by the use of matched resistor pairs, in which the effective resistance in the circuit is the difference between two resistors. This difference can be more tightly controlled than with one resistor alone.

Ion implantation also provides the analog circuit designer with a tool for producing resistors with a higher degree of control. Many analog circuits feature thin film resistors to achieve the required precision. The growth and popularity of digital circuits is based on their ability to produce a set output every time, regardless of noise or signal variations in the circuit. If a 5-V output is required, the digital circuit will deliver exactly that level.

Digital circuits, however, do not respond as fast as linear circuits. The term used in electronics is *real-time response*. In some applications, such as airplane controls, real-time response is mandatory. Recent developments in digital circuit speed is facilitating the encroachment of digital circuitry into this traditional use of analog circuits. A major advantage of digital circuits over analog circuits is in general-purpose computers. Analog circuits are more difficult to design to respond to a general range of problems. All modern general-purpose computers are based on digital circuits.

The most popular use for analog circuits is in amplifiers. They are designed in a variety of configurations, for many different applications. All have the same basic principle—the incoming signal or pulse is amplified. Audio circuits require amplification of a weak signal from the record tone arm or other input so as to produce the level required to operate a speaker.

The real-time aspect of analog circuits also makes them real-world circuits. Wherever there is a real-world measurement, such as temperature or movement, analog circuits are used. Even when the majority of the circuitry in a system is digital, analog circuits are often part of the interface with the outside world.

Most analog amplifier circuits are of the differential operational type. These circuits produce an output voltage amplified from and proportional to the difference of two input signals. Bipolar technology is favored for these circuits, because bipolar circuits are modular electric current devices and are better suited to the applications required of analog circuits.

The output signal of an analog device can have a "one-to-one" relationship with the input signal. These circuits are called *linear*. If the input is changed, the output changes in a linear manner. So many analog circuits are of linear design that the two terms are often used interchangeably. However, there are nonlinear circuits, such as those featuring a logarithmic relationship between the input and output.

Logic circuitry is built around the logic gate. A gate controls and directs passage through a barrier. The size and design of a gate influence the amount of passage allowed. For example, a room with many "in" doors and only one "out" door is a gate. Many people can enter the room, but their exit is restricted, because only one door is provided. This example gate can also be operated in reverse, allowing people to enter through only one door and leave through many.

Electronic logic gates perform similar functions but with electrical signals. In a circuit, they perform the necessary logic operation by the dictates of Boolean logic. A discussion of their incorporation in logic design is beyond the scope of this text. The point for this text is that gates, both analog and digital, are formed in a logic circuit by wiring together various components.

**Custom-semicustom logic.**   Using any of the logic gate approaches listed, hundreds of thousands of different logic circuits can be constructed. They vary from custom small-volume circuits to off-the-shelf standards. The bulk of logic circuits require some degree of customization. Several design and fabrication approaches are used to deliver custom and semicustom circuits to the customer at reasonable costs.

The approaches are:

1. Full custom

2. Standard circuit—custom gate pattern

3. Standard circuit—selective wiring gate

4. Programmable array logic

**Full custom.**   A full custom-designed logic circuit is specified by the customer, who pays for design and mask-making fees along with the fabrication costs. This approach is expensive and lengthy and is not geared for experimenting with different circuits in the design stage of a project. Custom-designed circuits are not cost effective in quantities of less than 100,000.

**Standard circuit—custom gate pattern.**   This fabrication process starts with a standard logic circuit design, but only the gates required for the particular application are formed during the fabrication process. The input, output, and other circuit sections are standard for a family of circuits.

**Standard circuit—selective wiring gate arrays.**   This system is similar to the custom gate approach but is based on a standard circuit design for most of the fabrication process. These circuits are built with a standard number of gates. This gate section is called the *array,* and the circuit is known as a *gate array.* Working with the basic design, customers can instruct the fabrication department to wire together only the gates required to produce their custom circuit logic function.

The result is faster turnaround time than full custom processes can achieve, and moderate cost. The cost per logic function of gate arrays is higher than that of a custom circuit produced in production quantities. The larger gate section required to allow many different circuit variations results in a larger chip. This larger chip size leads to a higher manufacturing cost per chip and/or a lower yield.

The wafers receive a common process up to the contact mask. The contact mask is customized to form contacts only to the gates required for the particular circuit. After metal deposition, only the gates with contacts are wired into the circuit. A variation of this process is to open the contacts in all the gates but use a customized metal mask that wires in only the wanted gates.

**Programmable array logic (PAL).**   Each of the three systems described requires the chip manufacturer to do the "customizing." This requirement can result in delivery or scheduling problems and generally forces the user to buy a minimum quantity of parts. Monolithic Memories, Inc., addressed this problem with the introduction of their PAL line of circuits in 1978. PAL stands for programmable array logic. MMI applied the programmable fuse technology used in memory products to logic circuits. The result was a field programmable (custom) logic circuit.

The concept is similar to the standard gate array, but, within the circuit, a fuse connects each logic gate into the circuit. The user programs the circuit by blowing the fuses at the unneeded gates, thus removing them from the circuit.

**Memory circuits**

Around 1960, industry forecasters began predicting that solid state memory circuits would overtake the traditional core memory. The advantages of solid circuits were their reliability, small size, and faster speed. This prediction was made every year until the early 1970s, when solid state memories finally did surpass core memory. The major factor that prolonged the life of core memory was lower cost.

Metal-oxide semiconductor (MOS) is the favored transistor structure for memory circuits. During the 1960s, however, the cleanliness requirement for MOS processing was not reliably available. High-yield MOS processing also requires accurate alignment and clean thin gate oxides. These processes were not fully developed in those years. The resultant low process yields kept MOS memories more expensive than core memories.

With process improvements and improved silicon gate structures, complementary metal-oxide semiconductor (CMOS) is the memory technology of choice. Some bipolar memories are favored for their fast speed and switching capabilities. While logic circuits can be (and are) made in MOS technology, most MOS circuits produced are memories, with the majority incorporated into computers. They are also used in microprocessor-based products, which require auxiliary memory chips. There are two principal types of memory circuits: volatile and nonvolatile (Fig. 17.6).

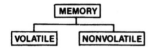

**Figure 17.6**  Memory circuit types.

**Nonvolatile memories.**   A nonvolatile memory device is one that does not lose its stored information when it loses its power. An example of this is a compact disk (CD), which is an information-storage device. If power to the record player is lost, the songs are not lost from the record itself. A number of nonvolatile memory circuits are listed in Fig. 17.7.

**ROM.**   In integrated circuits, the ROM design is the principal nonvolatile circuit. ROM stands for *read-only memory*. The sole function of this type of circuit is to give back precoded information. The information required in the circuit is specifically designed into the chip memory array section during fabrication. Once the chip is made, the stored information is a permanent part of the circuit.

Other memory types have *read and write* capability. That is, they can receive and store information from an input device (keyboard, magnetic tape, floppy disk, and so on). A phonograph record is a nonvolatile ROM device. A magnetic tape is an example of a nonvolatile device with both read and write capabilities, because information can be erased and rerecorded.

In a calculator, the constants and the rules required for the math operations are available in the ROM section. ROM circuits, like logic circuits, number in the hundreds of thousands. Although there are many standard types, the industry also produces many custom ROM circuits. The choices offered to the user in selecting a standard or custom chip are similar to those available with logic circuits. The user can buy a standard circuit, specify a variation on a standard basic circuit, design a total custom circuit, or buy a PROM, EPROM, or EEPROM.

**PROM.**   PROM stands for *programmable read-only memory.* A PROM is the memory equivalent of a PAL. Each memory cell is connected

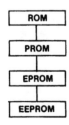

**Figure 17.7**  Nonvolatile memory.

into the circuit through a fuse. Users program the PROM to their own memory circuit requirements by blowing fuses at the unwanted memory cell locations. After programming, the PROM is changed to a ROM, and the information is permanently coded in the chip, where it becomes a read-only memory.

**EPROM.** For some applications, it is convenient to change the information stored in the ROM without having to replace the whole chip. EPROM (erasable programmable ROM) chips are designed for this use. The erasable feature is built in with the use of MMOS (memory MOS) transistors detailed in Chapter 16. The transistors can be selectively charged (or programmed), and they hold the charge for a long time—which holds the information in a nonvolatile fashion. Programming is by the mechanism of hot electron injection. When reprogramming is required, the charge in the transistors can be drained off (erasing the memory) by shining ultraviolet light on the chip. Reprogramming of the chip takes place by removing it from the circuit and putting in new memory information with an external programming machine. A typical EPROM can be reprogrammed up to ten times.

**EEPROM.** The next level of convenience in memory design is the ability to program and reprogram the chip while it is in a socket in the machine. This convenience is available with the EEPROM (or E2PROM), standing for *electronically erasable PROM*. Programming and erasing take place by pulses from the outside that place charges in selected memory cells or drain the charges away. Programming is by the same mechanism used for EPROMs, hot electron infection. Charge is drawn from the memory cell be a mechanism called Fowler-Nordheim tunneling. This connivance comes at the expense of a larger memory cell size and a commensurate reduction in chip density.

**Flash memories.** A flash memory is a form of EEPROM. It is a one-transistor cell design,[1] and like EPROMs can be programmed and reprogrammed while in their sockets in a system. Additionally, blocks, or the entire array, can be erased at one time.

**Volatile memories.** Semiconductor circuit and computer design involves the constant evaluation of trade-offs. In the case of memory, nonvolatile memory provides protection against power loss, but these memories are frequently slow and not very dense. More importantly, none of the circuits described above has a write capability, an essential feature in operating a computer. New information, such as a change in pay status, must be conveniently entered into the computer and stored temporarily while the new check is being written. Memory must also be easily erasable so the computer can quickly process new information or accept a completely new program. Several memory circuit designs are used to produce fast and high-density memory circuits. Both are of the volatile type; that is, when power is lost to the chip, all the stored information is lost; information presented on a computer screen, and not saved, is eligible for loss if the power to the computer goes off.

**RAM.**   One type of circuit used for high-density memory (Fig. 17.8) storage is random-access memory, or RAM. "Random" refers to the ability of the computer to directly retrieve any information stored in the circuit. Unlike a serial memory, the RAM design allows the chip to find the exact information asked for, wherever it is located in the computer memory. This feature allows faster information retrieval and makes the RAM the principal memory circuit in computers.

**Dynamic random-access memory (DRAM).**   DRAMS come in two principal designs: dynamic and static (Fig. 17.8). A dynamic memory design called a DRAM, for dynamic RAM, is used in great quantities in computer memories. The memory cell design is based on only one transistor and a small capacitor (Fig. 17.9). The information is stored in it by a charge built up in the capacitor. Unfortunately, the charge drains away very rapidly. To combat this problem, the memory information must be re-inputted to the circuit on a constant basis. The term for this function is *refresh*. The refreshing of the circuit occurs many thousands of times per second. Dynamic RAMS are vulnerable to both power loss and interruption, and to problems with the refreshing circuit.

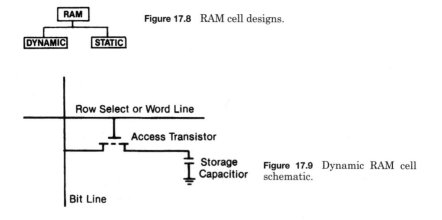

**Figure 17.8**  RAM cell designs.

**Figure 17.9**  Dynamic RAM cell schematic.

The goal of DRAM design is small-cell design for high-density and closely spaced components with small and thin parts for speed. These requirements have driven DRAM design and processing to the highest levels of the technology. All the advantages available by advanced, state-of-the-art equipment and processing are applied to DRAM circuits. This fact make them the industry's leading-edge circuit.

**Static random-access memory (SRAM).**    Static random-access RAM (SRAM) memories are based on a cell design that does not need a refresh function. Once the information is put into the chip, it will stay as long as the power remains on. This is accomplished with a cell containing several transistors and capacitors (Fig. 17.10). The information is stored as conditions of the transistors are alternately on or off. Information can be read and written with a SRAM cell much faster than with a RAM design, since transistors can be switched faster than capacitors can be charged and drained. The penalty paid for this lesser degree of volatility and speed is loss of space. The larger cell design makes static memories less dense than DRAMs.

Memory capacity is measured by the number of bits that can be stored. A 1k RAM has a capacity of 1024 bits of information; 1024 is a power of 2. A 64k RAM actually has a capacity of 65,536 bits of information. RAM capacity has expanded rapidly, with larger megabit memories (64 and higher) expected to be produced in quantities with presently identified technology. Each step upward in RAM capacity places greater pressure on wafer processing and yield improvement. The nature of the semiconductor chip business is exemplified by the 64k RAM, introduced by IBM in 1977. The chips were soon available in the merchant market, priced at over $100 per circuit. By 1985, competition and yield improvements had lowered the price to under $1 per circuit!

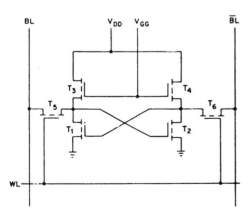

Figure **17.10**  Static RAM cell schematic.

**Ferroelectric memories (FERAM and DRAM).** Ferroelectric materials (see Chapter 16) used in capacitors create memories that operate faster than conventional SiCMOS technology capacitors.[2] The challenge is the integration of this nonsilicon technology into a standard silicon process.

## Redundancy

Redundancy is the inclusion of extra circuit components in the design. If one or more of the components don't work, others are available that do. The trade-off for redundancy is larger chip size. Also, extra circuitry is required within the main circuit to detect the functioning and nonfunctioning components and direct the selection of a functioning component. Although this approach to higher yield has been discussed for years, it hasn't yet become a mainstay of circuit design. This is due to the problem of locating the working and nonworking components and wiring the working ones into the circuit.

## The Next Generation

Since the 1950s, the semiconductor industry has maintained a rather constant rate of product evolution. In general, there has been a new product generation every three years.[3] Within each generation, the density of memory chips increased four times and logic chips two to three times. Every two generations (six years), the feature size decreased by a factor of two, with chip area and package pin count increasing by a factor of two.

Predicting the future is always difficult, but there are identified end points for chip circuits as we now know them. The SIA technology roadmap[4] profiles development to the 50-nanometer feature size level by the year 2012. At that size, chip densities of 16G (billion) bits for memory and 20M (million) gates for logic. Wafer diameters climb to 200 to 400 mm, and die sizes are in the 1000 mm$^2$ range (1.2 inches per side) or larger.

Reaching this level will require development of X-ray lithography and low-resistance metals, low-leakage contacts, and ultraclean materials and processes. Planarization techniques will be strained as the number of metal interconnect levels climbs to six or seven. Perhaps the major challenge is reliable, stable gate oxides in the 35-Å range. Below 35 Å, carriers can tunnel through the layer, causing leakage problems. These technical problems must be overcome in a manufacturing atmosphere that is increasingly expensive and that requires automation and extensive measurement and process control programs.

Circuits coming out of the advanced technology will be a mixture of increased performance existing circuits and new circuits. Certainly, DRAMs will continue to increase in memory capacity, and microprocessors will become more powerful and faster. New areas for chip development are speech recognition, expert systems, and the continuing "microchipization" of automobiles and most power-consuming products (control and conservation).

## Review Questions

1. Name the two principal solid-state circuit types.
2. Name the two principal logical circuit types.
3. Give an example of a digital electrical device.
4. Which of the following describes a PROM circuit?

   *a.* Factory programmable

   *b.* Field programmable

   *c. a* and *b*

   *d.* Not programmable

5. List the three functions in a microprocessor circuit.
6. What does RAM stand for?
7. What is the difference between a DRAM and SRAM?
8. How does a serial memory device read back information? Give an example.
9. Why are DRAM circuits considered the leading-edge ULSI circuits?
10. Express the number "25" in binary code.

## References

1. M. McConnell, et al., "An Experimental 4-Mb Flash EEPROM with Sector Erase," *IEEE Journal of Solid State Circuits,* vol. 26, no. 4, April 1991.
2. R. Jones, "Integration of Ferroelectric Nonvolatile Memories," *Solid State Technology,* October 1997, p. 201.
3. C. Hu, "MOSFET Scaling in the Next Decade and Beyond," *Semiconductor International,* June 1994, p. 105.
4. Semiconductor Industry Association, Palo Alto, CA, 1992.

# 18

# Packaging

## Overview

After wafer sort, the chips on the wafer surface are completed and the electrically functioning ones identified. The chips will be incorporated into individual protective packages, mounted with other components in hybrid or multichip modules (MCMs), or connected directly onto a printed circuit board. This chapter describes the packages and processes used for chip protection and to connect the chip to electrical connections which, in turn, allow integration of the circuit into a larger electronic system.

## Objectives

Upon completion of this chapter, you should be able to:

1. List the four functions of a semiconductor package.
2. List the five common parts of a package.
3. Recognize and identify the major package designs.
4. List and describe the major packaging process flows.

## Introduction

After wafer sort, the chips are still part of the wafer. For use in a circuit or electronic product, they must be separated from the wafer and, in most cases, put in a protective package. They may also be mounted onto the surface of a ceramic substrate as part of a hybrid circuit, put into a large package with other chips as part of a multichip module (MCM), or be connected directly on board a printed circuit, chip-on-board, or direct chip attach (COB or DCA) (Fig. 18.1). All three options share some common processes. The packaging process, in addition to protecting the chip, provides an electrical connection system allowing the chip to be integrated into an electronic system, and it provides environmental protection and heat dissipation. This series of processes is known variously as *packaging, assembly,* or the *back-end process.* In the packaging process, the chips are called *dies* or *dice.*

Over the years, semiconductor packaging has lagged wafer fabrication in process sophistication and manufacturing demands. The advent of the VLSI/ULSI era in chip density has forced a radical upgrading of chip packaging technology and production automation.

Higher-density chips require more input connections (I) and more output connections (O). These are referred to as the I/O count or simply the pin count. The SIA *Technology Roadmap* (ITRS) projects pin counts increasing into the 3000 range by 2007 (Fig. 18.2). The IRTS lists pin count, cost, chip size, thickness, and temperature considerations as the primary physical drivers of packaging technology. As solid-state circuits have found more applications, the need for special package designs has increased. Higher pin counts have led to the adoption of bump/flip chip technology. Size and speed considerations have driven the use of chip scale packages in consumer products, such as cell phones and hand-held products. The harsh environments of space, automotive use, and military applications require special packages, processing, and testing to ensure high reliability in the field. These packages, processes, and tests are referred to as *hi-rel.* The other chips and packages are referred to as *commercial* parts.

No longer is packaging the stepchild of the semiconductor industry. Many feel that, eventually, packaging will be the limiting factor on the

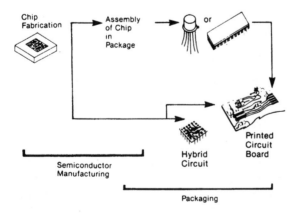

**Figure 18.1**   Chip packaging options.

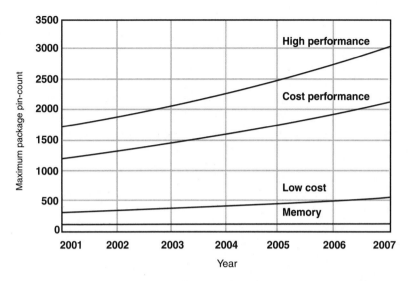

**Figure 18.2**  Pin-count predictions. (*Courtesy of* Semiconductor International.)

growth of chip size. For the time being, however, much effort is going into new package designs, new material development, and faster and more reliable packaging processes.

## Chip Characteristics

Throughout this text, many characteristics of discrete devices and integrated circuits have been mentioned. Several of them have a direct bearing on package design and the packaging processes (Fig. 18.3). The chip density (integration level) determines the number of connections required, with higher-density chips having larger surface areas and more bonding pads. The trend to larger chips has resulted in the need for thicker and larger-diameter wafers. These factors have caused changes in die separation processes, package design, and the need for wafer thinning.

In previous chapters, it was pointed out that the functioning of the chip components (transistors, diodes, capacitors, resistors, and fuses)

- Integration Level
- Wafer Thickness
- Dimensions
- Environmental Sensitivity
- Physical Vulnerability
- Heat Generation
- Heat Sensitivity

**Figure 18.3**  Chip characterizations affecting the packaging process.

can be altered by various contaminants. Primary among them are chemicals such as sodium and chlorine. Additionally, other chemicals can attack the chip layers, and environmental factors (e.g., particulates, humidity, and static) can ruin chips or change their performance. Other concerns are the influence of light and radiation impinging on the chip surface. Some chips are extremely light or radiation sensitive. This factor is considered in the selection of package materials and processing. A dominant chip characteristic is the extreme vulnerability of its surface to physical abuse. The surface components are only a small distance down into the wafer surface, and the surface wiring can be as thin as 1 micron.

These environmental and physical concerns are addressed in two ways. First is the passivation layer deposited near the end of the fabrication process. This may be a hard layer based on silicon dioxide or silicon nitride. Often, passivation layers are doped with boron, phosphorus, or both to increase their protective properties. Alternatively, it may be a soft layer such as a polyimide (Fig. 18.4). The second method of protecting the chip is provided by a package.

Another chip characteristic of importance to the package design and material is heat generation. Chips used in high-power circuits and highly integrated circuits can generate enough heat to actually damage themselves and the circuit. Package design includes heat dissipation factors. Heat is also an important parameter in packaging processes, with packaging process temperatures limited to 450°C. Above this temperature, the aluminum and silicon contacts on the chip can form an alloy in the wafer surface that causes electrical shorts.

## Package Functions and Design

There are four basic functions performed by a semiconductor package. They are to provide:

1. A substantial lead system
2. Physical protection

- Silox
- Vapox
- Pyrox
- Glassivation Layer
- PSG
- BSG
- PBSG

Figure 18.4 Passivation layer names.

3. Environmental protection

4. Heat dissipation

**Substantial lead system**

The primary function of the package is to allow connection of the chip to a circuit board or directly to an electronic product. This connection cannot be made directly, due to the thin and fragile metal system used to interconnect the components on the chip surface. The metal leads are typically less than 1.5 µm thick and often only 1 µm wide. The thinnest wires available are 0.7 to 1.0 mils in diameter, which is many times larger than the surface wiring. Solder balls used in bump connection technology are about 100 µm in diameter. This difference in wiring sizes is the reason why the chip wiring terminates in the larger bonding pads.

Even though the wires are larger, at 1 mil in diameter, they too are very fragile. This fragility is overcome by a more substantial electrical lead system that serves as the connection of the chip to the outside world by traditional leads, pins, or the balls used in grid array packages (Fig. 18.5). The lead system is an integral part of the package.

**Physical protection**

The second function of the package is the physical protection of the chip from breakage, particulate contamination, and abuse. Physical protection needs vary from low, as in the case of consumer products, to very stringent, as in the case of automobile circuits, space rockets, and military uses. The protection function is accomplished by securing the chip to a die-attachment area and surrounding the chip, wire bonds, and inner package leads with an appropriate enclosure. The size and eventual use of the chip dictate the choice of materials for the enclosure and the design and size of the package.

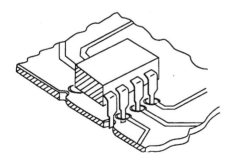

**Figure 18.5** DIP through-hole assembly.

## Environmental protection

Environmental protection of the chip from chemicals, moisture, and gases that may interfere with the chip functioning is provided by the package enclosure.

## Heat dissipation

Every semiconductor chip generates some heat during operation. Some generate large quantities. The package enclosure materials serve to draw the heat away from most chips. Indeed, one of the factors in choosing a package material is its thermal dissipation property. The chips that generate large quantities of heat require additional consideration in the package design. This consideration will influence the size of the package and will often require the addition of metal heat-dissipating fins or blocks on the package.

## Overview of Packaging Operations

In this section, a simple overview of the primary packaging process is presented.

### Cleanliness and static control

The chips are vulnerable to contamination during their entire lifetime. While assembly areas are not generally required to maintain the same cleanliness levels as fabrication areas (see Chapter 4), cleanliness is important. The table in Fig. 18.6 lists common static-control practices used in packaging areas. High-rel areas, particularly, demand higher cleanliness levels. In fact, many companies are finding that halfway contamination control programs are doomed to failure. Consequently, more assembly areas are practicing very stringent controls, especially for people-generated particles and chemicals.

An environmental danger that is most serious in packaging areas is static. In the fabrication cleanrooms, static is controlled primarily to prevent the attraction of particles to the wafer surface. This is also a concern in a packaging area. But the greatest concern is *electrostatic*

- HEPA Filters/VLF Air
- Smocks, Hats, Shoe Covers
- Finger Cots or Gloves
- Filtered Chemicals
- Sticky Floor Mats
- Static Control

**Figure 18.6** Contamination control practices.

*discharge,* or *ESD.* Static charge can build up to levels of tens of thousands of volts. If this voltage is suddenly discharged onto a chip surface, it can easily destroy a portion of the circuit. MOS gate structures are particularly vulnerable to ESD.

Every packaging area making high-density chips should have an active antistatic program (Fig. 18.7). Those assembling military parts will need to have one as a condition of getting the contract. The antistatic program is based on operators wearing grounding straps and nonstatic smocks; the use of antistatic carrier materials; moving work by lifting rather than sliding; and grounded equipment, work surfaces, and floor mats. Static is also reduced by the placement of ionizers in nitrogen and air blow-off guns (Fig. 18.8) and in the path of air coming out of HEPA filters.

### Basic processes

In wafer fabrication, the wafers pass many times in and out of four basic operations (layering, patterning, doping, and heat processing). In packaging as well, there are also several basic operations (Fig. 18.9). However, packaging is a once-through process. Each of the major processes is required only once. As in fabrication, the exact order of the operations is determined by the package type and other factors. Each process may or may not be used in a particular process, and each is

- Wrist Ground Straps
- Nonstatic Garments
- Antistatic Materials
- Grounded Equipment
- Grounded Work Surfaces
- Grounded Floors or Floor Mats

**Figure 18.7**  Static control practices.

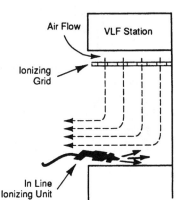

**Figure 18.8**  Static control techniques.

| | |
|---|---|
| • Backside Preparation | • Preseal Inspection |
| • Die Separation | • Packaging Sealing |
| • Die Pick | • Plating |
| • Die Attach | • Lead Trim |
| • Inspection | • Marking |
| • Bonding | • Final Tests |

**Figure 18.9**   Basic packaging operations.

customized to the particular chip and package requirements. There are three primary methods of connecting the chip to the package: wire bonding, bump (ball) technology, and tape automated bonding (TAB). The process flow(s) based on wire bonding are presented first, followed by the variations required of the bump technique and TAB. The details of the various operations and their optional processes are explained in more detail in the remainder of the chapter.

**Backside preparation.**   At the end of the fabrication process, some wafers have to be thinned (wafer thinning) to fit in the package and/or to remove backside damage or junctions. Wafers whose die are going to be attached to the package by a gold-silicon eutectic will receive a deposited layer of gold (backside gold).

**Die separation.**   The wafer is separated into individual die by sawing or scribe-and-break techniques.

**Die pick and place.**   The functioning die identified at wafer sort are picked from the separated wafer and placed in carriers (Fig. 18.10).

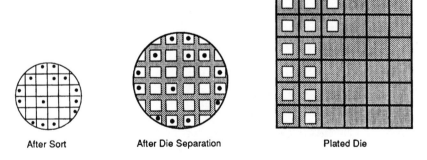

After Sort            After Die Separation            Plated Die

**Figure 18.10**   Die separation to plate.

**Inspection.**    The dies are optically inspected for edge integrity, contamination, and cosmetic defects.

**Die attach.**    Each die is attached into the die-attach area of the package by either a gold-silicon eutectic layer or an epoxy adhesive material (Fig. 18.11).

**Lead bonding.**    Thin wires are bonded between the chip bonding pads and the inner leads of the package. Bonding is also done by bumps of metal on the bonding pads or by the TAB technique.

**Preseal inspection.**    The bonded die is optically inspected in the package. Criteria checked for are the alignment of the die in the package, proper bond placement, contamination, die-attach quality, and bonding quality.

**Plating, trimming, and marking.**    The outer package leads are plated with an additional layer of conductive metal to improve their solderability into a printed circuit board. Near the end of the process, each package with outer leads will go through a trimming operation to separate the leads from supports. The marking operation is performed to code important information (Fig. 18.12) on the outside of the package enclosure.

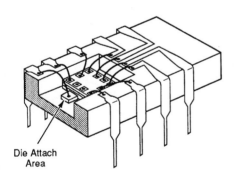

Die Attach
Area

**Figure 18.11**  Die attach area of package.

ORA6-6348
1987-3156

**Figure 18.12**  Marked package.

**Final test.**    A series of final tests, including electrical and environmental, are performed on each packaged chip to ensure quality. Some packages will receive a "burn-in" test to detect early failures.

### Common package parts

The four functions of a chip package are accomplished through the use of a wide variety of package designs. However, most packages have five common parts. They are as described below.

**Die-attachment area.**    In the center of every package is an area where the chip is securely attached into the package. This *die-attachment area* may have an electrical connection that serves to connect the back of the chip to the rest of the lead system. A major requirement for this area is absolute flatness so as to intimately support the chip in the package (Fig. 18.11).

**Inner and outer leads.**    The metal lead system chip is continuous from the die-attach cavity to the printed circuit board or electronic product. The system inner connections are called the *inner leads, bonding lead tips,* or *bond fingers.* The inner leads are generally the narrowest portion of the lead system. The leads become progressively wider, finally ending outside the package. These portions of the lead system are called the *outer leads* (Fig. 18.13). Most of the lead systems are composed of one continuous piece of metal. One exception is the side-brazed package. In this package construction method, the outer leads are brazed onto the interior leads. Two different alloys are used for the outer lead system—either an iron-nickel alloy or a copper alloy. The iron-nickel alloy is desirable for its strength and stability, while copper is used for its electrical and heat-conduction properties.

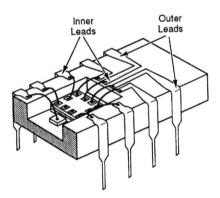

**Figure  18.13** Inner and outer leads.

**Chip/package connection.**    The chip is electrically connected to the lead system of the package with bonding wires, balls, or other on-chip connectors (Fig. 18.14).

**Enclosures.**    The die-attach area, bonding wires, and inner and outer leads constitute the electrical parts of the package. The other part is the *enclosure* or *body.* This is the part that provides the protection and heat-dissipation functions. These functions are achieved by several different techniques and package designs as described in the sealing section. The completeness of the seal falls into two categories: hermetic and nonhermetic (Fig. 18.15).

Hermetic sealing results in a package that is impervious to the penetration of moisture and other gases. Hermetic seals are required for chips operating in harsh and demanding environments such as in rockets and space satellites. Metal and ceramic enclosures are the preferred materials for hermetically sealed packages.

Nonhermetically sealed packages are adequate for most consumer applications such as computers and entertainment systems. This sealing system provides good and adequate environmental protection of the chip, except in the most demanding situations. A better term for this type of enclosure sealing method would be *less hermetic.* Nonhermetic packages are composed of epoxy resins or polyimide materials and are generally referred to as *plastic packages.*

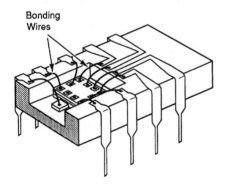

Bonding
Wires

**Figure 18.14**    Bonding wires.

Hermetic
- Metal
- Ceramic

Nonhermetic
- Epoxy Resins
- Polyimides

**Figure 18.15**    Package sealing designations.

## Packaging Processes

### Prepackaging wafer preparation

After the final passivation layer and an alloy step in wafer fabrication, the circuits are complete. However, one or two additional processes may be performed on the wafer before transfer to packaging. These steps (wafer thinning and backside gold) are optional, depending on the wafer thickness and the particular circuit design.

**Wafer thinning.**   The trend to thicker wafers presents several problems in the packaging process. Thicker wafers require the more expensive complete saw-through method at die separation. While sawing produces a higher-quality die edge, the process is more expensive in time and consumption of diamond-tipped saws. Thicker die also require deeper die attach cavities, resulting in a more expensive package. Both of these undesirable results are avoided by thinning the wafers before die separation.

Another situation requiring wafer thinning is electrical in nature. If the wafer backs are not protected as the wafers go through the dopant operations in fabrication, the dopants will form electrical junctions in the wafer back, which may interfere with good conduction in the back contact that is required for the circuit to operate correctly. These junctions may require physical removal by wafer thinning.

The thinning step generally takes place between wafer sort and die separation. Wafers are reduced to a thickness of 0.2 to 0.5 mm.[1] Thinning is done by the same processes (mechanical grinding and chemical-mechanical polishing-CMP) used to grind wafers in the wafer preparation stage (see Chapter 3). An alternate method is to protect the front of the wafers and chemically etch material from the back.

Wafer thinning is a worrisome process. In back grinding, there is the concern of scratching the front of the wafer and of wafer breakage. Since the wafer must be held down on the grinder or polishing surface, the front of the wafer must be protected and, once thinned, wafers are easier to break. In back etching, there is a similar need to protect the front of the wafer from the etchant. The protection can be provided by spinning a thick layer of photoresist on the front side. Other methods include attaching adhesive-backed polymer sheets cut to the wafer diameter. Stresses induced in the wafer by grinding/polishing processes must be controlled to prevent wafer and die warping. Wafer warping interferes with the die separation process (broken and cracked die). Die warping creates die attach problems in the packaging process.[2]

**Backside gold.**   Another optional wafer process is adding a layer of backside gold. A layer of gold is required on wafers that are going to be attached to the package by eutectic techniques (see the "Die attach" section). The gold is usually applied in the fabrication area (after back grinding) by evaporation or sputtering.

### Die separation

The chip-packaging process starts with the separation of the wafer into individual dies. The two methods of die separation are scribing and sawing (Fig. 18.16).

**Scribing.**  *Scribing,* or *diamond scribing,* was the first production die separation technique developed in the industry. It requires dragging a diamond-tipped scribe through the center of the scribe lines and separating the die by flexing the wafer. Scribing becomes unreliable in wafers over 10 mils thick.

**Sawing.**  The advent of thicker wafers has led to the development of sawing as the preferred die separation method. A saw consists of a wafer table with rotation capability, a manual or automatic vision system for orienting the scribe lines, and a diamond-impregnated round saw. Two techniques are used. Both start with the passage of the diamond saw over the scribe lines. For thinner wafers, the saw is lowered into the wafer surface to create a trench about one-third of the way through the wafer. The separation of the wafer into die is completed by the stress and roller technique used in the scribing method. The second sawing method is to separate the die by a complete saw-through of the wafer.

Often, the wafers for complete saw-through are first mounted on a flexible plastic film. The film holds the die in place after the sawing operation and aids the die pick operation. Sawing is the preferred die separation method due to the cleaner die edges and the fewer cracks and chips left on the sides of the die (Fig. 18.17).

### Die pick and place

After sawing, the separated die are transferred to a station for selection of the functioning die (non-inked). In the manual method, an op-

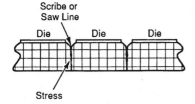

**Figure 18.16**  Scribe and saw separation.

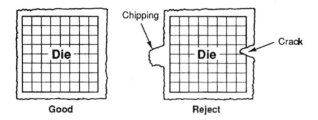

**Figure 18.17**  Die-separation results.

erator will pick up each of the non-inked dies with a vacuum wand and place it in a sectioned plate. Wafers that come to the station on the flexible film are first placed on a frame that stretches the film. The stretching separates the die, which aids the die pick operation.

In the automated version of this operation, a memory tape or disk that has the locations of the good die (from wafer sort) is loaded into the tool. A vacuum wand picks up good die and automatically places them in a sectioned plate for transfer to the next operation.

### Die inspection

Before being committed to the rest of the process, the die are given an optical inspection. Of primary interest is the quality of the die edge, which should be free of chips and cracks. This inspection also sorts out surface irregularities, such as scratches and contamination. Inspection may be manual with microscopes or automated with a vision system. At this step, the die is ready to go into a package.

### Die attach

Die attachment has several goals: to create a strong physical bond between the chip and the package, to provide either an electrical conducting or insulating contact between the die and the package, and to serve as a medium to transfer heat from the chip to the package.

A requirement is the permanency of the die-attachment bond. The bond should not loosen or deteriorate during subsequent processing steps or when the package is in use in an electronic product. This requirement is especially important for packages that will be subjected to high physical forces, such as in rockets. Additionally, the die attach materials should be contaminant-free and should not outgas during subsequent process heating steps. Lastly, the process itself should be productive and economical.

**Eutectic die attach.**    There are two principal methods of die attach: *eutectic* and *epoxy* adhesives. The eutectic method is named for the phenomenon that takes place when two materials melt together (alloy) at a much lower temperature than either of them separately. For die attach, the two eutectic materials are gold and silicon (Fig. 18.18). Gold melts at 1063°C, while silicon melts at 1415°C. When the two are mixed together, they start alloying at about 380°C. Gold is plated onto the die-attach area and alloys with the bottom of the silicon die when heated.

The gold for the die-attach layer is actually a sandwich. The bottom of the die-attach area is deposited or plated with a layer of gold. Sometimes, a preformed piece of metal composed of a gold and silicon mixture is placed in the die-attach area. When heated, these two layers, along with a thin layer of silicon from the wafer back, form a thin alloy layer. This layer is the actual bond forming the die-package attachment.

Eutectic die attach requires four actions. First is the heating of the package until the gold-silicon forms a liquid. Second is the placement of the chip on the die-attach area. Third is an abrasive action, called *scrubbing*, that forces the die and package surfaces together. It is this action, in the presence of the heat, that forms the gold-silicon eutectic layer. The fourth and final action is the cooling of the system, which completes the physical and electrical attachment of the chip and package.

Eutectic die attach can be performed manually or by an automated machine that performs the four actions. Gold-silicon eutectic die attach is favored for high-reliability devices and circuits for its strong bonds, heat dissipation properties, thermal stability, and lack of impurities.

**Epoxy die attach.**   The alternate die-attach process uses thick liquid epoxy adhesives. These adhesives can form an insulating barrier between the chip and package or become electrically and heat conductive with the addition of metals such as gold or silver. Polyimide may also be used as an adhesive. Popular adhesives are silver-filled epoxy for copper lead frames and silver-filled polyimide for Alloy 22 metal frames.[3]

Conducting
  • Gold/Silicon Eutectic
  • Metal-Filled Epoxy
  • Conducting Polyimide
Nonconducting
  • Epoxy Adhesive
  • Insulating Polyimide

**Figure 18.18**  Die-attach material matrix.

The epoxy process starts with the deposit of the epoxy adhesive in the die-attach area by dispensing the adhesive with a needle or screen printing it into the die-attach area. The die, held by a vacuum wand, is positioned in the center of the die-attach area. The second action is to force the die into the epoxy to form a thin uniform layer under the die. The final action is a curing step in an oven at an elevated temperature that sets the epoxy bond.

Epoxy die attach is favored for its economy and ease of processing, in that the package does not have to be heated on a stage. This factor makes the automation of the process easier. When compared to gold-silicon eutectic die attach, epoxy has the disadvantage of potential decomposition at the high temperatures of bonding and sealing operations. Epoxy die-attach films also do not have the bonding power of the eutectics.

Regardless of the attachment method used, there are several marks of a successful die attach. One is the proper and consistent alignment of the die in the die-attach area. Proper placement pays off in faster and higher-yield automatic bonding. Another desired result is a solid, uniform, and void-free contact over the entire area of the chip. This is necessary for mechanical strength and good thermal conduction. One evidence of a uniform bond is a continuous joint or "fillet" between the die edge and the package. The final mark of a good die-attach process is a die-attach area free of flakes or lumps that can come loose during use and cause a malfunction.

### Die-to-package bonding

Once the die and package are attached, they go to the bonding process. This is perhaps the most critical of all the assembly operations. Three techniques provide the critical chip/package connection: wire bonding, bump/flip-chip, and TAB. In wire bonding, up to hundreds of wires must be perfectly bonded from the bonding pads to the package inner leads. In bump/flip chip, the bonding pad/package connection is a solder ball. Tape automated bonding (TAB) system is a process that bonds the lead frame leads directly to the die bonding pads in one step.

### Wire bonding

The wire bonding procedure is simple in concept. A thin (0.7 to 1.0 mil) wire is first bonded to the chip bonding pad and spanned to the inner lead of the package lead frame. The third action is to bond the wire to the inner lead. Last, the wire is clipped and the entire process repeated at the next bonding pad. While simple in concept and procedure, wire bonding is critical because of the precise wire placement and electrical contact requirements. In addition to accurate placement, each and every wire must make good electrical contact at both ends, span between the pad and inner lead in a prescribed loop without kinks, and be at a safe distance from neighboring wires. Wire loops in conventional packages are 8 to 12 mils, while those in ultra-thin packages are 4 to 5 mils.[4] Distances between adjacent wires are referred to as the *pitch* of the bonding.

Wire bonding is done with either gold or aluminum wires. Both are highly conductive and ductile enough to withstand deformation during the bonding steps and still remain strong and reliable. Each has its advantages and disadvantages, and each is bonded by different methods.

**Gold wire bonding.**    Gold has several pluses as a bonding wire material. It is the best known room-temperature conductor and is an excellent heat conductor. It is resistant to oxidation and corrosion, which translates into an ability to be melted to form a strong bond with the aluminum bonding pads without oxidizing during the process. Two methods are used for gold bonding. They are *thermocompression* and *thermosonic*.

Thermocompression bonding (also known as *TC bonding*) starts with the positioning of the package on the bonding chuck and the heating of the chip and package to between 300 and 350°C. Chips that are going to be enclosed in an epoxy molded package are processed through die attach and bonded with the chip on the lead frame only. The bonding wire is fed out of a thin tube called a *capillary* (Fig. 18.19). An instantaneous electrical spark or small hydrogen flame melts the tip of the wire into a ball and positions the wire over the first bonding pad. The capillary moves downward, forcing the melted ball onto the center of the bonding pad. The effect of the heat (thermal) and the downward pressure (compression) forms a strong alloy bond between the two materials. This type of bonding is often called *ball bonding*. After the ball bond is complete, the capillary feeds

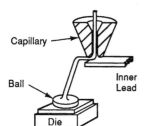

**Figure 18.19**   Gold ball bonding.

out more wire as it travels to the inner lead. At the inner lead, the capillary again travels downward to where the gold wire is forced by the heat and pressure to melt onto the gold-plated inner lead. The spark or flame then severs the wire, forming the ball for the next pad bond. This procedure is repeated until every pad and its corresponding inner pad are connected.

Thermosonic, gold ball bonding follows the same steps as thermocompression bonding. However, it can take place at a lower temperature. This benefit is provided by a pulse of ultrasonic energy that is sent through the capillary into the wire. This additional energy is sufficient to provide the heat and friction to form a strong alloy bond.

The majority of production gold wire bonding is done on automatic machines that use sophisticated techniques to locate the pads and span the wire to the correct inner lead. The fastest bonding machines can perform thousands of bonds per hour. There are two major drawbacks to the use of gold bonding wires. First is the expense of the gold. Second is an undesirable alloy that can form between the gold and aluminum. This alloy can severely reduce the conduction ability of the bond. It forms a purplish color and is known as the "purple plague."

**Aluminum wire bonding.**    Aluminum wire, while not having the conduction and corrosion-resistance properties of gold, is still an important bonding wire material. A primary advantage is its lower cost. The second advantage is that the bond with the aluminum bonding pad is a monometal system and thus less susceptible to corrosion. Also, aluminum bonding can take place at lower temperatures than gold bonding, making it more compatible with the use of epoxy die-attach adhesives.

The bonding of the aluminum follows the same major steps as gold wire bonding. However, the method of forming the bond is different. No ball is formed. Instead, after the aluminum wire is positioned over the bonding pad, a wedge (Fig. 18.20) forces the wire onto the pad as a pulse of ultrasonic energy is sent down the wedge to form the bond. Af-

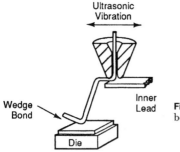

**Figure 18.20** Aluminum wedge bonding.

ter the bond is formed, the wire is spanned to the inner lead where another ultrasonic-assisted wedge bond is formed. This type of bonding is known variously as *ultrasonic* or *wedge bonding*.

After this bond, the wire is cut. At this point in the process, a major difference between the bonding of the two materials occurs. In gold bonding, the capillary moves freely from pad to inner lead, to pad, and so forth, with the package in a fixed position. In aluminum wire bonding, the package must be repositioned for every single bonding step. The repositioning is necessary to line up the pad and inner lead along the direction of travel of the wedge and wire. This requirement places an additional difficulty on the designers of automatic aluminum bonding machines. Nevertheless, most production aluminum bonding is done on high-speed machines.

### Bump/flip-chip bonding

Wire bonding presents several problems. There are electrical resistances associated with each bond. There are minimum height limits imposed by the required wire loops. There is the chance of electrical performance problems or shorting if the wires come to close to each other. Plus, the wires require an individual bonding step at both the chip bonding pad and at the package lead. Perhaps the biggest problem results from the increasing number of connections (pin count) needed to operate larger circuits. Chip designers simply run out of space to locate the required number of connections around the periphery of the chip. These issues are addressed by replacing wires with a deposited metal *bump* on each bonding pad. The bumps are also called *balls*, as in naming packages using bump/flip-chip processes as ball grid arrays (BGAs). This bonding method allows chip design with bonding pads located both along the edge of the die as well as in the interior of the die (Fig. 18.21). These locations place the bump closer to the chip circuitry, increasing signal processing speed. Connection to the package is made when the chip is flipped over and the bump soldered to a corresponding package inner lead (Fig. 18.22) on a package or printed circuit board. IBM calls their version of this technology controlled collapse chip connection (C4).[5]

This process leaves the die suspended above the package surface. Physical stresses and strains are absorbed by the soft solder bump. Additional stress tolerance is provided by filling the gap with an epoxy filling, called and underfill.

Bump connection technology starts in the wafer fabrication process. Wafers are processed through the usual metallization, passivation, and bonding pad patterning processes. The last patterning process leaves an opening in the passivation layer over the bonding pads.

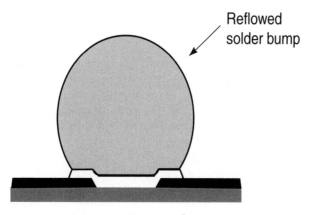

**Figure 18.21**  Reflowed solder bump. [5]

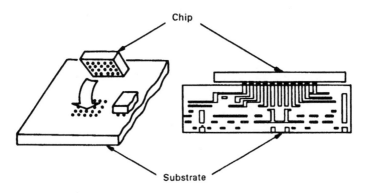

**Figure 18.22**  Flip-chip joining. *(Source:* Microelectronics Handbook, *by Tummala and Rymaszewski.)*

A number of process flows are available to form the solder bumps on the bonding pad. The process described below is an example.

**Sputter deposit intermetal stack.**   Lead/tin solder balls are the preferred "bump" material. However, an intermetal layer (or stack) is required between the bonding pad and the solder ball to prevent the lead from diffusing into the aluminum pad and to aid adhesion of the solder ball onto the pad. Various metal stacks are used, including chrome-copper-gold (Cr-Cu-Au), titanium-nickel (Ti-Ni), and plain copper (see "Alternative process").

**Patterning step of bump location.**   A patterning step covers the die surface with resist, leaving openings over the bonding pads and surrounding dielectric. The resist layer is thick enough to accommodate enough solder to form a ball of sufficient volume to provide structural support and lower electrical resistance between the chip and package or substrate.

**Deposit of intermediate layer stack.**    The intermetal materials are evaporated or sputtered through the openings on the pad.

**Deposit lead/tin solder.**    Deposit of the lead/tin solder can be by electroplating or evaporation. If electroplating is used, a seed layer is deposited before the electroplating. Lead/tin is used to lower the melting point of the solder.

**Remove resist.**    The resist is removed, leaving a bump of lead/tin connected to the bonding pad.

### Alternative Process

Figure 18.23 shows an alternative process. It starts with a blanket sputter deposition of a Ti/Ni or copper. After a patterning step over the bonding pad, there is another deposit of Ni, followed plating of the solder plug. The system is completed to with removal of the resist and etching away the intermetal materials left on the surface. The end result is similar to the first process—a ball of lead/tin solder positioned on the bonding pad.

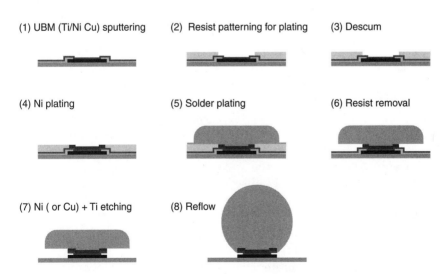

(1) UBM (Ti/Ni Cu) sputtering      (2) Resist patterning for plating      (3) Descum

(4) Ni plating      (5) Solder plating      (6) Resist removal

(7) Ni ( or Cu) + Ti etching      (8) Reflow

**Figure 18.23**  Alternative bump process. *(Courtesy of Future Fab International.)*

## Transfer to Packaging Area

Once these wafer fabrication processes are complete, the wafer is transferred to the packaging operation.

### Reflow

A heating of the wafer in the 350°C range in hydrogen enhances the electrical contact to the intermetal stack. Also, surface tension pulls the solder deposit into a spherical shape, much like surface tension forms spherical soap bubbles.

### Die separation and die pick and place

These two steps are the same as in the wire bonding process. Typically, the die are affixed to a flexible tape on a reel that allows automatic handling in the subsequent processes.

### Alignment of die to package

Packages used in flip-chip systems are usually ceramic or organic. *Organic* is the term used for packages based on the same materials used for circuit boards. Sometimes they are referred to as "plastic" packages. The first step is flipping the die upside down and aligning it to the package such that the bumps on the die are positioned over the corresponding leads on the package.

### Attachment to package (or substrate)

In the presence of a solder flux, the die/package combination is heated in an oven to melt the solder ball to the package lead.

### Deflux

A cleaning process removes excess flux from the surface.

### Underfillment

An epoxy is introduced under the corners of the die soldered to the package. When heated, the epoxy is drawn by capillary action underneath the entire chip. After a curing step it provides a "cushion" to absorb additional stresses on the system. The package designs using the bump/flip-chip technique are discussed later on in this chapter.

### Encapsulation

Typically, a molding compound is deposited over the upside-down chip to provide environmental and physical protection. The combination of the substrate, die, and protective top layer form the functions of a tradition package.

**Tape automated bonding.** Tape automated bonding (TAB) is used to wire up chips when extreme package thinness is required, such as in credit-card-size radios. The TAB process starts with formation of the lead system on a flexible strip of tape. Various methods are used to form the lead system. The metal for the system is deposited on the tape by sputtering or evaporation. Formation of the lead system is either by mechanical stamping or patterning techniques similar to the fabrication patterning process. The result is a continuous tape containing many individual lead systems. For the bonding operation (Fig. 18.24), the chip is positioned on a chuck, and the tape is moved by sprockets until one of these lead systems is positioned exactly over the chip. In this position, the inner leads of the system should be positioned over the bonding pads of the chip.

The bond is completed with a tool known as a *thermode*. The thermode is faced with a flat diamond surface and is heated. The thermode is moved downward, first contacting the inner leads. It continues downward with enough pressure to force the inner leads onto the bonding pads. The heat and the pressure are regulated to cause a physical and electrical bond between the two. Large chips require a larger TAB bonding area. For these chips, the thermode is faced with a synthetic diamond.

TAB bonding is also used in conjunction with package bonding. The advantages of TAB are speed, in that all of the bonds to the chip are made in one action, and the ease of automation offered by the tape and sprocket system.

**Preseal inspection**

After bonding, a number of steps are required to complete the packaging operation. The packaging steps described follow the traditional wire bonding and individual package process flow. Most of them have to be performed at some point in the bump/flip-chip and tape auto-

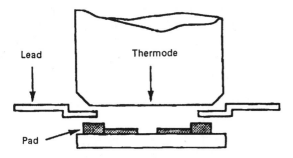

**Figure 18.24** Tape automated bonding. *(Source:* Microelectronics Handbook, *by Tummala and Rymaszewski.)*

mated bonding processes. An important step in the traditional chip-packaging process is the preseal or precap inspection, sometimes called *third optical inspection,* which takes place after the bonding step. The inspection is performed to provide feedback on the quality of the operations already performed. It also is performed to reject packaged chips that may represent reliability hazards when the chip is in operation in the field.

While there are many levels of inspection criteria, they fall into two main categories: commercial and high-reliability. The commercial inspections are given to chips and packages destined for use in commercial systems and are derived from the internal quality levels of the producing company in conjunction with its experience and customer specifications. The high-reliability specifications are derived from a set of government standards identified as "Mil-Standard 883." Commercial-level inspections screen the bonded chips for die-attach quality, correct placement of the bonds on the bonding pads and inner leads, the shape and quality of the ball or wedge bond, and the general condition of the chip surface in regard to contamination, scratches, etc. Mil-Standard 883 covers the same general issues as the commercial inspections but to more stringent requirements. In particular, this standard also specifies criteria for the chip surface, including pattern alignment, critical dimensions, and surface irregularities, such as small scratches, voids, and small defects. These criteria serve to reject bonded chips that may malfunction in the rigorous conditions encountered in space and military operation.

### Sealing techniques

After the bonded chip passes the optical inspection, it is ready for sealing in a protective enclosure. Several methods are used to achieve the enclosure of the chip. The method chosen depends on whether the seal must be hermetic or nonhermetic and which type of package is to be used. The principal sealing methods use welded seals, soldered seals, glass-sealed lids, CERDIP package construction, molded epoxy enclosures, and a sealing layer (see "Blob top") directly on a die bonded directly to a flip-chip package or printed circuit board (Fig. 18.25).

Hermetic
- Welding
- Soldered Lid
- Glass-Sealed Lid or Top

Nonhermetic
- Epoxy Molding
- Blob Top

**Figure 18.25** Package sealing methods.

**Metal can.**    If the package is a metal can type, a hermetic seal is achieved by welding the flanged lid to a matching flange on the base of the package.

**Premade packages.**    Premade ceramic packages are sealed by one of two methods, metal or ceramic lids (Fig. 18.26). Packages made for metal lids have a ring of gold around the top of the die-attach cavity, called the *seal ring*. Placed on top of the seal ring area is a performed piece of gold-tin solder. The metal lid is clamped in position over the seal ring and placed on the belt of a conveyor furnace. As the clamped package passes through the furnace, the lid and package are soldered together to form a hermetic seal. The sealing takes place at a temperature range of 320 to 360°C in a pure nitrogen atmosphere. If the package is to receive a ceramic lid, the procedure is similar. The part of the ceramic lid that contacts the base, outside the die attach area, is coated with a layer of low-melting-point glass. The hermetic seal is completed as the package passes through a conveyor furnace. This sealing takes place at a temperature of about 400°C in an atmosphere of clean, dry air.

**CERDIP packages.**    A completed CERDIP package results in a hermetic seal around the chip and bonding wires. This seal is accomplished with glass, similar to the ceramic seal on the premade packages.

In the case of the CERDIP package, the inner metal lead system is buried in a glass layer. The ceramic top of the package system has a cavity (Fig. 18.27). The underside of the lid, outside the cavity area, is coated with a layer of low-melting-point glass. The lid is placed over the base and clamped. The seal is accomplished as the assembly is passed through a conveyor furnace or placed in an oven. In the furnace or oven, the glass melts, fusing the base and top together. The CERDIP glass sealing system is used to seal DIP and flat packs. These latter packages are known as *Cerpacks* and *Cerflats*.

**Molded epoxy enclosures.**    The fourth major method of enclosure, *epoxy molding,* produces the plastic package (Fig. 18.28). The resultant

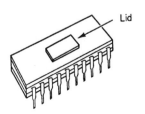

**Figure 18.26**  Premade ceramic package.

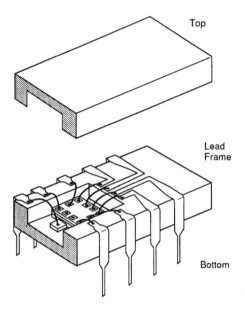

Top

Lead
Frame

Bottom     **Figure 18.27**  CERDIP parts.

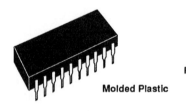

**Figure 18.28**  Molded CERDIP.

Molded Plastic

seal, while protecting the die from moisture and contamination, is not classified as hermetic. However, there exists a considerable amount of research into the development of improved epoxy materials to create better enclosures. The major advantages of epoxy molded enclosures are weight, low material cost, and manufacturing efficiency.

This sealing method follows a different process flow. The die is attached and bonded to a lead frame containing a number of lead systems (Fig. 18.29). After the preseal inspection, the lead frames are transferred to the molding area. The frames are placed on a mold mounted in a transfer molding machine. The molding machine is in turn charged with pellets of the epoxy material, which have been previously softened by a radio frequency heater. Inside, the pellets are forced by a ram into a liquid state. The ram then forces the liquid around the die on the lead frames, forming an individual package around each lead frame. After the epoxy sets in the mold, the frames are removed and placed in an oven for final curing.

**Figure 18.29**  Lead frame.

## Lead plating

An important feature of the completed package is the finish on the package leads. Most package leads are coated with lead-tin solder, tin plate, or gold plate. The plating serves several important functions.

First is the solderability of the leads into a circuit board. The additional metal finish improves the lead solderability, resulting in a more reliable electrical connection of the package and the printed circuit.

The second benefit of the lead finish is that it protects the leads from oxidation or corrosion during periods of storage prior to mounting on the circuit board. The third benefit of lead plating is the protection of the leads from corrosive agents in the packaging and printed-circuit board mounting processes. These agents include solder flux, corrosive cleaners, and even tap water. The plating continues to protect the leads during their lifetime of use.

**Electrolytic plating.**    Plated layers, such as gold and tin, are applied by electrolytic procedures. The packages are mounted on racks with each lead connected to an electric potential. The racks are placed in a tank containing a plating solution. Next, a small current is passed between the packages and an electrode in the tank. The current causes the particular metal in the solution to plate onto the leads.

**Tin-lead solder.**    Tin-lead solder layers are applied either by dipping the packages into a pot of the molten solder or by a wave soldering technique. This latter technique offers the advantage of good control of the layer thickness and provides a shorter exposure of the package to the molten solder.

### Plating process flows

Metal cans, side-brazed DIP packs, and pin grid arrays have their leads plated before starting into the packaging process. CERDIP and plastic packages go through the plating process after the sealing steps.

### Lead trimming

One of the last steps in the package assembly process is trimming away excess material from the leads. The outer leads of DIP and flat-pack packages are made with a tie-bar (Fig. 18.30). This bar keeps the leads from becoming bent during the packaging process. At the end of the process, the package goes through a simple trimming machine that simultaneously trims away the tie-bar and trims the leads to the proper length.

Plastic package lead frames have an additional piece of material. It is a bridge of metal close to the package body that functions as a dam to prevent the liquid epoxy material from running into the lead area (Fig. 18.31). The dam is cut away from the frame with a series of precise cutting tools. After the dam is cut away, the packages move to another station on the cutter where the frame is separated into individual packages. If the package is a surface-mount type, the leads will be bent into the required shape.

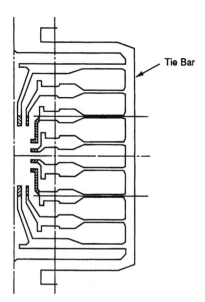

Tie Bar

**Figure 18.30**  Tie-bar.

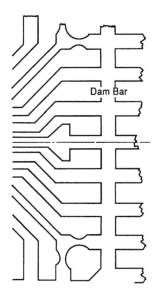

Dam Bar

**Figure 18.31**   Lead frame dam.

## Deflashing

Plastic packages receive an additional process, called *deflashing*, which is required to remove excess molding material from the package enclosure. Deflashing is done by either dipping the packages in a chemical bath followed by a rinse or by a physical abrasion process. Physical deflashing is done in a machine similar to a sandblaster that uses plastic beads as the abrasive.

## Package marking

Once completed, a package must be identified with key information. Typical information coded on the package is the product type, device specifications, date, lot number, and where it was made. The main methods of marking are ink printing and laser inscription. Ink printing has the advantage of good adherence to all of the package materials.

The composition of the ink is chosen for permanence in the eventual operating environment of the device. The ink is applied by an offset printer followed by a curing step. Curing is done by oven drying, room-temperature air drying, or by ultraviolet light. Laser printing is especially well suited to plastic packages. The mark is permanently inscribed in the package surface and can provide good contrast on the dark packages. Additionally, laser marking is fast and noncontaminating, since no foreign material is added to the package surface, and no curing step is required. A drawback to laser marking is the difficulty

of changing the mark if a wrong code was used or the status of the device is changed. Regardless of the marking method, all marks must
meet the requirements of legibility (especially on smaller packages)
and permanence when exposed to harsh environments.

### Final testing

At the conclusion of the packaging process, the completed package is
put through a series of environmental, electrical, and reliability tests.
These tests vary in type and specifications, depending on the customer
and use of the packaged devices. The tests may be performed on all of
the packages in a lot or on selected samples.

### Environmental tests

Environmental tests are performed to weed out leaking and defective
packages. Defects detected are loose chips, contaminants and particles
in the die-attach cavity, and faulty bonding. This testing series starts
with a stabilization bake to drive off any volatile substances in the
package. A typical bake is at 150°C for 24 hours.

The first environmental test is *temperature cycling*. The packages
are loaded into a chamber and cycled between two temperature extremes. The number of cycles may reach several hundred. The high
and low temperatures of this test vary with the device use. Commercial parts receive a narrower temperature range than hi-rel parts. The
hi-rel cycle range is –25 to 125°C. During the cycling, any weakness in
the seal, die attachment, or bonding will be aggravated and detected
in later electrical tests.

A second environmental test is *constant acceleration.* In this test,
the packages are accelerated in a centrifuge (Fig. 18.32) that creates a
force as high as 30,000 times the pull of gravity on the Earth (30,000
*g*'s). During the acceleration, loose particles in the package, poorly attached dies, and weakly attached bonds are stressed so that they will
be detected at the final electrical tests.

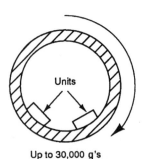

Units

Up to 30,000 g's

Figure 18.32   Acceleration test.

Leaks in the package enclosure are detected by two tests. *Gross leak* testing (Fig. 18.33) is conducted by submerging the packages in a hot liquid. The heated liquid raises the temperature of the package and forces trapped gases in the cavity to escape. The escaping gases are observable as bubbles rising in the liquid. The chamber has a transparent side, allowing an operator to observe the bubbles. Smaller (or fine) leaks are detected by using tracer gases. For this check, helium is pumped under pressure into a chamber containing the packages. If the package has small leaks, the gas will be pumped into the package cavity. The gas is detected as it escapes through the small leaks by a machine known as a mass spectrometer, which can identify the escaping gas. An alternate fine leak test uses the radioactive gas krypton-85. It too is pumped through any leaks into the package under pressure. Detection of any krypton-85 in the package is by a device similar to a Geiger counter.

### Electrical testing

The purpose of the wafer fabrication and packaging processes is to provide to the customer a specific semiconductor device that performs to specific parameters. Thus, one of the last steps is an electrical test of the completed unit to verify that it performs to specifications. The tests are similar to the wafer sort tests. The overall objective is to verify that the good chips identified at wafer sort have not been compromised by the packaging process.

First, there is a series of parametric tests. These electrical tests check the general performance of the device or circuit and ensure that it meets certain input and output voltage, capacitance, and current specifications. The second part of the final test is called the *functional test*. This test actually exercises the specific chip functioning. Logic chips are put through logic tests, and memory chips are exercised in their data storage and retrieval capabilities. The equipment used to conduct the final test is electrically similar to that used in the wafer-sort operation. The electrical tests are performed by a computer-con-

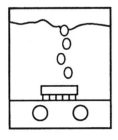

**Figure 18.33**  Gross leak bubbles.

trolled tester that directs the sequences and levels of the parametric and functional tests. The packages are connected to the tester through sockets; the socket unit is known as the *test head*. The packages are inserted into the test head manually or by an automatic unit known as a *handler* (Fig. 18.34). This handler may be mechanical or robotic, depending on the speed and complexity of the operation.

### Burn-in tests

The last of the tests is the optional burn-in test(s). The reason it is optional is that, although it is required for all high-reliability device lots, it may or may not be performed on commercial devices. The test requires the insertion of the packages in sockets and mounting in a chamber with temperature-cycling capability. During the test, the circuits are temperature cycled while under an electrical bias.

The burn-in test is intended to stress the electrical interconnection of the chip and package and drive any contaminants in the body of the chip into the active circuitry, thus causing failure. This test is based on data that indicate that chips prone to these types of failures actually malfunction in the early part of their lifetime. By conducting burn-in tests, the early failures are detected. The devices passing the test are statistically more reliable.

### Package Process Flows

It should be obvious that there is no universal packaging process flow. The package construction technique and lead-plating requirements dictate the flow. The table in Fig. 18.35 illustrates typical flows for three of the more common package types. As reported bump/flip-chip bonding techniques follow a completely different process flow.

### Package/Bare Die Strategies

Integrating a chip into an electronic product or system requires an electrical subsystem between the chip and the system. Two main ap-

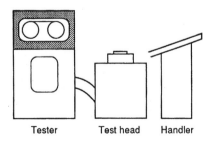

**Figure 18.34**  Final test.

Tester        Test head      Handler

| Process Step | Premade Ceramic | CERDIP | Epoxy Molded |
|---|---|---|---|
| Die Separation | X | X | X |
| Pick and Plate | X | X | X |
| Die Attach | X | X | X |
| Wire Bonding | X | X | X |
| Preseal Inspect | X | X | X |
| Lid Sealing | X | X | |
| Epoxy Molding | X | | |
| Deflashing | X | | |
| Lead Plating and Trim | | X | X |
| Marking | X | X | X |
| Environmental Test | X | X | X |
| Final Electric Test | X | X | X |
| Burn-In | X | X | X |

**Figure 18.35**  Table of packaging process flows.

proaches have emerged: individual packaging and bare die techniques (Fig. 18.36). There are two principal bare die techniques: wire bonding and bump/flip-chip. The wire bonding technique starts with adhesion of the die, face up, on the printed circuit board (PCB) or substrate. Each bonding pad is wire bonded to a corresponding electrical lead on the board. The bump/flip chip technique uses the same technology described to connect die directly to a printed circuit board surface.

### Package/PCB connections

On the package side, there are four primary techniques used to connect the package to a printed circuit board (PCB). They are through-hole, surface mount, TAB, and ball (bump) technology. Through-hole connections feature straight pins on the package, which are inserted into corresponding holes in the printed circuit board (Fig. 18.37). A newer method is *surface mount,* also referred to as *SMD.* This method

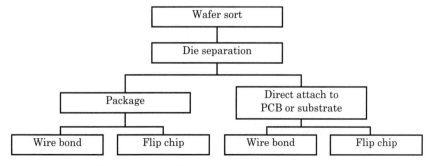

**Figure 18.36**  Post-die-separation options.

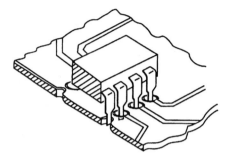

**Figure 18.37** DIP through-hole assembly.

features packages that have their leads bent into a J shape or bent outward to allow direct soldering to the board surface (Fig. 18.38). Some SMD packages do not have leads; rather, they terminate in metal traces hugging the package body. They are known as *leadless packs*. For inclusion on a circuit board, they are inserted into chip carriers, which in turn have the leads that attach to the printed circuit board. Last, there is the bump technology used to connect die to packages, adapted to connecting packages to a PCB.

Tape automated bonding (TAB) has two uses. One is bonding the chip bonding pads directly to the lead frame (see section on bonding). TAB is also a technique for bonding the outer leads directly to the PCB.

### Package Design

Up to the early 1970s, most chips ended up in either a metal package, known as a "can," or in the familiar dual in-line package (DIP). The trends in chip size and integration levels and new electronic products with special packaging requirements (smart cards) have driven the development of new packages and strategies.

### Metal cans

Metal cans are cylindrically shaped packages with an array of leads extending through the base (see Fig. 18.39). The chip is attached to

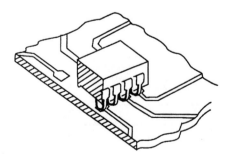

**Figure 18.38** Surface-mount device.

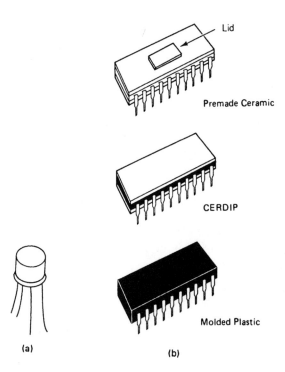

Lid

Premade Ceramic

CERDIP

Molded Plastic

(a)

(b)

**Figure 18.39**   (*a*) Metal can and (*b*) DIP packages.

the base and wire bonded to posts that are connected to the leads. The lid and base have matching flanges that are welded together to create a hermetic seal. These packages are designated by numbers, with the T0-3 and T0-5 being the most common. Metal cans are used to package discrete devices and small-scale integrated circuits.

The DIP is probably the most familiar package design. It features a thick sturdy body with two rows of outer leads coming out of the side and bending downward. DIPs are constructed by three different techniques (Fig. 18.39). Chips designed for high-reliability use will be packaged in a premade ceramic DIP. The package is formed with a solid body of ceramic with the leads buried in the ceramic. The die attachment area is a cavity recessed into the body. The hermetic seal is completed by a soldered metal lid or a glass-sealed ceramic lid.

Another approach to the DIP is the CERDIP, which stands for ceramic DIP. This type of package is composed of a bottom ceramic base with the lead frame held firm in a glass layer. The chip is attached to the base and wire bonded to the lead frame. The hermetic seal is completed with a ceramic top glass sealed to the base. CERDIP construction is used for a number of package types. The vast majority of DIPs are made by the epoxy molding technique. In this technique, the chip is attached to a lead frame and then wire bonded. After bonding, the frame is placed in a molding machine and the package is formed around the chip, wire bonds, and inner leads.

### Pin grid arrays

Larger chips, with more leads, have outgrown the DIP configuration. The pin grid array is a package designed for larger chips. It features a premade "sandwich" with the outer leads coming out of the bottom of the package in the form of pins (Fig. 18.40). The chip is attached in a cavity that is formed in either the top or bottom of the body, usually using bump/flip-chip technology. Connections to the chip cover the entire chip area, unlike most chips with connections restricted to bonding pads around the periphery of the chip. Ceramic PGAs are hermetically sealed with a soldered metal lid.

### Ball grid arrays (BGAs) or flip-chip ball grid arrays (FCBGAs)

BGAs have the same body shape as PGAs. Instead of pins on the package bottom, there is a series of solder bumps (balls) used to connect the package to the PCB (Fig. 18.40). This is essentially the same technology used to connect die to packages.

The effect is to lower the package profile and weight as well as providing higher pin counts by using the whole chip surface for bonding pads. Balls (or bumps) also bring the aspect absorbing stresses created from the thermal expansion differences between the package and the PCB.

(a)

(b)

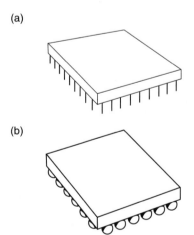

**Figure 18.40** (*a*) Pin grid array. and (*b*) ball grid array.

## Quad packages

While pin grid array packs are a convenient design for larger chips, their ceramic construction is expensive compared to molded epoxy packages. This consideration led to the development of the "quad" package. A quad (short for quadrant) pack is constructed by the epoxy molded technique but has leads coming out of all four sides of the package (Fig. 18.41).

## Thin packages

New products such as smart cards require thin packages. Several techniques are used to make thinner packages. They are called flat packs (FPs), thin small outline packages (TSOPs), small outline IC (SOIC), or ultra-thin packages (UTPs). Flat packs are constructed by the same techniques used to form DIP packages. These packages are designed with flatter height profiles and have their leads bent out to the side of the package (Fig. 18.42). Ultra-thin packages have total body heights in the 1 mm range. There are also quad flat packs (QFPs).

## Chip-scale packages

In the world of ICs, the perfect package is no package. It is recognized that any package brings with it electrical resistance, weight, the opportunity to degrade the circuit performance, and cost. Overall, the smaller the package, the cheaper the cost of packaging and the benefit of higher densities. Chip-scale packages meet this need. They are simply packages with dimensions within 1.2 times the die size.[6] The

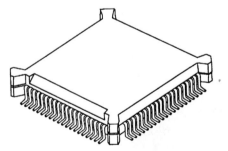

**Figure 18.41**   Quad package.

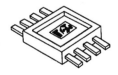

**Figure 18.42**   Flat pack.

challenges are to provide adequate mechanical and environmental protection for easy connection to printed circuit boards.

General design approaches favor flip-chip technology with ball grid arrays and blob top protection. The march to smaller packages and more reliable electrical connections has lead to the micro-ball grid array, or µBGA.

### Lead on chip (LOC)

The LOC package, intended for large die, has the bonding pads arranged down the middle of the chip (Fig. 18.43). Package leads sit on a cushion over the chip surface.

### Bare die techniques and blob top

Increased reliability with higher density and faster circuits presents an ongoing goal and challenge. Reliability is addressed in hybrid circuits. Speed and higher density comes with the elimination of the individual die package. Fewer links in the chain reduce resistances and (in some cases) shorten the length carriers have to travel, increasing speed. This strategy, called *bare die strategy,* is used in hybrid, multichip modules, and chip on board technology.

The most direct use of bare die is to bond them directly to the printed circuit board. Bonding techniques include all the ones used to bond chips into packages. Protecting the chip after connection to the board is by *blob top protection.* The protection is accomplished by covering the chip and bonds with a blob of epoxy resin material (Fig. 18.44).

The material is similar in properties to that used to mold the plastic packages. Blob top coverings are used with TAB and other packaging schemes.

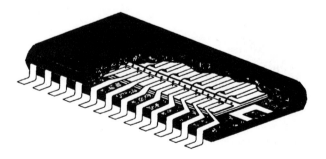

**Figure 18.43** Lead on chip (LOC). *(Source:* Semiconductor International, *May 1993.)*

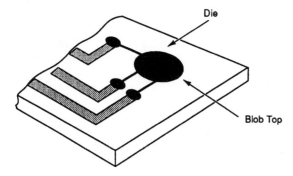

**Figure 18.44**  Blob top.

## Hybrid circuits

Hybrid circuitry is an old technology, long favored for use in military and harsh environments. Hybrid circuits consist of a substrate on which standard electrical and semiconductor devices are mounted. The devices are connected by conductive or resistive thick-film paths on the surface of an insulating substrate. These paths are formed by silk screening inks containing a proper filler or from thin films evaporated or sputtered onto the surface and patterned by photolithography techniques. The term *hybrid* refers to the mix of solid state and conventional passive (resistors, capacitors) electrical devices present in the same circuit. Most hybrid substrates are ceramic. High-performance hybrids may have AlN, SiC, or diamond substrates used alone or in combinations.[7]

Hybrid circuits offer the advantages of structural rigidity and low leakage between devices due to the insulating property of the ceramic. They can provide a circuit with a mix of CMOS, bipolar and other components and offer functions not available in ASIC circuits.[8]

On the downside, they generally have a much lower density than integrated circuits and have a higher cost.

## Multichip modules (MCMs)

Mounting individual chip packages on a PCB presents several problems. A chip package is several times the area of the chip taking up space on the board. Circuit resistance is increased by the individual resistances of all the package pins, and the electrical path lengths are multiplied by the number of chips and package leads. Each of these problems is reduced by mounting several chips on the same substrate. The technology is similar to hybrid circuits, but thick film screened components are not, or rarely, used.

Three types of MCMs have evolved. MCM-L (laminated) is similar to advanced laminated printed circuit boards, using copper foil conductors on plastic dielectrics. MCM-C (ceramic) are more like hybrid circuits. The substrate is a cofired ceramic with thick film conductors. MCM-D (deposited) uses ceramic, metal, or silicon substrates use deposited thin metal conductors.

### The known good die (KGD) problem

In the individual package process, a final test assures quality of the completed product. If the chip has gone bad or the process was faulty, the entire chip and package is discarded. But the cost is much higher when putting bare die into hybrid, MCM, and printed circuit boards. These vehicles are more expensive to build and carry other expensive chips or components.

One option is to rely on the results of the wafer-sort test to certify die performance. Unfortunately, wafer sort does not include environmental tests or long-term reliability tests. However, performing a final test on a bare die is difficult if not impossible.

### Package Type/Technology Summary

There are thousands of individual package types, and there is no uniform system of identifying them. Some are named by their design (DIP, flat packs, and so on), some are named by their construction technique (molded, CERDIP, and so forth), and others are named by their use, such as SMDs. When trying to understand a package type, keep in mind the three considerations: design type, construction technique, and use.

Into the future, wire-bonded/back-connected chip techniques will be overwhelmed by pin counts numbering in the hundreds. While exact packaging techniques are hard to project, the feeling is that flip-chip, ball grid arrays (BGAs), and chip scale packages (CSPs) will be the package designs of future.[9]

### Review Questions

1. Name the four functions of a chip package.
2. What is the function of marking dies with ink dots?
3. Name the five parts of a package.
4. Define a hermetic package.
5. Give an example of a hermetic and nonhermetic package.

6. List, in order, the six major steps in the packaging process.

7. Describe the process and material used for eutectic die attachment.

8. Describe the principal thermosonic and ultrasonic bonding techniques.

9. What is a DIP package and how is it formed?

10. Name two methods of package marking.

## References

1. H. Hinzen and B. Ripper, "Precision Grinding of Semiconductor Wafers," *Solid State Technology*, PennWell Publishing, August 1993, p. 53.
2. F. I. Blech and D. Dang, "Silicon Wafer Deformation after Backside Grinding," *Solid State Technology*, PennWell Publishing, August 1994, p. 74.
3. L. Plummer, "Packaging Trends," *Semiconductor International*, Cahners Publishing, January 1993, p. 33.
4. R. Iscoff, "Ultrathin Packages: Are They Ahead of Their Time?" *Semiconductor International*, Cahners Publishing, May 1994, p. 50.
5. R. Tummala and E. Rymaszewski, *Microelectronics Packaging Handbook*, Van Nostrand Reinhold, New York, 1989.
6. T. DiStefano and J. Fjelstad, "Chip-scale Packaging Meets Future Design Needs," *Solid State Technology*, April 1996, p. 82.
7. N. Nguyen, "Using Advanced Substrate Materials with Hybrid Packaging Techniques for Ultrahigh-Power ICs," *Solid State Technology*, PennWell Publishing, February 1993, p. 59.
8. R. Iscoff, "Will Hybrid Circuits Survive?" *Semiconductor International*, Cahners Publishing, October 1993, p. 57.
9. J. Baliga, "Package Styles Drive Advancements in Die Bonding," *Semiconductor International*, June 1997, p. 101.

# Glossary*

**acceptor** An impurity that causes semiconducting materials to accept valence electrons, thereby leaving "holes" in the valence band. The holes act like carriers of positive charge, referred to as P type.

**aligner (align and expose)** A process tool used to align wafers and masks or reticles and expose the photoresist with a UV or other radiation source.

**alignment** Refers to the positioning of a mask or reticle with respect to the wafer.

**alignment marks** Targets on the mask and wafer used for correct alignment.

**alloy** (1) A compound composed of two metals. (2) In semiconductor processing, the alloy step causes the interdiffusion of the semiconductor and the material on top of it, forming on ohmic contact between them.

**aluminum (Al)** The metal most often used in semiconductor technology to form the interconnects between devices on a chip. It can be applied by evaporation or sputtering.

**amorphous** Materials with no definite arrangement of atoms, e.g., plastics are amorphous.

**angstrom** A unit of length, an angstrom (Å) is one ten-thousandth of a micron ($10^{-4}$ μm) or 100,000,000 Å = 1 cm.

**anistropic** An etch process that exhibits little or no undercutting.

**anneal** A high-temperature processing step (usually the last one), designed to minimize stress in the crystal structure of the wafer.

**antimony (Sb)** A Group V element that is an N-type dopant in silicon. It is often used as the dopant for the buried layer.

**arsenic** A Group V element that is an N-type dopant in silicon.

**assembly** The series of operations after fabrication in which the wafer is separated into individual chips and mounted and connected to a package.

**atmospheric oxidation** A process of oxidation of silicon carried out at atmospheric pressure. The equipment used for thermal oxidation is the same as that used for thermal diffusion. It is composed of four subassemblies: a reactant source cabinet, a reaction chamber, a heating source, and a wafer holder.

---

*Source: Glossary terms are abstracted from Beverly Griggs, Anne Miller, and Peter Van Zant, *Semiconductor Terminology—Graphic Glossary of Terms,* Semiconductor Services, Redwood City, CA, 1989.

**atomic force microscope (AFM)**   A microscope for profiling wafer surfaces by plotting the output of a spring-balanced probe moved over the surface.

**atomic number**   A number assigned to each element equal to the number of protons (therefore the number of electrons) in the atom.

**atomic particles**   The parts of an atom: electrons, protons, and neutrons.

**base**   (1) The control portion of an NPN or PNP junction transistor. (2) The P-type diffusion done using boron that forms the base of NPN transistors, the emitter and collector of lateral PNP transistors, and resistors.

**binary notation**   A way of representing any number using a power of 2 (the digits 0 and 1).

**bipolar transistor**   A transistor consisting of an emitter, base, and collector, whose action depends on the injection of minority carriers from the base by the collector. Sometimes called NPN or PNP transistor to emphasize its layered structure.

**bi-MOS**   A circuit containing both bipolar and MOS transistors.

**boat**   (1) Pieces of quartz or metal joined together to form a supporting structure for wafers during high-temperature processing steps. (2) A Teflon® or plastic assemblage used to hold wafers during wet processing steps.

**boat puller**   A mechanical arrangement to push a boat loaded with wafers into a furnace and/or withdraw it at a fixed speed.

**BOE**   See *buffered oxide etch*.

**bonding pads**   Electrical terminals on the chip (generally around the periphery) used for connection to the package electrical system.

**boron (B)**   The P-type dopant commonly used for the isolation and base diffusion in standard bipolar integrated circuit processing.

**boron trichloride (BCl$_3$)**   A gas that is often used as a source of boron for doping silicon.

**bubbler**   An apparatus in which a carrier gas is "bubbled" through a heated liquid, causing portions of the liquid to be transported with the gas, e.g., a carrier gas (nitrogen or oxygen) is bubbled through deionized water at 98 to 99°C on its way to the oxidation tube.

**buffered oxide etch**   A mix of hydrogen fluoride (HF) and ammonium fluoride (NH$_4$F) used to allow oxide etching to occur at a slow, controlled rate.

**bump connection technology**   A structure of metal bumps formed on the bonding pads, allowing connection of the chip to a package when the chip is flipped over.

**buried layer**   The N+ diffusion in the P-type substrate done just prior to growing the epitaxial layer. The buried layer provides a low-resistance path for current flowing in a device. Common buried layer dopants are antimony and arsenic.

**can**  A metal package used for connecting a chip to a printed circuit board with from three to five leads.

**capacitor**  A discrete device that stores electrical charge on two conductors separated by a dielectric.

**capacitance**  Electrical charge storage capability.

**capacitance-voltage plot (C/V plot)**  A plot that provides information on the amount of mobile ionic contamination present in the oxide.

**carrier gas**  An inert gas that will transport atoms or molecules of a desired substance to a reaction chamber.

**centistokes**  Units used to measure viscosity; centipoise divided by density.

**channel**  A thin region of a semiconductor that supports conduction. A channel may occur at a surface or in the bulk, essential for the operation of MOS-FETs and SIGFETs. In cases where channels are not part of the circuit design, their presence may indicate contamination problems or incomplete isolation.

**channeling**  A phenomenon in which an ion beam will penetrate into the crystal planes of the wafer. Preventing channeling is accomplished by cutting the wafer "off orientation." The effect is to tilt the crystal planes relative to the beam direction.

**charge carrier**  A carrier of electrical charge within the crystal of a solid-state device, such as an electron or hole.

**chemical etching**  Selective removal of material by means of liquid reactants. The precision of the etch is controlled by the temperature of the etchant, the time of immersion and the composition of the acid etchant.

**chemical mechanical polishing (CMP)**  A wafer flattening and polishing process that combines chemical removal with mechanical buffing. Used for polishing/flattening wafers after crystal growing and wafer planarization during the wafer fabrication process.

**chemical vapor deposition**  See *CVD*.

**chip**  Die or device, one of the individual integrated circuits or discrete devices on a wafer.

**chrome**  A metal often used in mask fabrication to form the layer in which the circuit pattern is generated.

**circuit board**  See *printed circuit board.*

**circuit layout**  The calculation of the physical device dimensions required to produce the required electrical parameters. Vertical dimensions determine CVD and doping thickness specifications. Horizontal dimensions determine the wafer pattern dimensions and are the basis for a scaled drawing of the finished circuit (composite drawing).

**class number**  Number of contaminant particles in a cubic foot of air.

**cleanroom**   An area in which semiconductor device fabrication takes place. The cleanliness of the room is highly controlled so as to limit the number of contaminants to which the semiconductor is exposed.

**clear field mask**   A mask on which the pattern is defined by the opaque areas.

**cluster tool**   Several process stations or tools served by one loading–unloading chamber and wafer-transport system.

**CMOS (complementary field-effect transistor)**   N- and P-channel MOS transistors on the same chip.

**collector**   Along with the emitter and base, one of the three regions of the bipolar type of transistor.

**collimated light**   Light in which the rays are parallel; used for gross visual inspection of surfaces.

**composite drawing**   A scaled drawing of the finished circuit.

**conductivity**   The ability of materials to conduct electricity (measured in siemens for conductance or ohms for resistance).

**conductor**   A material that has low resistivity and high conductivity.

**contact**   The regions of exposed silicon that are covered during the metallization process to provide electrical access to the devices.

**contact aligner**   An aligner tool that clamps the wafer and mask into a tight contact before the resist exposure cycle.

**contact mask**   The step at which holes are put into the wafer layers to allow the metal layer to reach down to the doped silicon substrate.

**contamination**   A general term used to describe unwanted material that adversely affects the physical or electrical characteristics of a semiconductor wafer.

**copper (Cu)**   Metal used to connect semiconductor devices on a chip surface. Generally used with a dual-damascene patterning process.

**critical dimensions (CDs)**   The widths of the lines and spaces of critical circuit patterns as well as the area of contacts.

**cryogenic pump**   A vacuum pump that can produce a vacuum to the $10^{-10}$ torr range, the same level as the vacuum of space. It does not require forepumps or cold traps and is faster than other types of vacuum pumps.

**crystal**   A material in which the atoms are arranged in structured groups called *unit cells*.

**crystal defects**   Vacancies and dislocations in a crystal that influence the electrical performance of a circuit.

**crystal orientation**   The orientation of the primary crystal plane, expressed as Miller indices.

**crystal planes** The planes in the semiconductor crystal structure along which the die must be aligned so as to prevent "ragged" die edges when the wafer is separated into individual die.

**CUM yield** See *fabrication yield.*

**current** A measure of the number of charged particles passing a given point per unit time.

**curve tracer** A piece of electrical test equipment that displays the characteristics of a device visually on a screen.

**CVD (chemical vapor deposition)** A method for depositing some of the layers that function as dielectrics, conductors, or semiconductors. A chemical containing atoms of the material to be deposited reacts with another chemical, liberating the desired material, which deposits on the wafer while by-products of the reaction are removed from the reaction chamber.

**Czochralski crystal grower** A type of crystal grower that uses a seed to pull a crystal from a crucible of molten material.

**dark field mask** A mask on which the pattern is defined by the clear portion of the mask.

**deep ultraviolet (DUV)** A light wavelength often used to expose photoresist; it has the advantage of an ability to produce smaller image widths.

**defect density** The density of defects per square centimeter on a chip.

**dehydration baking** A heating process by which wafer surfaces are restored to a hydrophobic condition by baking. Surface water is evaporated from the wafer at elevated temperatures.

**deionized (DI) water** Process water that is free of dissolved ions. Specification levels are generally 15 to 18 m$\Omega$ of resistance.

**depletion layer** The region in a semiconductor where essentially all charge carriers have been swept out by the electric field that exists there.

**deposition** Process in which layers are formed as the result of a chemical reaction in which the desired layer material is formed and coats the wafer surface.

**design rule** The minimum feature size of a circuit.

**develop inspection** The first inspection in the photomasking process consisting of measurement of critical dimensions and inspection for defects. It is done after development or development and hard bake if an automatic baking system is used.

**development** A photoresist processing step in which photoresist is removed from areas defined by the masking and exposure step of wafer fabrication.

**developer** Chemical used to remove areas defined in the masking and exposure step of wafer fabrication.

**device** A single-function component such as a transistor, resistor, or capacitor.

**DI water** Deionized water; purity of this water is measured by its resistivity, with the standard being 18 MΩ.

**diborane ($B_2H_6$)** A gas that is often used as a source of boron for doping silicon.

**die** One unit on a wafer separated by scribe lines; after all of the wafer fabrication steps are completed, die are separated by sawing; the separated units are referred to as *chips*.

**die bonding** Assembly step in which individual chips are attached to the package with conductive adhesives or metal alloys.

**die sort** See *wafer sort*.

**dielectric** A material that conducts no current when it has a voltage across it. Two dielectrics encountered in semiconductor processing are silicon dioxide and silicon nitride.

**diffusion** A process used in semiconductor production that introduces minute amounts of impurities (dopants) into a substrate material such as silicon or germanium and permits the impurity to spread into the substrate. The process is very dependent on temperature and time.

**diffusivity** The rate of movement or diffusion of dopants in a semiconductor.

**diode** Device that enables current flow in one direction but not in the other.

**DIP (dual in-line package)** A rectangular circuit package, with leads coming out of the long sides and bent down to fit into a socket.

**discrete device** A circuit having a single electrical function. Discrete devices include capacitors, resistors, transistors, diodes, and fuses.

**dislocation** A discontinuity in the crystal lattice, a type of crystal defect.

**DMOS (diffused MOS)** A transistor structure that features a narrow (channel length) separation between the source and drain. The channel length is created by two sequential diffusions through the same hole.

**donor** An impurity that can make a semiconductor N-type by donating extra "free" electrons; electrons carry a negative charge.

**dopant** An element that alters the conductivity of a semiconductor by contributing either a hole or electron to the conduction process. For silicon, the dopants are found in Groups III and V of the periodic table.

**dopant deposition** The first step in the diffusion process, in which the dopant atoms are diffused into the wafer surface.

**doping** The introduction of an impurity (dopant) into the crystal lattice of a semiconductor to modify its electronic properties—for example, adding boron to silicon to make the material P type.

**DRAM (dynamic random access memory)** Memory device for the storage of digital information. The information is stored in a "volatile" state.

**drain** Along with the source and gate, one of the three regions of a unipolar or field-effect transistor (FET).

**drive-in** Stage in diffusion where the dopant is driven deeper into the wafer.

**dry etch** See *plasma etch.*

**dry ox** The growth of silicon dioxide using oxygen and hydrogen, which form water vapor at process temperatures, rather than using water vapor directly.

**dry oxide** Thermal silicon dioxide grown using oxygen.

**dual damascene** A patterning process that first defines the required pattern in a trench in the top wafer surface followed by overfilling with a conductive metal. The overfill is removed, usually by a chemical-mechanical-polishing process, leaving the metal pattern within the trench.

**e-beam (electron beam)** An exposure source that allows direct image formation without a mask. An e-beam can be deflected by electrostatic plates and therefore directed to precise locations, resulting in the generation of submicron-size patterns.

**e-beam aligner** An aligner tool that exposes the resist-coated wafer surface by steering (writing) an electron beam across the wafer surface.

**e-beam evaporation (electron beam evaporation)** Phase change that uses the energy of a focused electron beam to provide the required energy to change solid metal or alloys from solid to gas.

**e-beam exposure system** A machine in which the image pattern is stored in a computer memory and used to control the electrostatic plates that, in turn, direct the e-beam, resulting in the generation of patterns without the use of reticles or photomasks.

**edge die** The incomplete die located on the edge of the wafer.

**EEPROM (electrically erasable PROM)** A memory circuit with the capability of data erasure and acceptance of new information by the application of an electrical pulse.

**electromigration** The diffusion of electrons in electric fields set up in the lead while the circuit is in operation. It occurs in aluminum and is exhibited as a field failure, not as a process defect. The metal thins and eventually separates completely, causing an opening in the circuit.

**electron** A charged particle revolving around the nucleus of an atom. It can form bonds with electrons from other atoms or be lost, making the atom an ion.

**ellipsometer** An instrument that uses laser light sources to measure thin film thickness.

**emitter** (1) The region of a transistor that serves as the source or input end for carriers. (2) The N-type diffusion usually done using phosphorus, which forms the emitter of NPN transistors, the base contact of PNP transistors, the N+ contact of NPN transistors, and low-value resistors.

**epitaxial**   (Greek for "arranged upon.") The growth of a single-crystal semi-conductor film upon a single-crystal substrate. The epitaxial layer has the same crystallographic characteristics as the substrate material.

**epoxy package**   See *molded package.*

**EPROM (erasable PROM)**   A memory circuit with the capability of data erasure and acceptance of new information.

**etch**   A process for removing material in a specified area through a wet or dry chemical reaction or by physical removal, such as by sputter etch.

**evaporation**   A process step that uses heat to change a material (usually a metal or metal alloy) from its solid state to a gaseous state with the result of the source being deposited on wafers. Both electron beam and filament evaporation are common in semiconductor processing.

**exposure**   Method of defining patterns by the interaction of light or another form of energy with photoresist that is sensitive to the energy source.

**fabrication**   Integrated circuit manufacturing processes.

**fabrication yield**   The percentage of wafers arriving at wafer sort compared with the number started into the process.

**feature size**   The minimum width of pattern openings or spaces in a device.

**FET (field-effect transistor)**   A transistor consisting of a source, gate, and drain, whose action depends on the flow of majority carriers past the gate from the source to drain. The flow is controlled by the transverse electric field under the gate. See *unipolar transistor.*

**field oxide**   The region on an electrical device where the oxide serves the function of a dielectric.

**final test**   The final assembly step in which the packaged die is put through its last electrical test.

**flash memory**   An EPROM or EEPROM with the capability of block erasure of data in the memory array.

**flat zone**   The highly temperature-controlled region of a tube furnace.

**flip-chip joining**   A chip or package connection process where "bumps" of connecting metal are formed on the chip surface and the chip is "flipped" over for soldering to the package.

**four-point probe**   A piece of electrical test equipment used to determine the sheet resistivity of a wafer.

**furnace**   A piece of equipment containing a resistance-heated element and a temperature controller. It is used to maintain a region of constant temperature with a controlled atmosphere for the processing of semiconductor devices.

**fuse**   A circuit component that can be blown to allow a desired memory cell or logic gate to be programmed.

**gallium arsenide (GaAs)**  The most common of compound semiconductor materials. It has the advantage of producing higher-speed devices than those produced using silicon as a substrate.

**gate**  Along with the source and drain, one of the three regions of the unipolar or field-effect transistor (FET or MOS).

**gate array**  Type of integrated circuit made up of an arrangement of interconnected gates used to provide custom functions.

**gate oxide (gate ox)**  The thin oxide that causes the induction of charge, creating a channel between source and drain regions of a MOS transistor.

**germanium**  Semiconducting material used in the manufacture of crystal diodes and of early transistors.

**HEPA filter (high-efficiency particulate attenuator)**  A filter constructed of fragile fibers in an accordion-folded design that allows a larger filtering area at an air velocity low enough for operator comfort. This filter permits a filtering efficiency of 99.99 percent.

**hexamethyldisilizane (HMDS)**  Primer used to promote photoresist adhesion.

**high-pressure oxidation**  Oxidation carried out at high pressure (10 to 20 atm) to reduce the amount of heat or time required. The reaction chamber for this process must be constructed of stainless steel to safely contain the pressure.

**hole**  (1) The absence of a valence electron in a semiconductor crystal. Motion of a hole is equivalent to motion of a positive charge. (2) A "hole" in a surface layer created by the photomasking process.

**hybrid integrated circuit**  A structure consisting of an assembly of one or more semiconductor devices and a thin-film integrated circuit on a single substrate, usually of ceramic.

**hydrofluoric acid (HF)**  An acid used to etch silicon dioxide; often diluted or buffered before it is used.

**hydrogen ($H_2$)**  A gas used in semiconductor processing primarily as a carrier gas for high-temperature reaction steps such as epitaxial silicon growth.

**hydrophilic**  Affinity toward water (water-loving); a hydrophilic surface is one that will allow water to spread across it in large puddles.

**hydrophobic**  Aversion to water; a hydrophobic surface is one that will not support large pools of water. The water is pulled into droplets on the surface. These surfaces often are termed "dewetted."

**hydroscopic**  Attracts and absorbs water.

**integrated circuit**  A circuit in which many elements are fabricated and interconnected on a single chip of semiconductor material, as opposed to a "nonintegrated" circuit, in which the transistors, diodes, resistors, and so on, are fabricated separately and then assembled.

**integration level** The range of total component count in a die. Varies from SSI (small-scale integration, less than 50 components) to ULSI (ultra-large-scale integration, over 1,000,000 components).

**intrinsic semiconductor** An element or compound that has four electrons in its outer ring (i.e., elements from Group IV of the periodic table or compounds of Group III and V).

**ion** An atom that has either gained or lost electrons, making it a charged particle (either negative or positive).

**ion beam milling** A dry etching method that uses an ion beam. Argon atoms are ionized and accelerated toward a wafer. The exposed areas are removed through a sputtering action.

**ion implantation** Introduction of selected impurities (dopants) by means of high-voltage ion bombardment to achieve desired electronic properties in de- fined areas.

**interconnect** See *lead*.

**isolation diffusion** Diffusion step resulting in P-N junctions surrounding the areas to be separated.

**isotropic etching** Refers to the etching of the photoresist both downward and to the side.

**JFET (junction field-effect transistor)** Device in which voltage is applied to a terminal to control current between the source and drain regions.

**junction** The interface at which the conductivity type of a material changes from P type to N type or vice versa.

**killer defect** A defect that causes the failure of a device or circuit.

**lateral diffusion** The diffusion of dopants from side to side every time the wafer is heated near the diffusion temperature range.

**layering** A process by which thin layers of different materials are grown on, or added to, the wafer surface.

**lead** A metal strip on the wafer surface.

**LED (light-emitting diode)** A semiconductor device in which the energy of minority carriers in combining with holes is converted to light. Usually, but not necessarily, constructed as a P-N junction device.

**lift-off process** A material removal process in which the material is deposited into a hole in a photoresist layer and the pattern defined where the photoresist layer is removed (lifted off) the surface.

**light field mask** See *clear field mask*.

**lithography** Process of pattern transfer; when light is utilized, it is termed photolithography; when patterns are small enough to be measured in microns, it is referred to as microlithography.

**local oxidation of silicon (LOCOS)**   A MOS surface isolation scheme in which silicon dioxide is grown around islands of silicon nitride, which is in turn removed to leave oxide-free areas for the formation of the circuit components.

**low-pressure CVD (LPCVD)**   A CVD system and process performed in a low-pressure range.

**LSI (large-scale integration)**   Refers to chips with between 5000 and 100,000 components each.

**majority carrier**   The mobile charge carrier (hole or electron) that predominates in a semiconductor material—for example, electrons in an N-type region.

**mask**   A glass plate covered with an array of patterns used in the photomasking process. Each pattern consists of opaque and clear areas that, respectively, prevent or allow light through. Masks are aligned with existing patterns on silicon wafers and used to expose photoresist. Mask patterns may be formed in emulsion, chrome, iron oxide, silicon, or a number of other opaque materials.

**masking**   See *patterning*.

**memory**   The storing of data or information.

**metal mask**   The step at which an island of conducting material is left on the wafer surface.

**metalorganic CVD (MOCVD)**   A VPE process that uses halides and metalorganic sources.

**micrometer**   One-millionth of a meter ($10^{-6}$ m); abbreviation is μm.

**micron**   Same as micrometer.

**microchip**   See *chip*.

**Miller indices**   A numerical system of three numbers used to identify the orientation of planes in a crystal.

**mini-environment**   An environment that maintains wafer cleanliness by storing, transporting, and loading or unloading wafers in small, clean enclosures.

**minority carrier**   The nonpredominant mobile charge carrier in a semiconductor—for example, electrons in a P-type region.

**MMOS (memory MOS)**   A nonvolatile memory device structure that enables information to be retained during power shutdown.

**mobile ionic contaminant**   Wafer contaminants that are electrically charged particles, which can cause electrical failures in a wafer or device.

**molded package**   A package of epoxy or other polymer material that is molded around the chip and package lead system.

**molecular beam epitaxy (MBE)**   An evaporative deposition process capable of extreme control of the deposition process.

**molecule**   Smallest quantity of a substance that retains the properties of that substance.

**monochromatic light**   Light of a single wavelength.

**MOSFET**   A field-effect transistor containing a metal gate over thermal oxide over silicon.

**MSI (medium-scale integration)**   Refers to chips with between 50 and 5000 components each.

**multichip module (MCM)**   A package containing two or more IC chips connected by a thin-film metal system.

**multilayer resist process**   An image-resolution process that uses two or more layers of photoresist.

**nanometer (nm)**   length unit = $1 \times 10^{-9}$ meter.

**nanotechnology**   Processes and materials used to build semiconductor devices and other structures with nanometer dimensions.

**negative resist**   Photoresist that remains in areas that were not protected from exposure by the opaque regions of a mask while being removed by the developer in regions that were protected. A negative image of a mask remains following the develop process. A "clear" or "light" field mask is most often used with negative resist.

**nitric acid (HNO$_3$)**   A strong acid often used to clean silicon wafers or etch materials.

**nitridation**   The formation of silicon nitride by the high-temperature exposure of a silicon surface to nitrogen.

**nitrogen (N$_2$)**   A gas that seldom reacts with other materials. It is often used as a carrier gas for chemicals in semiconductor processing.

**NMOS**   N-channel MOS; type of MOSFET in which the channel is negative during conduction.

**nonvolatile memory circuit**   A memory circuit that retains its data when power to the chip is lost.

**N-type**   A semiconductor material in which the majority of carriers are electrons and therefore negative. N-type dopants in silicon are Group V elements, in which the fifth outer electron is free to conduct current.

**NPN transistor**   A transistor that has a base of P-type silicon sandwiched between an emitter and a collector of N-type silicon.

**Ohm's law**   A relationship between resistance, voltage, and current; $R = V/I$.

**oil diffusion pump**   A type of high-vacuum pump that uses evaporated hot oil particles to "push" chamber particles out of the system.

**overall yield**   The percentage of functioning packaged chips from a wafer related to the number of die mapped onto the wafer. Overall yield is the product of fabrication yield, sort yield, and assembly yields.

**oxidation**  The growth of oxide on silicon when exposed to oxygen. This process is highly temperature dependent.

**oxidation reaction chamber**  A chamber in which oxidation takes place. Quartz or silicon carbide tubing is used to make an oxidation reaction chamber due to their thermal resistance and purity.

**oxide**  See *silicon dioxide.*

**oxide etching**  An etching process that uses acid—usually hydrofluoric acid (HF). The acid must be buffered for the reaction to proceed at a rate slow enough to be controlled. Buffered oxide etch (BOE) is often used.

**oxygen ($O_2$)**  Gas used to combine with silicon to form silicon dioxide.

**package**  Protective container for semiconductor chip having electrical leads.

**packaging yield**  The percentage of packaged die passing the final tests compared to the number of good die that entered the packaging process.

**passivation**  Sealing layer added at the end of the fabrication process to prevent deterioration of electronic properties through chemical action, corrosion, or handling during the packaging processes. The passivation layer, usually silicon dioxide or silicon nitride, protects against moisture or contamination.

**patterning**  A process in which the pattern in a reticle or photomask is transferred to a wafer resulting in the identification of areas to be doped or selectively removed.

**pellicle**  A thin film of an optical-grade polymer that is stretched on a frame and secured to a mask or reticle. This solves the problem of airborne dirt collecting on the mask and acting as an opaque spot. During the exposure, the dirt is held out of the focal plane and does not "print" onto the wafer.

**phosphine ($PH_3$)**  A gas that is often used as a source of phosphorus for doping silicon.

**phosphorus (P)**  The N-type dopant commonly used for the sinker and emitter diffusions in standard bipolar integrated circuit technology.

**phosphorus oxychloride ($POCl_3$)**  A liquid that is often used as a source of phosphorus for doping silicon.

**photomasking**  See *patterning.*

**photoplate**  A coated mask blank before imaging.

**photoresist**  The light-sensitive film spun onto wafers and "exposed" using high-intensity light through a mask. The exposed (or unexposed, depending on its polarity) photoresist is dissolved with developers, leaving a pattern of photoresist that allows etching to take place in some areas while preventing it in others.

**pinhole**  A small, undesired hole in the photoresist or in the opaque region of a mask or reticle.

**pin grid array (PGA)**   A large chip package with many leads coming out of the entire bottom surface of the package.

**planar structure**   A flat-surfaced device structure fabricated by diffusion and oxide masking, with the junctions terminating in a single plane.

**plasma**   High-energy gas made up of ionized particles.

**plasma-enhanced CVD (PECVD)**   A CVD system and process using plasma energy to drive the deposition.

**plasma etch**   A dry-etch process using reactive gases energized by a plasma field.

**plastic package**   See *molded package.*

**plug (via plug)**   A metal (generally a refractory metal) deposited in a connecting via hole between the conducting layers of a multilayer metal system.

**PMOS (P-channel MOS)**   Type of MOSFET in which the channel is positive due to conduction achieved by holes.

**PNP**   Semiconductor crystal structure consisting of an N-type region sandwiched between two P-type regions, as commonly used in bipolar transistors.

**polycide MOS gate**   A MOS gate structure composed of a sandwich of silicon dioxide topped with a layer of polysilicon that is covered with a refractory metal silicide.

**polycrystalline silicon (poly)**   Silicon composed of many crystal unit cells randomly arranged.

**polymer**   A complex organic chemical compound made up of repeating units.

**positive resist**   Photoresist that is removed in areas that were not protected from exposure by the opaque regions of a mask while remaining after develop in regions that were protected from exposure. A positive image of the mask remains following the develop process. A "dark field" mask is used most often with positive resist.

**post exposure bake (PEB)**   A baking step performed after resist exposure to reduce standing wave effects.

**predeposition (predep)**   The process step during which a controlled amount of a dopant is introduced into the crystal structure of a semiconductor.

**primer chemical**   A chemical that enhances the adhesion of a desired layer (in semiconductor technology, the layer is usually photoresist).

**programmable read-only memory**   See *PROM.*

**projection alignment**   An exposure system in which the image on the mask is projected onto the wafer. This results in little mask or photoresist damage and has about the same productivity as the contact method. For LSI and VLSI, projection alignment is standard.

**projection aligner**   An aligner tool that projects the mask or reticle image over a distance onto the wafer.

**PROM (programmable read-only memory)**   A technology in which fuses are used in every memory cell, and selected fuses are blown so as to program the chip with user-specified information.

**proximity aligner**   An aligner tool that holds the wafer and mask or reticle a small distance apart during the resist exposure cycle.

**P-type**   Semiconductor material in which the majority carriers are holes and therefore positive. P-type dopants in silicon are Group III-A elements.

**quartz**   Commercial name for silicon dioxide formed into glass products. Because of its high temperature resistance, quartz is used in many processing steps in integrated circuit fabrication.

**RAM (random access memory)**   Device that temporarily stores digital information.

**rapid thermal oxidation (RTO)**   An oxidation process performed in a rapid thermal processing (RTP) tool.

**rapid thermal processing (RTP)**   A single-wafer processing tool that uses high-intensity lights or other sources to heat and cool the wafer in milliseconds.

**RCA clean**   A multiple-step process to clean wafers before oxidation; named after RCA, the company that developed the procedure.

**reactive ion etching (RIE)**   An etching process that combines plasma and ion beam removal of the surface layer. The etchant gas enters the reaction chamber and is ionized. The individual molecules accelerate to the wafer surface. At the surface, the top layer removal is achieved by the physical and chemical removal of the material.

**reactor**   (1) A piece of equipment used for the deposition of a layer of material used in semiconductor processing. Common types of reactors are epitaxial reactors, vapox reactors, and nitride reactors. (2) See *plasma etcher.*

**resistivity**   A measure of the resistance to current flow in a material. A function of the attraction between the outer electrons and inner protons of a material. The more tightly bound the electrons, the greater the resistivity.

**resolution capability**   The minimum feature size capability of a photolithography process or tool.

**reticle**   An exposure mask with only a portion of complete die pattern.

**rinse**   The removal of wet etchants or developers from the wafer. This process results in stopping the etching or developing processes and removing the active chemical from the surface. There are several different methods of rinsing: overflow rinsing, spray rinsing, dump rinsing, and spin-rinse dryers.

**ROM (read-only memory)**   A memory circuit containing permanent data with no capability of accepting new information.

**salicide MOS gate**   A polycide MOS gate structure sequenced in the process to self-align to the source or drain. See *polycide MOS gate.*

**scanning electron microscope (SEM)**   Microscope used to magnify images as much as 50,000 times by means of scanning with an electron beam. The impinging electrons cause electrons on the surface to be ejected. The ejected electrons are collected and translated into a picture of the surface.

**scribe lines**   Lines used to separate die on a wafer. The wafer will be sawed along the scribe lines, resulting in individual chips.

**self-aligned gate**   A MOS structure that allows the direct alignment of the source or drain to the gate without a photoresist alignment step.

**semiconductor**   An element such as silicon or germanium, intermediate in electrical conductivity between the conductors and the insulators, in which conduction takes place by means of holes and electrons. Common single-element semiconductors are Si (silicon) and Ge (germanium); a common compound semiconductor is GaAs (gallium arsenide).

**sheet resistance**   A measurement of resistance with dimensions of ohms per centimeter squared that shows the number of N-type or P-type donor atoms in a semiconductor.

**side diffusion**   see *lateral diffusion.*

**silicon (Si)**   The Group IV element used for fabricating diodes, transistors, and integrated circuits.

**silicon dioxide (SiO$_2$)**   A nonconducting layer that can be thermally grown or deposited on silicon wafers. Thermal silicon dioxide is commonly grown using either oxygen or water vapor at temperatures above 900°C.

**silicon gate MOS**   A MOS gate structure with a layer of polysilicon on top of a thin layer of silicon dioxide.

**silicon nitride (Si$_3$N$_4$)**   A nonconductive layer chemically deposited on wafers at temperatures between 600 and 900°C. When it is the final layer in the process, it protects devices against contamination.

**single crystal**   Refers to substances that have all unit cells arranged in a definite and repeated fashion as opposed to polycrystalline materials, which have unit cells randomly arranged.

**slope etching**   Controlled undercutting; an etch strategy in which the sides of the contact holes are purposely overetched so as to reduce the shadow effect of the sidewall and the resultant thinning of the film.

**soft baking**   A heating process used to evaporate a portion of the solvents in resist. The term "soft" describes the still soft resist after baking. The solvents are evaporated to achieve two results: to avoid retention of the solvent in the resist film and to increase the surface adhesion of the resist to the wafer.

**solid-state electronics**   Designation used to describe devices and circuits fabricated from solid materials such as semiconductors, ferrites, or thin films, as distinct from devices and circuits making use of electron tube technology.

**source**   Along with the gate and drain, one of the three regions of a unipolar or field-effect transistor (FET).

**spinning**   A technique in which the photoresist is spun onto the wafer, resulting in a typical photoresist layer 0.5 μm thick with an allowable thickness variation of 10 percent.

**spin rinse dryer**   A machine that automatically rinses and dries wafers by spinning them in cassettes around a central axis.

**sputtering**   A method of depositing a thin film of material on wafer surfaces. A target of the desired material is bombarded with radio-frequency-excited ions that knock atoms from the target; the dislodged target material deposits on the wafer surface.

**spectrophotometer**   An analytical instrument used to collect interference measurements that are used to calculate film thickness.

**spreading resistance**   A technique used for measuring the dopant concentration profile in a wafer.

**SSI (small-scale integration)**   Refers to chips with between 2 and 50 components each.

**standard mechanical interface (SMIF)**   A system that allows the mating of portable clean wafer boxes (called *pads*) to the clean microenvironment loading stations of process tools.

**standing wave effects**   A vertical resist exposure pattern that follows standing waves set up in the resist layer by constructive interference of the exposure light reflecting off the wafer surface.

**static RAM (static random access memory)**   Fast read-write memory cell based on transistors.

**steam oxide**   Thermal silicon dioxide grown by bubbling a gas (usually oxygen or nitrogen) through water at 98 to 100°C.

**step and repeat**   An operation in which the pattern on the reticle is transferred to the mask or wafer. The photoresist-coated mask blank (chrome, emulsion, or iron oxide) or wafer is placed on an *x-y* stage, and the reticle pattern is repeatedly imaged until the entire surface is filled with the reticle pattern.

**step coverage**   The ability of new layers to evenly cover steps formed in the existing wafer layers.

**stepper**   An aligner tool that aligns and exposes one (or a small number) of die at a time. The tool "steps" to each subsequent die on the wafer.

**stripping**   Removal process; usually refers to photoresist.

**subcollector**   See *buried layer.*

**substrate**   The underlying material upon which a device, circuit, or epitaxial layer is fabricated.

**sulfuric acid (H₂SO₄)**   A strong acid often used to clean silicon wafers and to remove photoresist.

**susceptor**    The flat slab of material (usually graphite) upon which wafers are held during high-temperature deposition processes such as epitaxial growth or nitride deposition.

**tape automated bonding (TAB)**    A chip-to-package connection process in which the package leads are formed on a flexible tape, and all the lead fingers are bonded to the chip in one action.

**target**    The material to be sputtered during the sputtering process.

**test die**    Die on a wafer that appear to have a different pattern from most others. These contain test devices created by the same processes at the regular die; however, the devices on these die are designed on a larger scale to allow in-process quality control.

**TCE (trichloroethylene)**    A solvent used for wafer and general cleaning.

**tetraethylorthosilicate (TEOS)**    A chemical source for the deposition of silicon dioxide.

**thermal diffusion**    A process by which dopant atoms diffuse into the wafer surface by heating the wafer in the range of 1000°C and exposing it to vapors containing the desired dopant.

**thermal oxide**    On silicon semiconductor devices, an oxide fabricated by exposing the silicon to oxygen at high temperatures. The resulting interface has low levels of ionic impurities and defects (surface states).

**thermocouple**    A device to measure the temperature in a furnace of a reactor. It is made by welding two wires together at a point. Heat generates a voltage between the two materials that is proportional to the temperature.

**torr**    Pressure unit; international standard unit replacing the English measure, millimeters of mercury (mmHg).

**transistor**    A semiconductor device that uses a stream of charge carriers to produce active electronic effects. The name was coined from the electrical characteristic of "transfer resistance."

**tube**    (1) See *furnace*. (2) A cylindrical piece of quartz with fittings on one or both ends. It is placed in a furnace to provide a contamination-free and controlled atmosphere.

**undercutting**    See *isotropic etching*.

**ULSI (ultra-large-scale integration)**    Refers to chips with more than 1,000,000 components each.

**ultraviolet (UV) light**    A portion of the electromagnetic spectrum from 250 to 500 nm. High-pressure mercury sources emit UV light for photoresist exposure.

**unipolar transistor**    A transistor such as an FET whose action depends on majority carriers only.

**vacancy**    (1) A position in the crystal for an atom which is empty. (2) A type of crystal defect.

**vacuum**   A low-pressure condition.

**vapor phase epitaxy (VPE)**   An epitaxial deposition system that can combine several source gases to deposit compound semiconductors.

**vapor priming**   A technique in which primer is applied in a vapor state such that the wafer never comes in contact with any possible contamination in the liquid or, in the case of HMDS, any particles of hydrolyzed HMDS.

**via**   Vertical opening filled with conducting material used to connect circuits on various layers of a device to one another and to the semiconducting substrate; serves same purpose as "contacts."

**viscosity**   The qualitative measure of liquid flow. Viscosity measurements are made by measuring the force required to move an object through the liquid. It is a measurement of "internal friction."

**VLF hood**   A workstation with vertical laminar airflow to keep particulate levels low.

**VLSI (very-large-scale integration)**   Refers to chips with between 100,000 and 1,000,000 components.

**volatile memory circuit**   A memory circuit that loses its data when power to the chip is lost.

**voltage**   The force applied between two points causing charged particles (and hence current) to flow.

**wafer**   A thin, usually round, slice of a semiconductor material from which chips are made.

**wafer fabrication**   The series of manufacturing operations in which the circuit or device is put in and on the wafer.

**wafer flat**   Flat area(s) ground onto the wafer's edges to indicate the crystal orientation of the wafer structure and the dopant type.

**wafer sort**   The step after wafer fabrication during which the electrical parameters of integrated circuits are tested for functionality. Probes contact the pads of the circuit to conduct the test, leading to the name "prober" for the equipment that performs electrical tests on each die site of completed wafers.

**wafer sort yield**   The number of functioning die at wafer sort as compared to the total number of die started; typically, the lowest major yield point for integrated circuits.

**wire bonding**   An assembly step in which thin gold or aluminum wires are attached between the die bonding pads and the lead connections in the package.

**X-ray aligner**   An aligner tool that uses X-rays and a mask to expose resist-coated wafers.

**X-ray exposure system**   Imaging system using X-rays as the exposure source. Due to their short wavelengths, X-rays exhibit no detrimental diffraction effects.

**yield**   A percentage used in the semiconductor industry that indicates the amount of finished products leaving a process as compared to the amount of product entering that process.

# Index

## A

acceleration tube, 349
accumulative wafer-fabrication,
    140–145
acids, 46–47
acoustic wave devices, 547
adiabatic cooling, 251
adiabatic expansion, 429
aerial image, 285
air quality, 97–103
airborne molecular contaminants, 95
alignment (wafer), 231–243
alkalis, 46–47
alloying, 403
alternating phase shift mask [alter-
    nating aperture phase shift mask
    (AAPSM)], 296–297
alternative transistor designs,
    531–532
aluminum, 405
aluminum-copper alloy, 406
aluminum-silicon alloys, 405
amorphous silicon deposition, 53,
    392–394
amplification, 2
anisotropic etching, 260
annealing (doping), 79, 180, 357
annular ring illumination, 298
antireflective coatings (ARCs),
    301–303
atmospheric pressure CVD systems
    (APCVD), 372–377
atomic force microscopy (AFM),
    451–452, 462–464
atomic layer deposition (ALD),
    381–382
atomic number, 27–29
atomic structure, 26–28
auger electron spectroscopy (AES),
    465
autodoping, 390

automated guided vehicles (AGVs),
    496
automatic defect detection, 460
automatic spinners, 225–227
automation (wafer fabrication),
    494–509
autoprofiling, 173

## B

back end of the line (BEOL), 16, 85,
    273
backside plating, 67, 416, 590–591
backstreaming, 427
bacterial contaminants (see contami-
    nants)
baking, 227–231
ball bonding, 585–586
ball grid arrays (BGAs), 587, 604
ballroom layout, 102
bare die techniques, 606
barrel etchers, 266
barrel radiant-heated PECVD, 381
barren radiant-induction-heated
    APCVD 374–375
barrier layer, 403
barrier metals, 406–407
barrier/liner deposition, 412
bases, 46–47
batch versus single-wafer process-
    ing, 500–501
beam focus (ionization), 351
beam scanning (ionization), 351–352
Bell, Alexander Graham, 448
Bi-MOS, 543–544
binary notation, 555–557
bipolar circuit formation, 534–538
bipolar structure, 17–19
bipolar transistors, 472
bird's beak, 185, 538–539
blob top, 606
Bohr atom, 26–27, 41
book-to-bill ratio, 490

Boole, George, 555–556
borosilicate glass (BSG), 397
Bose-Einstein (wafer sort), 152–153
breakdown voltage, 471–472
bridges (mask), 146
bubbler, 130, 176, 186
Buckminster Fuller, 555
budget, 280
bump/flip-chip bonding, 587–589, 601
buried layers, 391
burn-in tests, 600

**C**

C/V plot, 475–477
capacitance-voltage plotting, 474
capacitors, 30, 517–519
capacitor electrodes, 415
capillary force, 122
carrier mobility, 37
carriers, 33
cavitation, 250
cellulose acetate (AC), 299–300
centipoises, 215
CERDIP packages, 593–594, 603
channeling (doping), 358
chemical cleaning solutions, 124–125
chemical contaminants, 95
chemical mechanical polishing
    (CMP), 42, 310–316, 414–415
chemical purity, 48
chemical vapor deposition (CVD), 74,
    162, 173, 366, 369–381
chemical, contamination, 115–116
chemically amplified (CA) resist, 286
chemical-mechanical polishing
    (CMP), 304
chip and wafer size, 7–10
chip characteristics, 571–572
chip cost, 11
chip uses, 12
chip, 16
chip-on-board (COB), 17
chip-scale packages, 605
chlorine-added oxidation, 187
circuit density (die), 150
circuit design, 79–82
cleaning techniques, 127–128
cleanroom, 101–109, 119–120
closed-loop control-system automa-
    tion, 497

clustering, 495
CMOS epitaxy, 391
cold-wall induction, 374
collimated, 234
color-active matrix liquid crystal dis-
    plays (AMLCDs), 548–550
columnar poly, 393
complementary MOS (CMOS) cir-
    cuits, 19–20, 540–543
composition (wafer), 467
computer-aided process planning
    (CAPP), 499
computer-integrated manufacturing
    (CIM), 498–499
concentration/depth profile, dopant,
    441
conductors
    CVD, 399
    generally, 29–34, 531–532
    metallization, 405–412
    multilevel metal schemes, 403–407
    single-level metal, 402–403
conformal layer, 305–306
constant acceleration, 598
contact aligners, 237
contact holes, 402
contaminants, 92–96
contamination
    generally, 454–462, 465–467
    personnel generated, 111–113
    problems, 95
    sources, 97–98
continuous conduction-heated
    APCVD, 376
contrast effects (mask), 286
contrast enhancement layers (CEL),
    318
contrast threshold, 318
copper (metallization), 409
critical dimension (CD), 280, 452
crusting, 273
cryogenic cleaning, 128
cryogenic pumps, 429–430
crystals
    defects, 59–60, 357
    generally, 53
    growth, 56–60
    orientation, 55–56, 62–63
    quality, 59–60
cum fab yield, 141–143, 151

CVD metallization, 425–426
Czochralski, 56–57

**D**

Dark-field inspection, 457
defect density (die), 10, 150
defect detection, 454–462
deflashing, 597
deflux, 590
dehydration baking, 218–219
densification, 396
density, 44–45
deposited films (CVD), 385
deposited semiconductors (CVD),
    385–386
deposition
    doping, 333–338
    metallization, 416
depth of focus/field, 293–294
DESIRE process, 307–308
develop inspect, 254–257, 277
device dielectric (silicone dioxide),
    162
device electrical measurements,
    467–477
device processing yield, 95–96
device reliability, 95–96
devitrification, 172
diameter grinding, 62–64
diamond semiconductors, 40
diatomic molecules, 42
dichlorosilane source chemistry, 388
die
    area, 150
    attach, 582–584
    generally, 16
    sort, 87–89
    wafers, 147–148
dielectrics
    constant, 399
    generally, 30, 369
    isolation, 536–537
    wearout, 269
die-to-database system, 462
die-to-package bonding, 584
differential Hall effect (DHE), 442
differential oxidation, 169
diffraction effect, 234, 288
diffused MOS (DMOS), 529
diffusion (doping), 326–329

diffusion limited reaction, 164
diffusivity, 333–334
dimensional control, 233
diode sputtering, 423
diodes, 3, 470, 519–521
DIP packages, 603
DIP through-hole assembly, 573
direct writing, 242, 292
discrete devices, 4
dislocations, 61
dissolution inhibitor systems, 207
doping
    concentration, 355–359
    polysilicon, 414
    semiconductors, 32–37
    sources, 337–341
    silicon dioxide, 397
doping barrier (silicone dioxide), 161
double masking, 298
downstream plasma processing, 269
downstream strippers, 276
DRAM, 8, 12–13
drawback (spinners), 225
drive-in oxidation, 341–345
dry (or plasma) development, 252
dry cleaning, 128
dry etching, 264–273
dry oxidation, 165–166
dry oxygen, 185–186
drying, 133–134
dryox (*see* dry oxidation)
dual in-line package (DIP), 602
dual-damascene process, 309–310,
    411
dyed resists, 318
dynamic dispense, 223–224

**E**

e-beam gun, 418
edge bead removal, 225
edge die (wafer), 147
electrical conduction, 29
electrical sort, 87–89
electricity, static, 108–110
electrochemical plating (ECP),
    412–414
electromagnetic spectrum, 209
electromigration, 406
electron beam aligners, 242
electron beam scanning, 291–292

electron conduction, 33
electron cyclotron resonance (ECR), 267
electron flood gun, 354
electron spectroscope for chemical analysis (ESCA), 466
electron-beam projection lithography (EPL), 295
Electronic Numeric Integrator and Calculator, 2–3
Electroplating (see electrochemical plating)
electrostatic beam scanning, 352
elements, 26–28
elephant tube (wafer), 188
ellipsometers, 446–447
emissivity, 182
encapsulation, 590
end cropping, 62
end station (ionization), 353
energy-sensitive polymers, 204–205
engineered substrates, 39
engineered wafers, 69
ENIAC, 2–3
environmental, safety, and health (ESH), 501
epitaxial
    film, 388–389
    layer, 18, 366, 385
    process, 391
    silicon, 386
epoxy molding, 593
equipment (wafer fabrication), 491–494
equipment, cleaning, 110–111, 118
error budget, 280
error function, 334–335
etch
    definition, 319–320
    photoresist, 258–262
    profile control, 321
etchback planarization, 309
etching, 536
eutectic alloying, 406
eutectic die attach, 582
evaporation, 74
excimer lasers, 289
exponential model (wafer sort), 153–154
exposure (wafer), 231–243

exposure response curve, 210
exposure, photoresist, 285–287

**F**

fab floor layout, 500
factory-level automation (wafer fabrication), 497–501
Fairchild camera, 4
Faraday cup, 354
feature size, 6–7
Federal Standard 209E, 97–99
FEOL, 315
ferroelectric materials, 40
field effect transistor (FET), 17, 524–531
field emission devices (FEDs), 550
field oxide (FOX), 162, 540
filar dimension measurement, 452
film parameters (layer deposition), 367–369
first-fail basis, 257
flat panel displays (FPDs), 548
flat, grinding, 63
flip-chip ball grid arrays (FCBGAs), 604
flip-chip bonding, 587
float zone technique, 56–60
flood guns, 350
focused ion beams, 289
four-point probe (resistance), 438–441
fringes, 445
front end of the line (FEOL), 16, 85, 120, 273
functional test, 599
furnaces (oxidation), 170–173
fuses (metallization), 415–416
fuses, 531–532

**G**

gallium arsenide, 144, 384, 395, 545
gallium, 37–38, 58
gas source MBE or GSMBE, 383
gas, contamination, 116–117
gate array, 560
gate oxide integrity (GOI), 121, 442
gate width, 6–7
Gauss, Karl Friedrich, 502–504
germanium, 37–39

gettering, 67, 397
global planarization, 310–311
gravity, specific, 44
green fabs, 501
grinding (wafer), 62–68
gross leak testing, 599
Grove, Robert, 420
growth defects, 61

**H**

hard bake, 211, 252–254
heteroepitaxial, 386
heterostructures/heterojunctions, 39
hexamethyldisilazane (HMDS),
    220–221
high-aspect-ratio patterns, 368
high-density plasma CVD (HD-
    PCVD), 381
high-k dielectrics, 30, 369, 399
high-pressure oxidation, 184
high-pressure water cleaning, 123
hi-rel, 570
histograms, 502
hole conduction, 33–35
hole flow, 36
homoepitaxial, 386
hoods, 100–101
horizontal conduction-convection-
    heated LPCVD, 378
horizontal conduction-heated
    APCVD, 376
horizontal laminar flow (HLF), 101
horizontal vertical-flow PECVD, 379
horizontal-tube induction-heated
    APCVD, 373–374
Horni transistor, 4–5
hot plates, 228–230
hybrid circuits, 607
hydrido-organic siloxane polymers
    (HOSP), 410
hydrogen ions, 46
hydrogen reduction, 53
hydrophilic, 218–219
hydrophobic, 218–219
hydroxyl ion, 165

**I**

image reversal (planarization), 317
image shearing dimension measure-
    ment, 452

immersion, 249
index of refraction, 216
induction (silicone dioxide), 162
ingots, 56
inorganic acids, 47
inorganic residue, 124
insulated field effect transistor
    (IFET), 19
insulators and dielectrics (CVD), 395
integrated circuits, 4
integrated device manufacturers
    (IDMs), 14
integrated image processing,
    254–258
integrated-circuit formation, 533–537
integration level, 7–8
interconnects (*see* leads)
interdielectric layers (IDL), 315
intermediate metal dielectric (IMD),
    409–410
intermetallic dielectric layer (IDL or
    IMD), 403
*International Roadmap for Semicon-
    ductor Technology (IRST)*, 7, 10,
    21, 22, 48, 52, 95, 145, 284, 410,
    483, 570
intrinsic semiconductors, 31
inventory control (wafer fabrication),
    506
inversion voltage (*see* threshold volt-
    age)
ion beam etching, 271
ion implant
    doping, 326–329
    evaluation, 359
    implantation, 78–79, 345–355,
        361–362
    masks, 354–355
ion milling (*see* sputtering)
ion pumps, 42, 430
ionization chamber, 347
ISO 9000, 508
ISO Global Cleanroom Standards
    (ISO 14644–2), 119–120
isotropic etching, 260

**J**

Josephson effect, 544
junction
    depth, 448–449

doping, 326–332
field-effect transistors (JFETS), 530
isolation, 534
transistor, 17–19
jungle, 176
just-in-time (JIT) inventory control,
507–508

**K**
kerf, 64
Kilby circuit, 4–5
killer defects, 92–93, 145, 151, 153,
462
kinematic viscosity, 215
known good die (KGD) problem, 608

**L**
laminar gas flow, 178
latch-up (CMOS epitaxy), 391
latent image, 246
lateral diffusion, 330
lattice, 54
layer thickness measurements (wafers), 443–445
lead on chip (LOC), 606
lead plating, 595
lead trimming, 596
leadless packs, 602
leads, 10, 402, 578
leakage current, 471
lift-off process (etching), 319320
light-sensitive polymers, 204–205
line edge roughness (LER), 286
line organization (wafer fabrication),
508–509
line width measurements (see critical
dimensions)
linear stage (oxidation), 163
liquid encapsulated Czochralski
(LEC), 58
lithography, 8, 243, 285–286
loading, wafer, 188–190
local oxidation of silicon (LOCOS),
185, 398, 537
logic circuits, 557–560
low-k dielectric materials, 30, 369,
399, 409–410
low-pressure chemical vapor deposition (LPCVD), 377–381, 414,
425–426

**M**
magnetron sputtering, 424
manometer, 45
manual spinners, 225
Marangoni drying, 134
mask defects, 145–146
masking, 82–83, 198–202, 278–280
mass analyzing/ion selection, 348
mass flow meter, 176–177
material safety data sheet, 48
matter, 42–45
maximum solid solubility (doping),
333–334
mechanical scanning, 353
melt, 57–58
memory circuits, 557, 560–566
memory MOS (MMOS), 530
mercury (Hg) spectrum, 210
metal cans, 603
metal film uses, 415–416
metal lines (see leads)
metal removal (MR), 198
metal semiconductor field-effect transistor (MESFET), 531–532
metallic ions, 94
metallization process, 402
metalorganic CVD (MOCVD), 373,
384–385
metal-oxide semiconductor (MOS)
structure, 7, 30
gate electrodes, 415
generally, 17, 185, 191, 533–534
LOCOS isolation, 538
MOSFET gate, 525, 532
silicone dioxide, 162
transistors, 473
metrology, 436
microelectromechanical systems
(MEMS), 545
micro-environments, 103–104
microlithography, 198
microloading, 270
micromachines, 547
microprocessors, 557
microwave baking, 230
Miller indices, 55–56
mini-environments, 103–104
mini-fabs, 483
mobile ionic contaminants (MICs), 94

molecular beam epitaxy (MBE), 373, 383
momentum (ionization), 349
momentum transfer, 271, 420
monocrystalline, 54–55
Moore's law, 7, 243, 482
moving-arm dispense, 224
moving-belt infrared ovens, 230
MSI/bipolar integrated circuit, 87
multichip module (MCM), 16–17, 607
multilayer resist/surface imaging, 304
multimeter, 438
Murphy (wafer sort), 152–153

**N**

nanoanalysis era, 436
nanotechnology, 22–23
negative binomial (wafer), 152–153
negative resist development, 247
neutral beam trap (ionization), 351
next generation lithography (NGL), 285
nitridation, thermal, 194–195
nitrocellulose (NC), 299–300
Novolak, 205
Noyce, 4–5, 6
N-P junction, 329–330
NPN transistors, 523
N-type dopants, 328–332
N-type junction, 17
nucleation (layer deposition), 370
numerical aperture (NA), lens, 293–294

**O**

off-axis illumination, 294
Ohm's law, 438
ohmic, 405
oil diffusion pumps, 428–429
open-loop control, 183
optical critical dimension (OCD), 453
optical proximity corrected/correction (OPC), 297
optical resists, 204–207
optoelectronics, 546
organic acids, 47
organic residue, 124
out-diffusion, 390

ovens, convection, 228
overlay budget, 236, 280
overshoot, 173
oxidation
    automated, 190–191
    cycles, 191–193
    generally, 165–167
    process (wafer), 143
    wafer, 69
oxide impurities, 168
oxide removal (OR), 126, 198
oxide rupture (BVox), 442
oxygen passivation (OP), 117
ozonated chemistries, 127

**P**

packaging
    bare die strategies, 600
    design, 602
    functions and design, 572–578
    generally, 89
    marking, 597
    operations, 574–578
    parts, 578–590
    process flows, 600
    process, 590–597
    wafer, 69
pad mask, 84
pad oxide, 539
paddle, 188
pancake induction-heated APCVD, 375–376
parabolic stage (oxidation), 164
parallel downflow rinser, 130
parametric testing, 468
Pareto chart, 504–505
particle contaminants, 92–93
particulate removal, 122–123
passivation layer, 84
pattern shift, 389
patterning, 74, 198
pellicles, 298–300
periodic table, 27–28
Perkin Elmer light projection, 239
phase shift masks (PSMs), 296–297
phase shifting, 296–297
phenol-formaldehyde novolak resin structure, 205
phenolic organic strippers, 274
phosphorus silicate glass (PSG), 397

phosphorus-doped oxides, 168
photoacoustic, 448
photolithography, 198
photomasking, 198–202, 217–221
photoresist
    application (spinning), 221
    chemistry, 202–204
    development 246–252
    generally, 300
    performance, 207–214
    physical properties, 214–217
    process advances, 304
    removal, 121
photosolubilization, 205
physical measurement methods (wafers), 443
physical vapor deposition (PVD), 74, 366, 420
picked wafers, 188
pile-up
    dopants, 168
    drive-in oxidation, 343
pin grid arrays
pinholes (resist), 211
planar plasma etching, 267–268
planar technology, 4
planarity, 314–315
planarization, 11, 67, 303–304
planes, 18, 55–56
planetary wafer, 420
plasma descum, 251
plasma enhanced chemical vapor
    deposition (PECVD), 373, 379
plating process flows, 596
plug/plug filling, 408
PNP transistors, 523
point defects, 61, 145
point-of-use filters, 259
poisson (wafer sort), 152–153
polarization, 446–447
polishing
    pads, 313
    rates, 314
    wafers, 67–68
polycide-gate MOS, 528
polycrystalline, 54–55
polyimide planarization layers,
    308–309
polymerized resist, 247–248

polymethylmethacrylate (PMMA),
    305–306
polysilicon
deposition, 392
generally, 53
oxidation, 168
positive resists, 212–214, 247
post-CMP clean, 315
post-exposure bake (PEB), 241, 303
post-ion implant and plasma etch
    stripping, 276
postoxidation, 193–194
prebake (*see* hard bake)
pre-etch bake (*see* hard bake)
preseal inspection, 591–592
pressure, 45–46
priming, 220–221
printed circuit board (PCB), 601–602
process chemicals, 40–41
process control techniques (wafer),
    144
process defects (wafer), 145
proportional band controllers, 173
proximity aligners, 237
proximity effect, 286–287
P-type dopants, 328–332
P-type junction, 17
puddle development, 251
pumps, vacuum, 426–431
pyrogenic steam oxidation, 165
Pyrox, 395

**Q**

quad packages, 605
quality control (ASQC), 508
quality control and certification—ISO
    9000, 508
quartz and contamination, 118
quartz, 172–173

**R**

radiation hardened, 38
radiation, 227
rail guided vehicle (RGV), 496
ramp furnaces, 180
ramping, temperature, 174
rapid thermal oxidation (RTO),
    182–184

rapid thermal processing (RTP), 180–182
Rayleigh constant, 293
RC constant, 409–410
RCA clean, 126–127
reaction chamber (oxidation), 172
reactive ion etching (RIE), 272
reduction steppers, 239
redundancy, 566
reflectance, 453
reflectometry (see spectrophotometers)
reflow (planarization), 317
refraction, 216
refractory metals, 407–409
registration capability, 233
reoxidation (reox) (see drive-in oxidation)
resist coverage, 223
resist light scattering, 300
resist stripping, 273
resistance (wafer electrical measurements), 437
resistivity, 30–33
resistors, 31, 469–470, 514–517
resists, optical, 204–207
resolution capability, 207, 233
reticle defects, 145–146
reticle, 82–83
retrograde wells, 361, 543
rinsing, 129–131
roughing pumps, 426
rupture voltage (see oxide rupture)

S

safe delivery system (SDS), 347
salicide-gate MOS, 529
same-type doping, 330
saw, wafer, 65
scaled transistor designs (see alternative transistor designs)
SCALPEL, 295–296
scanning auger microanalysis (SAM), 466
scanning capacitance microscopy (SCM), 451–452
scanning electron microscope (SEM), 450, 458
scanning projection aligners, 238
scatterometry, 463–465

scavenger (furnace), 175
Schottky barrier bipolar transistors, 524
scratch layer, 84
scumming, 251
seal ring, 593
sealing techniques, 592
secondary ion mass spectrometry (SIMS), 442, 450–451
seed deposition (metallization), 412
seeds (wafer sort), 152–153
selective epitaxial silicon, 391
selectivity (etching), 261
self-aligned silicon gate (SAG) structure, 321
self-aligned structure, 320
semiconducting compounds, 37
Semiconductor Equipment and Materials International (SEMI), 499, 508
semiconductor history, 1–4
semiconductor production materials, 14–16, 37
semiconductor silicon preparation, 52
semiconductor-device formation, 513–517
Semiconductor Industry Association (SIA) Roadmap [see International Roadmap for Semiconductor Technology (IRST)]
shape metrology, 453
sheet resistance, 439
side diffusion, 330, 345
silane source chemistry, 388
silicon dioxide (SiO2), 18, 395
silicon gate MOS, 526–528
silicon nitride, 397–398
silicon on insulator (SOI) isolation, 394–395, 544
silicon on sapphire (SOS), 394–395
silicon planar processing, 18
silicon tetrachloride source chemistry, 386–388
silicon wafers, 53
silicon, 37–39
silicone dioxide layer uses, 160–163
silicon-gate MOS technology, 83–87, 414
Silox, 395
silylation/DESIRE process, 307–308

single crystalline, 54–55
single-chamber planar PECVD, 380
single-wafer versus batch processing, 500–501
slip, 61–62, 389
slurry, 313
snow cleaning, 128
soft bake, 211, 219, 227–231
soft-contact machines, 237
solar cells, 546
solid state discrete devices, 4, 14
solids content (resist), 214
solid state integrated circuit, 557–560
solid state thermal diffusion, 330, 333
solvents, 46–47, 276
sonic-assisted cleaning/rinsing, 132
source cabinet (furnace), 175
space charge forces, 352
spectral response characteristic, 209
spectrophotometers, 446
spin priming, 220
spinning, 220–226
spot defects, 145
spray development, 250
spreading resistance probe (SRP), 450–451
sputter deposition/sputtering (PVD), 419–425
sputter etching, 271
sputtering, 74, 588
stack thickness (wafer), 467
stacking faults, 389
standard mechanical interface (SMIF), 103–104
standing-wave effect, 303
static dispense spin process, 222–223
static, 108–110
station yields, 141
statistical process control (SPC), 501–509
steam oxidation, 165–166
step and scan aligners, 241
step-and-repeat reduction system, 239
steppers, 238
strain gauges, 545
stripping, chemical, 274–276
stylus, 447–448

subcollectors, 391
subject contrast, 286–287
substrate reduction, 425
substrates, 69
subsurface reflectivity, 300
suckback (spinners), 225
superconductors, 544–550
surface concentration, 334–335
surface dielectric (silicone dioxide), 161–162
surface passivation (silicone dioxide), 160
surface profileometers (see stylus)
surface roughness, 120
surface tension, 134, 215
surface topography, 294
SWAMI, 539
switching, 2
synchrotron radiation, 291

**T**

tape automated bonding (TAB), 576, 591
target chamber (ionization), 353
TC bonding (see thermocompression bonding)
teardrop transistor, 4–5
temperature control (oxidation), 173
temperature cycling, 598
temperature scales, 43–44
temperature sensing, 547
test wafers, 436
testing, 598–600
tetraethyl orthosilicate (TEOS), 396
tetramethylammonium hydroxide (TMAH), 248
thermal budget, 181
thermal diffusion, 78–79
thermal flow (masking), 211
thermal nitridation, 194–195
thermal oxidation mechanisms, 160, 163–180
thermocompression bonding, 585
thermode, 591
thermosonic, 585
thin packages, 605
third optical inspection, 592
threshold voltage, 361, 475
tie-bar, 596

time of flight secondary ion mass spectrometry (TOF–SIMS), 467
top surface imaged (TSI), 308
Torricelli, 45
transfer resistor (*see* transistor)
transistor operational analogy, 521
transistor, 3, 521–532
transmission electron microscope (TEM), 460
transport-limited reaction, 164
trapping, 67
trench isolation, 540
triboelectric charging, 109
trichlorosilane, 53
trilevel resist process, 307
tube furnaces, 178–179
tungsten, 425
tunnel/bay concept, 102
tunneling, 544
turbomolecular pumps, 431
twinning, 61

**U**

ULSI/VLSI patterning, 284
ultrahigh vacuum CVD (UHV/CVD), 379
ultraviolet and visible spectrum, 209
undercutting, 260
underfillment, 590
underpass conductors, 532
unit cell, 53
unpolymerized resist, 247–248
unyielded die cost, 488–490

**V**

vacuum
  baking, 231
  evaporation, 417–419
  generally, 45–46
  pumps, 426–431
  tube, 2
van der Pauw structure, 359–360
van der Waals force, 122
vapor density, 45
vapor phase decomposition/atomic absorption spectroscopy (VPD–AAS), 467
vapor phase epitaxy (VPE), 373, 382
vapor priming, 220–222

Vapox, 395
vertical laminar flow (VLF) station, 100
vias, 403
virtual leak, 423
viscosity, 214, 217
VLSI circuit fabrication, 417
VLSI/ULSI patterning, 284

**W**

wafer
  assembly and test of, 155
  breakage, 144
  charging, 350
  doping, 75–76, 167
  electrical measurements, 437–442
  fabrication, 14, 72–83, 484–491
  generally, 16, 52–53
  isolation technology (WIT), 103–104
  layering, 74–76, 77
  marking, 65–66
  orientation, 167
  overall process yields, 155–156
  patterning, 74–76, 78
  planes, 73
  preparation, 62
  priming, 219–220
  quality, 59–60
  scrubbers, 123
  size, 22, 147–150, 590
  slicing, 64
  sort, 16, 72, 87–89
  throughput, 233
  warping, 144, 174
wafer-cleaning station, 187
wafer-delivery automation, 496
wafer-loading, 187, 494–495
wafers in process (WIP), 506–507
wafer-sort yield factor, 146–149, 152–154;
water, contamination, 113–115
wet development methods, 249
wet etching, 261–264
wet oxidation, 165
wet process/cleaning, 124
wire bonding, 584–589
work function, 526

**X**

X-ray aligners, 241, 289–291

**Y**

yield limiters, 142

yield measurement point, 140–143

yielded die cost, 488–490

**Z**

zeta potential, 122